Advanced Welding Technology in Metals

Advanced Welding Technology in Metals

Editors

João Pedro Oliveira
Zhi Zeng

MDPI • Basel • Beijing • Wuhan • Barcelona • Belgrade • Manchester • Tokyo • Cluj • Tianjin

Editors
João Pedro Oliveira
Universidade NOVA de Lisboa
Portugal

Zhi Zeng
University of Electronic Science and Technology of China
China

Editorial Office
MDPI
St. Alban-Anlage 66
4052 Basel, Switzerland

This is a reprint of articles from the Special Issue published online in the open access journal *Metals* (ISSN 2075-4701) (available at: https://www.mdpi.com/journal/metals/special_issues/advanced_welding_technology_metals).

For citation purposes, cite each article independently as indicated on the article page online and as indicated below:

LastName, A.A.; LastName, B.B.; LastName, C.C. Article Title. *Journal Name* **Year**, *Volume Number*, Page Range.

ISBN 978-3-0365-5675-8 (Hbk)
ISBN 978-3-0365-5676-5 (PDF)

Contents

Article

A Comparative Study of Analytical Rosenthal, Finite Element, and Experimental Approaches in Laser Welding of AA5456 Alloy

Hamidreza Hekmatjou [1,*], Zhi Zeng [2], Jiajia Shen [3], J. P. Oliveira [3,*] and Homam Naffakh-Moosavy [1]

1 Department of Materials Engineering, Tarbiat Modares University, Tehran P.O. Box 14115-143, Iran; h.naffakh-moosavy@modares.ac.ir

2 School of Mechanical and Electrical Engineering, University of Electronic Science and Technology of China, Sichuan 221116, China; zhizeng@uestc.edu.cn

3 UNIDEMI, Department of Mechanical and Industrial Engineering, NOVA School of Science and Technology, Universidade NOVA de Lisboa, 2829-516 Caparica, Portugal; j.shen@campus.fct.unl.pt

* Correspondence: h.hekmatjou@modares.ac.ir (H.H.); jp.oliveira@fct.unl.pt (J.P.O.); Tel.: +351-21-294-96-18 (J.P.O.)

Received: 21 February 2020; Accepted: 24 March 2020; Published: 27 March 2020

Abstract: The thermal regime and microstructural phenomenon are studied by using finite-element (FE) modelling and the analytical Rosenthal equation during laser welding of aluminum alloy 5456 (AA5456) components. A major goal is to determine the merits and demerits of this analytical equation which can be an alternative to FE analysis, and to evaluate the effect of imperative assumptions on predicted consequences. Using results from the analytical and numerical approaches in conjunction with experiments, different physical features are compared. In this study, the results obtained from experiments in terms of melt pool shapes are compared with the predicted ones achieved from the numerical and analytical approaches in which the FE model is more accurate than the Rosenthal equation in the estimation of the melt pool dimensions. Furthermore, as to the partially melted zones, the estimations achieved from the numerical modeling are more genuine than ones from the analytical equation with regards to the experimental results. At high energy density, near keyhole welding mode, the reported results show that experimental melt widths are supposed to be narrower than the fusion widths estimated by the analytical solution. The primary explanation could be the influence of thermal losses that occurred during convection and radiation, which are neglected in the Rosenthal equation. Additionally, the primary dendrite arm spacing (PDAS) estimated with the numerical modeling and the analytical Rosenthal solution is comparable with the experimental results obtained.

Keywords: laser welding; numerical finite-element modeling; analytical Rosenthal equation; thermal regime; microstructure; AA5456

1. Introduction

Laser beam welding has numerous merits owing to high welding velocity, low heat input, high flexibility, high efficiency of production, narrow weld width and large weld depth in comparison to conventional welding processes, such as those based on the electric arc. Thus, the use of laser welding in different applications such as high pressure and vacuum vessels, crane fabrication, and in the marine and aerospace industries is currently a reality. For instance, pulsed Nd:YAG laser is of great importance because the laser parameters, that is, average power, pulse duration, and frequency, are controllable. As for aluminum and its alloys, they are comprehensively consumed in different industries such as aerospace and automobile owing to their high specific strength and corrosion resistance in comparison to other metallic alloys. Among different kinds of aluminum alloys, 5456 aluminum alloy (AA5456), which is a wrought and non-heat treatable aluminum alloy belonging to the 5xxx Al-Mg series, is focus of attention due to the high amount of magnesium (approximately 5 percent), which improved the material weldability [1].

During welding, process control is fundamental to obtain defect-free joints. All aluminum alloys encompass specific characteristics, namely, low absorption of the laser beam, volatile elements such as magnesium in 5456 aluminum alloy, tendency to create oxide films, and a propensity to form constituents with low melting points [2,3]. Therefore, some defects including excessive porosity, blowholes, and hot cracking can occur in laser welded aluminum alloys [4]. These kinds of flaws can be seriously damaging to the mechanical properties of the welds. To achieve a sound weld of aluminum alloys via laser welding, considering the relationships between microstructure and mechanical properties, determining optimum parameters in the process for different aluminum alloys is of paramount importance. It has been reported that hot cracking phenomenon is eliminated during laser welding of 5456 aluminum alloy by performing preheating just before welding [1]. Researchers have found that increasing pulse overlapping factor induces an increase in Mg loss so that a tendency to formation of excessive solidification cracking occurs [5]. With all these results considered, it is obvious that the solidification process and optimization of laser parameters are determining factors in the performance of laser welds and the resulting microstructures. As far as these imperative concepts are concerned, it is of paramount importance to evaluate the temperature regime during this particular process.

Experimental results play an integral role in characterizing laser welded components; however, these results could be restricted by the technique used, and the time-consuming aspect may also adversely affect the whole process. On the other hand, in order for the significant behavior in laser welding to be investigated and determined, mathematical models are additional ways. It should be considered that numerical and analytical solutions are two major approaches to obtain such insights into the effect of different process parameters. To predict the thermal features of welding, Rosenthal's analytical model can be widely utilized and has a potential to give an abrupt estimation [6]. However, some assumptions need to be performed when using the Rosenthal equation in order that this analytical solution would be applicable in welding process; nevertheless, this will raise concerns about the results' accuracy obtained from this equation. On the other hand, the amount of assumptions used in numerical models is often fewer than the analytical solution, which causes to be more genuine. In contrast to this idea, the calculation time needed for numerical models is drastically longer than that of analytical ones.

It has been reported that the Rosenthal equation was already applied to determine which conditions would induce lack of fusion circumstance in different alloys [7]. They found a prediction of blowhole formation induced by lack of fusion under various processing conditions. In another research, the primary dendrite arm spacing (PDAS) was estimated via a processing map [8]. Wang et al. [9] have conducted the phase-field model to evaluate the dendritic growth of grains along the melt pool boundary. Remano et al. [10] have utilized a finite-element (FE) model to investigate molten and solidified Inconel 718 during the laser-based additive manufacturing. In this research, they observed that the fusion width was overestimated when using high energy densities. In this regard, the absorptivity of laser beam by the substrate was determined using a factor which needs to be performed in numerical modeling so that the predicted results became comparable with experimental ones. Artinov et al. [11] developed a 3D model to calculate the heat source and predict the thermal behavior in fusion welding. Furthermore, they accomplished a relationship between numerical results and experimental ones for weld pool geometries and heat transfer. Satyanarayana et al. [12] have applied the convection mode of heat transfer and Marangoni stresses in the fusion zone and have also calculated heating and cooling rates with which they would then use to study the microstructure of fusion and heat affected zones during laser welding of ZR-1%NB alloy. In another study [13], a finite element model was developed to investigate hot cracking phenomena during laser welding of 6xxx aluminum alloys. It was shown that the heat flux vector field achieved from the simulation is in a good agreement with the grain orientation observed in the experiments. As to laser welding of 5xxx and 6xxx Al series, research works have been conducted using thermomechanical finite element models to study the temperature-dependent characteristics of the joints so that optimized laser parameters would be obtained [14]. Moreover, an irreversible melting state variable into the thermomechanical simulation was presented to evaluate how the melting state influences the mechanical and thermal features during laser welding of Al alloys [15]. Based on previous studies, numerical modeling and analytical solutions are both capable to produce satisfactory results during the laser welding process. However, they should be investigated separately, and there has been no comprehensive evaluation of these two routes yet, especially for laser welding of the 5456 aluminum alloy.

All in all, this research aims to describe both analytical and numerical solutions and compare them with regards to laser welding of AA5456 using a pulsed Nd:YAG laser. Based on the previous studies in this regard, it should be mentioned that the continuous mode of laser equipment has been widely used such that a comprehensive study in order to show the comparison between numerical and analytical solutions in laser welding of AA5456 using a pulsed Nd:YAG laser is still lacking. To do so, some important parameters like fusion dimensions, partially melted zone thickness, temperature gradient, cooling rate, solidification rate, and microstructure such as primary dendrite arm spacing (PDAS) were predicted and compared using analytical and numerical approaches in addition to experiments. Moreover, the predicted results were affected by the assumptions performed in the numerical modeling; therefore, these effects were elaborated. Furthermore, to validate these predicted results, some experimental results were obtained and compared to them. By doing so, it is expected that the knowledge of the correlation between determining parameters in the process and laser welded results is enhanced. Additionally, in order to investigate whether the analytical Rosenthal equation can supplement or replace the numerical modeling, the capability and limitations of this equation are studied in this research.

2. Numerical Modeling

In order for the temperature profile to be simulated during laser welding, a finite-element model was prepared. In this case, thermophysical properties of the material, the distribution mode of the laser beam focused on the substrate, and radiated heat losses in conjunction with heat losses through the convection are considered. For this purpose, COMSOL Multiphysics software version 5.4 developed by COMSOL Inc. (Stockholm, Sweden) is utilized to carry this numerical simulation out. Simulation circumstances are similar to processing parameters performed in laser welding, including a mean power of 50–80 W, a spot diameter of 0.5 mm, a frequency of 10 Hz, a duration of 4 ms, and a welding velocity of 2 mm/s.

2.1. Material Properties and Methods

Sheets of aluminum alloy 5456 with a thickness of 5 mm were used as the base metal in this investigation. The length and the width of these aluminum alloy sheets were 100 mm and 50 mm, respectively. Table 1 illustrates the chemical composition of this alloy. In order for the dendritic growth of grains within the fusion zone to be investigated, the scanning electron microscopy (SEM) was utilized.

Table 1. Chemical composition of samples (in wt.%).

Mg	Mn	Fe	Si	Cr	Cu	Ti	Al
4.7	0.66	0.22	0.09	0.09	0.01	0.03	Base

To weld AA5456 sheets in the bead-on-plate condition, a pulsed Nd:YAG laser with the model of IQL-10 was used. This laser apparatus was able to produce a maximum power of 80 W, a wavelength of 1064 nm, a focal length of 100 mm, and a spot diameter of 0.5 mm on the substrate. In the current study, the laser beam was collided with the substrate through the top surface. To protect the welding pool and its surroundings, argon gas at a flow rate of 30 l/min was applied linearly in the welding direction. Table 2 depicts the specific parameters which have been performed in the laser welding process.

Table 2. Laser welding parameters.

Specimen	Pulse Frequency (Hz)	Pulse Duration (ms)	Peak Power (kW)	Pulse Energy (J)	Average Power (W)	Heat Input (J/mm)
A1	10	4	1.25	5	50	25
A2	10	4	1.5	6	60	30
A3	10	4	1.75	7	70	35
A4	10	4	2	8	80	40

Thermophysical properties of the material such as density, specific heat, and thermal conductivity vary with temperature such that they should be calculated via polynomial coefficients method which is described below. These coefficients are depicted in Table 3 [16–18].

Table 3. Polynomial coefficients vary with temperature T for calculation of Density D, Specific Heat at constant pressure C_p, and Thermal Conductivity λ of Aluminum alloy 5456 [16–18].

Property y	Unit	Polynomial Coefficients $y=a+bT+cT^2$			Range T (K)	State	Reference
		a	b	c			
$D(T)$	$kg{\cdot}m^{-3}$	2717.683	−0.231	-	$293 \le T \le T_m$	s	[16,17]
$D(T)$	$kg{\cdot}m^{-3}$	2599.97	−0.27	-	$T_m \le T \le 1680$	l	[16,17]
$c_p(T)$	$J{\cdot}kg^{-1}{\cdot}K^{-1}$	787.73	0.457306	-	$293 \le T \le T_m$	s	[16,18]
$c_p(T)$	$J{\cdot}kg^{-1}{\cdot}K^{-1}$	1261	-	-	$T_m \le T \le 1491$	l	[16,17]
$\lambda(T)$	$W{\cdot}m^{-1}{\cdot}K^{-1}$	111.11	0.0888	-	$293 \le T \le T_m$	s	[16,18]
$\lambda(T)$	$W{\cdot}m^{-1}{\cdot}K^{-1}$	33.9	7.892×10^{-2}	-2.099×10^{-5}	$T_m \le T \le 1491$	l	[16,17]

As was mentioned above, thermophysical characteristics of the material are computed via polynomial coefficient. The value of each parameter was calculated as function of temperature using COMSOL Multiphysics; these diagrams are shown in Figure 1. In addition, in order for the phase change in the mushy zone to be considered, the heat capacity method is solved by encompassing the latent heat ($L = 290$ kJ·kg^{-1}) of the base metal [19]. By doing so, the modified specific heat can be calculated in the mushy zone between the solidus and liquidus temperatures of the material ΔT_m, which is about 67 K; in this region, average melting temperature is defined to be 877.5 K approximately [20] (Equation (1)).

$$C_p = \begin{cases} C_{p,sensible} \ for \ T < T_m - 0.5\Delta T_m \ or \ T > T_m + 0.5\Delta T_m \\ C_{p,\mathrm{modified}} = C_{p,sensible} + L/\Delta T_m \ for \ T_m - 0.5\Delta T_m < T < T_m + 0.5\Delta T_m \end{cases} \quad (1)$$

in this equation, c_p and T illustrate the specific heat and temperature, respectively.

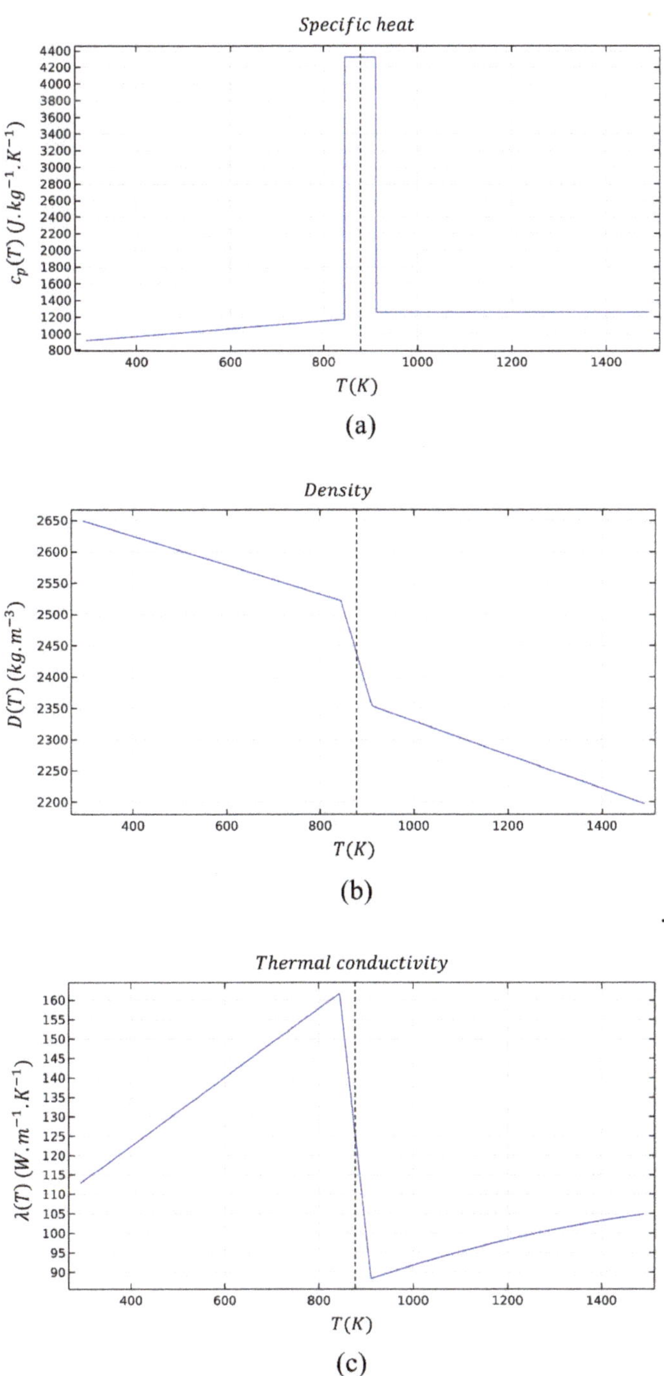

Figure 1. Variation of thermophysical properties for 5456 aluminum alloy (AA5456 with temperature: (**a**) specific heat; (**b**) density; and (**c**) thermal conductivity.

2.2. Thermal Modeling

The substrate with the width of 10 mm, the length of 10 mm, and the height of 2.5 mm is involved in the thermal modeling. In general, it should be mentioned that the dimension of actual substrate that is used for the laser welding process is larger than the domain considered for simulation. Nevertheless, the thermal model simply plays a role as a heat container in the welding process; the serious thermal involvement happens adjacent to the upper part of the substrate and decreases abruptly with shifting to the bottom [21]. It was verified with a domain size independence test that the calculation domain is appropriate to not influence the temperature regime during the welding process, while restricting the computational load. To do so, five kinds of domains were used to validate the domain size independence. These were $10 \times 10 \times 2.5$ mm^3, $8 \times 8 \times 2$ mm^3, $6 \times 6 \times 1.5$ mm^3, $4 \times 4 \times 1$ mm^3, and $2 \times 2 \times 1$ mm ^{3}, as domains 1 to 5, respectively. As shown in Figure 2, the temperature variation at point (5, 0, 2.5), as the initial point of laser beam with an average power of 80 W focused on the substrate, was extracted. The difference between these five domains is less than 1%. Therefore, domain 1 was selected for this calculation to reduce the computational cost and time. However, this issue should be taken into account that by increasing the heat input of laser beam, the molten depth can be larger than the height of the substrate determined in the thermal model. Therefore, if the molten depth becomes larger than the domain used in the thermal modeling, the height of the domain needs to be changed with regards to the molten depth. However, in order for the computational cost and time to be decreased with regards to the molten depth and width, which were the main concentration in the present study, a domain with 2.5 mm height was prepared. Additionally, natural convection within a liquid melt pool is neglected in the thermal calculation. By doing so, the molten pool temperature is supposed to be higher than its temperature in real experiments; however, this would affect the solidification process very little from the molten pool boundary in which phase transformation and heat conduction occur [22]. Equation (2) illustrates the distribution mode of laser beam power, which is considered to be Gaussian. In this equation, $(x_0,\ y_0)$ is the initial point of the laser beam focused on the substrate, which is equal to (5, 0). Moreover, it should be noted that a pulsed Nd:YAG laser was utilized in the present study; therefore, it has a pulse duration and frequency with which the on-time and off-time of the laser beam can be determined. According to the parameters in Table 2, the pulse duration and frequency are 4 ms and 10 Hz, respectively. This means the laser beam is activated for 4 ms and then deactivated for 96 ms. In this regard, there should be 10 pulses in a second of the laser welding process (the time length from a pulse to the next pulse is around 100 ms). To consider the pulse mode of the laser beam, a step function is defined in which its value is 1 when the laser is active; on the other hand, its value is 0 when the laser beam is off. This function φ, which is dependent on time of the welding process t, is shown below as Equation (3).

$$q(x, y, t) = \frac{2\lambda P\varphi(t)}{\pi r_0^2} exp\left\{\frac{-2\left[(x - x_0 - Vt)^2 + (y - y_0 - Vt)^2\right]}{r_0^2}\right\} \tag{2}$$

$$\varphi(t) = \begin{cases} 1 & t = D \\ 0 & t = f - D \end{cases} \tag{3}$$

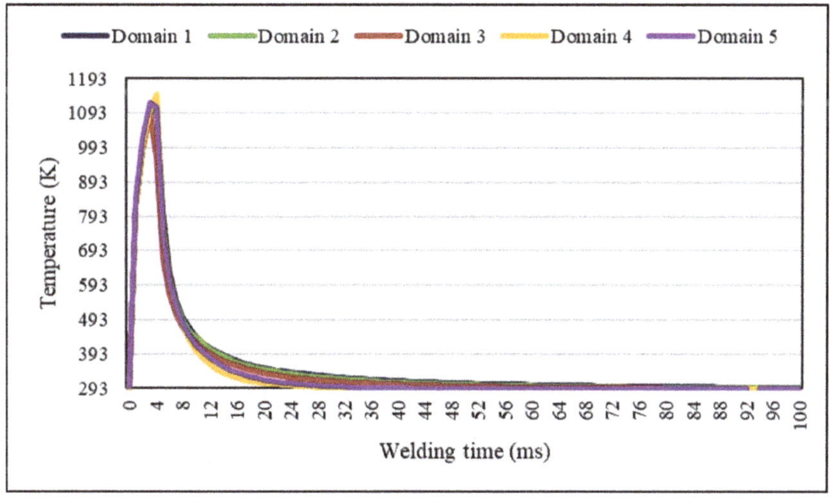

Figure 2. Domain size independence test.

In this equation, P is the laser power, λ is the absorption coefficient of the material, r_0 is the laser beam radius, V is the welding velocity, and $\varphi(T)$ is the function calculated via Equation (3), in which D and f are the pulse duration and frequency of the laser equipment, respectively. The absorptivity of aluminum alloy 5456 is reported to be 0.36 [23].

To calculate the amount of radiation losses during laser welding, Equation (4) is performed using a coefficient for radiative heat transfer through the top surface of the substrate:

$$h_{radiation} = \varepsilon\sigma\left(T^2 + T^2_{ambient}\right)(T + T_{ambient}) \tag{4}$$

In this equation, $h_{radiation}$ is supposed to be the coefficient for radiative heat transfer, σ is a constant coefficient known as the Stefan–Boltzmann constant, ε is the coefficient for emission of the substrate, and $T_{ambient}$ is considered to be the ambient temperature; these parameters are reported to be 5.67 $J/K^2{\cdot}m^4$, 0.022, and 293 K, respectively [24,25].

In the thermal model, the temperature of the bottom part of the substrate is kept at room temperature. Based on previous studies, it has been shown that the substrate temperature is not significantly affected by the temperature variation at the bottom part of the substrate [26]. Furthermore, there is an assumption in the thermal model in which the insulation aspect of side walls is taken into account. Figure 3a exhibits the thermal model, which is considered for simulation in this research. Moreover, the calculation domain used in this simulation is demonstrated in Figure 3b. It should be taken into account that the mesh used in the present study is extremely fine (530714 elements) to calculate the temperature-dependent parameters accurately. A tetrahedral-shaped mesh with the minimum element size of 0.002 mm and the element volume ratio of 0.06968 was used to investigate the thermal behavior of joints.

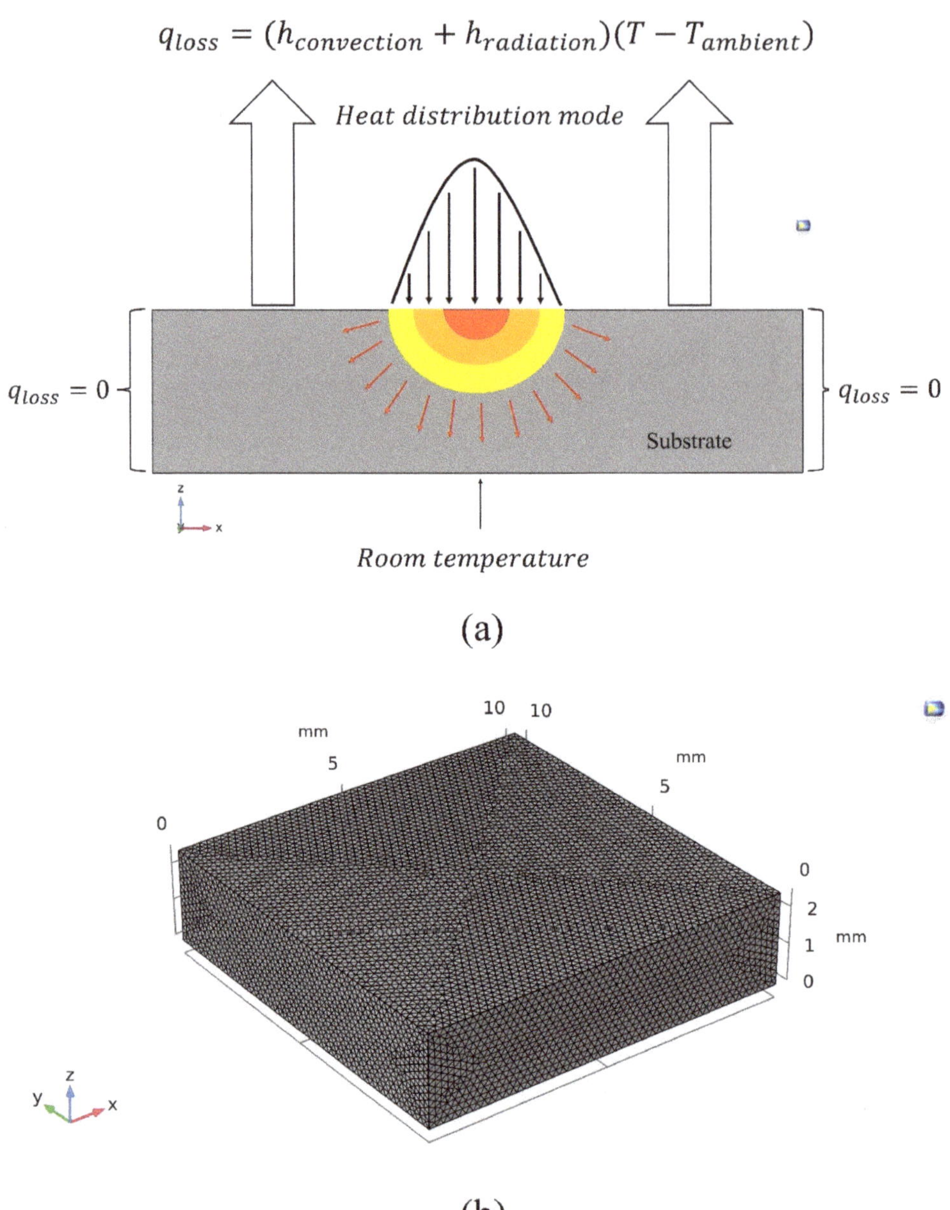

Figure 3. (**a**) Thermal model: (1) the distribution mode of laser heat source; (2) radiation and convection losses; (3) insulation of side walls; (4) the bottom part of the substrate remained at room temperature; (**b**) calculation domain used in the simulation.

3. Analytical Rosenthal Equation

Rosenthal [6] has proposed an analytical method to estimate the thermal characteristics of materials during fusion welding in conduction mode. Therefore, this equation can be used in laser welding processes undergoing conduction mode welding and not key hole, in order to understand the temperature-dependent behavior of materials during welding. The Rosenthal equation is a simple method to predict the thermal behavior of materials, which makes it applicable to various manufacturing processes such that time-dependent parameters like temperature gradient, cooling rate, and solidification rate are able to be calculated using this equation. For this purpose, there are some assumptions that have been taken into account to develop this particular equation, which are as follows [6]:

- Materials' properties such as specific heat at a constant pressure, density, and thermal conductivity are not the functions of temperature. In addition, the latent heat of materials is not considered when a phase transformation occurs between the solidus and liquidus temperatures of alloys, especially in the calculation of the specific heat in the mushy zone.
- The heat distribution mode on the substrate is considered to be quasi-stationary due to the consistency of welding velocity and power.
- A point heat source is selected in this analytical approach.
- Heat losses through radiation and convection are not taken into account. Furthermore, the natural convection within the substrate is not involved so that the heat produced in the substrate transfers by the conduction mode.

In order for the analytical Rosenthal equation to be applicable in the pulsed mode of a laser beam as a heat generator in the welding process, an additional assumption is made that the laser beam is active in the time range of its pulse duration 4 ms. To do so, Equation (3) should be performed directly in the Rosenthal equation. Therefore, the modified Rosenthal equation is achieved and shown in Equation (5):

$$T = T_0 + \frac{\lambda P \varphi(t)}{2\pi k r} exp\left[-\frac{V(r+\xi)}{2\alpha}\right] \tag{5}$$

According to Equation (5), T_0 is the base material starting temperature, k is the thermal conductivity, V is the welding velocity, α is the thermal diffusivity, and $\varphi(t)$ is the function that was previously described in Equation (3). It should be taken into account that in Equation (5), the welding is performed in the direction of the y-axis; so, to perform welding velocity and time in the equation, $y - Vt$ is substituted by ξ; r is the distance of any point on the substrate from the laser beam, which can be computed by $\sqrt{x^2 + \xi^2 + z^2}$. As discussed before, materials' properties are not the functions of temperature while using in the analytical solution. As a result, the properties mentioned in Table 4 are used in the analytical Rosenthal equation to predict the thermal behavior of the material in laser welding.

Table 4. Room-temperature thermal properties of aluminum alloy 5456 used in the analytical method.

Property	Value	Reference
Thermal conductivity, k	117 $W{\cdot}m^{-1}{\cdot}K^{-1}$	[27,28]
Density, ρ	2670 $kg.m^{-3}$	[27,28]
Specific heat, C_p	924 $J{\cdot}kg^{-1}{\cdot}K^{-1}$	[27,28]
Absorptivity, λ	0.36	[23]
Solid diffusivity, α	6.74×10^{-5} $m^2{\cdot}s^{-1}$	[17]

4. Results

In the present study, the results from experiments, the numerical modeling, and the analytical method are compared with regards to different energy densities. Equation (6) exhibits how the heat input can be calculated from the welding velocity and power. These parameters are of great importance because they affect the molten geometries and the heat transfer in the welding process. The calculated heat inputs in the present study are shown in Table 2.

$$E = P/V \tag{6}$$

Figure 4 demonstrates the heat transfer and a transverse-section molten pool achieved by the numerical modeling using an average power of 80 W, welding speed of 2 mm·s^{-1}, and welding time of 4 *ms*. By using this model, further calculations of the fusion depth and width, growth rate, and temperature gradient can be enabled, and thus, the microstructure of welds becomes available to be estimated using these temperature-dependent parameters. Furthermore, the dimension of the partially melted zone (PMZ) of welds can be determined using experimental, Rosenthal equation, and FE modeling results. The results obtained from the numerical modeling, the analytical method, and the experiments are compared. The temperature gradient and growth rate determined by the two routes (numerical and analytical) will be evaluated owing to the importance of these parameters in the prediction of microstructure. Moreover, the PDAS (primary dendrite arm spacing) obtained from two approaches will be investigated.

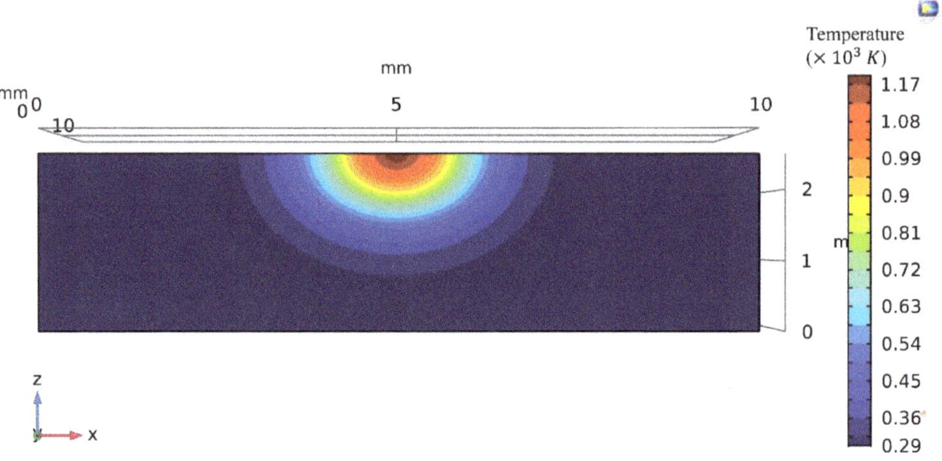

(a)

Figure 4. *Cont.*

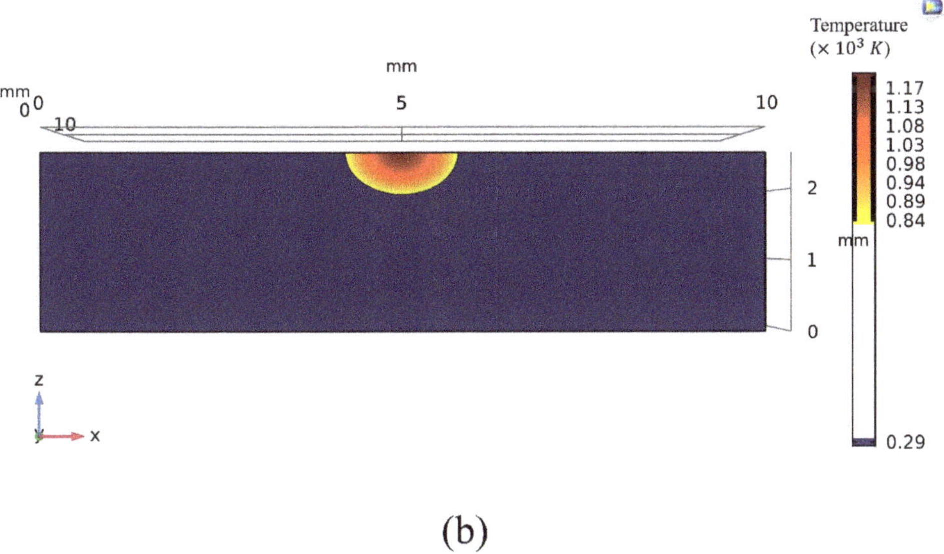

Figure 4. (**a**) Transverse-section (x-z) of the heat transfer; (**b**) molten pool achieved from the numerical modeling using an average laser power of 80 W, welding speed of 2 $mm.s^{-1}$, and absorptivity of 0.36 at $t = 4\ ms$.

4.1. Molten Pool Dimensions

According to the Rosenthal equation, the fusion width maximizes as $dx/d\xi = 0$. To compute the fusion width in the current study, a simple equation was developed which is derived from the Rosenthal equation. This particular equation enabled the calculation of the path length between the laser beam and any location on the melt pool at a specific temperature. By doing so, computing of the fusion width is possible, and then, the fusion depth can be determined as well. Regarding the welding velocity of 2 mm/s, the laser beam will move 8 μm in the direction of y-axis for 4 ms (the pulse duration of laser beam). The abovementioned criteria are shown as Equation (7).

$$\frac{2\alpha}{V} ln(r') + r' = -\frac{2\alpha}{V} ln\left(\frac{2\pi k(T - T_0)}{\lambda P}\right) - \xi \quad (7)$$

where r' is the path length between any location on the weld pool with a specific temperature and the laser beam; P is the welding power; T is the temperature of any particular location on the melt pool; T_0 is the initial temperature of the material, which is about 293 K; α is the thermal diffusion coefficient at 293 K; V is the welding velocity; k is the thermal conductivity of the material at 293 K; and λ is the absorption coefficient of the substrate. In order for the weld pool width to be calculated, Equation (8) is used which is written as follows:

$$w = 2r = 2\sqrt{(r')^2 - (V \times t)^2} \quad (8)$$

where r' is the path length between any location on the weld pool with a particular temperature and the heat generator, V is the welding velocity, and t is the welding time. It should be taken into account that the analytical Rosenthal equation creates a semi-circular molten pool in which the fusion width (w) is two times bigger than the fusion depth (d).

Experimental results in terms of the fusion width were compared with fusion widths, fusion depths, and partially melted zones thickness achieved by numerical modeling and analytical method.

To do so, Rosenthal equation and FE modeling were used in order to calculate these values with which the comparison of these results becomes available with regards to experimental results. It should be noted that the PMZ is the area in which solidification cracks can be produced during laser welding such that the study in terms of this particular area can be beneficial to predict the solidification microstructure. The parameters performed in the numerical modeling are similar to those utilized in experiments; these parameters are comprehensively mentioned in Section 2. In order to compare the validation of the numerical and analytical methods regarding experimental results, four distinct laser powers were selected in the present study.

Fusion widths obtained from experiments are compared with the estimated ones from the numerical modeling and the analytical method. This comparison is illustrated in Figure 5. The average fusion widths were measured from three different transverse sections perpendicular to the welding direction for each heat input. Based on these results, it is observed that fusion widths achieved from the numerical modeling are consistent with those from experiments up to an energy density of 30 J·mm^{-1}. By approaching higher energy densities, the predicted fusion widths tend to overestimate the average fusion widths obtained from experiments. However, there is a gap between the results from the analytical method and ones from experiments; obviously, fusion widths achieved from the Rosenthal equation are bigger than those from experiments. It is apparent that the results from Rosenthal equation are not matched with the experimental results due to some of selected assumptions described in Section 3. According to the FE models, natural convection within the molten pool is neglected so that in smaller melt pools, the temperature is overestimated, compared with genuine melt pools. Therefore, radiation losses are increased in smaller fusion welds. In contrast to this idea, based on the assumptions for the analytical Rosenthal equation, radiation losses are not involved in this method; therefore, fusion widths attained from the analytical method are larger than those from the numerical modeling and experiments.

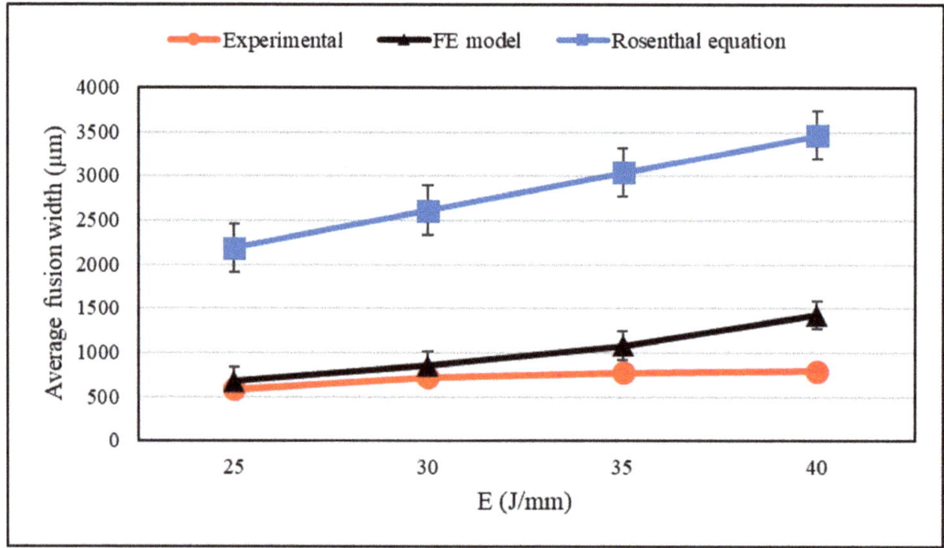

Figure 5. Comparison of the fusion width obtained from experiments, the numerical modeling, and the analytical Rosenthal equation.

Fusion depths obtained from experiments are compared with the estimated ones from the numerical modeling and the analytical method. This comparison is demonstrated in Figure 6. The average fusion depths were measured from three different transverse sections perpendicular to the welding direction for each heat input. According to these results, it is apparent that fusion depths achieved from the numerical modeling are consistent with those from experiments up to an energy density of 30 J·mm^{-1}. By approaching higher energy densities, the predicted fusion depths incline to overestimate the average fusion depths obtained from experiments. On the other hand, the predicted results from the Rosenthal equation seem to overestimate the average fusion depths attained from the experiments. The main reason may lie in the assumptions of the Rosenthal equation. As discussed above, there is no convection during the simulation of the welding process such that the results can be overestimated. Furthermore, the radiative losses might be too high especially for smaller melt pools; therefore, the predicted results might be inaccurate. In contrast, in the analytical method, radiation losses and natural convection are not considered with which the predicted results could be overestimated in comparison to the predictions from FE modeling and the experimental results.

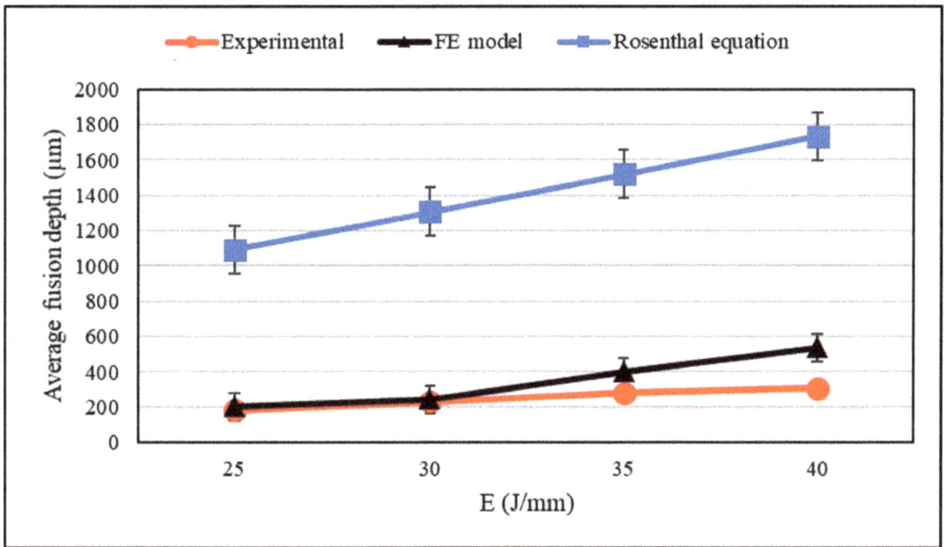

Figure 6. Comparison of the fusion depth obtained from experiments, the numerical modeling, and the analytical Rosenthal equation.

Figure 7 demonstrates a comparison made on the partially melted zones thickness obtained from the experiments, the FE modeling, and the Rosenthal equation as well as a cross-section of the molten pool, whose parameters were selected from A1 in Table 2, fitted by the FE model of the partially melted zone. As shown in Figure 6, the partially melted zone thickness obtained from the experiments increases when the heat input of the laser beam is augmented. Actually, the partially melted zone (PMZ) is an area formed between the eutectic (or solidus temperature for solutionized work pieces) and liquidus temperatures of materials [29]. It is reported that the energy density can only affect the PMZ length during welding [30]. According to the predicted results from the FE modeling, it is obvious that the partially melted zone thickness enhances with the heat input increment; however, these results overestimate the length of PMZ due to the simplifications described above, such as insulating of substrate walls and neglecting the existence of convection inside the melt pool. Regarding the results obtained from the analytical method, an overestimation of the experimental results can be seen from Figure 7. As described above, there were some assumptions performed to the Rosenthal equation with which the predicted results are not matched with the experimental ones in this regard.

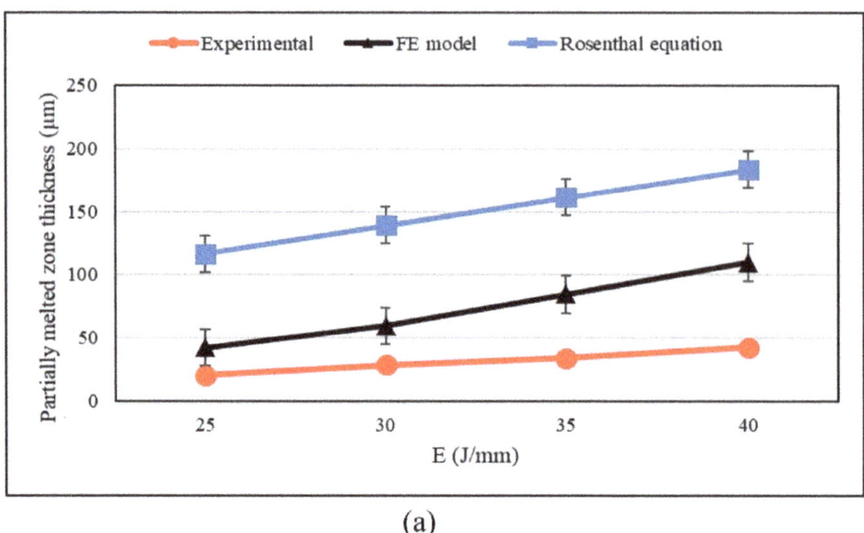

(a)

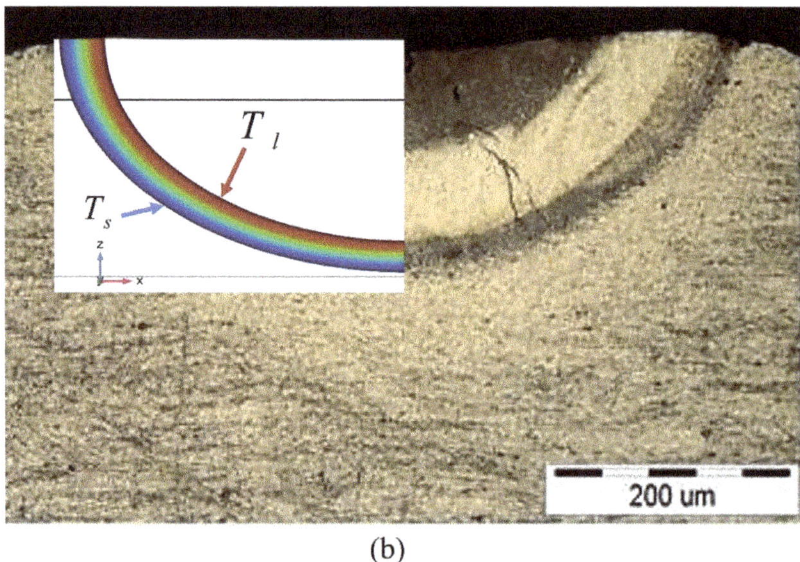

(b)

Figure 7. (**a**) Comparison of the partially melted zone thickness obtained from experiments, the numerical modeling, and the analytical method; (**b**) fitted partially melted zone from the finite-element (FE) model on the cross-section of the melt pool using an average laser power of 50 *W*, welding velocity of 2 mm·s^{-1}, and absorptivity of 0.36.

4.2. Predicted Temperature-Dependent Parameters

Both of the temperature gradient (G) and the growth rate (R) are of paramount importance for determining the microstructure of solidified welds. The analytical Rosenthal equation can be utilized to estimate these imperative parameters abruptly via derivation from Equation (5). The temperature gradients in the directions of the axis y- and the axis z- are computed using the following equations. Equations (9) and (10) demonstrate the temperature gradients in the direction of the axis y- and the axis z-, respectively; both of these equations can be utilized through a section in the y-z plane where x is equal to zero. Moreover, for any point in the transverse section of the molten pool at $x = 0$, the cooling rate can be computed using Equation (11).

$$\frac{\partial T}{\partial y} = \left[1 + \frac{y}{\sqrt{y^2+z^2}} + \frac{2\alpha y}{V(y^2+z^2)}\right]\left(-\frac{\lambda P}{2\pi k}\frac{V}{2\alpha}\frac{1}{\sqrt{y^2+z^2}}\right)exp\left[-\frac{V}{2\alpha}\left(y+\sqrt{y^2+z^2}\right)\right] \quad (9)$$

$$\frac{\partial T}{\partial z} = \left[1 + \frac{2\alpha}{V\sqrt{y^2+z^2}}\right]\left(-\frac{\lambda P}{2\pi k}\frac{V}{2\alpha}\frac{z}{y^2+z^2}\right)exp\left[-\frac{V}{2\alpha}\left(y+\sqrt{y^2+z^2}\right)\right] \quad (10)$$

$$\frac{\partial T}{\partial t} = \left[1 + \frac{y}{\sqrt{y^2+z^2}} + \frac{2\alpha y}{V(y^2+z^2)}\right]\left(\frac{\lambda P}{2\pi k}\frac{V^2}{2\alpha}\frac{1}{\sqrt{y^2+z^2}}\right)exp\left[-\frac{V}{2\alpha}\left(y+\sqrt{y^2+z^2}\right)\right] \quad (11)$$

From the work of Bontha et al. [31], temperature-dependent parameters, namely, temperature gradient, cooling rate, and growth rate vary considerably at different points inside the molten pool. It is of great importance that the fusion depth should be calculated by Equation (7) so that the temperature gradient and the cooling rate of solidified welds can be determined in this regard. In this case, the average fusion depth can be seen from Figure 5 as well. Using Equation (5), the points y and z for different heat inputs are computed from the bottom of the welds to the centerline. These values are utilized to investigate the heat transfer behavior within the molten pool using Equations (9)–(11). Then, the average amounts of the results obtained from the analytical method are compared for various locations. In order for the solidification rate within the melt pool to be determined, the values of temperature gradient and cooling rate achieved from the analytical Rosenthal equation are used in Equation (12) for each point in the melt pool.

$$R = \frac{1}{G}\frac{\partial T}{\partial t} = \frac{1}{\sqrt{\left(\frac{\partial T}{\partial y}\right)^2 + \left(\frac{\partial T}{\partial z}\right)^2}}\frac{\partial T}{\partial t} \quad (12)$$

As to the numerical modeling, it is also of paramount importance that the fusion depth of solidified welds needs to be determined to evaluate the heat transfer behavior within the molten pool. Then, temperature-dependent parameters including temperature gradient and cooling rate are calculated at different points within the melt pool. In order for this case to be clearer, conceive an example of a laser beam with an average power of 80 W, welding velocity of 2 mm.s^{-1}, and an absorption coefficient of 0.36. Performing these laser parameters, the obtained fusion depth using the numerical modeling is 576 µm. Temperature and temperature gradient, as thermal properties of the welded material, are computed in five different points from the bottom of the weld pools to the centerline: 76, 176, 276, 376, and 476 µm. By doing so, Figure 8a–c is attained in which the temperature and the temperature gradient in the direction of the axis z- are illustrated as functions of time and temperature, respectively. It is worthwhile to mention that the average amount of the temperature gradient is taken into account for the melting point and compared with the results obtained from the analytical method. For determining the solidification rate within the molten pool, the cooling rate is measured in the same way as the temperature gradient which was discussed before. Then, the temperature gradient and the cooling rate obtained from the numerical modeling are used in Equation (12), and ultimately, the values of the solidification rate are attained for different locations within the melt pool.

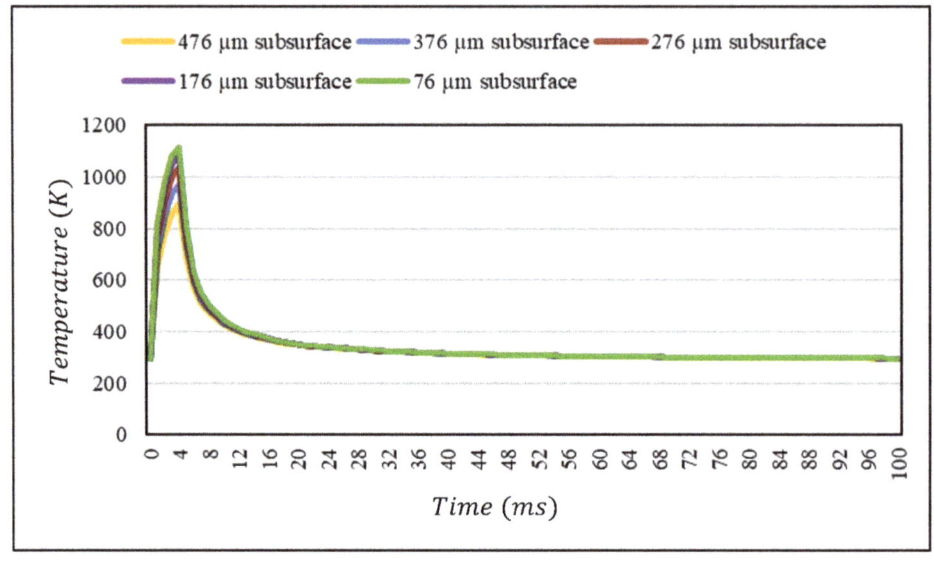

(a)

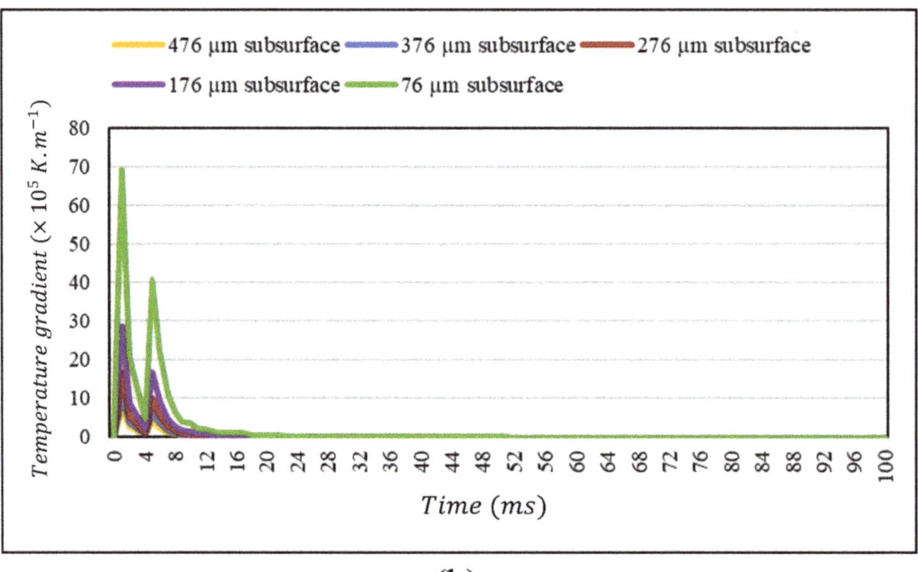

(b)

Figure 8. *Cont.*

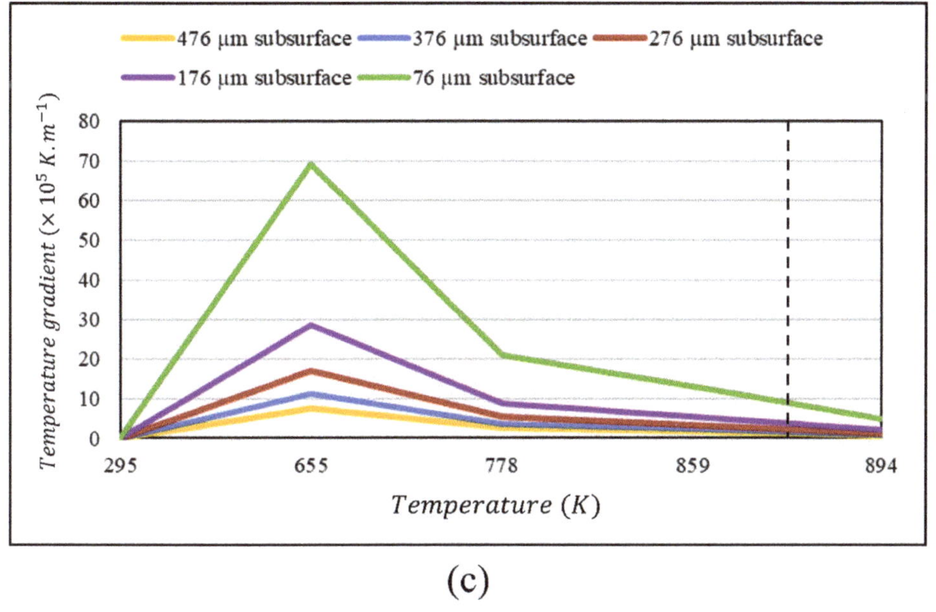

(c)

Figure 8. (**a**) Variation of temperature versus welding time for different points within the molten pool; (**b**) variation of temperature gradient versus welding time for different points within the molten pool; (**c**) variation of temperature gradient versus temperature for different points within the molten pool obtained from the numerical modeling using a laser beam with an average power of 80 W, welding velocity of 2 mm·s^{-1}, and absorption coefficient of 0.36.

Temperature gradients are determined for different heat inputs using the numerical modeling and the analytical method, and the results obtained from the numerical modeling are compared with those attained from the analytical Rosenthal equation as shown in Figure 9, where the welding velocity is constant and its value is about 2 mm·s^{-1}. As shown in Figure 9a, the temperature gradient is decreased by increasing heat input of the laser beam; this trend is the same for both approaches including the numerical modeling and the analytical method. The reason may lie in that by increasing the heat input of the laser beam, the molten pool becomes larger, which deteriorates the temperature gradient within the weld pool. According to the parameters used in this study, the results for temperature-dependent parameters such as temperature gradient, cooling rate, and growth rate obtained from the numerical modeling are slightly bigger than those from the analytical Rosenthal equation, as shown in Figure 9a–c.

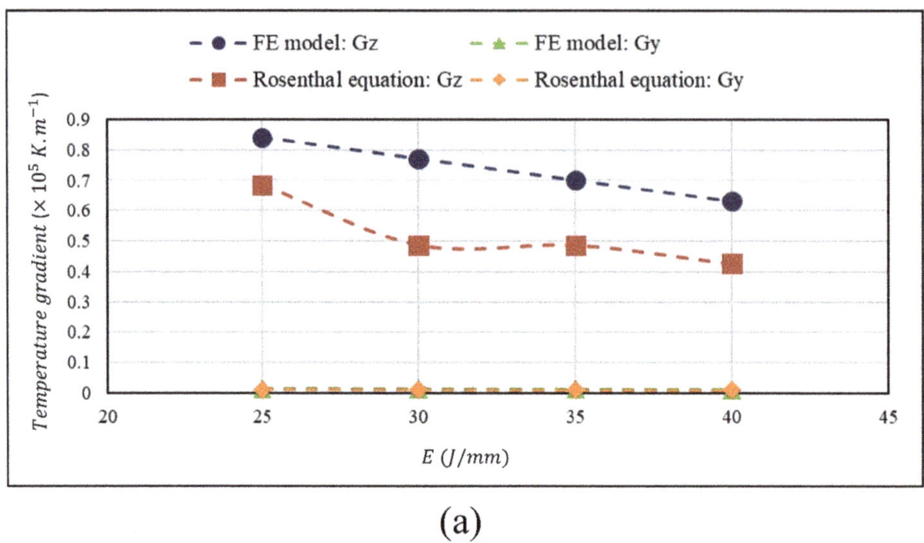

(a)

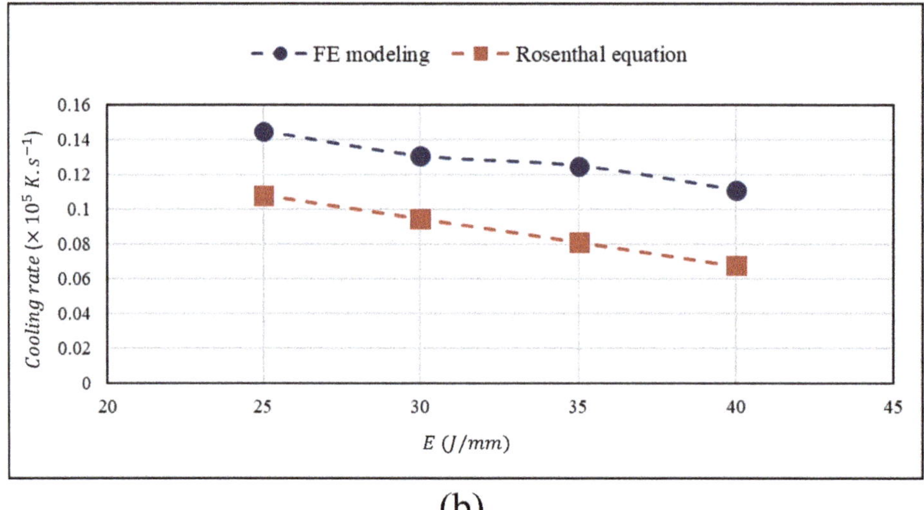

(b)

Figure 9. *Cont.*

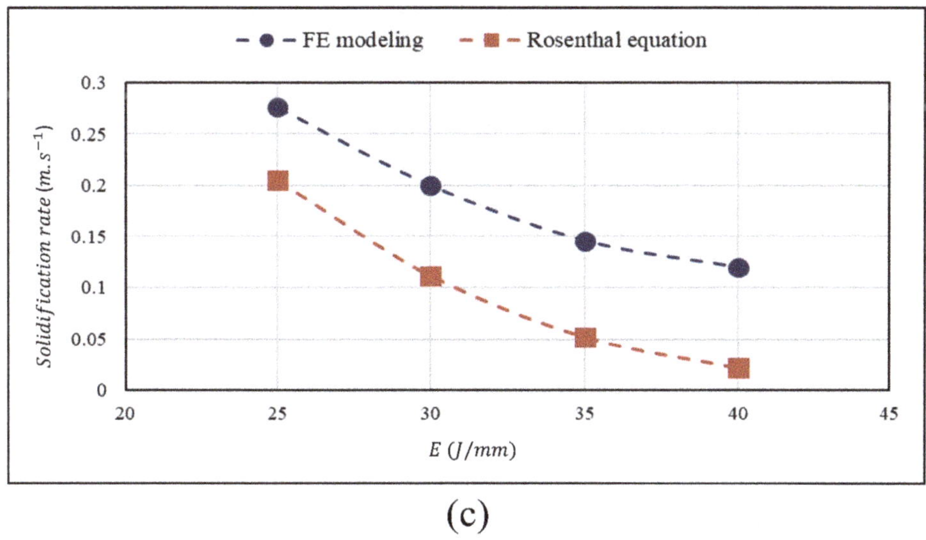

(c)

Figure 9. (**a**) Temperature gradient, (**b**) cooling rate, and (**c**) solidification rate achieved from the numerical modeling and the analytical method versus heat input of the laser beam and their comparison.

Additionally, as can be seen from Figure 9b, the difference between the values of cooling rates obtained from the numerical modeling and the analytical method is increased by enhancing the energy density of the laser beam. The major reason may lie in that by increasing the heat input, radiation loss is increased as well so that this plays an integral role in increasing the difference between the results achieved from the numerical modeling and the analytical method at higher energy densities. Furthermore, as to the analytical method, heat losses are not considered through the substrate such that the temperature is raised considerably within the molten pool which makes it bigger in comparison to the FE model. Therefore, the cooling rate attained from the analytical method is lower than that achieved from the numerical modeling. Moreover, the more heat input is increased, the more the cooling rate from the analytical method is decreased due to an increase in weld pool dimensions. In addition, as to the FE model, the thermal conductivity of the material is a function of temperature. Therefore, by approaching the temperature near to the melting point of the material, the thermal conductivity is increased so that this can help the cooling process in the simulation. However, this cannot be a positive point for the cooling rate achieved from the analytical method because materials properties are not dependent on the temperature of the material using in the analytical Rosenthal equation, as discussed before. Goldak et al. [32] conducted a research in terms of the cooling rate and a comparison of its value obtained from the numerical modeling, the analytical Rosenthal equation, and experiments during welding. Based on the results from the numerical modeling, the analytical method, and experiments, they reported that the cooling rate attained from the FE model was underestimated by 5%, while the cooling rate computed from the Rosenthal equation was overestimated by 41%. Quite contrary, the results obtained from this study differ from those achieved by Goldak et al. because the cooling rate attained from the numerical modeling is higher than that achieved from the analytical method in the present study due to the reasons explained above. It should be considered that Goldak et al. [32] used low-carbon steel as the base metal that has lower thermal conductivity at higher temperatures so that the cooling rate calculated by the numerical modeling became lower than that computed by the analytical Rosenthal equation.

While investigating the heat transfer behavior within the molten pool, it is noteworthy to mention that the initial temperature of the substrate is considered to be constant, which is equal to room temperature. However, in genuine environments, the accumulation of heating in the substrate may affect the initial temperature so that it cannot be considered to become constant. There are some factors that can influence on the initial temperature of the substrate, including a minute time length between pulses with which the time needed for the cooling process is decreased; a small welding path with which the temperature of the molten pool is raised considerably after each pulse; and small features in the geometry with which the heat conduction is become weak due to the reason that oxide films on the substrate have low thermal conductivity. Without considering the local temperature of the substrate, the analytical Rosenthal equation is not applicable to evaluate the heat transfer behavior of the welded material; nevertheless, particular geometries can be performed in the numerical modeling for investigation of the heat transfer during welding.

4.3. Microstructural Evaluation: Primary Dendrite Arm Spacing (PDAS)

Understanding the relationship between the laser parameters and the microstructure can require an understanding of the microstructural evaluation of the joints via the numerical and analytical studies. By doing so, the mechanical characteristics of laser welded components can be estimated using this invaluable knowledge. Primary dendrite arm spacing (PDAS) can be an appropriate microstructural instance with which the mechanical features of welded materials can be elaborated upon [9]. As a result, the correlation between the solidification process and primary dendrite arm spacing has become the attention of various researches [5,33,34]. The Kurz and Fisher (KF) model is one of the most traditional models with which the calculation of PDAS becomes available [34]. It should be mentioned that this model (Equation (13)) has been utilized in Al-Mg systems in which PDAS has reported to be accurate with regards to experimental results [5].

$$\delta = 4.3\left(\frac{\Delta T_0.D.\Gamma}{k_0}\right)^{1/4}.G^{-1/2}.R^{-1/4} \tag{13}$$

in this equation, G, which is known as temperature gradient, is the variation of temperature around a particular location at a given time; R, which is known as solidification rate or growth rate, is the travel speed of the solid-liquid interface at a given temperature; δ is the primary dendrite arm spacing (m). The values of ΔT_0, D, k_0, and Γ are intrinsic characteristics of the material which are provided in Table 5 [5,20].

Table 5. Material characteristics of the AA5456 used in the Kurz and Fisher (KF) model.

Feature	Amount	Reference
Solidification zone, ΔT_0	67 K	[20]
Gibbs–Thomson coefficient, Γ	1.3×10^{-7} K·m	[5]
Partition coefficient, k_0	0.48	[5]
Liquid diffusivity, D	10^{-8} m^2·s^{-1}	[5]

As shown in Figure 10, the variation of primary dendrite arm spacing (PDAS) obtained from the numerical modeling and the analytical method is illustrated versus the heat input of the laser beam, and a comparison between these results and experimental one has been made. Obviously, the G (temperature gradient) and R (solidification rate) values obtained from the numerical modeling and the analytical Rosenthal equation are different so that Equation (13) results in different values for PDAS attained from the numerical modeling and the analytical method. Wang et al. [35] have observed that both calculation approaches illustrate the same trend in which the PDAS is increased by enhancing heat input. It should be taken into account that the PDAS, which is predicted by the analytical method, is attained to be slightly larger than that from the numerical modeling. To observe the validation of the results obtained from the numerical modeling, an experiment was conducted using the same laser parameters as for the numerical method, and these results are compared with each other. Figure 10 shows the transverse-section of the laser welded material in which the laser parameters are selected from A2 in Table 2. As shown in Figure 11, the PDAS value is 1.029 μm at the heat input of 30 J·mm^{-1}. It is apparent that the results obtained from the numerical modeling and the analytical method are consistent with the result from an experiment which was done at the heat input of 30 J·mm^{-1}. Apparently, both of the numerical and analytical methods are capable to produce the primary dendrite arm spacing (PDAS), but it needs to have more experimental results to validate the numerical and analytical methods efficiently and compare their results to see which one is more accurate that the other one. Thus, there should be more experiments, specifically at various heat inputs, to verify the numerical modeling and the analytical method. Furthermore, with more available data, the determination of better methods will be attainable as well.

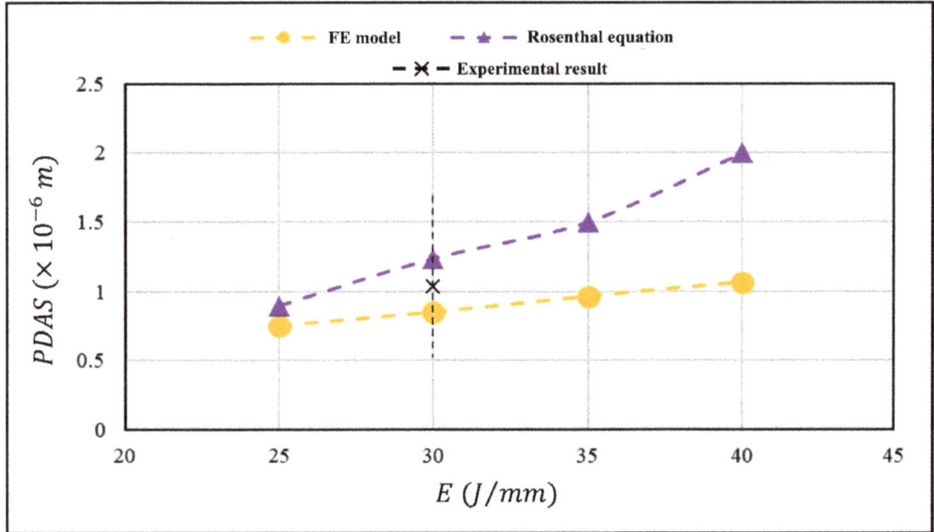

Figure 10. Variation of primary dendrite arm spacing (PDAS) obtained from the experiment, the numerical modeling, and the analytical method versus heat input.

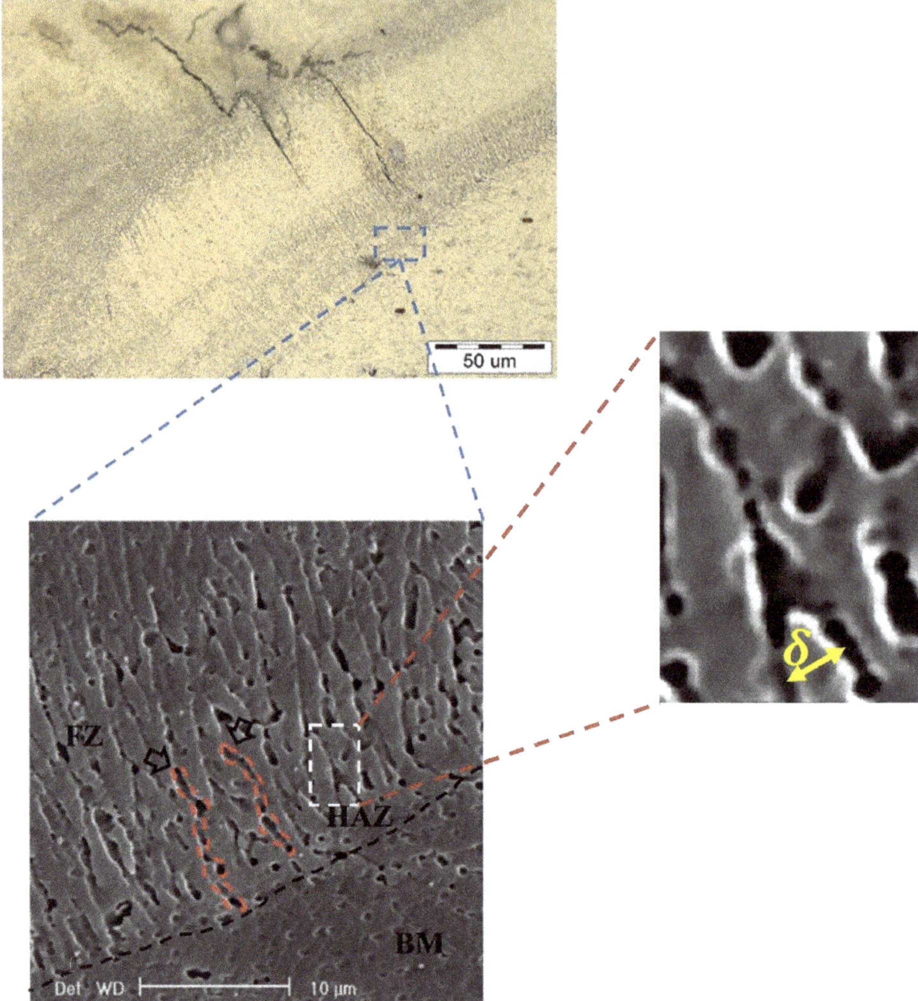

Figure 11. Measurement of dendrite arms taken by optical and scanning electron microscopies at the boundary of fusion zone using a laser with an average power of 60 W, welding velocity of 2 mm.s^{-1}, and absorption coefficient of 0.36.

5. Conclusions

In the present study, the heat transfer behavior of AA5456 as the base metal is investigated while welding by a pulsed Nd:YAG laser. To do so, the numerical modeling and the analytical method are performed, and the estimated results are compared with the experimental ones. According to the investigation, welding pool dimensions including fusion width, fusion depth, and PMZ thickness; temperature-dependent parameters such as temperature gradient, cooling rate, and growth rate; and primary dendrite arm spacing (PDAS) are determined for different heat inputs. Moreover, the results obtained from the numerical modeling and the analytical method are compared with the results from experiments to validate these methods as a confident tool in prediction of mechanical properties as well as the microstructure. The conclusions are mentioned below:

(1) The fusion width and depth obtained from the numerical modeling are consistent with the experimental results up to the heat input of 30 $J \cdot mm^{-1}$. However, by increasing the heat input, the discrepancy between the results from the numerical modeling and experimental ones becomes larger. The Rosenthal equation overestimates the melt pool dimensions. The reason may lie in the assumptions performed in the analytical Rosenthal equation in terms of neglecting heat losses through the substrate.
(2) With regards to the partially melted zone, the FE model is slightly larger than the experimental results at the heat input of 25 $J \cdot mm^{-1}$. However, by increasing the heat input, the discrepancy between the results achieved from the numerical modeling and the results from experiments becomes wider. On the other hand, the Rosenthal equation overestimates the partially melted zone in comparison to the FE model and the experiment; this could be a result of assumptions with which heat losses as well as latent heat of the material are supposed to be neglected.
(3) Numerical modeling produces temperature-dependent parameters including temperature gradient, cooling rate, and growth rate, which are higher than those obtained from the analytical method. Moreover, the discrepancy between the predicted results using the numerical modeling and the analytical method becomes larger by enhancing energy density.
(4) The primary dendrite arm spacing (PDAS) is measured using the numerical modeling and the analytical method. As a result, the values obtained from the analytical method are slightly larger than those from the numerical modeling. Furthermore, it is observed that both of the numerical and analytical methods predict the PDAS with somewhat accuracy at the heat input of 30 $J \cdot mm^{-1}$ regarding the experimental result. However, there should be more experiments to verify these methods as a confident tool for prediction of the microstructure in laser welded materials, especially at different heat inputs.
(5) All in all, the Rosenthal equation underestimates thermal results and overestimates microstructural results in comparison to the FE model. It should be taken into account that the discrepancy between the results attained from the analytical method and the numerical modeling becomes larger at higher heat inputs; therefore, there should be a restriction in utilizing the analytical Rosenthal equation at higher heat inputs to investigate the heat transfer behavior of laser welded materials due to the assumptions performed in this particular method. Thus, the numerical modeling can be widely used for investigation of heat transfer behavior of welded components at higher heat inputs, and its accuracy is more than the analytical method because it uses genuine materials' characteristics.

Author Contributions: H.H.: Investigation, formal analysis, writing of the paper; Z.Z.: Formal analysis, writing of the paper; J.S.: Investigation; J.P.O.: Formal analysis, writing of the paper; H.N.-M.: Formal analysis, supervision, writing of the paper. All authors have read and agreed to the published version of the manuscript.

Funding: This research was funded by Fundação para a Ciência e a Tecnologia (FCT–MCTES): Grant number: UIDB/00667/2020 (UNIDEMI).

Acknowledgments: J.S. and J.P.O. acknowledge Fundação para a Ciência e a Tecnologia (FCT–MCTES) for its financial support via the project UIDB/00667/2020 (UNIDEMI).

Conflicts of Interest: The authors declare no conflict of interest.

References

1. Hamidreza, H.; Naffakh-Moosavy, H. Hot cracking in pulsed Nd: YAG laser welding of AA5456. *Opt. Laser Technol.* **2018**, *103*, 22–32.
2. Sabina, C.; Casalino, G.; Casavola, C.; Moramarco, V. Analysis and comparison of friction stir welding and laser assisted friction stir welding of aluminum alloy. *Materials* **2013**, *6*, 5923–5941.
3. Cam, G.; Koçak, M. Progress in joining of advanced materials Part 2: Joining of metal matrix composites and joining of other advanced materials. *Sci. Technol. Weld. Join.* **1998**, *3*, 159–175. [CrossRef]

4. Zhao, H.; White, D.R.; DebRoy, T. Current issues and problems in laser welding of automotive aluminium alloys. *Int. Mater. Rev.* **1999**, *44*, 238–266. [CrossRef]
5. Beiranvand, Z.; Malekshahi, F.; Malek Ghaini, H.; Naffakh Moosavy, M.S.; Torkamany, M.J. Solidification cracking susceptibility in pulsed laser welding of Al–Mg alloys. *Materialia* **2019**, *7*, 100417. [CrossRef]
6. Rosenthal, D. Mathematical theory of heat distribution during welding and cutting. *Weld. J.* **1941**, *20*, 220–234.
7. Tang, M.P.; Chris, P.; Jack, L.B. Prediction of lack-of-fusion porosity for powder bed fusion. *Addit. Manuf.* **2017**, *14*, 39–48. [CrossRef]
8. Liang, Y.-J.; An, L.; Xu, C.; Pang, X.-T.; Wang, H.-M. Prediction of primary dendritic arm spacing during laser rapid directional solidification of single-crystal nickel-base superalloys. *J. Alloy. Compd.* **2016**, *688*, 133–142. [CrossRef]
9. Lei, W.; Wei, Y.; Zhan, X.; Yu, F.; Cao, X.; Gu, C.; Ou, W. Simulation of dendrite growth in the laser welding pool of aluminum alloy 2024 under transient conditions. *J. Mater. Process. Technol.* **2017**, *246*, 22–29.
10. John, R.; Ladani, L.; Sadowski, M. Laser additive melting and solidification of Inconel 718: Finite element simulation and experiment. *Jom* **2016**, *68*, 967–977.
11. Antoni, A.; Bachmann, M.; Rethmeier, M. Equivalent heat source approach in a 3D transient heat transfer simulation of full-penetration high power laser beam welding of thick metal plates. *Int. J. Heat Mass Transf.* **2018**, *122*, 1003–1013.
12. Satyanarayana, G.K.L.; Narayana, B.; Nageswara, R.M.S.; Slobodyan, M.A.E.; Kiselev, A.S. Numerical Simulation of the Processes of Formation of a Welded Joint with a Pulsed ND: YAG Laser Welding of ZR–1% NB Alloy. *Therm. Eng.* **2019**, *66*, 210–218. [CrossRef]
13. Wei, H.; Chen, J.S.; Wang, H.-P.; Blair, E.C. Thermomechanical numerical analysis of hot cracking during laser welding of 6XXX aluminum alloys. *J. Laser Appl.* **2016**, *28*, 022405. [CrossRef]
14. Wei, H.; He, Q.; Chen, J.-S.; Wang, H.-P.; Blair, E.C. Coupled thermal-mechanical-contact analysis of hot cracking in laser welded lap joints. *J. Laser Appl.* **2017**, *29*, 022412. [CrossRef]
15. He, Q.; Wei, H.; Chen, J.-S.; Wang, H.-P.; Blair, E.C. Analysis of hot cracking during lap joint laser welding processes using the melting state-based thermomechanical modeling approach. *Int. J. Adv. Manuf. Technol.* **2018**, *94*, 4373–4386. [CrossRef]
16. Mills, K.C. *Recommended Values of Thermophysical Properties for Selected Commercial Alloys*; Woodhead Publishing: Sawston, Cambridge, UK, 2002.
17. Leitner, M.; Leitner, T.; Schmon, A.; Aziz, K.; Pottlacher, G. Thermophysical properties of liquid aluminum. *Metall. Mater. Trans. A* **2017**, *48*, 3036–3045. [CrossRef]
18. Gaosheng, W.; Huang, P.; Xu, C.; Liu, D.; Ju, X.; Du, X.; Xing, L.; Yang, Y. Thermophysical property measurements and thermal energy storage capacity analysis of aluminum alloys. *Sol. Energy* **2016**, *137*, 66–72.
19. Comini, B.; Comini, G.; Fasano, A.; Primicerio, M. Numerical solution of phase-change problems. *Int. J. Heat Mass Transf.* **1973**, *16*, 1825–1832.
20. Muraca, R.F.; Whittick, J.S. *Materials Data Handbook: Aluminum Alloy 5456*; NASA Technical Report Server: Washington, DC, USA, 1 June 1972.
21. Salman, H.A.; Hubeatir, K.A.; AL-Kafaji, M.M. Modeling of Continues Laser Welding for Ti-6Al-4V Alloys Using COMSOL Multiphysics Software. *Eng. Technol. J.* **2018**, *36*, 914–918.
22. Hu, D.; Kovacevic, R. Modelling and measuring the thermal behaviour of the molten pool in closed-loop controlled laser-based additive manufacturing. *Proc. Inst. Mech. Eng. Part B J. Eng. Manuf.* **2003**, *217*, 441–452. [CrossRef]
23. Gaskell, D.R. *Introduction to the Thermodynamics of Materials*, 5th ed.; CRC Press: Boca Raton, FL, USA, 2008; pp. 66–91.
24. Bergman, L.T.; Incropera, F.P.; DeWitt, D.P.; Lavine, S.A. *Fundamentals of Heat and Mass Transfer*; John Wiley & Sons: Hoboken, NJ, USA, 2011.
25. Zhao, N.; Yang, Y.; Han, M.; Luo, X.; Feng, G.; Zhang, G. Finite element analysis of pressure on 2024 aluminum alloy created during restricting expansion-deformation heat-treatment. *Trans. Nonferrous Met. Soc. China* **2012**, *22*, 2226–2232. [CrossRef]
26. Patcharapit, P.; Onler, R.; Yao, S. Numerical and experimental investigations of micro and macro characteristics of direct metal laser sintered Ti-6Al-4V products. *J. Mater. Process. Technol.* **2017**, *240*, 262–273.

27. Ramasamy, S. CO_2 and Nd: YAG laser beam welding of 6111-T4 and 5754-O aluminum alloys for automotive applications. Ph.D. Thesis, The Ohio State University, Columbus, OH, USA, 1997.
28. Moon, D.W.; Metzbower, E.A. Laser beam welding of aluminum alloy 5456. *Weld. J.* **1983**, *62*, 53s–58s.
29. Rao, K.; Prasad, N.R.; Viswanathan, N. Partially melted zone cracking in AA6061 welds. *Mater. Des.* **2008**, *29*, 179–186.
30. Homam, M.; Mohammad-Reza, N.; Seyed, A.; Seyedein, H.; Goodarzi, M.; Khodabakhshi, M.; Mapelli, C.; Barella, S. Modern fiber laser beam welding of the newly-designed precipitation-strengthened nickel-base superalloys. *Opt. Laser Technol.* **2014**, *57*, 12–20.
31. Srikanth, B.; Klingbeil, N.W.; Kobryn, P.A.; Fraser, H.L. Effects of process variables and size-scale on solidification microstructure in beam-based fabrication of bulky 3D structures. *Mater. Sci. Eng. A* **2009**, *513*, 311–318.
32. John, G.; Chakravarti, A.; Bibby, M. A new finite element model for welding heat sources. *Metall. Trans. B* **1984**, *15*, 299–305.
33. Lu, S.-Z.; Hunt, J.D. A numerical analysis of dendritic and cellular array growth: The spacing adjustment mechanisms. *J. Cryst. Growth* **1992**, *123*, 17–34. [CrossRef]
34. Kurz, W.; Fisher, D.J. Dendrite growth at the limit of stability: Tip radius and spacing. *Acta Metall.* **1981**, *29*, 11–20. [CrossRef]
35. Xiaoqing, W.; Gong, X.; Chou, K. Review on powder-bed laser additive manufacturing of Inconel 718 parts. *Proc. Inst. Mech. Eng. Part B J. Eng. Manuf.* **2017**, *231*, 1890–1903.

Article

Influence of Tool Material, Tool Geometry, Process Parameters, Stacking Sequence, and Heat Sink on Producing Sound Al/Cu Lap Joints through Friction Stir Welding

Hongyu Wei [1], Abdul Latif [2], Ghulam Hussain [1,2,*], Behzad Heidarshenas [1] and Khurram Altaf [3]

1. College of Mechanical & Electrical Engineering, Nanjing University of Aeronautics & Astronautics, Nanjing 210016, China
2. Faculty of Mechanical Engineering, GIK Institute of Engineering Sciences & Technology, Topi 23460, Pakistan
3. Mechanical Engineering Department, Universiti Teknologi Petronas, Bandar Seri Iskandar, Perak 32610, Malaysia

* Correspondence: gh_ghumman@hotmail.com; Tel.: +92-938281026

Received: 28 June 2019; Accepted: 23 July 2019; Published: 8 August 2019

Abstract: The present study was focused on establishing guidelines for successful friction stir welding of Al alloys and Cu lap joints. Detailed investigations in respect to tool geometry, tool material, work-piece material, welding parameters, stacking sequence, and heat sink were carried out. The soundness of welded joints was tested through microscopic analysis and the lap shear test. The results revealed that the tungsten carbide (WC) tool with square-pin produced sound joints in terms of minimized defects and high strength. Further, the use of heat sink proved as an important pre-requisite when the stacking sequence was inversed (i.e., Cu-Al), and this stacking configuration in comparison with the Al-Cu stacking yielded weaker joints. The influence of the tool welding speed (F, mm/min) was found to depend upon the tool material. A range of tool welding speed (23.5–37.5 mm/min) worked well for the WC tool. However, only two values of welding speed (30 mm/min and 37 mm/min) were observed to be conducive when the tool material was HSCo (high-speed cobalt)-steel. Finally, it was concluded to employ the WC tool with square-pin, a welding speed of 30 mm/min, the rotational speed (S, rpm) of 1500 mm/min, and Al-Cu stacking sequence to successfully process the Al/Cu lap joints.

Keywords: friction stir welding; dissimilar metals; temperature; process conditions; guidelines

---

1. Introduction

Al-Cu bilayer sheet offers an attractive combination of high thermal and electrical conductivity and good corrosion resistance. This combination of metals is also comparatively cheaper than the standalone Cu sheet. Owing to these salient properties, the Al-Cu metal has found a host of applications in aerospace, chemical, transport, electronic, and power industries [1]. The most common applications include electrical connectors, power supply module, power LED, heat sinks, electromagnetic shielding, solder float, and radiators. Another interesting application of Al-Cu sheet is that it can be employed as a fire-resistant material [2]. Numerous conventional joining techniques such as brazing and laser welding are applied to join Al and Cu but it is challenging because of the difference in the physical and chemical properties of the metals and the tendency to form brittle intermetallic compounds (IMCs) during the formation of welded joints [3]. These IMCs can impair the mechanical and electrical properties of the joint [4].

The temperature of metals in solid-state joining, on the other hand, does not approach melting point and this fact minimizes the undesired formation of intermetallics in dissimilar joints. Friction stir welding (FSW) is relatively an innovative solid-state welding technique, whereby joining is realized by the stirring action of a pin-tool without applying any significant external heating. This aspect renders FSW a very competitive process for joining the dissimilar metals, as described in the literature [5–8]. Besides experimental analyses, numerical modeling has also significant contribution in establishing the fundamentals of FSW process. Luo et al. [9] worked on numerical modeling of AA2A14-T6 to visualize the material flow during the FSW process. They found that either high welding speed or low rotational speed could cause welding defects such as holes and cracks. Dialami et al. [10] performed numerical modeling to study the microstructure evolution of AZ31Mg alloy during FSW and established relation among grain size, strain rate, micro-hardness, and temperature.

Some researchers [11,12] have attempted to produce dissimilar joints of Al with other metals. Sharma et al. [13] utilized five different pin profiles including cylindrical, taper, cylindrical cam, taper cam, and square to produce butt joint between AA5754 and commercially pure Cu. They reported that square-pin offered better mechanical properties because it facilitated mixing at the nugget zone. Akbari et al. [14] examined the trend of mechanical properties with respect to the material position in dissimilar 7070 Al/Cu lap joint and showed that better welded joint quality achieved when Al was placed on the top of Cu. Bisadi et al. [15] studied the effect of FSW parameters on the microstructure and mechanical properties of Al 5083 and commercial Cu FSW lap joint and reported that very low and very high welding temperature can lead to several joint defects like channel and voids defects. Celik et al. [16] performed a similar investigation on Al/Cu butt joint and suggested that higher tensile strength is attributed to dispersion strengthening of fine Cu particles distributed over the Al material in the stir zone.

Karimi et al. [17] investigated the effect of tool material on the metallurgical and mechanical properties of dissimilar Al/Cu butt joints. They found that the tool with low thermal conductivity produced better welded joints. Çevik et al. [18] employed uncoated and TiN-coated X210Cr12 steel tools in order to fabricate 7075-T651 Al butt joints revealing that the uncoated tool produced favored results. Bozkurt et al. [19], on the other hand, observed opposite findings while butt welding of AA2124-T4 alloy with uncoated and CrN and AlTiN-coated HSS tools.

Although some scholars [14,15,20–23] have performed work on different aspects of FSW of Al/Cu lap joint, a common agreement has not yet arrived in certain respects. As an example, the effect of tool material has not been agreed upon. Furthermore, nature of the influence of various factors is associated with the type of weld (say butt or lap). Therefore, more investigations are required to acquire a thorough understanding on FSW of Al/Cu lap joints. Moreover, most studies furnish knowledge on one or two aspects of the process. A comprehensive study undertaking a range of important aspects/factors can provide useful insights to the user for successful welding. The present study is an attempt in this direction wherein the effects of a number of factors namely tool geometry, tool material, process parameters, stacking sequence, and heat sink are taken into account.

Two types of tool materials namely HSCo and WC; two thicknesses of materials (1.65 mm and 3 mm for Cu and 2.15 mm and 4 mm for Al); and two types of tool geometries namely round and square are employed. To successfully produce Al/Cu lap joints at varying of these variables, the welding speed and rotational speed are altered over a range. It is observed that the joining results (i.e., lap shear strength and defects) vary as either of the tool material, plate thickness, or tool geometry is changed. Further, the favorable range of feed and speed also experiences a change with a change in the rest of the conditions. The presented results can serve as a guideline to produce sound FSW Al/Cu lap joints.

2. Materials and Methods

For the current investigations, three different materials namely 1060 Al (thick), 2219 Al (thin), and pure Cu as listed in Table 1 and three different tools such as high-speed cobalt taper-pin (HSCo) tool (HSCo), high-speed cobalt (HSCo) square-pin tool, and tungsten carbide (WC) square-pin tool were employed as shown in Figure 1 and specifications listed in Table 2. The stress–strain curve of each of the materials is shown in Figure 2 and the mechanical properties are presented in Table 3. The blank from each material was cut to the size of 100 mm × 70 mm, as shown in Figure 3. The surface oxides from the plates were removed using the abrasive paper. The plates were firmly held in a fixture shown in Figure 4. The FSW was performed along the long dimension of plates (i.e., rolling direction) utilizing the BYJC vertical milling machine. The tilt angle was kept constant at 2° and the dwell time ranged from 20–35 s. After welding, the strength of lap joints was characterized by conducting lap shear tests on the Universal Testing Machine 5567 (Instron Corp., Norwood, MA, USA): The geometry of the test sample is shown in Figure 5. To examine the microstructure and defects, samples (size: 16 mm × 16 mm) were cut using a CNC EDM wire cut machine and were ground with 220, 320, 500, 800, 1000, 2400, and 4000 abrasive papers of Silicon Carbide (Figure 6). The samples were thoroughly observed with the TESCAN scanning electron microscope (TESCAN, Brno, Czech Republic) and OLYMPUS B061 optical microscope (OLYMPUS, Tokyo, Japan). OMEGA infrared thermometer (OMEGA, Norwalk, USA) was utilized during friction stir welding to measure the temperature at the center on the top surface of the upper plate as shown in Figure 4.

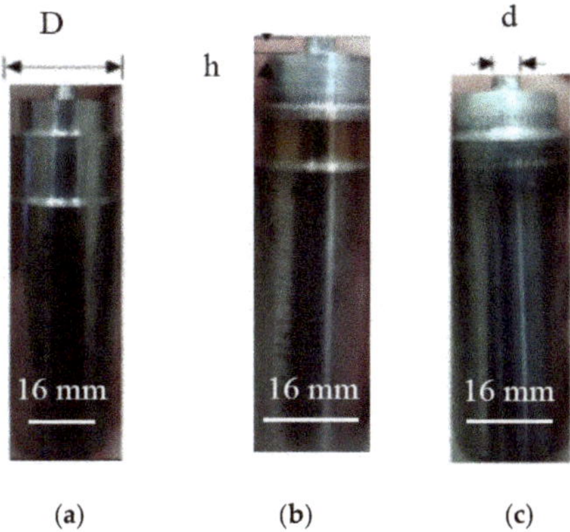

Figure 1. Tool geometries employed in the present study: (**a**) HSCo tapered tool; (**b**) HSCo squared tool; and (**c**) WC squared tool.

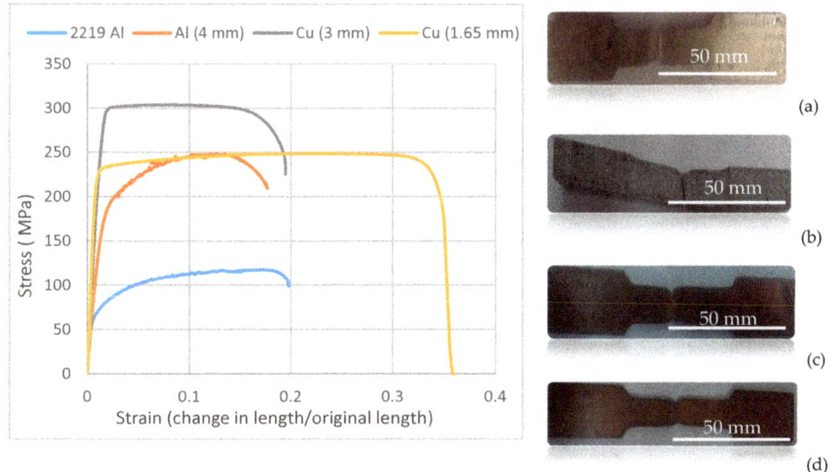

Figure 2. Stress–strain curve for: (**a**) 2.15 mm 2219 Al, (**b**) 4 mm 1060 Al, (**c**) 3 mm Cu, and (**d**) 1.65 mm Cu.

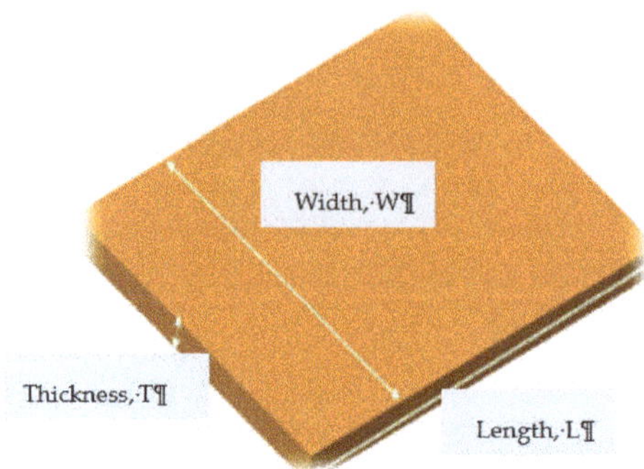

Figure 3. Schematics of friction stir welding (FSW) lap sample.

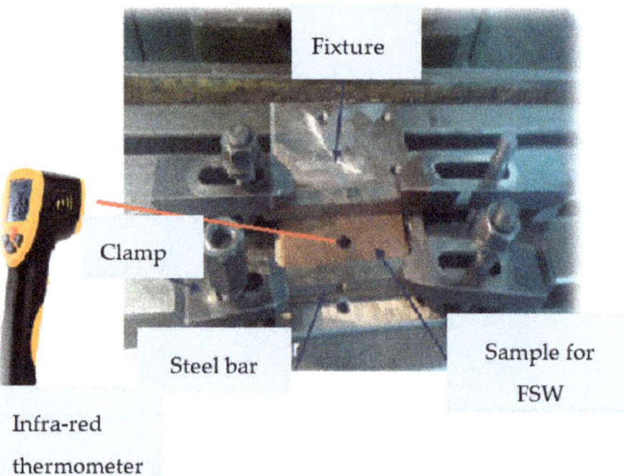

Figure 4. Setup for clamping FSW sample.

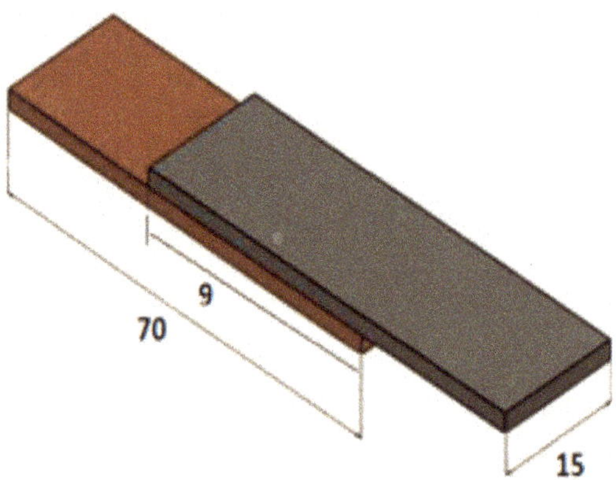

Figure 5. Lap shear sample.

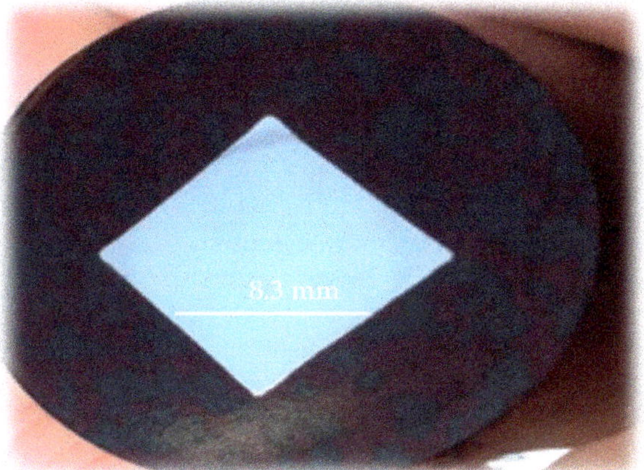

Figure 6. Microscopy and micro-hardness sample.

Table 1. Blank size utilized in the present study.

Material	Length (mm), L	Width (mm), W	Thickness (mm), T
1060 Al	100	70	4
2219 Al	100	70	2.15
Commercial Cu 1	100	70	1.65
Commercial Cu 2	100	70	3

Table 2. Specifications of FSW tools.

Tool	Shoulder Diameter (mm)	Pin Height (mm)	Pin Cross Section Dimension (mm)
HSCo Tapered Tool	16	3.2	Small diameter = 3 Large diameter = 5
HSCo Squared Tool	16	3.2	Squared shaped each side = 4
Carbide Squared Tool	16	3.2	Squared shaped each side = 4

Table 3. Mechanical properties of base metals.

Material	Yield Strength (MPa),YS	Ultimate Tensile Strength (MPa), UTS	%Elongation (mm/mm), e	Vicker's Micro-Hardness at 1 gm
1060 Al	181.72	248.66	17.69	81.12
2219 Al	60.42	117.97	19.75	52.02
Commercial Cu 1	219.21	248.85	36.56	107.71
Commercial Cu 2	291.18	303.86	19.48	112.21

3. Results and Discussion

3.1. Selection of Tool Geometry to Fabricate Dissimilar Al/Cu Lap Joints

Experiments were performed to select suitable tool geometry to successfully produce Al/Cu lap joints. Based on literature reports [24–26], two shapes of pins namely round tapered and square were utilized (Figure 1). The shoulder diameter in each case was 16 mm, the pin height was 3.2 mm, and the other details were as listed in Table 2. The commercial Cu sheet (1.65 mm thick) was placed on the top of the 1060 Al sheet (4 mm thick) to fabricate dissimilar lap joint. The welding was performed employing the tapered-pin tool made of high-speed cobalt (HSCo) steel (Figure 1a). The attempts were made by employing the rotation speed of 1500 rpm and varying the feed from 23.5 mm/min to 47.5 mm/min as listed in Table 4. As shown in Figure 7a and indicated in Table 4, surface cracks were generated while welding thus rendering all of the attempts to be unsuccessful. Moreover, the

HSCo tapered-pin tool was observed to have a shorter life in these tests wherein pin detachment led to complete failure of the tool.

The second experimental plan was launched to select appropriate tool geometry with the square-pin tool. As summarized in Table 5, the square-pin tool contrary to the tapered-pin tool apparently produced successful joints as observable from Figure 7b,c. However, this tool also encountered failure after a few tests as indicated in Figure 7c whereby a pin left inside the material can be noticed. The observation from these primary tests revealed that the square-pin tool produced comparatively successful joints because its squared corners while rotating in a circle generates waves that help in the breaking and mixing of materials [24]. However, the issue regarding the breakage of the tool-pin that probably occurred due to tool softening owing to heating while welding yet needed to be resolved. Therefore, further tests were performed by varying the tool material (WC) as detailed in the coming section. A test experiment was performed through the WC square-pin tool exactly on the same parameters (S 1500 rpm and F 36.5 mm/min) as applied for the HSCo square-pin tool to check the success of the welded joint (Cu-Al) and the strength of WC tool as shown in Figure 7d.

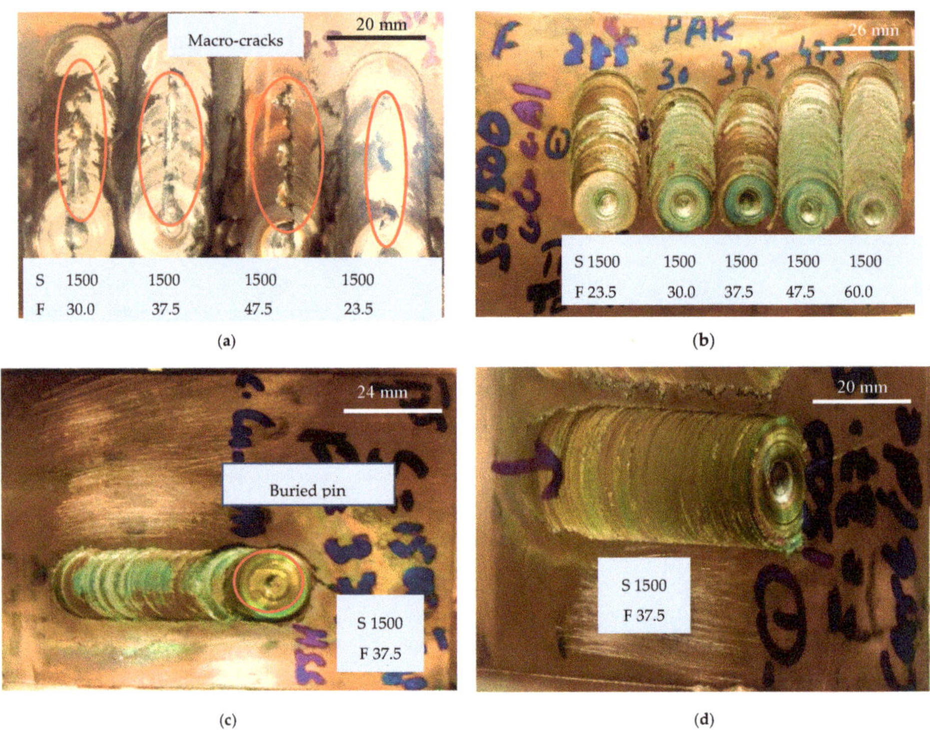

Figure 7. 1060Al-Cu lap joints produced; HSCo (**a**) tapered-pin tool, (**b**,**c**) square-pin tool, and (**d**) tungsten carbide (WC) square-pin tool.

Table 4. Experimental conditions employed to fabricate Cu 1-1060Al lap joints using HSCo-steel tapered-pin tool (T_D: 22 s).

Test no.	S, rpm	F, mm/min	Remark
1	1500	23.5	Crack as shown in Figure 7a
2		30.0	Crack as shown in Figure 7a
3		37.5	Crack as shown in Figure 7a
4		47.5	Crack as shown in Figure 7a

Table 5. Experimental conditions employed to fabricate Cu 1-1060Al lap joints using HSCo square-pin tool (T_D: 22 s).

Test No.	S, rpm	F, mm/min	Remark
1	1500	23.5	Successful as shown in Figure 7b
2		30.0	Successful as shown in Figure 7b
3		37.5	Successful as shown in Figure 7b
4		47.5	Successful as shown in Figure 7b
5		60.0	Successful as shown in Figure 7b

3.2. Selection of Tool Material to Fabricate Dissimilar Al/Cu Lap Joints

In an attempt to opt an appropriate tool material for successful welding of Al/Cu joints, two dissimilar lap joints of 2219 Al-Cu and Cu-1060 Al were fabricated employing two types of tool materials namely HSCo Steel and WC Carbide. The pin geometry in both cases was square because, as found above, this geometry offered better results in comparison to other considered ones. The complete set of conditions has been listed in Table 6.

Table 6. Fabrication of 2219 Al-Cu (1.65 mm) lap joint using WC and HSCo square-pin tool (T_D: 22 s).

Test No.	S, rpm	F, mm/min	Tool	Remarks
1	1500	23.5	WC	Successfully fabricated (Figure 8a)
2		30.0		Successfully fabricated (Figure 8a)
3		37.5		Successfully fabricated (Figure 8b)
4		47.5		Crack and hole occurred (Figure 8c)
5		60.0		Crack occurred (Figure 8d)
1	1500	23.5	HSCo	Crack occurred (Figure 8e)
2		30.0		Successfully fabricated (Figure 8e)
3		37.5		Successfully fabricated
4		47.5		Crack occurred

Lap shear tests were carried out in order to assess the mechanical performance of lap joints. The joints and their strength results are shown in Table 7. As observable, both 2219 Al-Cu and Cu-1060 Al dissimilar joints exhibit fairly good strength when welding was done utilizing the WC tool. Furthermore, the WC tool offers greater joint strength than the HSCo-Steel tool. For example, the joint strength of the 2219 Al-Cu joint is doubled when the tool material is altered from HSCo-steel to WC. The strength gain in the case of the Cu-1060 joint is even better (about 29 times). Figure 9 presents the temperature profiles recorded during welding. The peak temperature obtained with the WC tool is 250 °C and that obtained with HSCo-steel tool is 150 °C. This difference in the temperature is due to a fact that the WC material has higher conductivity (110 W/mK) than HSCo-steel (60 W/mK). As a result, the workpiece experiences greater heat input that in turn helps in proper mixing and defect free welding when the WC tool is employed. This can be witnessed from the SEM images shown in Figure 10 wherein joints produced with the WC tool are sound while those fabricated with the HSCo-steel tool suffer from defects like pores and voids. These shreds of evidence support why the WC tool produces sound joints with high strength.

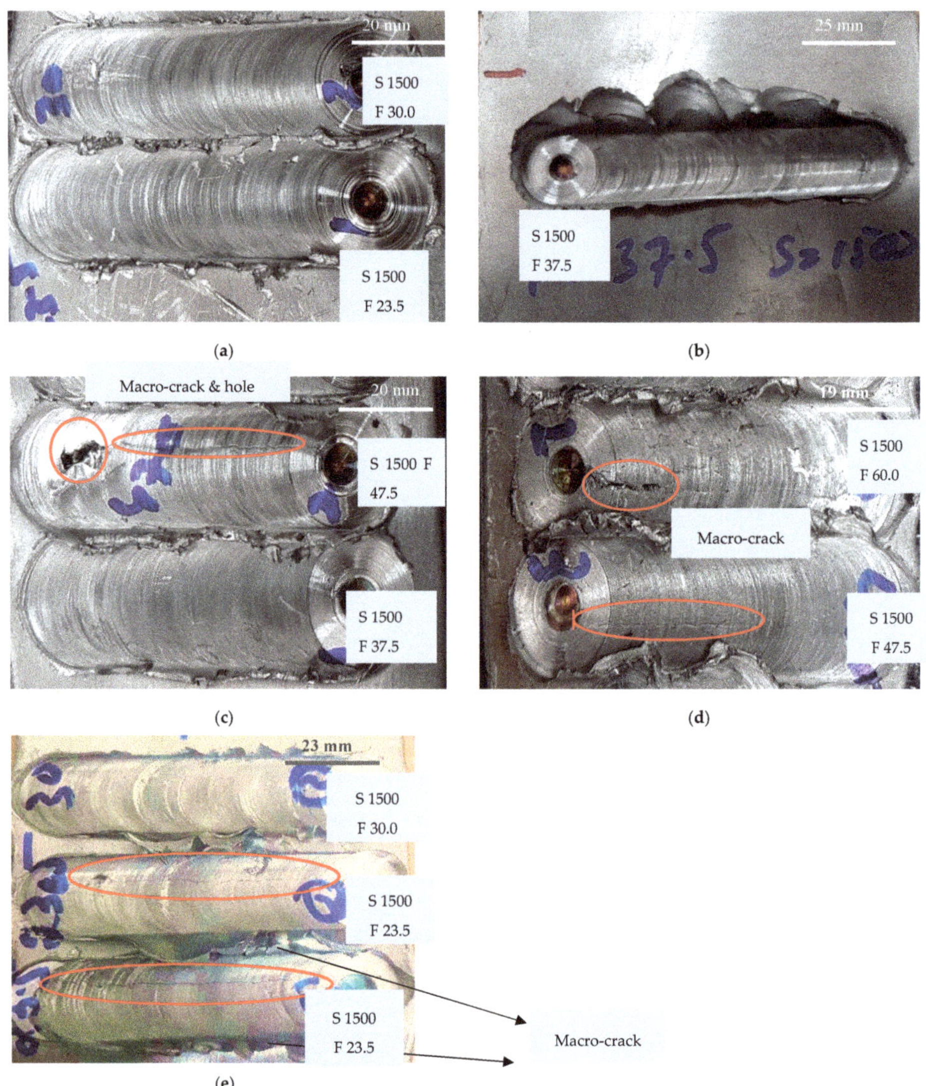

Figure 8. Dissimilar 2219 Al-Cu lap joints fabricated employing different tool materials: (**a**–**d**) WC square-pin tool at scale, (**e**) HSCo square-pin tool.

Table 7. Comparison of joint strength between dissimilar lap joints fabricated with WC and HSCo-steel tools.

S. #	Lap Joints	F mm/min	Joint Strength (MPa) by WC Tool	Joint Strength (MPa) by HSCo Tool
1	2219 Al-Cu	30.0	25.3	13.1

Table 7. *Cont.*

S. #	Lap Joints	F mm/min	Joint Strength (MPa) by WC Tool	Joint Strength (MPa) by HSCo Tool
2	Cu-1060 Al	37.5	17.8	0.58

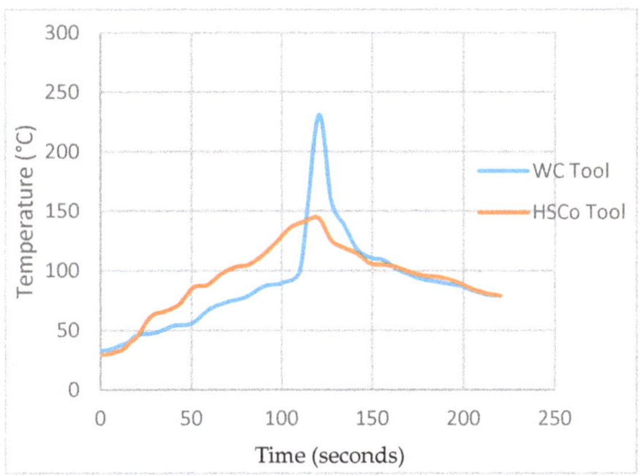

Figure 9. Effect of tool material on temperature profiles during FSW of 2219 Al-Cu joints with square-pin tools.

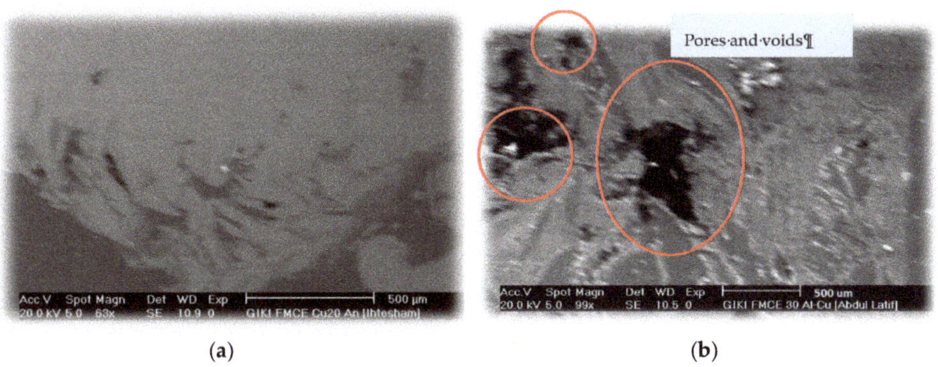

Figure 10. Scanning electron microscopy of 2219Al-Cu lap joint at F 30.0 mm/min using (**a**): WC tool, (**b**) HSCo square- tool.

3.3. Effect of Plate Positioning in Dissimilar Al/Cu Lap Joints

To compare the soundness of lap joints with respect to the position of the plates, initial experiments were performed keeping 2219 Al plate onto the Cu plate (1.65 mm). The joining was done using the WC tool and a range of welding speed as listed in Table 8. Successful joining in this stacking configuration was realized for the welding speed ranging from 23.5 mm/min to 37.5 mm/min (Figure 8 and Table 8). On the other hand, the entire range of joints remained unsuccessful when the Cu plate was stacked onto the Al plate (Table 9). In fact, the authors observed melting of the Al plate during welding as evidenced in Figure 11a,b. This observation points out that the Cu plate being very conductive radiated heat into the Al plate, and the heat was sufficient to cause melting of Al.

Table 8. Experiments for fabricating 2219Al-Cu lap joint using WC square-pin tool (T_D: 22 s).

Test No.	S, rpm	F, mm/min	Remarks
1		23.5	Successfully fabricated (Figure 8a)
2		30.0	Successfully fabricated (Figure 8a)
3	1500	37.5	Successfully fabricated (Figure 8b,c)
4		47.5	Crack occurred (Figure 8c,d)
5		60.0	Cracks occurred (Figure 8d)

Table 9. Experiments for fabricating Cu -2219Al lap joint using WC square-pin (T_D: 20 s).

Test No.	S, rpm	F, mm/min	Remarks
1	1500	23.0	Cracks and bubbles occurred (Figure 11a)
2	1500	37.5	Cracks hole and bubbles occurred (Figure 11b)
3	1500	60.0	Cracks occurred (Figure 11a)
4	1500	75.0	Cracks and bubbles occurred (Figure 11a)
5	950	30.0	Cracks occurred (Figure 11c)
6	750	30.0	Cracks occurred (Figure 11c)
7	600	30.0	Cracks occurred (Figure 11d)

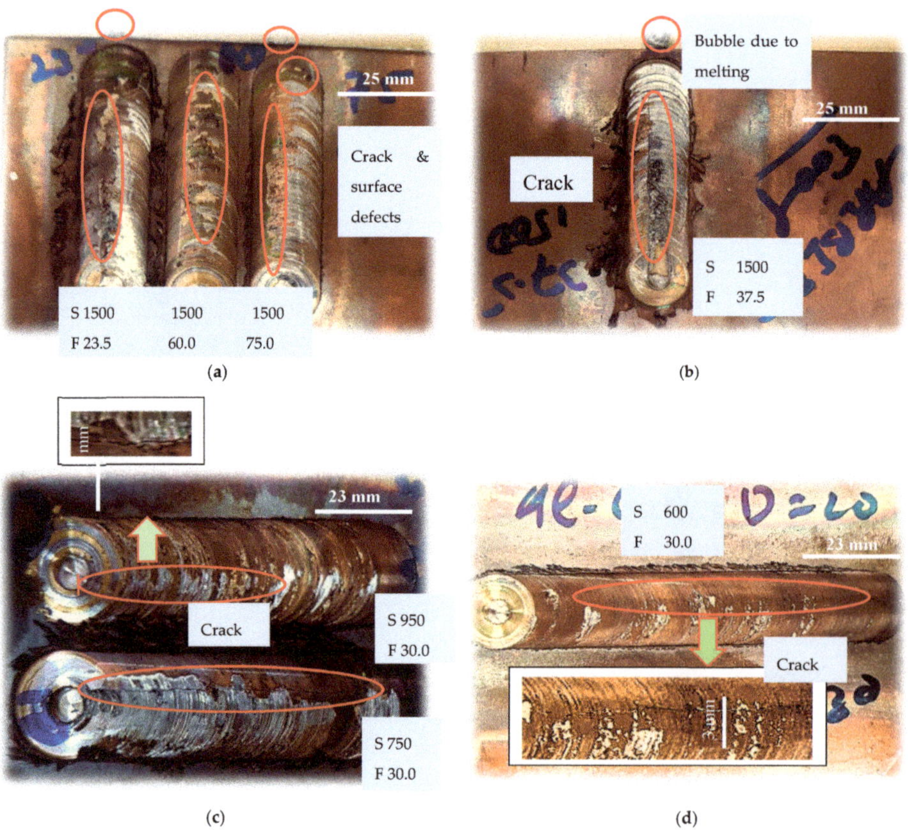

Figure 11. Dissimilar Cu-2219 Al lap joints fabricated by WC square-tool.

To reduce excessive heat generation, the rotational speed was decreased in steps from 1500 rpm to 600 rpm (Table 9). However, success could not be realized as the defects like cracks and holes occurred as observable from Figure 11c,d.

3.4. Role of Heat Sink in 2219 Al-Cu Lap Joints

In order to avoid melting of the Al plate in the Cu-Al stacking configuration found earlier, a number of trials were made with having the Cu plate as a sink under the Al plate. Two plates of Cu with thickness of 1.65 mm and 3 mm were utilized. The welding speed was varied over a range as listed in Table 10. The cracks and flakes of Cu were observed in many instances although the quality of the weld joint was significantly improved in terms of bubbles formation and cracking in comparison to the scenario(s) when no heat sink was used. The sound joint was achieved when thicker heat sink (3 mm Cu plate) was placed and the joining was performed at the rotational speed of 1500 rpm and welding speed of 30 mm/min as indicated in Table 10.

Table 10. Experiments to fabricate dissimilar Cu-2219 Al lap joint with WC square-tool using heat sink (T_D: 20 s).

S, 1500 rpm	Heat sink	F	Remarks
Test No.			
1	Cu 1	23.5	Cracks occurred (Figure 12a)
2		60.0	Surface defect and crack occurred (Figure 12a)
3		75.0	Small pore occurred (Figure 12b)
4	Cu 2	30.0	Successfully fabricated (Figure 12c)
5		37.5	Crack occurred (Figure 12c)
6		47.5	Pore occurred (Figure 12c)

Table 11 compares the effect of stacking sequence on the joint strength. It can be noticed that the Al-Cu sequence offers around 4 times greater strength than the Cu-Al stacking. This happens due to a fact that the peaks temperature in Al-Cu stacking is 220 °C and that in Cu-Al stacking is 160 °C (Figure 13), which promotes material mixing and defect-free joining. Further, the cooling rate in the former case is higher that promotes the joint strength [27,28]. Moreover, Akbari et al. [14] have reported that weaker compounds are formed when joining is done with the Cu-Al stacking sequence. These, besides defects, impair the interfacial strength.

Table 11. Effect of plate position on the strength of lap joint.

S. #	Lap Joints	F, mm/min	Joint Strength (MPa) by WC Tool
1	2219 Al-Cu	30.0	25.3
2	Cu-2219 Al	30.0	5.9

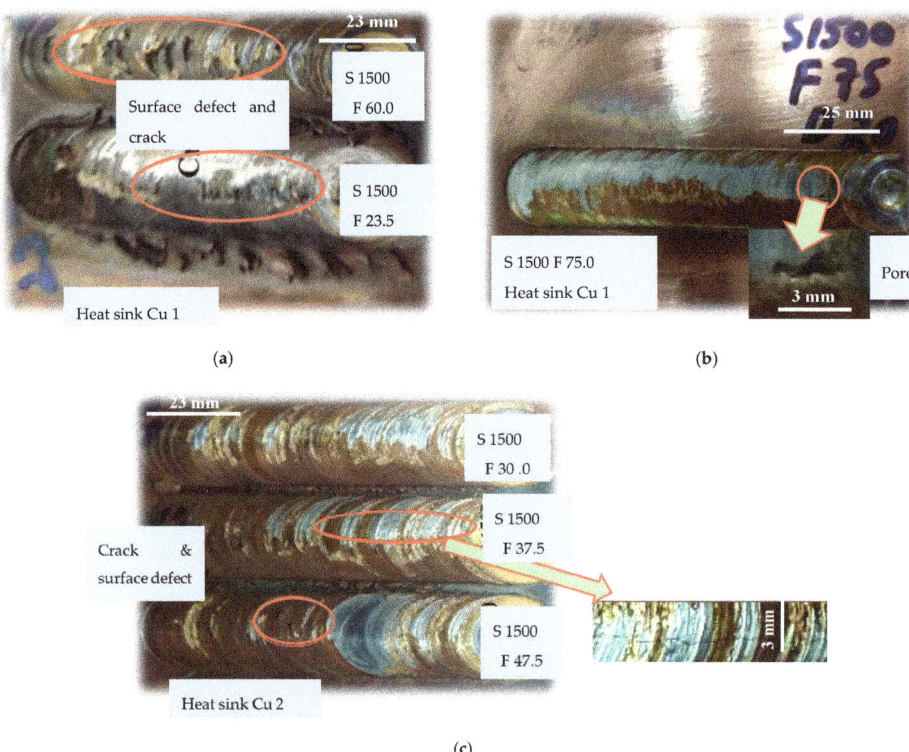

Figure 12. Redesign of dissimilar Cu-2219 Al lap joints fabricated by WC square-tool and using heat sink: (**a**,**b**) Cu 1 and (**c**) Cu 2.

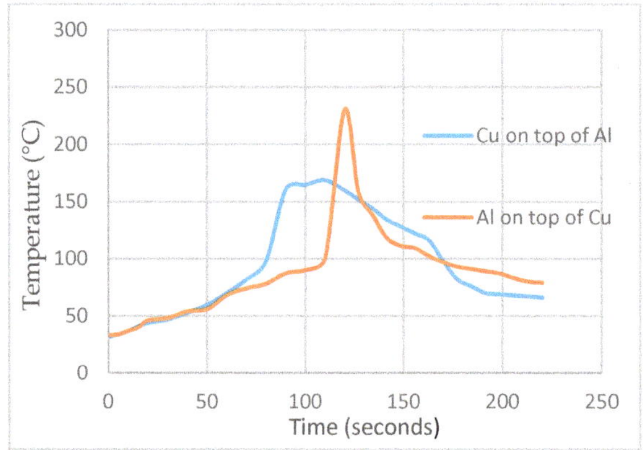

Figure 13. Temperature distribution curves of 2219 Al-Cu and Cu-2219 Al fabricated by WC at S 1500 rpm and F 30.0 mm/min.

4. Conclusions

In the present study, investigations were conducted in order to produce sound Al/Cu dissimilar lap joints through friction stir welding. For this purpose, a number of process conditions including tool-pin shape, tool material, stacking sequence, workpiece material, heat sink, rotational speed, and welding speed were altered. To know their soundness in terms of defects and strength, the joints were subjected to microscopic and mechanical tests. The following important findings can be drawn from the study:

1. The HSCo-steel tool with tapered-pin does not adequately mix the materials to realize a sound joint. The successful joining is observed when the square-pin tool is employed.
2. The WC tool forms superior joints (in terms of defects and strength) than the HSCo-tool, attributing to a reason that greater peak temperature is achieved with the former tool (say 250 °C vs. 150 °C in case of 2219 Al-Cu joints). The former tool offers joint strength of 25 MPa in comparison to 13 MPa offered by the HSCo-tool. Moreover, the WC tool exhibits greater life than the HSCo-Steel tool.
3. Stacking sequence is one of the key factors to have a successful Al/Cu lap joint. The joining is realized only when the Al-Cu configuration is arranged. With an inverse arrangement (Cu-Al), Al plate melts down. However, use of heat sink proves beneficial to radiate the heat and thus for the successful joining in the Cu-Al stacking configuration. Additionally, sound joint in this configuration is achieved with a particular set of conditions, i.e., heat sink: 3 mm thick Cu plate, feed rate: 30 mm/min, speed: 1500 rpm, tilt angle: 2°, tool: WC with square-pin.

Author Contributions: Conceptualization, A.L., H.W. and G.H.; methodology, A.L. and H.W.; formal analysis, A.L., H.W., G.H. and B.H.; investigation, A.L., H.W. and G.H.; writing—original draft preparation, A.L. and H.W.; writing—review and editing, A.L., G.H., K.A. and B.H.; visualization, A.L. and H.W.; supervision, G.H.

Funding: The necessary funding to realize this work in the research community was provided by Fundamental Research Funds for the Central Universities (Grant No. NS2015055), High-End Foreign Experts Project with Universities directly under the Administration of Ministries and Commissions of the Central Government (Grant No. 011951G19061), National Natural Science Foundation of China (Grant No. 51105202) and State Administration of Foreign Experts Affairs PR China and Ministry of Education PR China (111 project, Grant No. B16024), for which the authors are grateful.

Acknowledgments: The authors are thankful to GIK Institute of Engineering Sciences and Technology (Pakistan) for providing advisory and technical support.

Conflicts of Interest: The authors declare no conflict of interest.

Abbreviations

S	Rotational speed
rpm	Revolution per minute
F	Welding speed
mm/min	Millimeters per minute
Cu	Copper, Cu 1 thickness is 1.65 mm & Cu 2 thickness is 3 mm
Al-Cu	Aluminum plate placed on the copper plate
Cu-Al	Copper plate placed on the Al plate
Al/Cu	Either aluminum on top of copper or vice versa
HSCo	High-speed cobalt
WC	Tungsten carbide
T_D	Dwell time
T (°C)	Temperature in centigrade

References

1. Perrett, J.; Martin, J.; Threadgill, P.; Ahmed, M. Recent developments in friction stir welding of thick section aluminium alloys. In Proceedings of the 6th World Congress on Aluminium Two Thousand, Florence, Italy, 13–17 March 2007; pp. 13–17.
2. Ji, F.; Xue, S.; Dai, W. Reliability studies of Cu/Al joints brazed with Zn–Al–Ce filler metals. *Mater. Des.* **2012**, *42*, 156–163.
3. Kah, P.; Vimalraj, C.; Martikainen, J.; Suoranta, R. Factors influencing Al-Cu weld properties by intermetallic compound formation. *Int. J. Mech. Mater. Eng.* **2015**, *10*, 10. [CrossRef]

4. Galvão, I.; Oliveira, J.; Loureiro, A.; Rodrigues, D. Formation and distribution of brittle structures in friction stir welding of aluminium and copper: Influence of shoulder geometry. *Intermetallics* **2012**, *22*, 122–128. [CrossRef]
5. Shigematsu, I.; Kwon, Y.-J.; Suzuki, K.; Imai, T.; Saito, N. Joining of 5083 and 6061 aluminum alloys by friction stir welding. *J. Mater. Sci. Lett.* **2003**, *22*, 353–356. [CrossRef]
6. Cao, X.; Jahazi, M. Effect of tool rotational speed and probe length on lap joint quality of a friction stir welded magnesium alloy. *Mater. Des.* **2011**, *32*, 1–11. [CrossRef]
7. Xu, N.; Ueji, R.; Fujii, H. Dynamic and static change of grain size and texture of copper during friction stir welding. *J. Mater. Process. Technol.* **2016**, *232*, 90–99. [CrossRef]
8. Galvao, I.; Oliveira, J.; Loureiro, A.; Rodrigues, D. Formation and distribution of brittle structures in friction stir welding of aluminium and copper: Influence of process parameters. *Sci. Technol. Weld. Join.* **2011**, *16*, 681–689. [CrossRef]
9. Luo, H.; Wu, T.; Wang, P.; Zhao, F.; Wang, H.; Li, Y. Numerical Simulation of Material Flow and Analysis of Welding Characteristics in Friction Stir Welding Process. *Metals* **2019**, *9*, 621. [CrossRef]
10. Dialami, N.; Cervera, M.; Chiumenti, M. Numerical modelling of microstructure evolution in friction stir welding (FSW). *Metals* **2018**, *8*, 183. [CrossRef]
11. El Chlouk, Z.G.; Achdjian, H.H.; Ayoub, G.; Kridli, G.T.; Hamade, R.F. The Effect of Tool Geometry on Material Mixing during Friction Stir Welding (FSW) of Magnesium AZ31B Welds. In Proceedings of the TMS Middle East-Mediterranean Materials Congress on Energy and Infrastructure Systems (MEMA 2015), Doha, Qatar, 11–14 January 2015; pp. 235–242.
12. Patel, V.; Li, W.; Wang, G.; Wang, F.; Vairis, A.; Niu, P. Friction Stir Welding of Dissimilar Aluminum Alloy Combinations: State-of-the-Art. *Metals* **2019**, *9*, 270. [CrossRef]
13. Sharma, N.; Siddiquee, A.N.; Khan, Z.A.; Mohammed, M.T. Material stirring during FSW of Al–Cu: Effect of pin profile. *Mater. Manuf. Processes* **2018**, *33*, 786–794. [CrossRef]
14. Akbari, M.; Abdi Behnagh, R.; Dadvand, A. Effect of materials position on friction stir lap welding of Al to Cu. *Sci. Technol. Weld. Join.* **2012**, *17*, 581–588. [CrossRef]
15. Bisadi, H.; Tavakoli, A.; Sangsaraki, M.T.; Sangsaraki, K.T. The influences of rotational and welding speeds on microstructures and mechanical properties of friction stir welded Al5083 and commercially pure copper sheets lap joints. *Mater. Des.* **2013**, *43*, 80–88. [CrossRef]
16. Celik, S.; Cakir, R. Effect of friction stir welding parameters on the mechanical and microstructure properties of the Al-Cu butt joint. *Metals* **2016**, *6*, 133. [CrossRef]
17. Karimi, N.; Nourouzi, S.; Shakeri, M.; Habibnia, M.; Dehghani, A. Effect of tool material and offset on friction stir welding of Al alloy to carbon steel. *Adv. Mater. Res.* **2012**, *445*, 747–752. [CrossRef]
18. Çevik, B.; Özçatalbaş, Y.; Gülenç, B. Effect of tool material on microstructure and mechanical properties in friction stir welding. *Mater. Test.* **2016**, *58*, 36–42. [CrossRef]
19. Bozkurt, Y.; Boumerzoug, Z. Tool material effect on the friction stir butt welding of AA2124-T4 Alloy Matrix MMC. *J. Mater. Res. Technol.* **2018**, *7*, 29–38. [CrossRef]
20. Wiedenhoft, A.G.; Amorim, H.J.; Rosendo, T.S.; Strohaecker, T.R. Friction Stir Welding of Dissimilar Al-Cu Lap Joints with Copper on Top. *Key Eng. Mater.* **2017**, *724*, 71–76. [CrossRef]
21. Wiedenhoft, A.G.; de Amorim, H.J.; Rosendo, T.d.S.; Tier, M.A.D.; Reguly, A. Effect of Heat Input on the Mechanical Behaviour of Al-Cu FSW Lap Joints. *Mater. Res.* **2018**, *21*. [CrossRef]
22. Marstatt, R.; Krutzlinger, M.; Luderschmid, J.; Constanzi, G.; Mueller, J.; Haider, F.; Zaeh, M. Intermetallic layers in temperature controlled Friction Stir Welding of dissimilar Al-Cu-joints. *IOP Conf. Ser. Mater. Sci. Eng.* **2018**, *373*, 012017. [CrossRef]
23. Mubiayi, M.P.; Akinlabi, E.T. Friction stir welding of dissimilar materials between aluminium alloys and copper, An overview. In Proceedings of the World Congress on Engineering, London, UK, 3–5 July 2013; pp. 3–5.
24. Patel, J.B.; Bhatt, K.; Shah, M. Replacement of Tool-pin Profile and Simulation of Peak Temperature & Flow Stress during FSW of AA6061 Alloy. *IJISET* **2014**, *1*, 273–276.
25. Sahu, P.K.; Kumari, K.; Pal, S.; Pal, S.K. Hybrid fuzzy-grey-Taguchi based multi weld quality optimization of Al/Cu dissimilar friction stir welded joints. *Adv. Manuf.* **2016**, *4*, 237–247. [CrossRef]
26. Buffa, G.; De Lisi, M.; Sciortino, E.; Fratini, L. Dissimilar titanium/aluminum friction stir welding lap joints by experiments and numerical simulation. *Adv. Manuf.* **2016**, *4*, 287–295. [CrossRef]

27. Humphreys, F.J.; Hatherly, M. *Recrystallization and Related Annealing Phenomena*; Elsevier: Oxford, UK, 2012.
28. Nelson, T.; Steel, R.; Arbegast, W. In situ thermal studies and post-weld mechanical properties of friction stir welds in age hardenable aluminium alloys. *Sci. Technol. Weld. Join.* **2003**, *8*, 283–288. [CrossRef]

Article

Microstructure and Mechanical Properties of Resistance Heat-Assisted High-Power Ultrasonic Dissimilar Welded Cu/Al Joint

Huan Li [1,2,*] and Biao Cao [2]

1 School of Mechanical Engineering, Yangtze University, Jingzhou 434023, China
2 School of Mechanical and Automotive Engineering, South China University of Technology, Guangzhou 510640, China
* Correspondence: lihuan7@126.com

Received: 9 July 2019; Accepted: 7 August 2019; Published: 8 August 2019

Abstract: The Cu/Al dissimilar joint, welded by high-power ultrasonic welding technology, is still facing challenges despite the significant research attention it has attracted. In this work, the microstructure and mechanical properties of resistance heat-assisted high-power ultrasonic welding of Cu/Al are investigated, in order to obtain high-quality joints. The intermetallic compound (IMC) at the interface of hybrid welding is primarily composed of Al2Cu, and the additional resistance of heat reduces the thickness of this brittle IMC layer. The average shear stress for the joint prepared by hybrid welding is ~97 MPa, which is higher compared to the joint strength without resistance heat (90 MPa). Moreover, the duration of the hybrid welding process is shorter. Finally, the fracture of the hybrid weld is found to be a brittle–ductile hybrid mode.

Keywords: resistance heat-assisted ultrasonic welding; mechanical properties; fracture morphology; intermetallic compounds; microstructure

---

1. Introduction

Ultrasonic welding can produce welds faster and with less material loss than other common welding methods of dissimilar lap joints, such as laser welding (LW) [1,2] and friction stir welding (FSW) [3,4]. It also requires lower energy input than resistance spot welding (RSW) [5]. These advantages make this method suitable for welding dissimilar metals, such as copper and aluminum, which have high thermal and electrical conductivity and are widely used in the aerospace industry and battery packs [6,7]. Recently, the dissimilar copper and aluminum joint, welded by high-power ultrasonic welding (HPUSW) technique, is becoming more attractive due to its ability to join thicker sheets [8]. However, a major challenge of this technique is the thick continuous intermetallic compound (IMC) layer that is formed at the Cu/Al interface, which reduces the mechanical properties of the joint and leads to poor welding quality [9,10]. This issue inhibits the wider commercialization of this technique. Recently, interlayer metals, such as Al and Zn, were placed on the Cu/Al interface and were used to enhance the ultrasonic welding quality of the joint [11,12]. However, the metallurgical reaction at the specimen/interlayer interface is very complicated, making it difficult to understand the effects of the interlayer on the mechanical properties of the joints [13]. In addition, during the ultrasonic metal welding process, the metal interlayer will widen the heat-affected zone [14].

In order to improve the joint quality, several types of energy sources have been used to assist the ultrasonic welding, such as laser beam and resistance heat. Dehelean et al. [15] carried out hybrid ultrasonic-resistance welding of advanced materials and found that the resistance heat results in low weld strength. In our previous work, resistance heat-assisted low-power ultrasonic welding was proposed and it was found that the additional resistance heat can significantly increase the peak power of ultrasonic vibration and the welding strength [16]. Unfortunately, this welding method is only suitable for joining thin sheets. The finite element method was used to study the influence of additional resistance heat on the interface temperature and plastic deformation of the material, during ultrasonic welding [17]. However, the mechanical properties and microstructure of joints prepared by resistance heat-assisted high-power ultrasonic welding (RUSW) have not yet been reported. In this paper, the mechanical properties and microstructure of Cu/Al joints prepared by RUSW are studied, in order to obtain high welding quality.

2. The Principle of Hybrid Welding

The RUSW system includes a lateral-driven high-power ultrasonic welder and an inverter resistance power supply. The current and ultrasound vibrations act on the workpiece simultaneously. In RUSW, a high pressure creates sufficient contact between the upper and lower workpieces to ensure electrical conductivity between them. The ultrasonic waves from the sonotrode pass through the upper workpiece, causing local relative vibration at the upper/lower specimen interface and thus generating friction. Resistance, friction, and plastic deformation heat cause the interface temperature to rise rapidly, resulting in solid-state joining. Figure 1 shows the schematic diagram of the RUSW process.

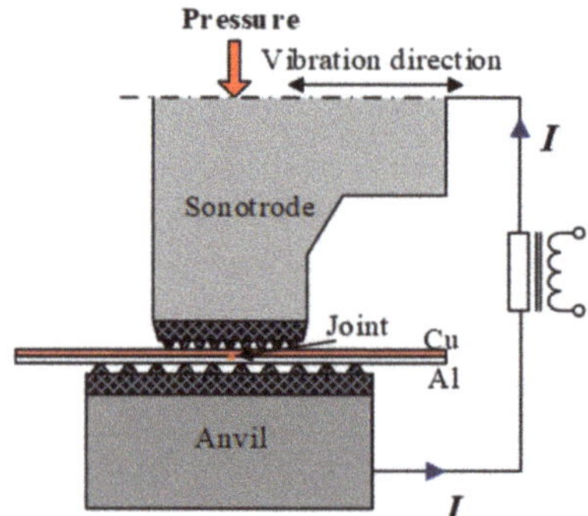

Figure 1. The principle of resistance heat-assisted high-power ultrasonic welding (RUSW).

In RUSW, the generated heat Q_{RUSW} includes one part from ultrasonic vibration Q_{USW} and another part from electrical resistance Q_{RSW}, as described below:

$$Q_{RUSW} = Q_{USW} + Q_{RSW}\,. \tag{1}$$

Ultrasonic vibration energy Q_{USW}, is converted into friction heat Q_f and plastic deformation heat Q_q [18], as shown in Equation (2):

$$Q_{USW} = Q_f + Q_q. \tag{2}$$

Q_{RSW} can be expressed as:

$$Q_{RSW} = I^2 \times R_{total} \times t \tag{3}$$

where I is the electrical current and t is the welding time. R is the total resistance providing thermal input during the RUSW process. Total resistance R involves seven components, as shown in Figure 2, and can be expressed as:

$$R_{total} = R_{st} + R_{c1} + R_{cu} + R_{c2} + R_{Al} + R_{c3} + R_{av} \tag{4}$$

where R_{st} is the resistance of the sonotrode, R_{Cu} is the resistance of Cu plate, R_{Al} is the resistance of Al plate, R_{av} is the resistance of the anvil, R_{c1} is the contact resistance between the sonotrode and the upper specimen, R_{c2} is the contact resistance between the upper and the lower specimen, and R_{c3} is the contact resistance between the anvil and the lower specimen.

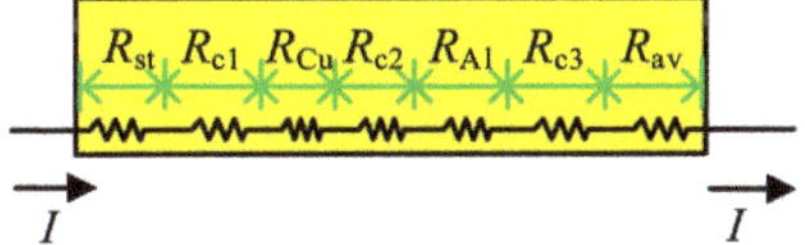

Figure 2. Schematic of the resistances in the RUSW system.

In RUSW, material softening arises from two sources: Ultrasonic softening and thermal softening [19]. Thus, the material softening rate in the welding process can be expressed as [18]:

$$\alpha = \alpha_{us} \times \alpha_T = \alpha_{us} \times \frac{\sigma_T}{\sigma_{T_0}} \tag{5}$$

where, α_{us} and α_T are the ultrasonic softening rate and thermal softening rate, respectively. σ_T and σ_{T0} are the yield stresses of material at temperature T and room temperature, respectively.

According to the above theory, the interaction between the ultrasonic vibration and the electrical resistance can be described as follows: When the ultrasonic vibration is applied to the RSW, the intensity of the material softening is increased, according to Equation (5). This benefits the breaking of the metal oxide film and promotes the resistance heat. When the resistance heat acts on the ultrasonic metal welding, it promotes the increase in interface temperature according to Equation (1), and thus promotes the metallurgical reaction on the interface. These interactions may potentially increase the weld strength of Cu/Al joints.

3. Experimental Details

The RUSW system included a lateral-driven, high-power ultrasonic welder (4.0 kW) and an inverter resistance power supply (Figure 3a). The high-power ultrasonic welder used a lateral spot welder (Telsonic M5000) with a vibration frequency of 20 kHz. A rectangular sonotrode tip with dimensions of 7 mm × 5 mm was used in this study (Figure 3b). To avoid the occurrence of arcs and marks on the Cu/Al interface, the shape of the sonotrode tooth was trapezoidal, instead of triangular. A resistance spot welder with a maximum electrical current supply of 4000 A was used. A zero-to-peak amplitude of 24 μm, current of 3900 A, clamping force of 1975 N and welding time of 0.2 s were selected for the hybrid welding. The interval between each sonotrode tooth was 0.9 mm. In order to avoid high current density occurring at the anvil/Al interface, the interval of the anvil tooth was 0.1 mm larger than that of the sonotrode tooth. For the conventional HPUSW, the welding time was 0.5 s while the clamping force was set to 1575–1975 N. The specimens were 6061-T6 aluminum alloy and pure copper, cut into 100 × 25 × 0.8 mm^3 pieces. The overlapping area of the specimens was 25 × 25 mm^2. In this work, prior to welding, the samples were ultrasonically cleaned with acetone to remove surface contaminants. The ultrasonic vibration direction (VD) was perpendicular to the length direction of the workpiece (Figure 3d).

The fracture surface morphology was observed by a JEOL JSM-7001F field emission gun scanning electron microscope (FEG SEM) (Jeol, Tokyo, Japan), equipped with an energy-dispersive X-ray spectrometer (EDS). X-ray diffraction (XRD) analysis was carried out using a PANalytical Empyrean (Malvern PANalytical, Almelo, The Netherlands) diffractometer to identify the phases of IMC at the welding interface. The XRD spectra were measured from 20° to 90°, with a step size of 0.02° and a scanning speed of 0.02 °/s. The tensile tests were performed using a Shimadzu AGS-X electronic testing machine (SHIMADZU, Kyoto, Japan). Tensile-shear tests were carried out at a tensile speed of 1 mm/min. At least three samples were tested for each process condition. The mechanical strength of the joints was evaluated according to tensile-shear strength, which was calculated by dividing the maximum tensile-shear force by the sonotrode area of 7 mm × 5 mm.

The temperature was measured using 0.1 mm K-type thermocouples, inserted through a semi-circular groove of radius 0.5 mm on the aluminum sheet surface. Thermocouple tip was placed at 1.5 mm from the center of the Cu/Al interface. Figure 3c,d show a schematic of the tensile-shear and thermocouple temperature measurement tests, respectively.

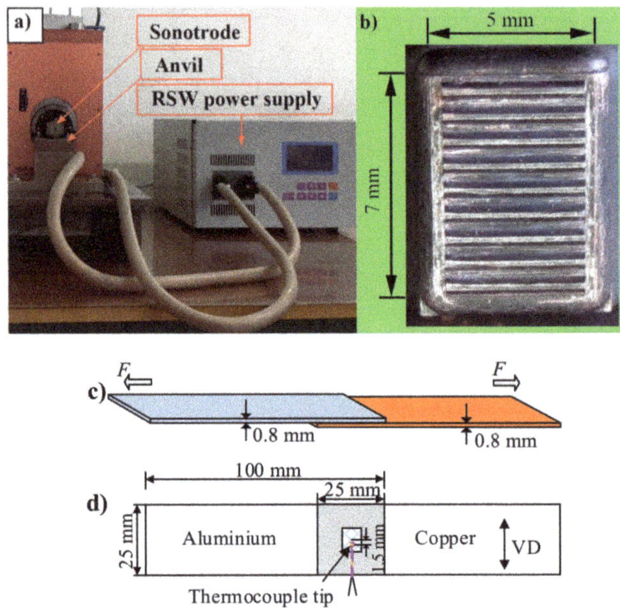

Figure 3. (**a**,**b**) The RUSW machine setup; (**c**) configuration of the tensile-shear, and (**d**) temperature measurement tests setup.

4. Results and Discussion

4.1. Interface Temperature

Figure 4 shows the measured temperature as the current changes from 0 to 3900 A. The results show that at the same welding time of 0.2 s, the measured peak temperature at the welding interface in RUSW is 445 °C, which is significantly higher than the interface temperature of 301 °C in conventional HPUSW. The peak temperatures with a current of 1300 A, 2600 A, and 3900 A at the welding time of 0.2 s are 328 °C, 380 °C, and 445 °C, respectively. This means that the temperature rises at a high gradient as the welding current increases, because the resistance heat *QRSW* has a quadratic relationship with the electric current, according to Equation (3). The higher interface temperature benefits interfacial diffusion, resulting in higher welding strength.

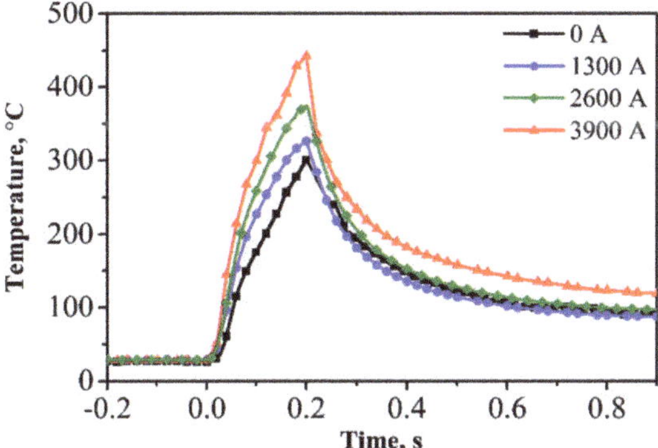

Figure 4. Measured temperatures for various currents.

Figure 5 shows the values of the interface temperature during hybrid welding and conventional HPUSW, at different welding moments. It is evident that the peak interface temperature in hybrid welding is similar to that in conventional HPUSW under the same conditions (welding time of 0.4 s, clamping force of 1975 N, welding time of 0.5 s, and welding pressure of 1575 N). This demonstrates that the additional resistance heat can speed up the HPUSW process.

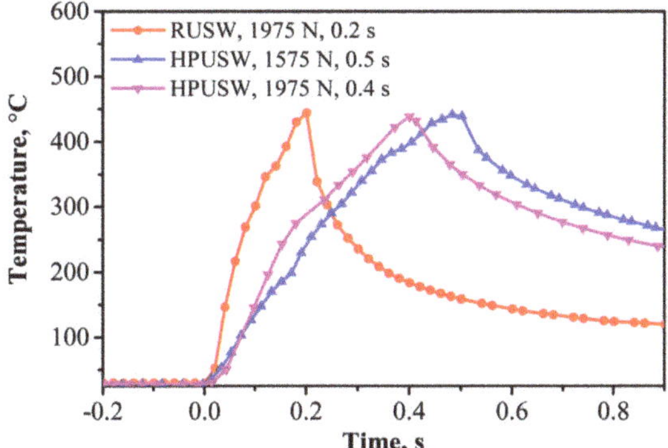

Figure 5. Measured temperature progress with welding time and pressure.

4.2. IMC Layer

During HPUSW of dissimilar joints, the thickness of the IMC layer determines the welding quality [20,21]. Figure 6a,b show backscattered images of the RUSW joint at $tw = 0.2$ s, $FN = 1975$ N, and HPUSW joint at $tw = 0.5$ s, $FN = 1575$ N, respectively. The thickness of the IMC layer in the joints prepared by hybrid welding and single ultrasonic welding is 1.5 μm and 2.2 μm, respectively. The interface temperature can be measured with good repeatability due to the proximity of the thermocouple to the center of welding spot and stable vibration amplitudes obtained under the same process parameters [22], which produces good repeatability in material vacancy concentration [23]. This results in good repeatability of the IMC thickness measurements. Although the interface temperature of the hybrid welding is the same as that of the conventional HPUSW, the duration of the hybrid welding is shorter, leading to a thinner IMC layer compared to the conventional HPUSW result.

The chemical composition of IMC in joints prepared by RUSW and conventional HPUSW was analyzed using EDS point analysis, and the results are shown in Figure 6c,d, respectively. The chemical composition at points A and B is 54.82 Al-45.12 Cu and 55.02 Al-44.98 Cu (by weight percentage (wt%)), respectively. This result demonstrates that the chemical composition of the IMC layer, in both the hybrid and conventional HPUSW methods, is primarily brittle Al2Cu alloy, according to the binary-phase diagram of Cu-Al [24]. In conclusion, the composition of the IMC layer is quite similar for joints prepared by both hybrid and conventional welding methods. However, the IMC layer is thinner in the hybrid welding approach due to the shorter duration of the welding process, producing better quality results.

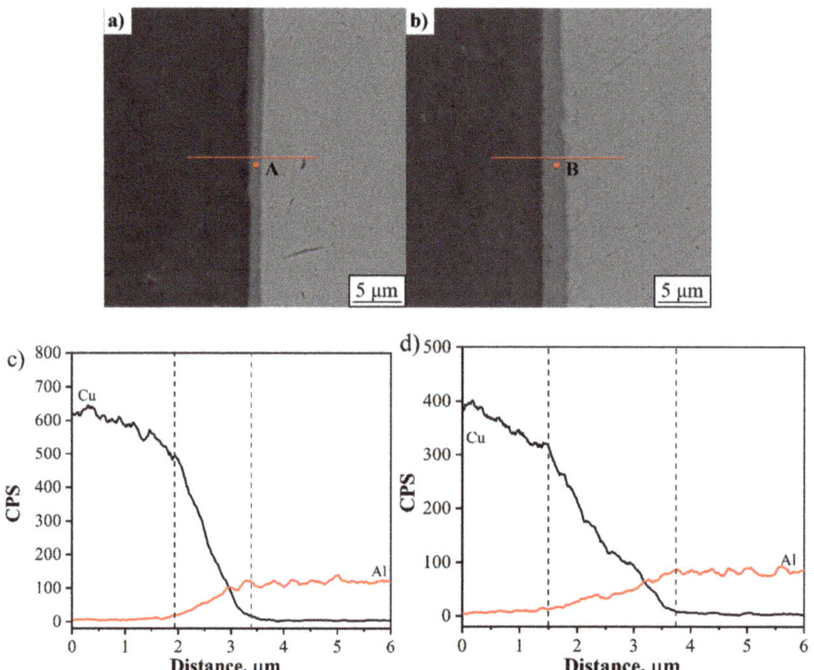

Figure 6. (**a**,**b**) Cu/Al interface in the RUSW and high-power ultrasonic welding (HPUSW) joints, respectively; (**c**,**d**) EDS results for the lines intersecting the points indicated in (**a**,**b**).

X-ray diffraction analysis was performed on the Al side of the fracture surface to further evaluate the phase composition of RUSW, as shown in Figure 7. The results show that the IMC formed during hybrid welding is primarily composed of Al_2Cu. Therefore, the XRD results are in agreement with the EDS results.

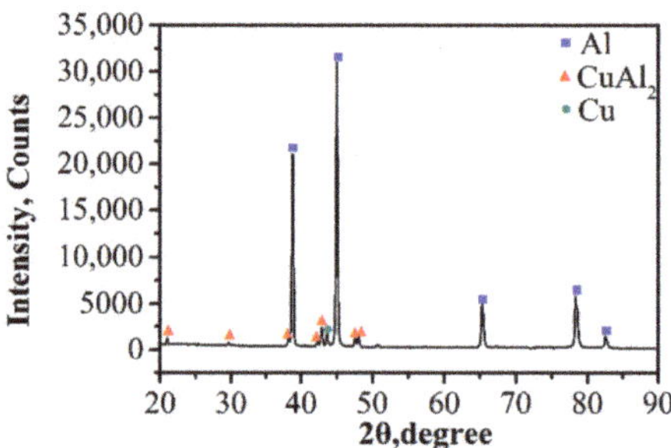

Figure 7. XRD patterns of the fracture surface.

4.3. Weld Cross-Section

The mechanical behavior was also affected by materials penetration. Figure 8a,b show the profiles of the weld cross-section, for the hybrid welded joint at tw = 0.2 s and for conventional HPUSW joint at tw = 0.4 s, under clamping force of 1975 N. The joint cross-section after hybrid welding has no obvious defects, whereas, in a conventional HPUSW joint, cracks appear on the copper, just under the tip of the sonotrode. This happens because the material is under high-stress concentration below the edge of the sonotrode, due to the action of long-term ultrasonic waves. In addition, in the conventional HPUSW joint, cracks propagate from the base metal to the welding interface under high-pressure action, which reduces the area of the welding and allows the aluminum alloy to leak from the crack and get attached to the sonotrode tip. This effect decreases the welding quality and may even cause damage to the sonotrode [25].

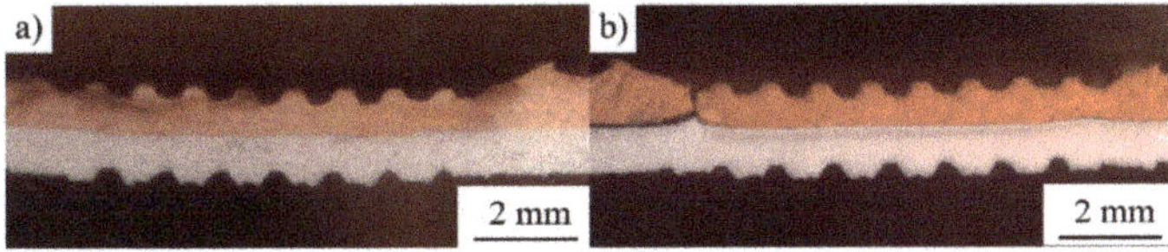

Figure 8. Profile of weld cross-section (**a**) RUSW at 0.2 s, and (**b**) HPUSW at 0.4 s.

4.4. The Average Shear Stress

The experiments show that the average shear stress of the joint obtained with additional resistance heat reaches 97 MPa and elongation extends to about 1.2% (Figure 9a). This value is significantly higher compared to the joint prepared without resistance heat, under the same clamping force of 1975 N. In our previous work, the average shear stress of the RUSW joint was also slightly higher than the average shear stress of 90 MPa obtained in conventional HPUSW, at a pressure of 1575 N and welding time of 0.5 s, which are the welding conditions that produce the highest strength in HPUSW of Cu/Al joint [9]. This is attributed to the fact that the additional resistance heat can increase the interface temperature (Figure 4), reduce the thickness of the brittle IMC layer (Figure 6), and avoid cracks at the edges of the weld zone (Figure 8). The strength of the hybrid weld is also much higher than that of RSW, because of the lower interface temperature in RSW. As the resistance of the steel sonotrode is an order of magnitude higher than that of the workpieces, the resistance heat mainly occurs at the sonotrode/Cu interface, rather than at the Cu/Al interface.

It is well known that the repeatability in welding dissimilar materials is poor [6]. The poor repeatability could be related to the fact that IMC may not be uniformly distributed across the entire weld interface (Figure 6a,b) [26]. Moreover, differences in IMC layer thickness may exist in different regions of the weld. While there were issues with repeatability in this study, some general trends in weld strength were observed. Figure 9b plots the weld tensile strength against the current. Although there is a lot of scattering, the figure suggests that a high current can produce welds with increased tensile strength. The duration of hybrid welding is shorter than that of conventional HPUSW technique, and thus the period of ultrasonic excitation is shorter, reducing the risk of fatigue fracture [27].

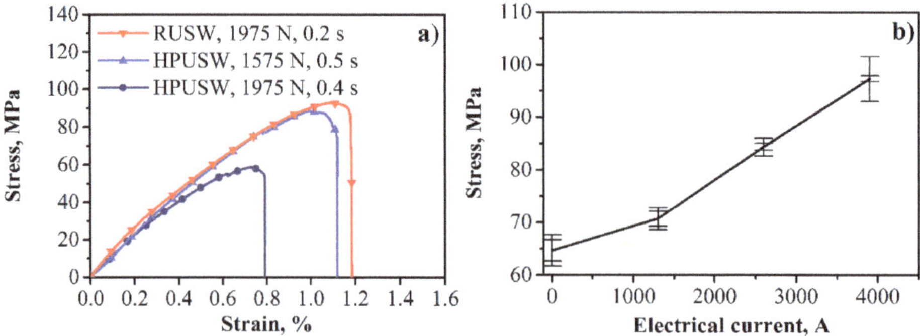

Figure 9. (**a**) The relationship between lap shear stress and strain; (**b**) the relationship between lap shear stress and electric current.

4.5. Fracture Morphology

Figure 10a,b show the macroscopic fracture morphology of the Cu/Al joint made by RUSW at 0.2 s and by conventional HPUSW at 0.4 s, under a clamping force of 1975 N on the Al side. Figure 10c is an enlarged view of Figure 10a, showing the microscopic topography of the fracture. It is observed that there are differences in fracture behavior. For the joint fabricated with RUSW, only welded regions are observed in the fracture surface. This demonstrates that the fracture of the hybrid welding joint occurs at the IMC of the Cu/Al interface, indicating high welding strength. In the case of a joint fabricated with HPUSW, some welded regions and scratched regions exist in the fracture surface (Figure 10b). There are two fracture models demonstrated in Figure 10c. From the enlarged view of the two regions (Figure 10e,f) from Figure 10c, it is obvious that a large number of dimples of different sizes are present in the c region, while there are some cleavage planes in the f region. These features indicate that the fracture of the RUSW is ductile–brittle type.

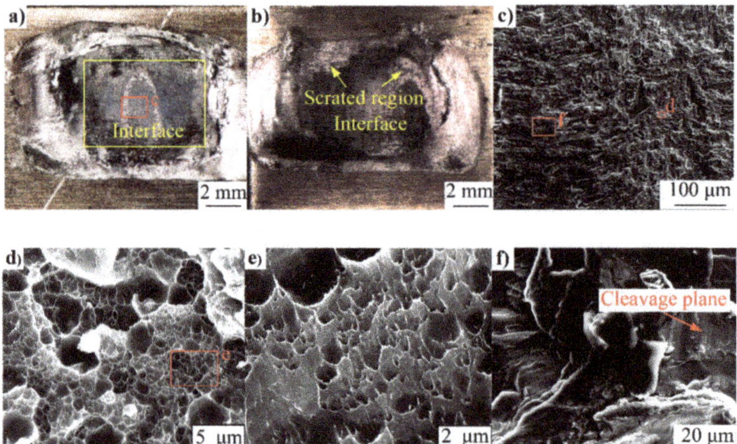

Figure 10. Fracture surface (**a**) RUSW at 0.2 s; (**b**) HPUSW at 0.4 s; (**c**) enlarge view of (**a**); (**d**) region d in (**c**); (**e**) enlarge view of (**d**); (**f**) region f in (**c**).

5. Conclusions

Dissimilar joints of pure copper and 6061-T6 alloys prepared by RUSW are investigated and compared to those prepared by the conventional HPUSW method. It is found that the additional resistance heat promotes the increase in interface temperature and reduces the thickness of the brittle IMC layer. The short duration of RUSW prevents the formation of cracks on the copper surface. As a result, the mechanical properties of the joint prepared by hybrid welding are significantly improved. Fracture surface morphology exhibits dimples and some cleavage planes, indicating a ductile–brittle hybrid fracture in RUSW.

Author Contributions: H.L. conceived, designed, and performed the experiments, and wrote the manuscript; B.C. contributed to the discussion of the experimental data, interpreted the results and revised the manuscript.

Funding: This work was supported by the National Natural Science Foundation of China (Grant No. 51175184).

Acknowledgments: The authors would like to acknowledge the kind assistance from Caiyou Zeng from the South China University of Technology.

Conflicts of Interest: The authors declare no conflict of interest.

References

1. Peng, H.; Jiang, X.; Bai, X.; Li, D.; Chen, D. Microstructure and mechanical properties of ultrasonic spot welded Mg/Al alloy dissimilar joints. *Metals* **2018**, *8*, 229. [CrossRef]
2. Pereira, A.B.; Cabrinha, A.; Rocha, F.; Marques, P.; Fernandes, F.A.; Alves de Sousa, R.J. Dissimilar Metals Laser Welding between DP1000 Steel and Aluminum Alloy 1050. *Metals* **2019**, *9*, 102. [CrossRef]
3. Eslami, N.; Hischer, Y.; Harms, A.; Lauterbach, D.; Böhm, S. Influence of Copper-Sided Tin Coating on the Weldability and Formation of Friction Stir Welded Aluminum-Copper-Joints. *Metals* **2019**, *9*, 179. [CrossRef]
4. Eslami, N.; Hischer, Y.; Harms, A.; Lauterbach, D.; Böhm, S. Optimization of process parameters for friction stir welding of aluminum and copper using the taguchi method. *Metals* **2019**, *9*, 63. [CrossRef]
5. Zhou, K.; Yao, P. Overview of recent advances of process analysis and quality control in resistance spot welding. *Mech. Syst. Signal Process.* **2019**, *124*, 170–198. [CrossRef]
6. Liu, G.; Hu, X.; Fu, Y.; Li, Y. Microstructure and mechanical properties of ultrasonic welded joint of 1060 aluminum alloy and T2 pure copper. *Metals* **2017**, *7*, 361. [CrossRef]
7. Liu, J.; Cao, B.; Yang, J. Effects of vibration amplitude on microstructure evolution and mechanical strength of ultrasonic spot welded Cu/Al joints. *Metals* **2017**, *7*, 471. [CrossRef]
8. Zhang, C.; Chen, D.; Luo, A. Joining 5754 automotive aluminum alloy 2-mm-thick sheets using ultrasonic spot welding. *Weld. J* **2014**, *93*, 131.
9. Yang, J.; Cao, B.; He, X.; Luo, H. Microstructure evolution and mechanical properties of Cu–Al joints by ultrasonic welding. *Sci. Technol. Weld. Join.* **2014**, *19*, 500–504. [CrossRef]
10. Zhao, Y.; Li, D.; Zhang, Y. Effect of welding energy on interface zone of Al–Cu ultrasonic welded joint. *Sci. Technol. Weld. Join.* **2013**, *18*, 354–360. [CrossRef]
11. Ni, Z.L.; Ye, F.X. Weldability and mechanical properties of ultrasonic joining of aluminum to copper alloy with an interlayer. *Mater. Lett.* **2016**, *182*, 19–22. [CrossRef]
12. Balasundaram, R.; Patel, V.K.; Bhole, S.D.; Chen, D.L. Effect of zinc interlayer on ultrasonic spot welded aluminum-to-copper joints. *Mater. Sci. Eng. A* **2014**, *607*, 277–286. [CrossRef]
13. Macwan, A.; Kumar, A.; Chen, D. Ultrasonic spot welded 6111-T4 aluminum alloy to galvanized high-strength low-alloy steel: Microstructure and mechanical properties. *Mater. Des.* **2017**, *113*, 284–296. [CrossRef]
14. Ni, Z.; Zhao, H.; Mi, P.; Ye, F. Microstructure and mechanical performances of ultrasonic spot welded Al/Cu joints with Al 2219 alloy particle interlayer. *Mater. Des.* **2016**, *92*, 779–786. [CrossRef]
15. Dehelean, D.; Oanca, O.; Toma, C.; Dorohoi, C.; Budau, V.; Craciunescu, C. Advanced materials joining using a hybrid ultrasonic-electric resistance technique. *J. Optoelectron. Adv. Mater.* **2010**, *12*, 1935–1941.
16. Yang, J.; Cao, B. Investigation of resistance heat assisted ultrasonic welding of 6061 aluminum alloys to pure copper. *Mater. Des.* **2015**, *74*, 19–24. [CrossRef]
17. Li, H.; Cao, B.; Yang, J.; Liu, J. Modeling of resistance heat assisted ultrasonic welding of Cu-Al joint. *J. Mater. Process. Technol.* **2018**, *256*, 121–130. [CrossRef]
18. Li, H.; Cao, B.; Liu, J.; Yang, J. Modeling of high-power ultrasonic welding of Cu/Al joint. *Int. J. Adv. Manuf. Technol.* **2018**, *97*, 833–844. [CrossRef]
19. Lee, D.; Cai, W. The effect of horn knurl geometry on battery tab ultrasonic welding quality: 2D finite element simulations. *J. Manuf. Process.* **2017**, *28*, 428–441. [CrossRef]

20. Samanta, A.; Xiao, S.; Shen, N.; Li, J.; Ding, H. *Atomistic Simulation of Diffusion Bonding of Dissimilar Materials Undergoing Ultrasonic Welding. The International Journal of Advanced Manufacturing Technology*; Springer: London, UK, 2019; pp. 1–12.
21. Das, A.; Masters, I.; Williams, D. *Process Robustness and Strength Analysis of Multi-Layered Dissimilar Joints Using Ultrasonic Metal Welding. The International Journal of Advanced Manufacturing Technology*; Springer: London, UK, 2019; Volume 101, pp. 881–900.
22. Zhao, J.; Li, H.; Choi, H.; Cai, W.; Abell, J.A.; Li, X. Insertable thin film thermocouples for in situ transient temperature monitoring in ultrasonic metal welding of battery tabs. *J. Manuf. Process.* **2013**, *15*, 136–140. [CrossRef]
23. Hu, T.; Zhalehpour, S.; Gouldstone, A.; Muftu, S.; Ando, T. A method for the estimation of the interface temperature in ultrasonic joining. *Metall. Mater. Trans. A* **2014**, *45*, 2545–2552. [CrossRef]
24. Tan, C.; Jiang, Z.; Li, L.; Chen, Y.; Chen, X. Microstructural evolution and mechanical properties of dissimilar Al–Cu joints produced by friction stir welding. *Mater. Des.* **2013**, *51*, 466–473. [CrossRef]
25. Badamian, A.; Iwamoto, C.; Sato, S.; Tashiro, S. Interface Characterization of Ultrasonic Spot-Welded Mg Alloy Interlayered with Cu Coating. *Metals* **2019**, *9*, 532. [CrossRef]
26. Bhamji, I.; Preuss, M.; Moat, R.; Threadgill, P.; Addison, A. Linear friction welding of aluminium to magnesium. *Sci. Technol. Weld. Join.* **2012**, *17*, 368–374. [CrossRef]
27. Carboni, M.; Annoni, M. Ultrasonic metal welding of AA 6022-T4 lap joints: Part II–Fatigue behaviour, failure analysis and modelling. *Sci. Technol. Weld. Join.* **2011**, *16*, 116–125. [CrossRef]

Article

Process and Parameter Optimization of the Double-Pulsed GMAW Process

Ping Yao [1], Kang Zhou [2,*] and Shuwei Huang [1]

[1] College of Electromechanical Engineering, Guangdong Polytechnic Normal University, Guangzhou 510635, China; ypsunny@163.com (P.Y.); hsw515087011@163.com (S.H.)
[2] School of Mechatronical Engineering, Beijing Institute of Technology, Beijing 100081, China
* Correspondence: zhoukang@bit.edu.cn or zhoukang326@126.com; Tel: +86-130-1108-3682

Received: 20 August 2019; Accepted: 10 September 2019; Published: 15 September 2019

Abstract: The double pulsed gas metal arc welding (DP-GMAW) process has been effectively employed to realize joining of steel plates and obtain weld bead surfaces with high quality fish scale ripples. In this work, a DP-GMAW process based on robot operation using the latest *twinpulse* XT DP control technology was employed to join the stainless-steel base plates. Four key operational parameters, which were robot welding speed, twin pulse frequency, twin pulse relation and twin pulse current change in percent, were selected to be input elements of orthogonal experimental design, which included nine experiments with three levels. To accurately understand the performance and process of weld bead obtained from DP-GMAW operation based on robot operation, the appearance observation and key shape parameters measurement, microstructure analysis, tensile and hardness testing, as well as stability analysis of the electrical signals, were conducted. Correlation analysis showed that the grain size was significantly correlative to the toughness and hardness. Then, to obtain quantitative evaluation results, fuzzy comprehensive evaluation (FCE) was employed to provide quality evaluation of weld beads from the above experiments. The influential levels of the key operational parameters on the appearance, grain size and FCE scores, and corresponding physical analyses, were respectively presented. In addition, optimal parameters combinations for obtaining weld beads with optimal appearance, grain size, and the highest FCE scores of weld bead quality were respectively provided according to the range analysis of the results from orthogonal experimental design. This work can provide an effective analysis method of influential levels of key operational parameters on the performance of the weld bead, optimal operational parameters combination seeking method, and quantitative quality evaluation method for the DP-GMAW process, which can improve the process optimization and increase the production efficiency, both in academic research and actual industrial production.

Keywords: double-pulsed; robot operation; microstructure; fuzzy comprehensive evaluation; orthogonal experimental design

---

1. Introduction

The double-pulsed gas metal arc welding (DP-GMAW) process is a mature arc welding operation technology that is prevalently employed in modern industrial manufacturing occasions. Though this process is designed based on the traditional pulsed-GMAW (P-GMAW), it has some significant merits when compared to the P-GMAW, such as the DP-GMAW process can reduce the porosity incidence [1] and improve the solidification cracking susceptibility [2]. Also, the process has better gap bridging ability [3], and better ability to control the mode of droplet transfer than those of the P-GMAW process [4]. This new technique is an effective variation of the traditional P-GMAW process, in which the pulsing current aiming to metal transfer control is overlapped by a thermal pulsation [5], which induces changes of temperature and stress of the welding pool [6]. Hence, it has been paid more and more attention in academic research and actual industrial production areas in recent years.

The difference between P-GMAW and DP-GMAW processes is the current waveform. During the DP-GMAW process, the current waveform is composed of a rhythmic thermal pulse phase (TPP) and a thermal base phase (TBP) [7], which have different frequencies and amplitudes, and the sum of durations of the two phases is equal to a thermal period (TP) [8]. Because of the existence of double pulses, it was also been named the twin-pulsed GMAW process in some published literature. To assure a stable welding process and obtain weld beads with high quality, researchers and scholars have put a lot of efforts to change the waveforms. The original waveform was a usual types of square waveforms where strong current and weak current were alternately appeared during the process. Under the circumstance, arc quenching may appear when actual currents switch between TPP and TBP, because the wire feeding equipment may not be able to catch up with the variation of the current pulse with adequate speed due to the mechanical inertia [8]. To alleviate the sudden changes of the current amplitude differences in these two phases, corresponding improvements have been revealed, such as trapezoid waveform, or sinusoidal waveform [8,9]. By means of the changes, the changes of the currents were replaced by gradual switches, which can improve the stability of the arc to achieve a stable droplet transfer [10]. Apart from arc quenching, splashes, short-circuit or open-circuit of the electrical system may frequently occur during the process if the parameters setting and matching are improper. All of these phenomena should be carefully considered during system design and operation in order to decrease the cost of production and improve the actual efficiency.

According to the principle of DP-GMAW, there are many parameters requiring proper setting during the process, such as the thermal period and corresponding frequency, twin pulse current change and duty cycles in two phases, and so on. Also, currently the welding robot has been employed more and more in manufacturing, and the robot traveling speed is also an important parameter. Hence, this is a typical multi-parameter system, and how to obtain an optimal operational parameters combination for achieving satisfactory performance is a challenge work for all users. In practical application, testing and justifying each parameter on the welding quality can cost so much and cannot be accepted in the majority of occasions.

No matter which type of welding technology is employed, quality estimation or evaluation is so important. For example, for resistance spot welding, the tensile-shear strength of the weld can be used for evaluating the welding quality [11]. For pulsed GMAW products, which is the weld bead, the quality involves more elements, such as crack, appearance, penetration, microstructure, and so on [12]. In general, these different elements should be properly combined to yield one reliable quality criterion. Nowadays, with developing computer technology, artificial intelligent (AI) technology has demonstrated many achievements. In welding research area, AI technology has been also employed, such as in quality estimation of resistance spot welding [11], or in optimal parameter prediction in double-wire-pulsed metal inert gas (MIG) arc welding [13]. As for the quality evaluation of arc welding products, many previous contributions have paid a lot attention to it. Casalino et al. [14] employed neural networks to establish a relation between process parameters and geometry of the molten zone of the welds, and then used a fuzzy C-means clustering algorithm to evaluate the quality. Wu et al. [15] used a Kohonen network to monitor the welding process and evaluate the quality in a GMAW process. The inputs were the probability density distribution (PDD) of the welding voltages and the class frequency distribution (CFD) of short circuiting times, and the network can recognize and classify the undisturbed and intentionally disturbed GMAW experiments. It can be noticed that the AI technology can be proper in estimating the quality of weld bead and exert remarkable effects.

This work aimed to explore how to obtain an optimal parameters matching in order to obtain the weld bead with satisfactory quality, while researching the influential levels of the key operational parameters on the different performances of the weld beads. The DP-GMAW process involves various input parameters, and the quality of weld bead also includes a lot of evaluation criteria. To achieve preliminary goals, two main contents have been included in this work. The first was the optimal parameters machining. Among various operational parameters, few key operational parameters were selected to do experiments. To achieve the desired effects and decrease experimental complexity, orthogonal experimental design, which is an important experimental design method to explore the system effects typically involving multiple factors and multiple levels [16], was employed. This experimental design method can reduce the workload and involves corresponding methods of analyzing the experimental results, so to yield more reliable conclusions [17].

After employing orthogonal experimental design to obtain weld beads using different operational parameter combinations, an appropriate quality evaluation method can be used to estimate the experimental results. Because the quality of weld bead involves various elements, the quality evaluation method should also be a multi-input system. In addition, to clearly reflect the evaluation results, the output is better when quantitatively presented. In this work, considering the characteristics of weld beads and application of current AI technology, fuzzy comprehensive evaluation (FCE), which was an effective evaluation method based on the fuzzy sets and fuzzy mathematics, was chosen to conduct the quality evaluation for the weld beads. FCE was introduced in the 1960s, and has become an effective multi-factor decision-making tool for comprehensive evaluations so far. During the actual application, combining with the expert experiences, this method can make a full and comprehensive refection on the evaluation criteria and the influence factors of fuzziness, and produces evaluation results closer to the actual situation [18]. It has been used in a lot of different areas, such as in power policy making [19], teaching and education performance evaluation [20], motion performance evaluation of autonomous underwater vehicle [21], water resources carrying capacity [22], distinct heating system evaluation [23], real estate investment risk research [24], quality assessment for compressed remote sensing images [25], and other relative areas.

In this work, the DP-GMAW process based on an industrial robot operation was conducted, the objective was seeking optimal operational parameters combination in order to obtain weld bead with satisfactory quality, and obtaining the influential levels of different operational parameters on the performances of the weld bead. During the process, according to principle and operational characteristics of this process, some key operational parameters were selected to design orthogonal experiments, and then the FCE method was employed to do quality evaluation according to relative experimental results and obtained optimal operational parameters combinations. Advanced experimental designing methods and quantitative quality evaluation methods were effectively combined in this work, and the contribution can serve the current DP-GMAW process improvement and parameter optimization.

2. Operational Characteristics of the DP-GMAW Process

The DP-GMAW process involves a lot of typical operational parameters. The current pulses in TPP and TBP have different frequencies. The thermal period (TP) denotes the duration of two phases. Another important operational parameter, twin pulse frequency (TPF), which is a reciprocal of the TP, can describe the speed of current waveform adjustment between the two double pulses. Also, the TPF can determine the number of fish scale ripples. It can be noticed that this process is a low frequency modulation based on high frequency phases (TPP and TBP), in other words, the low frequency current pulsation or the thermal pulse is superimposed on a pulsed current for active metal transfer control and weld pool stirring [26]. During the process, TPF can reflect the varying speed of the strong set and weak pulses set, and each set may include up to 10–20 high frequency pulses, the maximum frequencies of the pulses may achieve is 100 Hz. Hence, in a general case, the value of TPF is below 5 Hz. The welding currents were switched between two high frequencies with TPF of the switch frequency. The frequency in TPP is higher than that in TBP, and the pulse set in TPP is called a strong pulse set, which is signed as *PulseS*, and the pulse currents are switched between base current I_{bs} and peak current I_{ps} while the pulse set in TBP is called as weak pulse set, which is signed as *PulseW*, the corresponding base current and peak current are respectively I_{bw} and I_{pw}. The durations of TPP and TBP were, respectively, T_s and T_w. In addition, the proportion of the time of thermal pulse phase, which is T_s, in one thermal period TP, is called twin pulse relation, which can be mathematically described in Equation (1):

$$D_T = T_s/TP \tag{1}$$

where this parameter is denoted as D_T in this work.

The strong pulse set is to control the droplet transfer for obtaining enough welding penetration, in a general case, one pulse corresponds to one droplet, while the function of the weak pulse set is obtaining a series of regular pulses to stir the weld pool [27], and one TBP corresponds to one weld pool. In addition, the average current in TPP is I_{avs}, while the average current in TBP is I_{avw}, and the average current in one thermal period is marked as I_{av}. In general, the I_{av} is a preliminary arc welding setting value during the process. Under this circumstance, the regular fish scale grain can form if all the parameters are properly set. Figure 1 shows the schematic and main operational parameters of the DP-GMAW process.

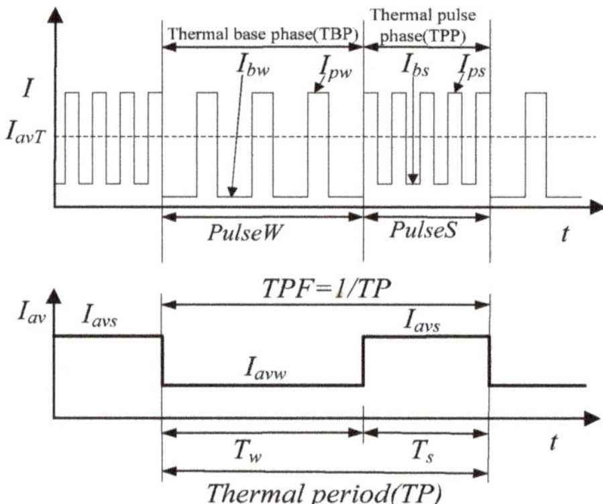

Figure 1. Schematic presentation of the double-pulsed gas metal arc welding (DP-GMAW) process.

In this figure, TP is the sum of T_w and T_s. Apart from the relations shown in the Figure 1, twin pulse current change, whose value is half of the subtraction between mean value of currents in TPP and TBP, is also an important parameter during the process because it can reflect the current variation in between two phases. In this work, to clearly reflect the current adjustment, this parameter can be described combining the I_{av} and in percent format as in the following equation:

$$I_\Delta = \frac{I_{avs} - I_{avw}}{2I_{av}} \times 100\% \tag{2}$$

where I_Δ is called twin pulse current change in percent. Moreover, the relations between I_Δ, and I_{avs} as well as I_{avw}, can be also be derived as follows:

$$\begin{cases} I_{avs} = I_{av} \times (1 + I_\Delta/100) \\ I_{avw} = I_{av} \times (1 - I_\Delta/100) \end{cases} \tag{3}$$

During the welding process, proper combination of the operational parameters of the pulse sets in TPP and TBP, and other process parameters, are the utmost important for improving the welding quality and obtaining weld bead with satisfactory fish scale ripples. During the traditional DP-GMAW operation process, various relative operational parameters cannot be accurately adjusted and matched one by one. In this work, the latest *twinpulse* XT DP control technology developed by the LORCH Company was employed. This process has two significant features:

1. The pulse frequency in TPP is so high, which can achieve 100 Hz, on the other hand, the pulse frequency in TPB can also achieve 30 Hz, and both of these two pulse frequencies were higher than those of traditional DP- GMAW process.
2. The peak current and base current in TPP and TPB are unchanged when the average current I_{av} in one thermal period is unchanged. The process control can be conducted through only adjusting average current I_{av} and the twin pulse relation D_T, then the parameters about *PulseW* and *PulseS* can correspondingly vary and need not to be adjusted, so that the number of operational parameters during the process which are required to be adjusted can be significantly decreased. Using this new control technology, the operational parameters of the arc welding process can be more effectively set and matched.

According to the above introduction about the DP-GMAW process, it can be noticed that there are various operational parameters included in the DP-GMAW process, and to obtain the weld bead with satisfactory quality, all the operational parameters should be carefully and seriously considered. Employing *twinpulse* XT DP control technology can decrease the setting complexity and improve the control performance, because this new control technology can use a few key operational parameters to control two waveforms, some important process parameters, such as peak current, base current, frequency, duty cycle, peak current, base current, peak time, base time, did not individually set. Therefore, realizing the proper control of energy delivery into the base plate was so convenient. Then in the next section, corresponding experiments can be conducted to produce weld beads with different qualities, and then the influential levels of different operational parameters on the selected performance of weld bead can be seriously explored after combining selected quality evaluation method.

3. Experimental Design of the DP-GMAW Based on Robot Operation

3.1. Experimental Platform

In this work, corresponding experimental design had been conducted. The experimental platform was composed of a FANUC Robot M-10IA industrial robot (FANUS Corporation, Oshino-mura, Yamanashi Prefecture, Japan), a LORCH S-RobotMIG arc welding machine (Lorch Schweißtechnik GmbH, Im Anwänder, Auenwald, Germany), a wire feeder machine, a welding torch and other auxiliary equipment. During the experimental process, the industrial robot controlled the welding speed, the current waveforms in TPP and TBP were controlled by the LORCH arc welding machine. In addition, one self-designed robot welding multi-signals collection and analysis instrument, which was based on the USB-6363 data acquisition card developed by NI (National Instruments) company (Austin, TA, USA), can be utilized to synchronously collect and analyze the current, voltage, arc sound signals. Figure 2 presented corresponding experimental instruments.

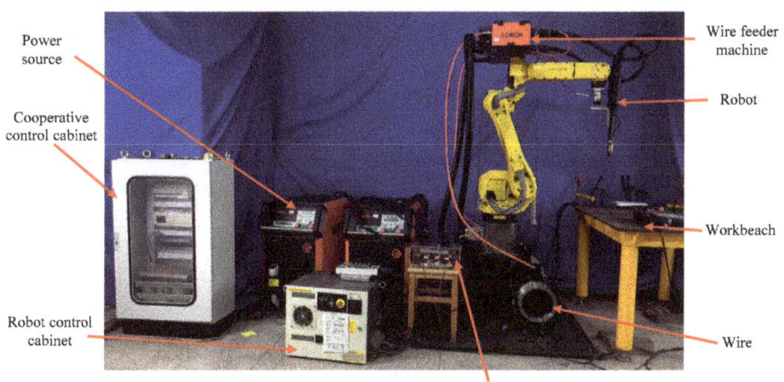

Figure 2. Experimental instruments and equipment.

Where two cabinets were used in the experiment. The robot control cabinet was responsible for the robot control operation, while the cooperative control cabinet was responsible for coordinately controlling the DP-GMAW process and robot operation.

3.2. Experimental Conditions and Methods

3.2.1. Experimental Conditions

The experimental conditions used in this work was follows: the base plates are the stainless steel 304, whose tensile strength was 520 MPa, the diameter of the used welding wire was 1.2 mm, and the material was stainless steel 316L. The shielding gas was composed of 98% pure argon and 2% CO_2 (15 L/min flow), the length of wire extension was 12 mm, and flat surfacing welding was used. The size of the base plate was 250 mm × 100 mm × 3 mm. The material characteristics of the base plate and the welding wire was shown in Table 1.

Table 1. Material characteristics of the base plate and welding wire.

Materials	C	Si	Mn	Cr	Ni	S	P	N	Mo
304	≤0.08	≤1	≤2	18–20	8–10.5	≤0.03	≤0.03	≤0.1	-
316L	≤0.03	≤1	≤2	16–18	10–14	≤0.03	≤0.045	-	2–3

To accurately reflect the effects of the different operational parameters on the welding quality, the base plate should be carefully preprocessed: the surface has been processed by angle grinder to eliminate the oxides, and then washed by special alcohol. After the surface of base plate was clear and dry enough, the welding action can be taken.

3.2.2. The Design of Orthogonal Experiments

Orthogonal experimental design is a powerful tool to deal with the system with multiple input parameters. The method can pick some typical parameters combinations from full possible combinations to conduct the experiments. According to corresponding analyses for the experimental results to comprehensively figure out the full experimental situation, and then obtain an optimal parameters combination. This method can reduce the number of the experiments and instruct to seek an optimal parameters combination for a special system.

To further explore the effects of some key operational parameters during the DP-GMAW process on the quality or other performances of the weld bead, according to the principle and characteristics of this welding process combined robot operation, four key operational parameters, which were robot welding speed V_R, twin pulse frequency TPF, twin pulse relation D_T and twin pulse current change in percent I_Δ, were chosen to design orthogonal experiments. The average welding current I_{av} during the process was set to 80 A, and the electrode inclination angle was 86° using backward inclination mode in this work. After serious preliminary analyses and process experiments, the value ranges of some main parameters can be confirmed during the ranges which the welding action can be normally conducted. Each chosen operational parameter corresponded to three levels as shown in Table 2, and then a detailed orthogonal experimental design, whose form was $L_9(3^4)$ with a four-element-three-level, can be depicted as shown in Table 3.

Table 2. Parameters (elements) and levels.

Level	V_R (cm/min)	TPF (Hz)	D_T (%)	I_Δ (%)
1	20	1	30%	30%
2	30	2	40%	40%
3	40	3	50%	50%

Table 3. The program of the orthogonal experimental design.

Index	V_R (cm/min)	TPF (Hz)	D_T (%)	I_Δ (%)
L1	20	1	30%	30%
L2	20	2	40%	40%
L3	20	3	50%	50%
L4	30	1	40%	50%
L5	30	2	50%	30%
L6	30	3	30%	40%
L7	40	1	50%	40%
L8	40	2	30%	50%
L9	40	3	40%	30%

Hence, to sufficiently explore the effects of different operational parameters on the quality or other performances of weld bead, nine experiments with different effective parameters combinations should be conducted.

3.2.3. Quality Evaluation of the Weld Bead

After orthogonal experimental design using chosen key operational parameters with different levels, for different experimental results, an accurate quantitative quality evaluation should be employed. In this work, the FCE method was introduced to quantitatively evaluate the quality of weld bead. Because this evaluation method can provide quantitative evaluation results based on some input conditions, and the quality of weld bead involves various different aspects including the appearance and shape, microstructure, and process stability of electrical signals, hardness and tensile performance, and so on, these quality criteria can be inputs of the FCE model using fuzzy mathematical algorithm, and then quantitative scores can be obtained.

FCE used the principle of fuzzy logic to evaluate the targets. There were two important procedures, the first was making evaluation using single element, while the second was making comprehensive evaluation using all elements. The detailed evaluation steps are shown in many published contributions [18,23,25]. In the next section, FCE can be employed in quantitatively estimating the quality of weld bead after orthogonal experiments.

4. Experimental Results and Analyses

A series of corresponding experiments were conducted based on the orthogonal experiment design as shown in Tables 2 and 3. Then, to deeply understand the welding quality, some relative measurement, observation and testing, which included appearance, shape parameters measurement, microstructure observation, tensile and harness testing, stability of electrical signals, can be sequentially conducted.

4.1. Appearance and Measurement of the Weld Beads

First, the macroscopic appearances, and cross-sections of weld beads of L1–L9 are shown in Table 4. In the table, the photos of the cross-sections of the weld beads, which were obtained using a microscopy with 20 times magnification, were provided. It can be observed that the overall appearance of the weld bead of L1 was unsatisfied, the grains of the fish scale ripples were rough, and there were some unfused parts appearing in the edge of the bead. The appearance of weld bead of L2 was still unsatisfied, the widths and heights of the ripples were non-uniform, as well as the weld bead was not straight and some curved bead existed. The bead of 2–5 cm was so thick, but the thickness in 7 cm was suddenly decreased, which meant that some sudden transitions occurred during the process. In addition, arc pits appeared in terminate of the bead. The quality of weld bead of L3 was very low and the appearance was irregular. Sudden transitions appeared in 10–15 cm, and serious splashes occurred in 13 cm, also, the shape located in the terminate was seriously irregular. The weld bead of L4 was regular and successive without any arc interruption or short-circuit occurring, however, the grain of the fish scale ripples was a bit rough and some unfused ripples appeared. The weld bead of L5 has high quality, the bead was straight and regular, and both of the penetration and height were reasonable. Moreover, compacted fish scale ripples are shown in the bead, and no obvious splash appeared during the process. The weld bead of L6 was very thin and high, and fish scale ripples were so obvious. However, the edge of the bead was irregular, and a few splashes occurred during the process. The weld bead of L7 was successive, but some irregular humps in the bead can seriously affect the mechanical performances. Both of the weld beads of L8 and L9 were regular, the overall shape of the bead was thin and high, though some splashes occurred during the process. According to the cross-sections of all the weld beads, the penetration of the weld beads in L1–L3 was so large that full penetrations appeared, while the widths of bead were irregular with a bit large. The penetration of weld bead of L4 was small, while the penetration of weld bead of L5 was 60% of the whole thickness of the base plate, which showed the penetration was proper. In addition, the penetrations of weld beads of L6, L8 and L9 were a bit small, while the penetration of weld bead of L7 can reach half of the whole thickness of the base plate.

Table 4. The appearances and cross-sections of weld beads of L1–L9.

No	Weld Bead	
	Appearance	Cross-Sections
L1		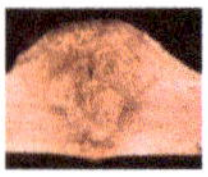
L2	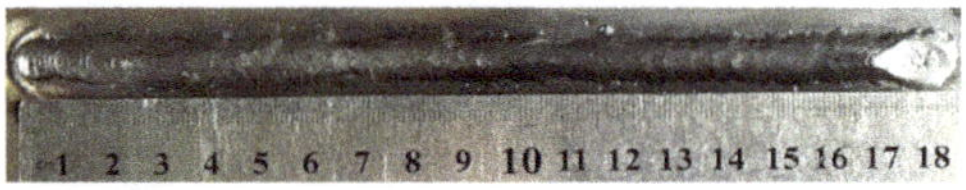	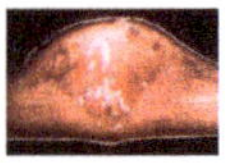
L3		
L4		
L5		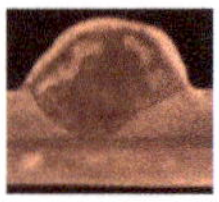
L6		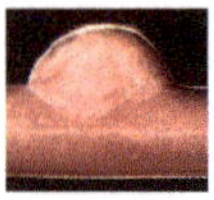
L7	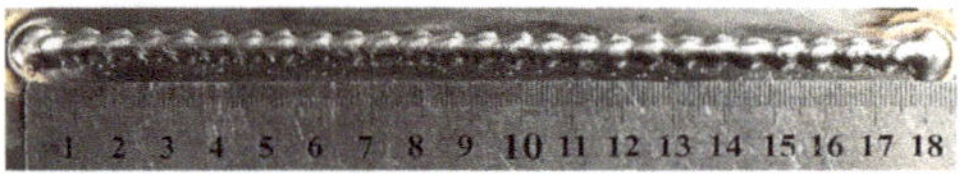	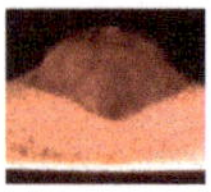
L8		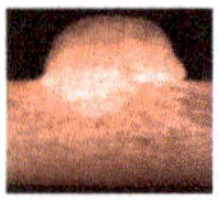
L9		

Furthermore, to further understand the morphology of the weld bead, detailed measurements for the shape of the weld bead were also conducted. Relative key measurement parameters of the shape can be seen in Figure 3.

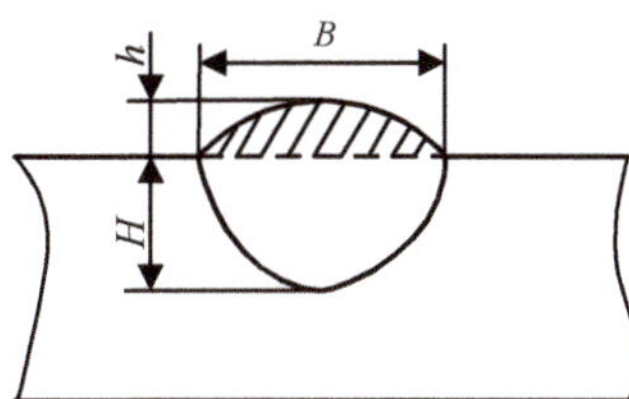

Figure 3. Schematic of shape parameter measurement for a weld bead.

Where the key shape parameters included bead width *B*, bead height *h*, penetration *H*. Then for the above nine weld beads, corresponding measurements were also conducted; the results are shown in Table 5.

Table 5. Measurement of the weld bead (mm).

Weld Bead	L1	L2	L3	L4	L5	L6	L7	L8	L9
h	2.16	2.335	2.54	1.955	2.065	2.13	1.595	1.995	2.1
H	>3	>3	>3	1.16	1.58	1.09	1.33	0.98	1.17
B	8.7	9.26	9.8	6.75	6.78	4.76	6.56	5.16	4.11

Combined with Tables 4 and 5, it can be noticed that the bead height *h* of L1 was proper, while the same item of L3 was relatively larger. However, the bead height of L3 was irregular with some heaves appearing on the surface; in addition, the transition was also not smooth. Under this circumstance, the stress may concentrate, which may deteriorate the load bearing capacity of the joints, and the cracks may appear when special loads were applied, which can affect the appearance of the weld bead. In addition, the penetration of weld beads of L1–L3 were so large, even full penetrations appeared, and the values of the bead width were also large enough. According to the observation of the cross-sections of these three beads, the forming situation was also not satisfactory. The bead height and bead width of L4 and L5 were proper, however, the penetration *H* of weld bead of L4 was small. The value of bead height *h* of weld bead of L6 was a bit larger, this was because both of the penetration and the bead width were relatively small, which showed the thin and high weld bead as shown in Table 4. The reason for this phenomenon was that the heat delivery was not sufficient during the process. The weld bead of L7 was hump type bead, which had irregular bead height *h*, and the stress may concentrate under this situation. The bead heights of L8 and L9 were proper, but the values of penetrations were small. Especially for the weld bead of L9, the bead width *B* was very small, so the overall bead appeared high and thin, therefore the forming situation was not well enough.

4.2. Materials Characteristic Performance Analyses of the Weld Beads

4.2.1. Analysis of Microstructure for the Weld Beads

The grain size is very important for the material characteristics, especially for the metals. The influence of grain size on the materials characteristics actually resulted from the influence of the grain boundary surface area. A smaller grain size denotes a bigger grain boundary surface area, and larger influence on the materials characteristics. In general, smaller grain size means higher strength and stiffness, as well as higher toughness and ductility for the metals in room temperature. This is because a plastic deformation can occur in more grains when the grain sizes are smaller, which can make the deformation more equal. Moreover, smaller grain size means more overall grain boundary surfaces, and boundaries are more complex, which can make the combination of different grains may more compacted. Under this circumstance, the propagation and growth of the cracks may be effectively prevented, which means that the metal has good strength and toughness.

To explore the materials characteristics, above nine weld beads can be observed by Phenom pro-electron microscopy (Thermo Fisher Scientific Inc, Waltham, MA, USA), which has 2000 multiplication times. The observations can be seen in Figure 4. In addition, to clearly present the grain sizes of the weld beads, the grain length was introduced to describe the grain size. In general, smaller grain length meant smaller grain size. For these nine weld beads, the corresponding grain length were 10.8, 10.38, 11.74, 9.64, 7.94, 9.00, 8.44, 6.53 and 6.75 μm from the weld beads from L1 to L9. It can be noticed that all the microstructures were cellular grain structure. The grain sizes of the weld bead of L1–L3 were larger, while the grain sizes of weld beads of L4–L6 were relatively smaller. In addition, the grain size of weld bead of L4 was a little larger than that of L5 and L6, and the weld beads of L7–L9 have the smallest grain sizes.

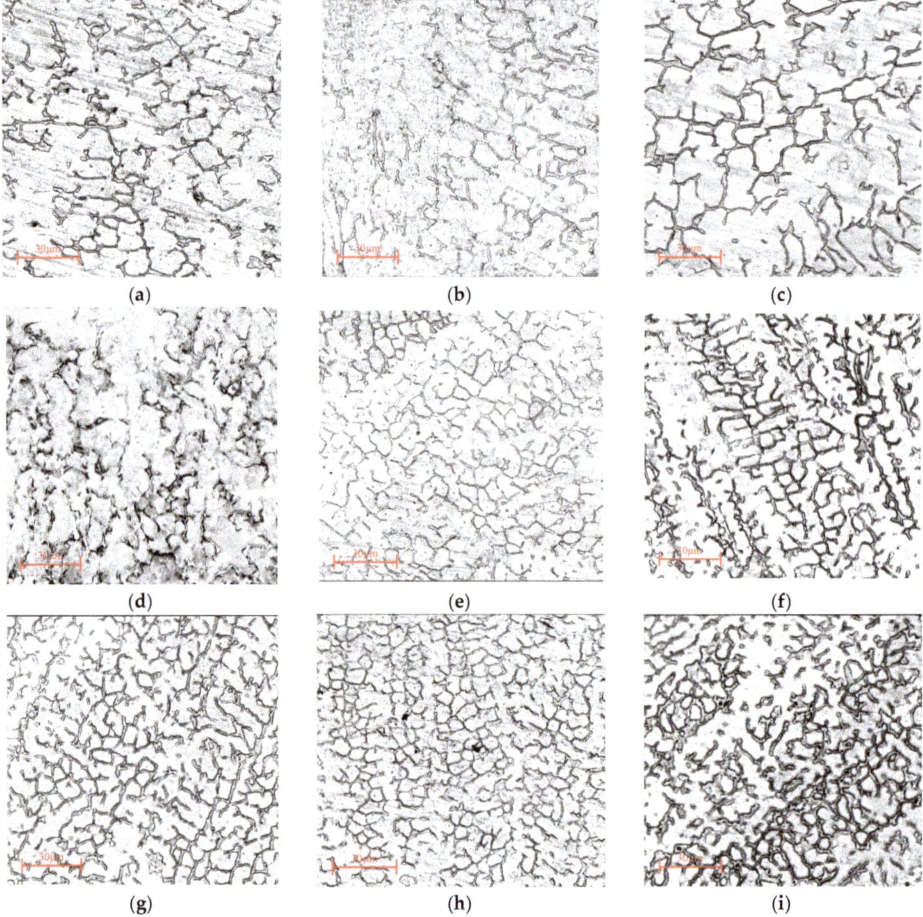

Figure 4. Microstructures of the weld bead observed by electron microscopy. (**a**) The weld bead of L1, (**b**) the weld bead of L2, (**c**) the weld bead of L3, (**d**) the weld bead of L4, (**e**) the weld bead of L5, (**f**). the weld bead of L6, (**g**) the weld bead of L7, (**h**) the weld bead of L8, (**i**) the weld bead of L9.

4.2.2. Performance Testing for the Weld Beads

To comprehensively obtain the performances of the weld beads of L1–L9, key performance tests can also be conducted in this work, in this part, tensile and hardness testing were carried out.

The first was the tensile testing, which was employed to test the axial toughness of the weld bead at room temperature (20 °C). The instrument used the WA-600 hydraulic universal testing machine (Yangzhou Jiangdu Open Source Test Machinery Factory, Yangzhou, China) to explore the toughness of the weld beads. Two terminals of each weld bead were seriously processed in order to conveniently hold, and then each specimen can be stretched to completely fracture. The species and experimental results are shown in Figure 5.

It can be noticed that the species of L1–L8 were fractured at the base plate which located in terminals of the weld beads, only for the weld bead of L9, the fracture occurred in the middle of the weld bead, and this bead had the longest elongation. The elongation of each specimen is recorded in Table 6.

Combining the observation and analysis results about microstructure of the weld beads in preceding part, the results in Table 6 showed that the weld beads which had larger grain sizes had stronger toughness, and as the grain sizes decreased, the toughness was weaker and weaker.

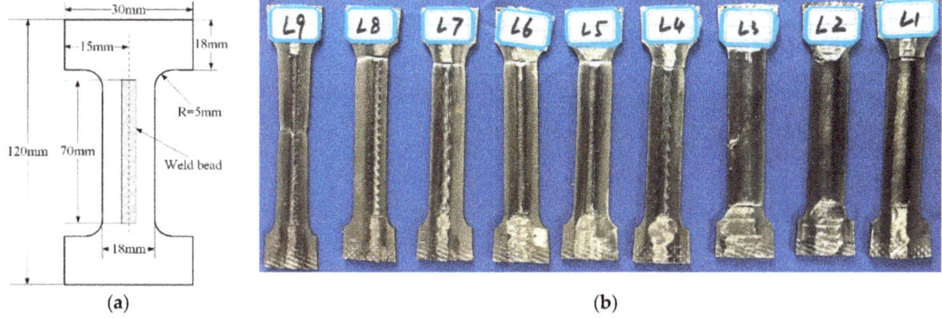

Figure 5. (**a**) Geometry of your tensile samples, (**b**) Tensile tests of the weld beads.

Table 6. Elongation of each specimen (mm).

Weld Bead	L1	L2	L3	L4	L5	L6	L7	L8	L9
Elongation	12.51	10.12	12.07	16.55	14.17	14.51	17.01	19.16	28.44

Apart from the tensile testing, the hardness testing was also conducted. The measuring instrument used the SCTMC HR-150DT electric Rockwell hardness tester (Shanghai Shangcai Testermachine CO., LTD, Shanghai, China) to test the harness for the weld bead. To accurately reflect the performance of the weld bead, five testing points were selected from the cross-section of each weld bead to conduct the hardness testing. The first testing point was in the center of the cross-section, and other four testing points were evenly distributed. A Cartesian axis was used, the origin was the center, and the horizontal axis was a line which went through the origin and parallel the surface of the base plate, while the vertical axis was a line which also went through the origin and vertical to the horizontal axis. Then, the four testing points were distributed in the four quadrants and the distance between the point and the edge the bead was 1.5 mm, as shown in Figure 6, and the applied external force on each point was 980.7 N.

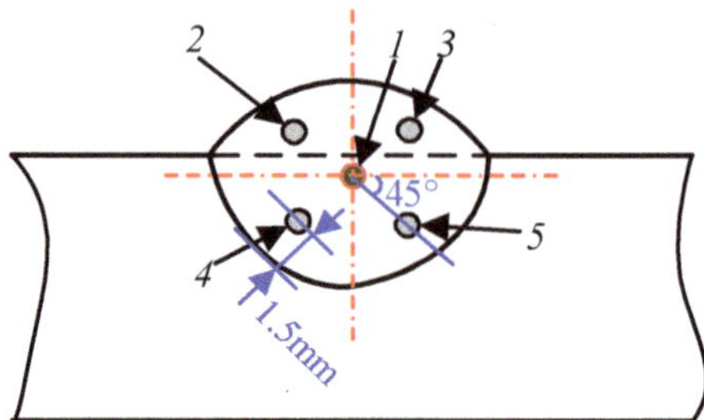

Figure 6. Testing points in the weld bead during hardness testing.

Where the dotted lines were axes. Then both maximum and minimum values in the five output values were rejected, and the final result was the mean value of the other three values for each specimen, corresponding results for nine weld beads are shown in Table 7.

Table 7. Hardness test of weld beads.

Weld Bead	L1	L2	L3	L4	L5	L6	L7	L8	L9
HRB	77	76.33	76.17	78.83	78.5	78	80.83	79.83	83

Where HRB is the abbreviation of Hardness Rockwell B, which is a commonly used criterion to describe the hardness of the weld bead using stainless steel. According to the testing results, it can be observed that each three tests formed one array, and there were three arrays in the experiments: the first array was from L1 to L3, the second array corresponded to L4 to L6, while the last array referred to L7 to L9. The overall hardness was increasing from the first to the third array. The harness values from L1 to L3 were very small and the values were so approaching. The average value of the second array was about 78.5, which was a little bigger than that of the first array. The third array had the biggest HRB values, whose average value was about 80. The HRB of the base plate was 73, which denoted that the hardness of these weld beads was higher than that of base plate. It can be concluded that the variation characteristic of the hardness was approximately the same as that of grain size.

4.2.3. The Correlation Analysis

According to above analyses, it is noticed that the toughness and hardness of the weld beads were relative to the grain size. To analyze the relations between grain size, and toughness as well as the hardness, corresponding correlation analyses had been also conducted.

Firstly, the nine weld beads can be ranked according to their grain sizes as shown in Figure 4. The weld bead with the smallest grain size can be marked as "Excellent", while the weld bead with the largest grain size was marked as "Poor", the other weld beads can be considered as "Good" and "Medium". Also, because the ranking values, toughness and hardness values had different value ranges, the normalized data processing must be conducted, in other words, the grain sizes, elongations and HRB values in the tensile testing were processed in between 0 and 1. The final results can be seen in Figure 7.

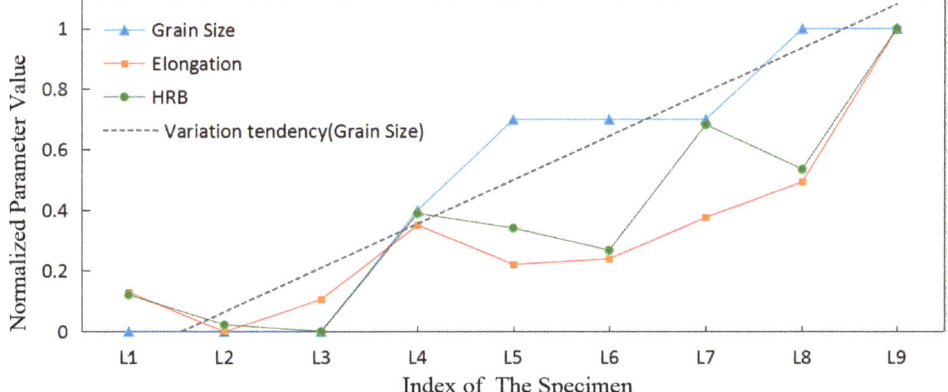

Figure 7. The variation tendency of grain size, toughness and hardness.

The vertical axis was the normalized parameter value whose range was between 0 and 1, the horizontal axis denoted the weld bead of L1–L9. It can be seen that the toughness, hardness and grain size of the weld beads had the approximately the same variation tendency.

Then, the correlation coefficients of above relation can also be calculated following the Equation (4), and the larger value of correlation coefficient denotes a greater degree of correlation.

$$r(X,Y) = \frac{Cov(X,Y)}{\sqrt{Var[X]Var[Y]}} \tag{4}$$

where *r(X,Y)* referred to the correlation between *X* and *Y*, *Cov(X,Y)* denoted that the Covariance of *X* and *Y*, and *Var*[*X*] and *Var*[*Y*], respectively, referred the variances of *X* and *Y*. The correlation coefficient between the ranking by grain sizes and the elongations of weld bead in the tensile test was calculated to be 77.56%, and the corresponding values between the ranking by grain sizes and hardness values was 84.35%, which can be considered as significant correlation. It can be concluded that a smaller grain size meant better toughness and higher hardness. Hence, in the following multi-parameter analysis, only considering the grain size should be enough.

4.3. Stability Analysis of the Electrical Signal

In this work, a one self-designed multiple-sensor-signal fusion system was employed to online collect the welding voltage and current signals. Using collected signals, the stability analysis of the welding current, which was closely relative to the stability of welding process and quality of weld bead, was conducted. After sampling all of nine welding current signals, three current signals corresponding to different stabilities, which were L3, L5 and L7 according to above analyses, were selected to make detailed stability analyses, corresponding results are shown in Figures 8–10. Figure 8 shows the transient welding current waveform and corresponding possibility density function (PDF) results of welding current of L3. The PDF of the welding current, together with the transient welding current, were commonly employed methods to evaluate the stability of welding process [28,29]. As shown in Figure 8a, the larger values of welding current appeared in 37.7–37.8 s and 38–38.5 s; the values were above 350 A; this was because the welding currents were unstable, which can induce the metal transfer non-uniform, and short-circuits occurred so that the transient welding currents sharply increased. While in Figure 8b, two convex peaks appeared at 17 A and 36 A; this was because the base current in TBP and TPP were respectively 17 A and 36 A. another convex peak appeared at 316 A, which was the peak value of the pulse current. However, there are some current possibility densities distributed beyond 316 A; this was because the welding current was unstable, and big volume metal transfer may occur, which can also induce short-circuits and make the welding current sharply increase; this was why the welding current can be distributed from 316–400 A. Also, the welding current can be distributed from 0 to 17 A, the 0 A of the welding current denoted the open-circuit occurring. The appearance of short-circuit and open-circuit showed that the welding current was unstable.

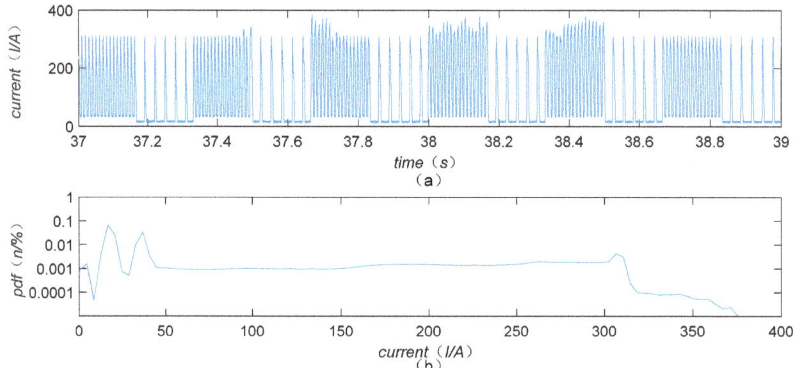

Figure 8. Welding current in L3 experiment, (**a**) Transient welding current waveform, (**b**) Possibility density function (PDF) of the welding current.

Corresponding figures of L5 are shown in Figure 9. The signal in Figure 9a was so regular, which denoted that the process had no short-circuit or open circuit. While in Figure 9b, the possibility density distributions were concentrative on three convex peaks, which corresponded to 21, 32 and 316 A. These three current values were the base currents and peak currents in TPP and TBP. The variation of possibility density distribution curve had large slopes, which meant the current sharply changed. In addition, there was no current distributed in 0 A and beyond 316 A, which also demonstrated that no short-circuit or open circuit occured, and the currents were smoothly switched between bases and peak values.

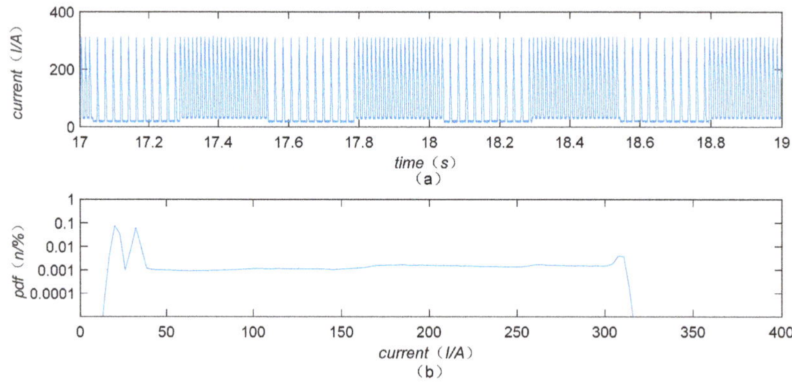

Figure 9. Welding current in L5 experiment, (**a**) Transient welding current waveform, (**b**) PDF of the welding current.

Figure 10 shows the corresponding analyses results of L7. The signals in Figure 10a were regular, which denoted that no short-circuit and open circuit occurring. While in Figure 10b, the current possibility density distribution concentrated on three convex peaks, which were 19, 35 and 316 A corresponding to the base currents and peak current in TPP and TBP. However, the slopes of variation curve were smaller than that in Figure 9b. In addition, there were small flat slopes at 19 A and 35 A, which showed that the current cannot be stable in these setting values. Moreover, the current possibility density increased from 175 to 316 A, the amplitude was about 0.001%, which meant the current had a small variation over this range.

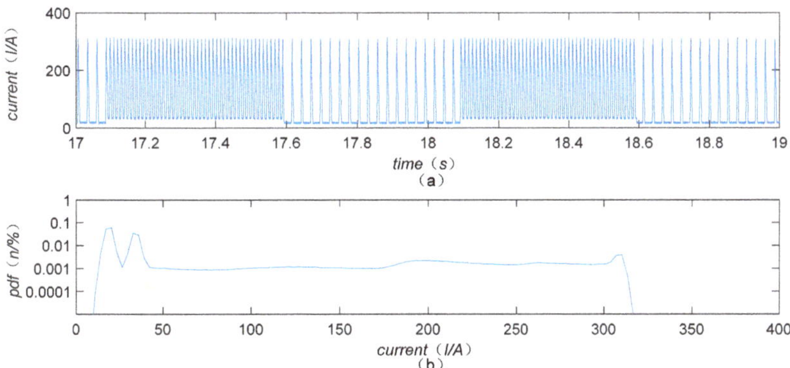

Figure 10. Welding current in L7 experiment, (**a**) transient welding current waveform, (**b**) PDF of the welding current.

According to detailed analyses and comparisons of Figures 8–10, it can be noticed that the welding current in Figure 8 was the most unstable, whose stability can be ranked as "Poor". The welding current in Figure 9 had the highest stability and can be ranked as "Excellent". The stability of welding current in Figure 10 was a little worse than that of Figure 9, and can be ranked as "Good". The stability of the welding currents in between that of Figure 8 and that of Figure 10 can be ranked as "Medium". In addition, in the above three experiments, it can be noticed that the peak values of the current pulses were the same, both in different experiments and in two current pulse sets, but the base values were different in different experiments and different current pulse sets. It was because the base values can be determined by the twin pulse frequency TPF and twin pulse current change, which were different in the above experiments.

4.4. Quality Evaluation of Weld Bead Using FCE

4.4.1. Evaluation Procedures

In this work, to employ the FCE method to evaluate the quality of weld bead, some steps were taken as follows.

Step1: Confirm the evaluation indexes elements set U for this work.

The most intuitive evaluation method for the weld bead is the appearance. In general, experienced welders can justify whether the process operational parameters are proper or not according to the appearance of the weld bead, and then make corresponding optimizations. Hence, the appearance was an important evaluation index element. In addition, the grain structure and size can determine the microstructure properties of the weld bead, therefore, the grain size was also an important index element. Moreover, some abnormal phenomena, such as short-circuiting, arc interruption and splashes, can induce the instability of the electrical signals, so the stability of the electrical signals can also be helpful for evaluating the quality of weld bead. Furthermore, the key shape measurement parameters of the weld bead were bead width, bead height and penetration. The bead height and penetration should follow a certain relation, hence, a new variable, which was bead height coefficient, was introduced to assist the evaluation, the bead height coefficient can be derived using a quotient between bead width and bead height (B/h). Also, the penetration and bead width should be individual evaluation index elements. Hence, a comprehensive evaluation index element set can be confirmed as shown in Equation (5):

$$U = \{\text{Appearance, grain size, stability of electrical signal, bead height coefficient, penetration, bead width}\} \tag{5}$$

Step2: Establish the evaluation indexes weights set W.

Considering the different levels of importance of above confirmed evaluation index elements, after consulting experienced experts and examined relative materials, corresponding weights set was shown in Equation (6):

$$W = \{0.3, 0.3, 0.2, 0.06, 0.07, 0.07\} \tag{6}$$

Step3: Establish the evaluation results set V.

According to the experience of welding researches, hierarchy was used in this work, the quality evaluation of the weld bead can be divided into four grades, which are shown in Equation (7).

$$V = \{\text{Excellent, Good, Medium, Poor}\} \tag{7}$$

To provide accurate qualitative evaluation results, the four grades can correspond to mathematical values, as shown in Table 8.

Table 8. Mapping from grades to mathematical values.

Grade	Excellent	Good	Medium	Poor
Value	100	85	70	50

In addition, the fuzzy operator used in this work was M (·,+) [21].

4.4.2. The Foundation and Evaluation Results of FCE

The foundation of quality evaluation for weld bead in this work can be seen in:

1. Weld bead had regular and a beautiful straight appearance, the fish scale ripple was compact, and no other obvious drawbacks, which can denoted that the weld bead had excellent appearance. According to the experimental observations, the appearance can be divided into four grades.
2. Smaller grain sizes denoted high quality of the weld bead, which can be also divided into four grades based on grain size.
3. Other four grades can also be divided according to the stability analyses of electrical signals, which included stability analyses of transient welding current and possibility density distribution.
4. The bead height coefficient whose value approached 4 denoted that the weld bead had high quality, while a proper penetration was that the value was beyond a half and approached 60% of the width of the base plate. At last, the bead width of the weld bead with high quality should be the width of base plate adding 3–4 mm. About these three criteria, four grades were divided for each criterion during the evaluation.

Hence, all of these five evaluation index elements should be divided into four grades, which used the division described in Equation (7) as shown in Step 4. Then using FCE algorithm, quantitative evaluation scores can be obtained. Corresponding evaluation for each criterion and final quantitative evaluation scores can be seen in Table 9.

Table 9. Evaluation results of FCE.

Index	Appearance	Grain Size	Stability of Electrical Signal	Bead Height Coefficient	Penetration	Bead Width	Scores
L1	Medium	Poor	Poor	Excellent	Poor	Medium	60.4
L2	Medium	Poor	Poor	Good	Poor	Medium	59.5
L3	Poor	Poor	Poor	Good	Poor	Medium	53.5
L4	Medium	Medium	Medium	Good	Good	Excellent	74.05
L5	Excellent	Good	Excellent	Good	Excellent	Excellent	94.6
L6	Good	Good	Medium	Medium	Medium	Poor	77.6
L7	Poor	Good	Good	Excellent	Good	Excellent	76.45
L8	Medium	Excellent	Good	Medium	Medium	Good	83.05
L9	Medium	Excellent	Good	Poor	Medium	Poor	79.4

As shown in Table 9, not only the individual evaluate criterion for weld bead was provided, but also the overall quality of weld bead can also be qualitatively presented, these results can also be used in the orthogonal experimental analyses in the following part.

4.5. Experimental Analyses

After obtaining the quality evaluation using FCE, which comprehensively considered various elements to accurately obtain the detailed analyses about how the process operational parameters affected the various elements which were important for the quality evolution of the weld bead, the orthogonal experimental analyses method was used in this work.

4.5.1. The Influence Analysis of Appearance on the Weld Bead

The first analysis target was the appearance, and the analysis was based on evaluation results in Table 9 and mapping between grades and mathematical values shown in Table 8. The analyses parameters selected the four key operational parameters, which were robot welding speed V_R, twin pulse frequency TPF, twin pulse relation D_T and twin pulse current change in percent I_Δ. Each parameter had three different levels. For each parameter in each level, the corresponding $Score_{avg}$ can be calculated following Equation (8):

$$Score_{avg}(i,j) = \frac{\sum_{k=1}^{3} Score(i,j,k)}{3} \quad (8)$$

where i was the sequence of the operational parameter, i = 1, 2, 3 and 4, and j was the sequence of the level, j = 1, 2 and 3. For each parameter with one level, there were 3 scores in the experiment, these different scores were marked as k. The values of i, j and k were different under different situations. The value of *Score* can be obtained combing Tables 8 and 9. Then, the influence analysis of four operational parameters on appearance is shown in Table 10.

Table 10. Influence analysis of four operational parameters on appearance.

Level	$Score_{avg}$			
	V_R	TPF	D_T	I_Δ
1	63.333	70	75	80
2	85	80	70	75
3	70	68.333	73.333	63.333
Range value D	21.667	11.667	5	16.667

Where the range value D can be calculated based on Equation (9):

$$D(i) = \max_{j=1}^{3}(score_{avg}(i,j)) - \min_{j=1}^{3}(score_{avg}(i,j)) \quad (9)$$

where the meanings of i and j were the same as those in Equation (8). It can be seen that the robot welding speed is the most influential factor on the appearance; the corresponding range value was 21.667, and the following influential factors were twin pulse current change in percent, twin pulse frequency and twin pulse relation. Hence, to obtain satisfactory appearance of weld bead, a proper combination of the robot welding speed and twin pulse current change in percent was very important. It was because the robot welding speed and twin pulse current change in percent can determine the amount of energy delivery into the base plate in one-unit length, and this energy delivery was a key factor which can determine the final appearance.

According to above analyses results, to obtain the weld bead with optimal appearance, the optimal level for V_R should be Level2, the optimal level for TPF should be Level2, the optimal level for D_T should be Level1, and the optimal level for I_Δ should be Level1. Those corresponding values were respectively 30 cm/min, 2 Hz, 30% and 30%. Using these optimal parameters, an experiment was conducted and the result can be seen in Figure 11. It can be seen that the weld bead was straight and regular, and the fish scale ripples were so compact. The bead width and bead height were also proper, and the forming was satisfactory, which was the same as that in former analyses.

Figure 11. The appearance of the weld bead using optimal parameters combination.

4.5.2. The Influence Analysis of Grain Size on the Weld Bead

Grain size is another important element affecting the quality of weld bead, corresponding analyses are shown in Table 11, based on the same calculation methods as that in the preceding part. It can be seen that the robot welding speed was still the most influential factor to determine the grain size. Higher robot welding speed denoted smaller grain size. When the robot welding speed was 40 cm/min, which corresponded to Level3 in the robot welding speed, the smallest grain size was obtained. It may be due to two reasons. The first was that when high robot welding speed was employed, the weld bead could be fast cooling, which induces the number of grains increasing and the size being reduced. The second reason was that the weld pool was stirred because a series of regular pulses with high and low frequencies were alternatively used during the DP-GMAW operational process, and then a high robot welding speed denoted more frequent stirring in the weld pool, which could deliver more extern energy into the liquid metal and energy fluctuations were more severe, which can also make the grain size smaller. Hence, to obtain a satisfactory microstructure of weld bead, the robot welding speed should be properly increased.

Table 11. Orthogonal analyses of the grain size of the weld bead.

Level	$Score_{avg}$			
	V_R	TPF	D_T	I_Δ
1	50	68.333	78.333	78.333
2	80	78.333	73.333	73.333
3	95	78.333	73.333	73.333
Range value D	45	10	5	5

According to above analyses results, to obtain the weld bead with the most proper grain size, the optimal level for V_R should be Level3, the optimal level for TPF should be Level2 or

Level3, the optimal level for D_T should be Level1, and the optimal level for I_Δ should be Level1. Those corresponding values were, respectively, 40 cm/min, 2 Hz or 1 Hz, 30% and 30%.

4.5.3. Orthogonal Experiment Analysis for the Results of FCE

Apart from the appearance and grain size, which were important elements used in FCE, the results of the FCE method can also be analyzed using orthogonal experiment analysis, in order to obtain the influential levels of key operational parameters on the quality of weld bead. The analysis method was the same as in the above two analyses about appearance and grain size, however, all of the *Scores* used the accurate score of FCE in Table 9 and the parameters referred to the Table 3. The corresponding analyses results can be seen in Table 12. It can be seen that the effects of different parameters on the quality of weld bead were so obvious. The most influential factor was the robot welding speed of the welding robot and a corresponding range value achieved 24.283, and then the following factors were twin pulse frequency and twin pulse current change in percent, while the last was the twin pulse relation, whose influential effect was very low. It was because the robot welding speed can influence the heat delivery for weld bead and cooling rate of the weld pool. To obtain a weld bead with satisfactory quality, the system must provide a proper combination of robot welding speed and twin pulse frequency, which should appropriately match the average welding current and thickness of the base plate.

Table 12. Orthogonal analyses of FCE scores of the weld bead.

Level	$Score_{avg}$			
	V_R	TPF	D_T	I_Δ
1	57.8	70.3	73.683	78.133
2	82.083	79.05	70.983	71.183
3	79.633	70.167	74.85	70.2
Range value D	24.283	8.883	3.867	7.933

According to the range analysis results, the optimal parameter combination in these nine experiments can be obtained as follows: robot welding speed was 30 cm/min of Level2, the twin pulse frequency was 2 Hz of Level2, the twin pulse relation was 50% of Level3, and the twin pulse current change in percent was 30% of Level1. These parameter combinations can correspond to the L5 in the above experiments. Also, another testifying experiment was conducted to repeat testify this parameter combination. The obtained weld bead is shown in Figure 12.

Figure 12. The morphology of the weld bead using optimal parameters combination.

It can be seen that the weld bead was regular with proper bead width. The appearance was bright with compact and smooth fish scale ripples. Also, the welding process was stable without short-circuit and interruption occurring. All of these characteristics meant that the weld bead with satisfactory quality can be obtained using this operational parameter combination.

5. Conclusions

In this work, the DP-GMAW process based on robot operation for stainless steel 304 was seriously considered; all of the operational procedures could be successfully accomplished, and stable welding processes and weld beads with satisfactory quality could be obtained. After introducing the principle and operational characteristics about this process, orthogonal experimental design based on four key operational parameters and corresponding performance testing, which included the appearance observation and key shape parameters measurement, microstructure analysis, tensile and hardness testing, as well as stability analysis of the electrical signals, were conducted. Then, the FCE method was employed to provide quantitative quality evaluation for the weld bead. According to the combined orthogonal experimental analyses about the appearance, grain size and FCE scores of quality of weld bead, an optimal operational parameter combination for each condition can be obtained. Based on these serious explorations, some important conclusions can be drawn:

1. The FCE method was employed in this work to evaluate the quality of weld bead, and during the evaluating process, the appearance, microstructure and key shape parameters measurement were comprehensively considered. This evaluating method can be helpful for realizing the quantitative evaluation for a weld bead.
2. The appearance of the weld can be mainly determined by robot welding speed and twin pulse current change in percent. The proper combination of these two operational parameters can achieve appropriate heat delivery into the weld bead in one-unit length, and then obtain a good appearance. In addition, the grain size was mainly determined by robot welding speed based on robot operation. Higher robot welding speed denoted the weld bead with smaller grain size, better toughness and hardness can be obtained with other operational parameters unchanged.
3. According to the orthogonal experimental analysis for the FCE scores of quality of weld bead, the most influential factor on the welding quality was the robot welding speed V_R, the following were twin pulse frequency TPF and twin pulse current change in percent I_Δ, and the last was the twin pulse relation D_T. In the experiments in this work, the optimum process parameters were that the V_R was 30 cm/min, the TPF was 2 Hz, the D_T was 50% and the I_Δ was 30%, which corresponded to the L5 in the experiments. Under this circumstance, the appearance was beautiful, and the obvious and bright fish scale ripples were obtained; in addition, the grain sizes were small enough, and the microstructure property was also satisfied when compared to that of other experiments.

This work can provide effective methods for analyzing the influential levels of key operational parameters on one or more performances of the weld, and then obtaining a corresponding optimal parameter combination. The work can improve the process parameter optimization and operational performance in the academic research or actual industrial production. In the future, corresponding works will probably continue to be applied in welding base plates of other materials, or other multi-parameter welding systems, and the influential levels of other operational parameters on the quality of weld bead will be further considered.

Author Contributions: There are three authors contributed to this manuscript. P.Y. conducted the experiments and analyzed the experimental results. K.Z. proposed the idea of the work, and originated the experiment, and write the paper. S.H. provided assisted works for the work, and calculated the fuzzy comprehensive evaluation scores and sorted out the experimental data, and conducted the microstructure observation experiments.

Funding: This research was funded by National Natural Science Foundation of China, China (Grant No: 51805099, 51605103), Science and Technology Planning Project of Guangdong Province, China (2017B090914005), Science and Technology Program of Guangzhou, China (201805010001), and Beijing Institute of Technology Research Fund Program for Young Scholars, China.

Conflicts of Interest: The authors declare no conflict of interest.

References

1. Mathivanan, A.; Devakumaran, K.; Kumar, A.S. Comparative Study on Mechanical and Metallurgical Properties of AA6061 Aluminum Alloy Sheet Weld by Pulsed Current and Dual Pulse Gas Metal Arc Welding Processes. *Mater. Manuf. Process.* **2014**, *29*, 941–947. [CrossRef]
2. da Silva, C.L.M.; Scotti, A. Performance Assessment of the (Trans) Varestraint Tests for Determining Solidification Cracking Susceptibility when using Welding Processes with Filler Metal. *Meas. Sci. Technol.* **2004**, *15*, 2215–2223. [CrossRef]
3. Yamamoto, H.; Harada, S.; Ueyama, T.; Ogawa, S. Development of Low-Frequency Pulsed MIG Welding for Aluminium Alloys. *Weld. Int.* **1992**, *6*, 580–583. [CrossRef]
4. Wang, L.L.; Wei, H.L.; Xue, J.X.; DebRoy, T. Special Features of Double-pulsed Gas Metal Arc Welding. *J. Mater. Process. Technol.* **2018**, *251*, 369–375. [CrossRef]
5. Sen, M.; Mukherjee, M.; Pal, T.K. Prediction of Weld Bead Geometry for Double Pulse Gas Metal Arc Welding Process by Regression Analysis. In Proceedings of the 5th International & 26th All India Manufacturing Technology, Design and Research Conference (AIMTDR 2014), IIT Guwahati, Guwahati, India, 12–14 December 2014.
6. Yi, J.; Cao, S.-F.; Li, L.-X.; Guo, P.-C.; Liu, K.-Y. Effect of Welding Current on Morphology and Microstructure of Al Alloy T-joint in Double-pulsed MIG Welding. *Trans. Nonferrous Metals Soc. China* **2015**, *25*, 3204–3211. [CrossRef]
7. Devakumaran, K.; Rajasekaran, N.; Ghosh, P.K. Process Characteristics of Inverter Type GMAW Power Source under Static and Dynamic Operating Conditions. *Mater. Manuf. Process.* **2012**, *27*, 1450–1456. [CrossRef]
8. Wang, L.; Xue, J. Perspective on Double-pulsed Gas Metal Arc Welding. *Appl. Sci.* **2017**, *7*, 894. [CrossRef]
9. Wang, L.; Heng, G.; Chen, H.; Xue, J.; Lin, F.; Huang, W. Methods and Results Regarding Sinusoid Modulated Pulse Gas Metal Arc Welding. *Int. J. Adv. Manuf. Technol.* **2016**, *86*, 1841–1851. [CrossRef]
10. Ghosh, P.K.; Dorn, L.; Kulkarni, S.; Hofmann, F. Arc Characteristics and Behaviour of Metal Transfer in Pulsed Current GMA Welding of Stainless Steel. *J. Mater. Process. Technol.* **2009**, *3*, 1262–1274. [CrossRef]
11. Zhou, K.; Yao, P. Overview of Recent Advances of Process Analysis and Quality Control in Resistance Spot Welding. *Mech. Syst. Signal Process.* **2019**, *124*, 170–198. [CrossRef]
12. Yao, P. Intelligent Control Strategies and Performance Evaluation of Integrated Double Wire Arc Welding Power Source. Ph.D. Thesis, South China University of Technology, Guangzhou, China, 2012.
13. Yao, P.; Xue, J.; Zhou, K. Study on the Wire Feed Speed Prediction of Double-wire-pulsed MIG Welding based on Support Vector Machine Regression. *Int. J. Adv. Manuf. Technol.* **2015**, *79*, 2107–2116. [CrossRef]
14. Casalino, G.; Hu, S.J.; Hou, W. Deformation Prediction and Quality Evaluation of the Gas Metal Arc Welding Butt Weld. *Proc. Inst. Mech. Eng. Part B J. Eng. Manuf.* **2003**, *217*, 1615–1622. [CrossRef]
15. Wu, C.S.; Polte, T.; Rehfeldt, D. Gas Metal Arc Welding Process Monitoring and Quality Evaluation using Neural Networks. *Sci. Technol. Weld. Join.* **2000**, *5*, 324–328. [CrossRef]
16. Yao, P.; Xue, J.; Zhou, K.; Wang, X.; Zhu, Q. Symmetrical Transition Waveform control on double-wire MIG welding. *J. Mater. Process. Technol.* **2016**, *229*, 111–120. [CrossRef]
17. Keppel, G. *Design and Analysis: A researcher's Handbook*; Prentice-Hall Inc.: Upper Saddle River, NJ, USA, 1991.
18. Xie, Q.; Ni, J.-Q.; Su, Z. Fuzzy Comprehensive Evaluation of Multiple Environmental Factorsfor Swine Building Assessment and Control. *J. Hazard. Mater.* **2017**, *340*, 463–471. [CrossRef] [PubMed]
19. Liang, Z.; Yang, K.; Sun, Y.; Yuan, J.; Zhang, H.; Zhang, Z. Decision Support for Choice Optimal Power Generation Projects: Fuzzy Comprehensive Evaluation Model based on the Electricity Market. *Energy Policy* **2006**, *34*, 3359–3364. [CrossRef]
20. Chen, J.-F.; Hsieh, H.-N.; Do, Q.H. Evaluating Teaching Performance based on Fuzzy AHP and Comprehensive Evaluation Approach. *Appl. Soft Comput.* **2015**, *28*, 100–108. [CrossRef]
21. Liu, Y.; Fang, P.; Bian, D.; Zhang, H.; Wang, S. Fuzzy Comprehensive Evaluation for the Motion Performance of Autonomous Underwater Vehicles. *Ocean Eng.* **2014**, *88*, 568–577. [CrossRef]
22. Meng, L.; Chen, Y.; Li, W.; Zhao, R. Fuzzy comprehensive evaluation model for water resources carrying capacity in Tarim River Basin, Xinjiang, China. *Chin. Geogr. Sci.* **2009**, *19*, 89–95. [CrossRef]
23. Wei, B.; Wang, S.-L.; Li, L. Fuzzy Comprehensive Evaluation of District Heating Systems. *Energy Policy* **2010**, *38*, 5947–5955. [CrossRef]

24. Zhang, M.; Yang, W. Fuzzy Comprehensive Evaluation Method Applied in the Real Estate Investment Risks Research. *Phys. Procedia* **2012**, *24*, 1815–1821.
25. Zhai, L.; Tang, X. Fuzzy Comprehensive Evaluation Method and Its Application in Subjective Quality Assessment for Compressed Remote Sensing Images. In Proceedings of the Fourth International Conference on Fuzzy Systems and Knowledge Discovery (FSKD 2007), Haikou, China, 24–27 August 2007.
26. Sen, M.; Mukherjee, M.; Pal, T.K. Evaluation of Correlations between DP-GMAW Process Parameters and Bead Geometry. *Weld. J.* **2015**, *94*, 265s–279s.
27. Bosworth, M.R.; Deam, R.T. Influence of GMAW Droplet Size on Fume Formation Rate. *J. Phys. D Appl. Phys.* **2000**, *33*, 2605–2610. [CrossRef]
28. Rehfeldt, D.; Schmitz, T.H. A System for Process Quality Evaluation in GMAW. *Weld. World* **1994**, *34*, 227–234.
29. Yao, P.; Zhou, K. Research of a Multi-Frequency Waveform Control Method on Double-Wire MIG Arc Welding. *Appl. Sci.* **2017**, *7*, 171. [CrossRef]

Article

Study on Short-Circuiting GMAW Pool Behavior and Microstructure of the Weld with Different Waveform Control Methods

Tao Chen [1], Songbai Xue [1,*], Bo Wang [2], Peizhuo Zhai [1] and Weimin Long [2]

[1] College of Materials Science and Technology, Nanjing University of Aeronautics and Astronautics, Nanjing 210016, China; taocmsc@nuaa.edu.cn (T.C.); zhaipz@nuaa.edu.cn (P.Z.)
[2] China Intelligent Equipment Innovation Institute (Ningbo) Co., Ltd., Ningbo 315700, China; wangbo4175@126.com (B.W.); brazelong@163.com (W.L.)
* Correspondence: xuesb@nuaa.edu.cn; Tel.: +86-8489-6070

Received: 11 November 2019; Accepted: 4 December 2019; Published: 7 December 2019

Abstract: In order to study internal relation among the behavior of the weld pool, the microstructure of weld bead and the waveform of short-circuiting gas metal arc welding (S-GMAW), a high speed photograph-images analysis system was formed to extract characteristics of weld pool behavior. Three representative waveform control methods were used to provide partly and fully penetrated weld pools and beads. It was found that the behavior of the weld pool was related to the instantaneous power density of the liquid bridge at the break-up time. Weld pool oscillation was triggered by the explosion of the liquid bridge, the natural oscillation frequencies were derived by the continuous wavelet transform. The change of weld pool state caused the transition of oscillation mode, and it led to different nature oscillation frequencies between partial and full penetration. Slags flow pattern could be an indication of the weld pool flow. Compared with the scattered slags on fully penetrated weld pool, slag particles accumulated on partially penetrated weld pools. The oscillating promoted the convection of the welding pool and resulted in larger melting width and depth, the grain size, and the content of pro-eutectoid ferrite in the weld microstructure of S235JR increased, the content of acicular ferrite decreased.

Keywords: short-circuiting gas metal arc welding; waveform control method; weld pool oscillation and flow; microstructure; high speed photograph; image processing; continuous wavelet transform

1. Introduction

Short circuit gas metal arc welding (S-GMAW) has various advantages such as low heat input, small heating area, and high thermal stability. Benefiting from advances in digital control technology, the power sources can control the voltage and current and output specific shapes of the arc curve which aim to handle the molten material transfer and control the spatter. Kah [1] proposed a classification of control techniques for S-GMAW: Natural metal transfer, current controlled dip transfer, and controlled wire feed short circuit mode. On account of the controlled wire feed short circuit mode represented by cold metal transfer welding (CMT) introducing external mechanical forces on the wire it will not be discussed here. Different shapes of the current and voltage waveform change the droplet transition, which leads to the different weld pool behavior, weld shape and microstructure.

Currently, there are a variety of S-GMAW waveform control methods on the market, which can be divided into two types. The first method is represented by Surface Tension Transfer (STT) [2] and Cold Arc process [3]. This type of methods reduces the circuit current at the beginning and end of the short circuit period to permit a smooth touch and break of the bridge of the molten metal, preventing spatter. The other type is represented by Cold MIG (Metal-Inert Gas Welding) process [1] and Low Spatter

Control (LSC) [4]. In this type of methods, considerable increase of the current gradient to accelerate the droplet detachment during the short-circuit period meanwhile the short circuit period, dramatically reduces the short-circuit period. The current in this period is reduced and occurs faster compared to a conventional short arc. However, the welding circuit of this process maintains partial current when the bridge breaks compared with the first type processes. Figure 1 compares the waveform of the conventional, the Cold Arc, and the LSC.

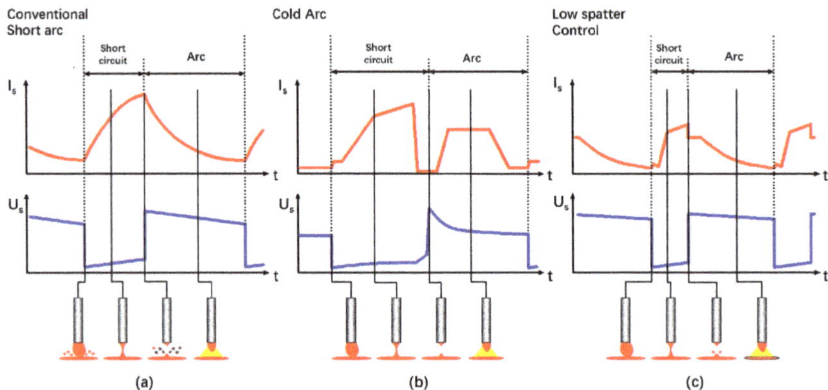

Figure 1. Comparison of (**a**) conventional, (**b**) Cold Arc, and (**c**) Low Spatter Control (LSC) waveform.

The behavior of weld pool is a direct reaction of weld pool state, the dynamic variation of weld pool has a great influence on the weld bead shape and microstructure. Weld pool behavior may contain sufficient information to understand the mechanisms of welding bead formation and control the stability of welding process [5–7], so it is of great significance to study the weld pool behavior of S-GMAW using different waveforms.

Many researchers have studied the weld pool behaviors in recent years. In gas tungsten arc welding (GTAW), for acquiring the amplitude and oscillation frequency of the weld pool, Yu Shi et al. [8] used line laser to illuminate the surface of weld pool. It is found that the oscillation frequency and amplitude of GTAW pool change abruptly in the process of partial penetration to full penetration. Liu [9] investigated the pulse frequency on fluid flow behavior of the weld pool in pulsed current GTAW. The result showed that weld pool oscillations triggered by pulse current lead to more heterogeneous nucleation sites, and the resonance between the movement of the weld pool and pulse current frequency greatly promotes grain refinement.

Compared with GTAW, there are complex interactions among arc plasma, droplet transfer, and pool behavior in GMAW. Richardson et al. [10] found that current pulses could not be used to trigger weld pool oscillation effectively for GMAW, the interactions between the transferred droplets and the weld pool can trigger the weld pool into oscillation. Tang et al. [11] developed a filter-reflection observation system to acquire the weld pool profile during double-pulsed gas metal arc welding process. It was found that the weld pool oscillation caused by low frequency pulse can effectively reduce the porosity and refine the weld structure.

To date, most investigations of weld pool behavior mainly focuses on GTAW and pulsed GMAW processes. However, no much research has been done in the area of weld pool characteristics with different S-GMAW waveform control methods. In this paper, a high speed photograph-images analysis system for weld pool observation was formed to capture the dynamic behavior of S-GMAW weld pool with the aim to reveal the internal relationship among the S-GMAW current waveform, the behavior of weld pool, the geometry of weld bead, and the microstructure.

2. Materials and Methods

2.1. Experimental System

The experimental system consisted of welding system, high-speed photography system, and welding electrical signal synchronous acquisition system, as shown in Figure 2.

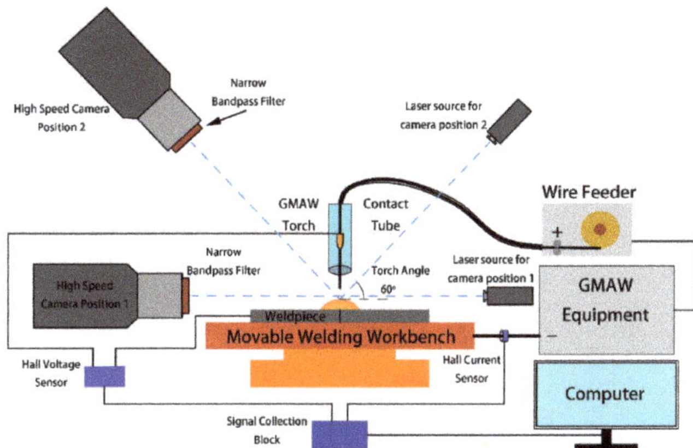

Figure 2. Schematic of high-speed photography system for weld pool observation.

EWM Phoenix 521 (EWM Hightec Welding GmbH, Mundersbach, Germany), EWM Cold Arc (EWM Hightec Welding GmbH, Mundersbach, Germany), and Fronius TPS5000 (FRONIUS, Pettenbach, Austria) were selected as power sources of welding system to provide needed waveforms. The welding position was PA(Flat position, as per ISO 6947). The movement of the workbench was controlled by servo motor. Current sensor, voltage sensor and signal acquisition card constituted the welding electrical signal synchronous acquisition system. Acquisition frequency of signal acquisition card is 1.5×10^6 Hz. The high-speed photography system was used to recorded the behavior of the weld pool.

In order to capture the side-view of welding pool during the welding process, the high speed photography camera was at position one of Figure 2, which is on the same horizontal plane as the welding test plate. Dynamic information about the weld pool oscillation from the high speed photography pictures was obtained by tracing the pool surface as a function of time. The shooting angle is perpendicular to the welding seam in the same plane. A light-emitting-diode (LED) was used as its excitation light source at position one, whose wave length was 850 nm, and continual output was 3 W. A laser source 850 nm near infrared filter was used to filter out strong arc during welding process. Acquisition frequency of position one is 10,000 Hz. The shooting picture is shown in Figure 3a. Positions of the camera and backlight source need to be changed to capture the contour and the flow behavior of the weld pool surface, as shown in position two of Figure 2. Different narrow-band filters were selected to obtain different information of the weld pool surface which have different spectral characteristics. The high speed camera was equipped with 850 nm near infrared filter to obtain the contour information of the weld pool with the laser shined by the same type laser source mentioned above at position two, as shown in Figure 3b. 650 nm near infrared filter was installed to obtain metal flow information of the weld pool with no laser shined, as shown in Figure 3c. Acquisition frequency of position two of Figure 2 was 2000 Hz. During video capturing, the camera was placed at an angle deviation of about 0–5°. In the analysis, the effect of these angles was not taken into account. However, based on the geometry employed, it is estimated that average errors are of the order of 1.5%.

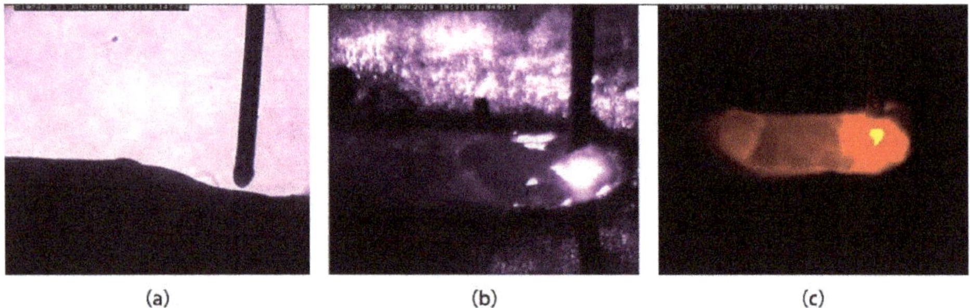

(a) (b) (c)

Figure 3. Pictures captured for different condition: (**a**) side-view shooting picture; (**b**) overhead shooting effect (850 nm near infrared filter was added); and (**c**) overhead shooting effect (650 nm near infrared filter was added).

2.2. Materials and Welding Parameters

In order to acquire both partly and fully penetrated weld pools under the same welding parameters, 2 mm and 4 mm of size 150 mm by 150 mm S235JR (1.0038) steel plate were selected as the base material, the weld bead was located in the middle of the plate. The length of the weld bead was 120 mm. The filler wire of ER70S-6 (G42) mild steel with a diameter of 1.2 mm was used for welding. The chemical composition of the base material and filler wire are given in Table 1. A mixture of 82% Ar + 18% CO_2 was used as a shielding gas, with a flow rate of 15 L/min. The travel speed was kept constant at 22 cm/min. The contact tip to workpiece distance (CTWD) was 20 mm. Welding conditions were selected that give an almost constant arc length, with an average voltage of approximately 20 V. The joint type was bead-on-plate. Welding parameters are listed in Table 2.

Table 1. Chemical compositions (in wt%) of base metal and filler wire (Fe balance).

Materials	C	Mn	Si	P	S	Ni	Cr	Mo	V	Other
S235JR	0.17	1.40	0.3	0.035	0.035	-	-	-	-	N 0.012
ER70S-6	0.06–0.15	1.40–1.85	0.80–1.15	0.025	0.035	0.15	0.15	0.15	0.03	Cu 0.5

Table 2. Welding parameters.

No.	Waveform	Wire Feed Rate (m/min)	Voltage(V)	Thickness (mm)	Penetration
1	Conventional	2.4, 2.7, 3.0, 3.3	19	4	Partial
2	LSC	2.4, 2.7, 3.0, 3.3	19	4	Partial
3	Cold Arc	2.4, 2.7, 3.0, 3.3	19	4	Partial
4	Conventional	3.0	19	2	Full
5	LSC	3.0	19	2	Full
6	Cold Arc	3.0	19	2	Full

2.3. Principle of Measurement

The image analysis was carried out using a computer program built with LabView to obtain the change of the height of weld pool surface and the diameter of liquid bridge neck. Direct information about the weld pool oscillation from the high speed photography pictures was obtained by tracing the height of reference point on the weld pool surface as a function of time. The processing processes are as follows: (1) the contour of weld pool and wire silhouette in high speed photography was extracted by image processing system to obtain the pixel coordinates of contour, as shown in Figure 4b. (2) Direct information about the weld pool motion from the high speed video pictures was obtained by tracing the pool surface as a function of time. For this purpose, a reference point was defined on the weld pool surface, as depicted in Figure 4b. The distance between the reference point and the center of wire was

1.8 mm (1.5 times wire diameter). The Y-coordinate of reference point was measured as a function of time. In this way the change of the position of the reference point during welding can be outlined and the trend of the pool motion can be revealed [10]. (3) The shortest distance between the white line and the red line was calculated to obtain the diameter of liquid bridge necking of liquid bridge during short-circuit period. The process flow is shown in Figure 4.

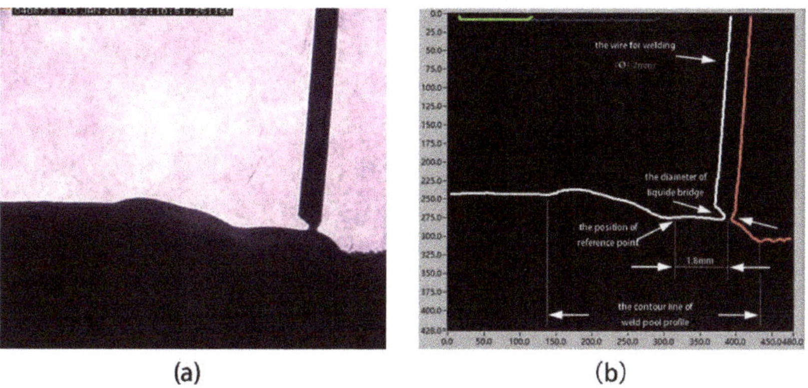

Figure 4. Principle of measurement: (**a**) original images; (**b**) extraction of the outline of weld pool and the smallest diameter of liquid bridge.

3. Results and Discussion

3.1. The Metal Transfer Process and Impact on the Weld Pool

During the short circuit period, the heat source of weld pool is mainly the resistance heat of the filler material and the molten filler material. Arc heat is the main heat source during the arc period. In order to compare the heating power of different waveform control methods to the weld pool, Equation (1) was used to calculate the welding power for all waveform control methods, Equations (2) and (3) were used to calculate the welding line energy on the base metal [12,13].

$$P_w = \frac{1}{t_a + t_s}\left[\left(\int_0^{t_a} u(t)i(t)dt\right) + \left(\int_0^{t_s} u(t)i(t)dt\right)\right] \tag{1}$$

$$Q_b = P_w * \eta_{\text{eff}} * (t_a + t_s) = P_w * \eta_{\text{eff}} * t_w \tag{2}$$

$$Q_{pl} = \frac{Q_b}{v * t_w} = \frac{P_w * \eta_{\text{eff}}}{v} \tag{3}$$

where $u(t)$ is the voltage curve during welding, $i(t)$ is the current curve, t_w is the welding time, t_{arc} is the burning-arc time, t_s is the arc-shorting time, P_w is the welding power during single a droplet transfer cycle, Q_b is the heat in the base material, η_{eff} is the thermal efficiency of the welding process, Q_{pl} is the heat power applied to the weld pool per unit length, and v is the welding speed.

The arc length of S-GMAW is short hence the heat losses to the surrounding atmosphere are low. The effective thermal efficiency is high and the η_{eff} of S-GMAW is 0.85 [14] which is higher than that of Pulsed GMAW and Spray GMAW.

Figure 5 shows the combination of voltage and current waveform and metal transfer process of different S-GMAW processes. In order to ensure the comparability of waveforms, the volume of the drops was similar at the time of waveforms acquisition. The arc current curve of traditional S-GMAW process is influenced by two factors: Inductance of the welding circuit and re-striking current. The re-striking current determines the peak current in the arc period. Then the current declined to the background current, this period was t_{arc1} as shown in Figure 5a. Inductance of the welding circuit determined the rate of current decline. The LSC process maintained the large current for a defined short period of time after the arc ignites to ensure that the arc had sufficient energy to heat the welding wire and the base material. Then the current decreased to the background current by the current control to regulate and initiate the next detachment, this period was t_{arc1} as shown in Figure 5b. As for Cold Arc process, the current was decreased dramatically to permit a smooth break of the bridge of the molten metal at the end of short-circuit period. After the arc had been stabilized, the current was raised for a defined short period of time, known as melt pulse, to heat the welding wire and the base material. Then the current decreased to the background current, this period was t_{arc1} as shown in Figure 5c. The average heating power to the base material and weld pool outlines of three waveforms is shown in Table 3.

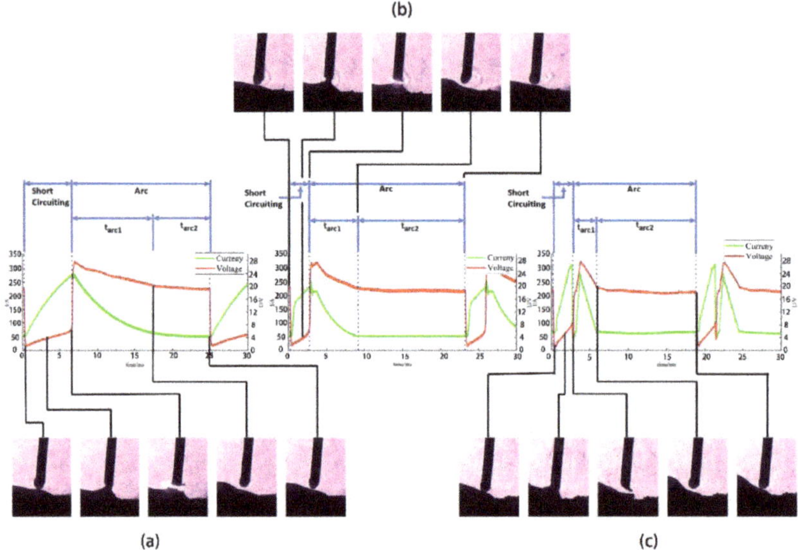

Figure 5. The droplet transition process: (**a**) conventional process; (**b**) LSC; and (**c**) Cold Arc.

Table 3. The effective heating power to the base material and weld pool outlines.

Waveforms	Plate Thickness/mm	Wire Feed Rate/m·min^{-1}	Effective Average Heating Power/KJ·m^{-1}	Pool Width/mm	Pool Length/mm
Conventional	4	3	409.915	5.6 ± 0.5	11.2 ± 1
LSC	4	3	344.656	5.5 ± 0.5	10.5 ± 1
Cold Arc	4	3	327.533	5.3 ± 0.5	9.6 ± 1

Figure 6 shows the profile of the weld pool during the transition period of a single molten drop. The first column of Figure 6 is the surface profiles of the weld pools at the short circuit stage, the second at the time when the liquid bridge exploded, the third at the arc stage, and the fourth at the short circuit stage of next droplet transition stage. The weld pools size was shown in Table 2. The area of the weld pool was measured by the Photoshop software, the border between the liquid and the solid was outlined manually, which could not be found by the software for tiny gray scale differences. The pool size can be obtained imprecisely by measuring the image, but the influence trend of waveform control mode on the pool size can be obtained under the same shooting condition. The results show that there was no obvious difference in the width of the weld pool, but there was a great difference in the length of the weld pool. These differences were directly related to the impact of electrical explosion at the end of short circuit.

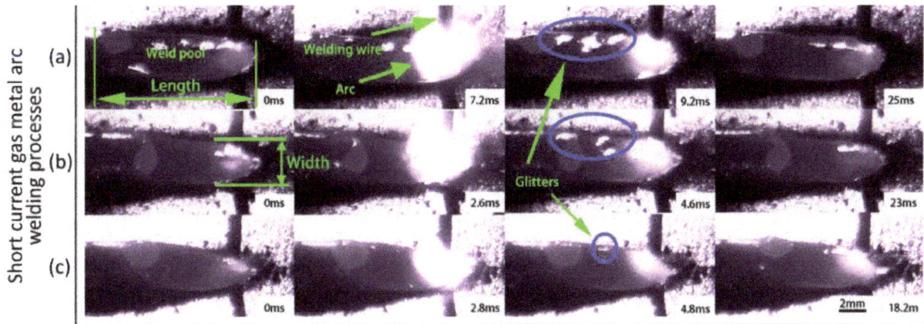

Figure 6. Variation of weld pool profile of short circuit gas metal arc welding (S-GMAW) under different waveforms (wire feed rate: 3 m/min, thickness of base plates: 4 mm): (**a**) conventional waveform, (**b**) LSC waveform, and (**c**) Cold Arc waveform.

As shown in Figure 6, strong arc light appeared at the moment of electric explosion. There was no obvious change in the size of the weld pool before and after the electric explosion, but there was obvious difference between the surface of different weld pools. The glitters in the blue circles of Figure 6 were due to backlight and weld pool surface, which was the mirror-like reflection. The weld pool fluctuation resulted in the change of surface curvature. The more violent the surface fluctuation of the weld pool, the greater the chance of mirror reflection and the more glitters there were. The surface of the weld pool with traditional waveform fluctuated the most, which was followed by LSC, and the weld pool of Cold Arc basically did not change. In the transition period of a single melt droplet, the energy carried by electric explosion was mainly propagated to the melt pool in the form of momentum, which changes the flow state of the metal inside the weld pool.

The resistance heat is the main factor that causes the liquid bridge explosion during short circuit period. Due to the highest resistance at the neck of the liquid bridge, it was the location where the electric explosion occurred. The instantaneous heat generation power per unit volume of the metal at the neck of the liquid bridge is calculated, the process is shown as follows:

$$R = \frac{\rho * dl}{\pi r^2} \tag{4}$$

$$P_h = I^2 * R = I^2 * (\rho * dl)/\left(\pi r^2\right) \tag{5}$$

$$P_v = \frac{P_h}{V} = \frac{I^2 * \frac{\rho * dl}{\pi r^2}}{\pi r^2 * dl} = \frac{I^2 * \rho}{\pi^2 * r^4} \tag{6}$$

where R is the resistance at liquid bridge neck, ρ is the resistivity of metal at liquid bridge, r is the radius of liquid bridge neck which was extracted by image processing system which was mentioned above, dl is the fluid bridge neck differential length, P_h is the thermal power of resistance at neck of liquid bridge, and P_v is the instantaneous power density of the liquid bridge. The electrical explosion is caused by overheating of the metal at the neck of the bridge. The diameter of the liquid bridge changes gently in a small area near the neck constriction whose volume can be replaced by a cylinder whose diameter is equal with the diameter of the neck of the liquid bridge. V is the differential volume of the fluid bridge neck length.

The image processing system was used to extract the diameter of the shrinking neck of the liquid bridge in the short circuit period. Figure 7 are the relationship curves that show the diameter of the neck of the liquid bridge along with time.

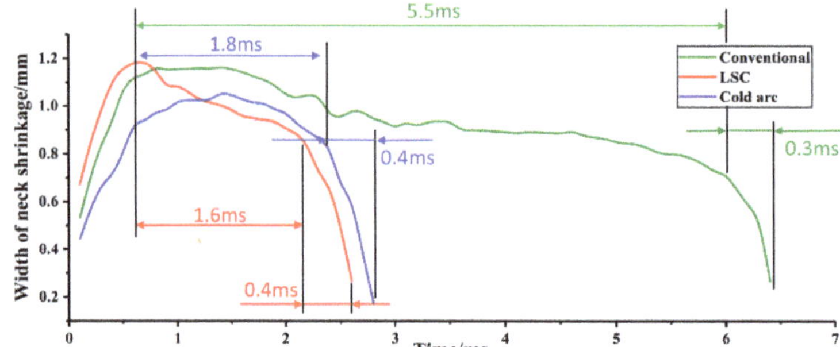

Figure 7. The diameter of the neck of the liquid bridge as a function of time.

It has been pointed out that the surface tension and electromagnetic pinch force are the main forces to make droplet transfer which have close relation with formation, destabilization, and break-up of short circuit liquid bridge. The curves in Figure 7 all showed a process of rapid rise, then stability, and finally rapid decline. The rapid decline stage of diameter was the process of destabilization and break-up of liquid bridge. The sharp slumping stage of three curves lasted nearly the same time as shown in Figure 7. At that time, the current in the welding loop were 280 A, 210 A, and 50 A in conventional, LSC, and Cold Arc, respectively, as shown in Figure 5. The results showed that electromagnetic shrinkage force had little effect on the duration of destabilization and break-up of short circuit liquid bridge. The difference of stability times of liquid bridges was obvious, which indicated that the rising rate of loop current in the short circuit stage can effectively promote the formation of neck of liquid bridge and greatly reduce the short circuit stage time.

Figure 8 shows the relationship curves of the instantaneous power density of liquid bridge neck with time under three waveform conditions. As shown in Figure 8, The curve of the instantaneous power density of the liquid bridge along with time was acquired by substituting the diameter of the shrinking neck of the liquid bridge and the current corresponding to it into Equation (6).

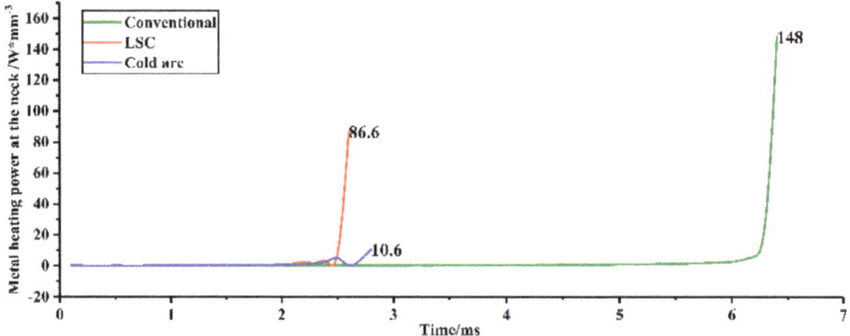

Figure 8. The instantaneous power density of the liquid bridge as a function of time.

It can be seen from Figure 8 that the instantaneous power density of the liquid bridge was extremely low during short circuit period for most of the time. The energy accumulated in a very short time before the liquid bridge explosion is the main factor influencing the impact of electric explosion. Therefore, the instantaneous power density of the liquid bridge during the burst can effectively measure the magnitude of the electric explosive impact force. The instantaneous power density of liquid bridge metal in cold arc power supply was relatively small. The instantaneous power density of liquid bridge metal at the end of short circuit in LSC process was about half of that of traditional process, and the electric explosion impact force was less than that of traditional process. The impact force of electric explosion determines the dynamic characteristics of weld pool. Figure 9 shows the probability density distribution of oscillation amplitude of weld pool:

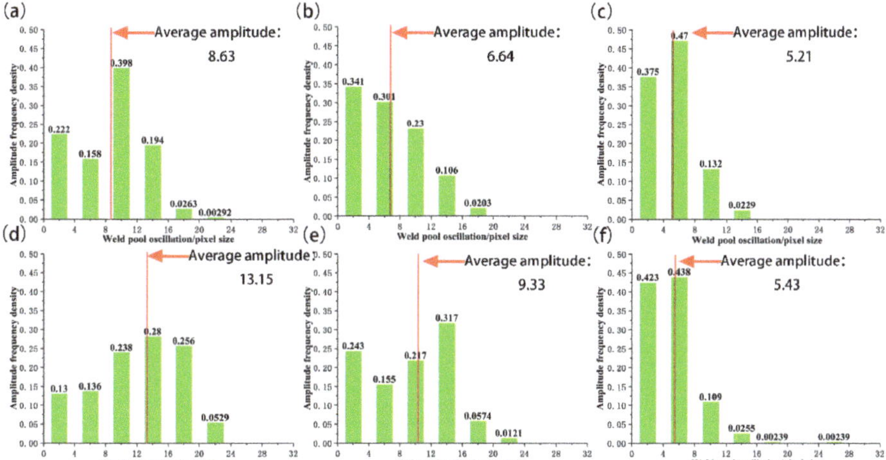

Figure 9. The probability density distribution of oscillation amplitude of weld pool: (**a–c**) are the amplitudes of the partial penetration pool of conventional process, LSC, and Cold Arc and (**d–f**) are the amplitudes of the full penetration pool of conventional process, LSC, and Cold Arc.

The amplitude of weld pool is proportional to the impact of electric explosion. The probability of large amplitude of weld pool in traditional process was greater than that of LSC and Cold Arc. The impact of electric explosion on the weld pool in Cold Arc process was very small, and the liquid level of the weld pool had no obvious fluctuation.

The results of the statistics of oscillation amplitude are in good agreement with Figure 6. The amplitude was proportional to the impact of surface traveling wave on the boundary of weld pool. This can explain the obvious difference in the length of weld pool with little difference in the width of weld pool. The oscillation amplitude of weld pool was affected by the state of weld pool. The full penetration pool had larger amplitude of the weld pool was affected by the state of the weld pool; compared with the partial penetration. When the weld pool was impacted, the bottom of the full weld pool was liquid metal level, which had little effect on the downward movement of metal flow. As a result, the amplitude of full penetration pool was larger than that of partial penetration pool under the same welding parameters.

3.2. Oscillation of Weld Pool

High speed photograph pictures showed that the liquid waves in S-GMA welding were triggered primarily by the electric explosion, not by the change in the arc pressure during the arc period [15]. The relationship curves the height of the reference point on the weld pool surface with time were obtained by using the method described in Section 2.3, as shown in Figure 10.

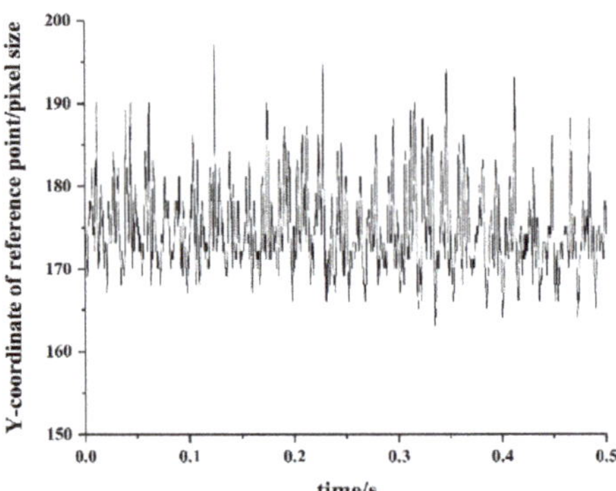

Figure 10. Oscillation signal extracted by high speed photograph-image analysis system.

The period of droplet transfer cycles fluctuated randomly within a range, and the oscillation curve of weld pool is a time-varying signal, which causes the oscillation frequency of weld pool change within a range. The Fast Fourier Transform Algorithm (FFT) could not extract the characteristics of weld pool oscillation. The Continuous Wavelet Transform (CWT) can analyze time-varying signals in time domain and frequency domain simultaneously. In this paper, Morlet continuous wavelet transform was applied to weld pool oscillation signals at different wire feeding speeds. Morlet wavelet base function is shown in Equation (7):

$$\psi_{a,b}(t) = \sqrt{a}exp\left(i\omega_0\frac{(t-b)}{a}\right)exp\left(-\frac{(t-b)^2}{2a^2}\right) \tag{7}$$

In the continuous wavelet transform, the scale vector a is associated with the central frequency and the support interval of the basis function, and the frequency of weld pool oscillation and its corresponding time frequency resolution can be obtained at any time. For a particular scale vector, the signal frequency allowed by the wavelet transform should be close to the corresponding frequency of the scale vector. Therefore, the continuous wavelet transform can clearly reflect the variation of oscillation frequency with time. In this experiment, wavelet transform is carried out on the acquired signal of melt pool oscillation, and the center frequency ω_0 of base function was three. The oscillation frequency range of traditional GMAW weld pool is below 300 Hz [16]. The scale vector a selected in this experiment was between 50 and 700, and the corresponding oscillation frequency identification range was 40–600 Hz. b is the duration of signal acquisition. The contour diagram of transform coefficient of signal reflects the energy density distribution of the signal in the time-scale plane. The energy of the signal is mainly concentrated around the wavelet-ridge-cure in the time-scale plane, from which the instantaneous frequency of the signal can be determined. Signal sampling frequency ($f_{\text{Sampling frequency}}$) was equal with the fps of high-speed photography, and the corresponding relationship between the oscillation frequency of weld pool ($f_{\text{Oscillation frequency}}$) and the scale vector a of wavelet-ridge-cure is as Equation (8):

$$f_{\text{Oscillation frequency}} = \frac{f_{\text{Sampling frequency}} \cdot \omega_0}{a} \tag{8}$$

Figure 10 is the contour diagram of continuous wavelet transform coefficient of weld pool oscillation signal of traditional S-GMAW process under different wire feeding speeds:

It can be seen from Figure 11 that the oscillation of weld pool of S-GMAW had significant periodicity. The relationship curve of the oscillation frequency with different wire-feed speeds is shown in Figure 12. When the wire feeding speed was 2.4 m/min, the weld pool volume was small, resulting high oscillation frequency of the weld pool. With the increase of wire feeding speed, the volume of weld pool increases, the propagation time of travelling wave on the surface of weld pool increased, and the oscillation frequency decreased.

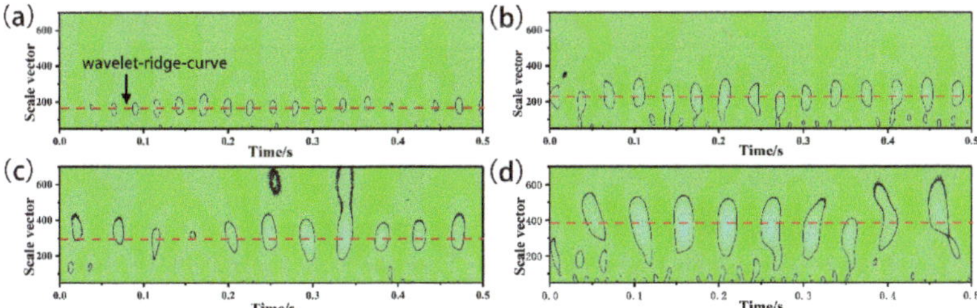

Figure 11. The contour diagram of transform coefficient of weld pool oscillation: (**a**) 2.4 m/min; (**b**) 2.7 m/min; (**c**) 3.0 m/min; and (**d**) 3.3 m/min.

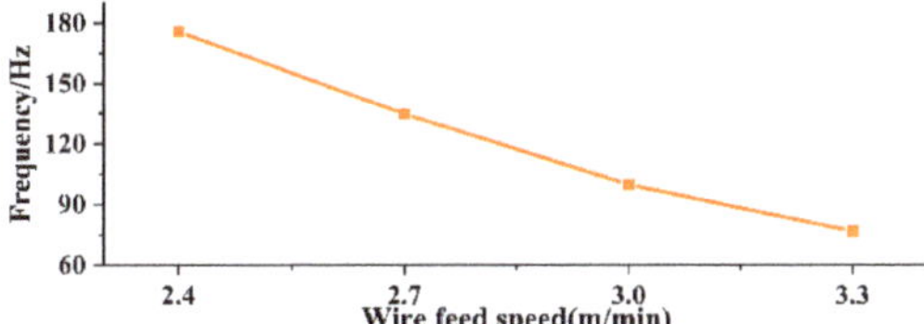

Figure 12. The oscillation frequency of weld pools with different wire-feed speeds.

Different waveforms and penetration states of weld pool led to different oscillation frequencies. Figure 13 shows the contour curves of the oscillation wavelet transform coefficients of the fusion and partial weld pools of three waveforms, and Table 4 shows the oscillation frequency statistics.

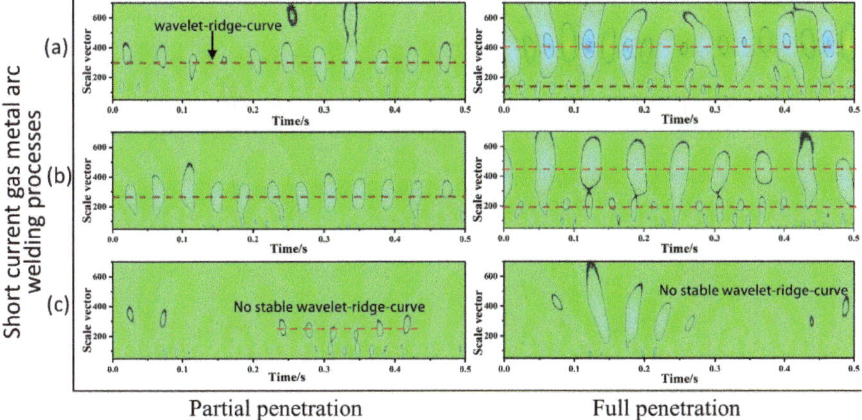

Figure 13. The contour curves of the oscillation wavelet transform coefficients of the fully and partly penetrated weld pools: (**a**) conventional process; (**b**) LSC; and (**c**) Cold Arc.

Table 4. The oscillation frequency statistics.

Waveform	Partial Penetration	Full Penetration
Conventional	100 Hz	75 Hz 200 Hz
LSC	112 Hz	68 Hz 165 Hz
Cold Arc	Not available	Not available

As can be seen from Figure 13, there was a significant difference in the oscillation frequency between the partly and fully penetrated weld pools. These pictures at both sides of Figure 13 are the contour diagram of the distribution of oscillation wavelet coefficients of partly and fully penetrated weld pool using different waveforms. Only one frequency occurred during the oscillation process of partly penetrated weld pool, while there were two characteristic frequencies on the oscillation spectrum of the fully penetrated weld pool. The difference between high frequency and low frequency was generally about 40 Hz, which indicated that there were two oscillation periods of different frequencies in the weld pool. Zacksenhouse [17] established a pool analysis model based on the stretch film theory and studied the oscillation frequency of the full penetration pool. In the full penetration pool, the vibration frequency is obviously lower than that of the partial penetration pool, and the amplitude of the oscillation of the fully penetrated weld pool is relatively larger than that of the partly penetrated weld pool due to the disappearance of the bottom constraint, which was consistent with the Figure 9.

In order to explain the two oscillation frequencies of the full penetration pool, the metal flow process in the pool should be considered, as shown in Figure 14. With the impact of electric explosion, the liquid weld pool flowed radially symmetrically with the arc axis. In the full penetration pool, axial flow occurred for the bottom of the weld pool was no longer supported by any solid material. The liquid in the middle of the weld pool can move vertically, while the liquid in the periphery of the weld pool was supported by the solid material and forced to flow laterally. It led to the fact that although the weld pool was in the state of full penetration, the traveling wave propagation process was still similar to that of non-molten penetration at the periphery of the weld pool. During travelling S-GMAW welding process, the penetration position was relatively small compared with the length of the weld pool, and most of the weld pool metal was still supported by solid metal at the bottom of the weld pool, the oscillation behavior was similar to partial penetration. Therefore, the full penetration pool had two characteristic oscillation frequencies: High frequency and low frequency.

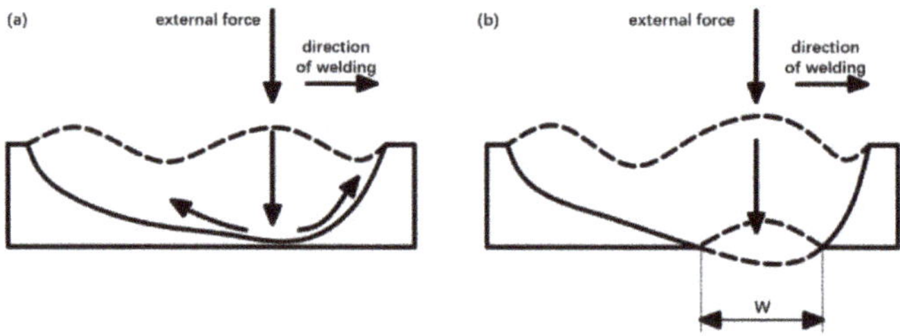

Figure 14. Oscillation mode of weld pool: (**a**) partial penetration; (**b**) full penetration.

The characteristics of the pool oscillation of three waveforms were different. As shown in Figure 13, No stable wavelet-ridge-cure occurred in the contour curves of the oscillation wavelet transform coefficients of Cold Arc, so the weld pools of Cold Arc had no stable oscillation frequency. The Cold Arc process reduced the current at the end of short-circuit stage, which greatly reduces the impact of electric explosion on the pool. At the same wire feeding speed, the weld pool oscillation frequency of LSC was slightly lower than that of conventional process, which was related to the size of weld pool and the electric explosive impact force at the end of short circuit, and the surface tension of weld pool was one of the factors that caused the difference of the frequency. Different waveforms led to different surface temperature of the weld pool, resulting in different surface tension of the weld pool. The surface tension of the weld pool metal is also one of the important factors affecting the oscillation frequency of the weld pool.

3.3. *Flow Behavior of Weld Pool*

To study the weld pool flow behavior, positions of the camera and backlight source need to be changed, as shown in position two of Figure 2. 650 nm near infrared filter were installed to obtain metal flow information of the weld pool. When active gas served as a shielding gas, alloying elements like silicon and manganese, which were present in the base metal and the wire, had a high affinity to react with oxygen and form silicon oxide and manganese oxide. These oxides accumulate on the surface of the weld pool and form slag [18]. the slags have a lower density than the molten metal and follow the flow pattern of the weld pool. Hence, slag flow pattern and accumulation location can disclose the weld pool flow behavior [19].

GMAW weld pool consists of the hot part of the weld pool and the cold part of the weld pool [19,20]. The hot part of the weld pool consisted of the area directly under the arc and the surrounding region, and the cold part of the weld pool is located behind the hot part of weld pool. According to Grong et al., the metal oxides in the high temperature zone of the weld pool exists in the form of metal oxide powder, which cannot aggregate into slags [20,21]. The metal oxides in the cold part of weld pool accumulate into blocks to form slags. Slag is a poor conductor of heat and prevents the red glow of the weld pool, which can block the light at a wavelength of 650 nm. The slag flow pattern can be clearly observed by using 650 nm polaroid as filter.

Figure 15 represented the frames from the high speed video to show the partly penetrated weld pool flow pattern and the slag accumulation location for conventional process, LSC, and Cold Arc, respectively. The white powder in the front and middle of the weld pool was the silicon oxide and manganese oxide particles, which are separated from the weld metal due to the strong turbulence in the weld pool in this part and pushed to the low temperature area of the weld pool under the action of the pool flow. The slags flow patterns behaviors of different processes showed significant difference.

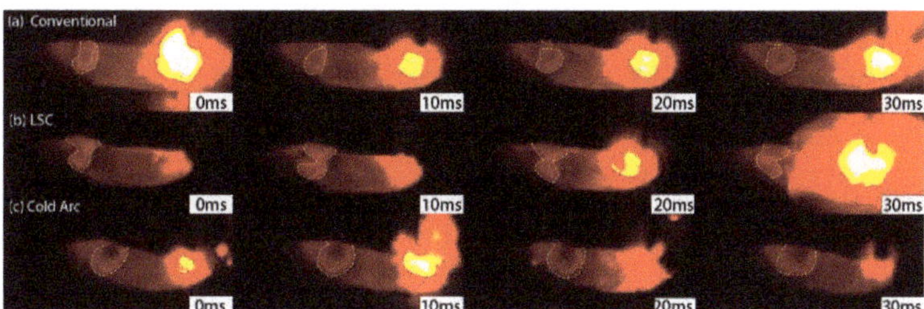

Figure 15. The partly penetrated weld pool flow pattern and the slag accumulation location: (**a**) Conventional; (**b**) LSC; (**c**) Cold Arc (the slag islands were outlined by white dotted lines).

The weld pool flow in partly the penetrated weld pool is explained with the assistance of Figure 16. When the weld pool is forced to flow downwards (by the external force), it is blocked by the base metal and is forced to flow to the back of the weld pool. At the back of the weld pool, the metal liquid flow will rebound off the solid metal interface and flow to the front of the pool. By the present experiment results in Section 3.1, the impact of the traditional S-GMAW process on the weld pool is the strongest of the three, and the bounced metal flow was the strongest which led to the formation of two spinning large slag islands. The impact of the LSC on the weld pool was smaller than the conventional process. The dumbbell shaped slag island in LSC was formed with co-extrusion of the bounced metal flow and the backward flow of metal on the surface of weld pool. Cold Arc process has little impact on the weld pool, because the bounced metal strength was negligible, so the metal oxides gathered into a single round island of slag.

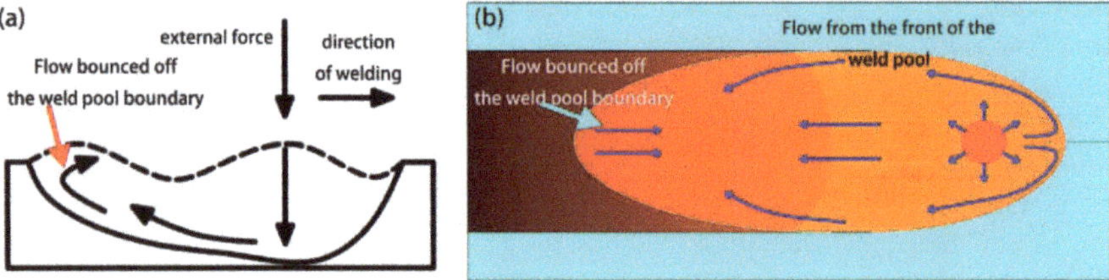

Figure 16. Schematic diagram of flow of partial penetrated weld pool: (**a**) inside. (**b**) surface.

As shown in Figure 17, compared with partial penetrated weld pool, the characteristics of the full penetrated weld pool flow pattern have the obvious differentiation, the full penetrated weld pools also have obvious low temperature zone and high temperature zone, but the metal oxides in the cold part of the weld pool did not gather and form huge slag islands, instead distributed at the back of the weld pool evenly. The bottom surface of the full penetrated weld pool can expand and contract with impact, and the energy was absorbed due to the existence of the free surface at the bottom, as shown in Figure 18. The liquid metal in the hot part of weld pool was not pressed, so the volume and length of the fully penetrated weld pool was larger than that of partly penetrated weld pool. For the fully penetrated weld pool, the slag at the end of the pool formed discrete scattered islands and did not gather into a large slag island, as was observed for the partial penetration case. This change was simply dependent on the surface wave of the weld pool.

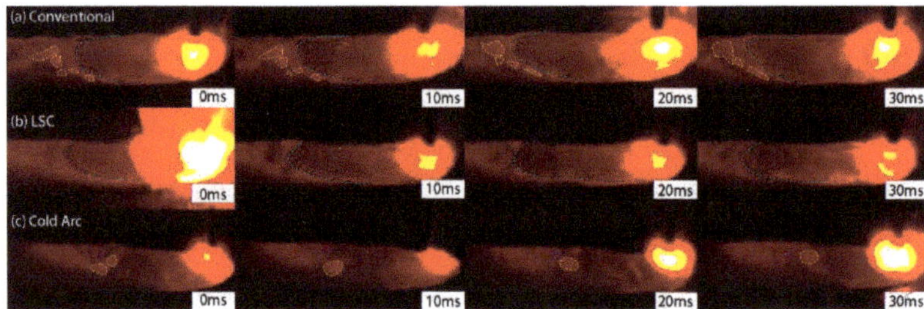

Figure 17. The fully penetrated weld pool flow pattern and the slag accumulation location: (**a**) Conventional; (**b**) LSC; (**c**) Cold Arc (the slags were outlined by dotted lines).

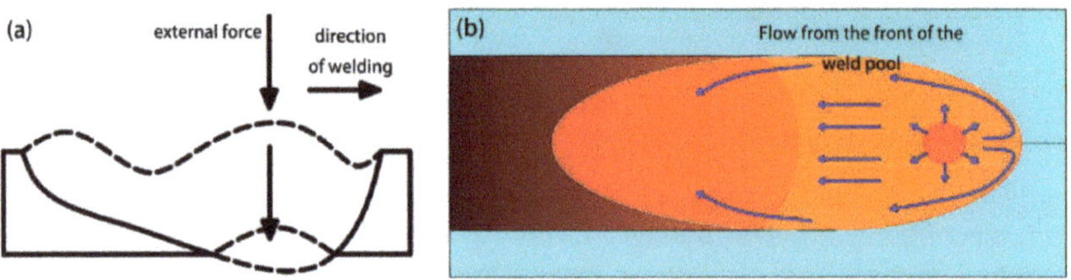

Figure 18. Schematic diagram of flow of fully penetrated weld pool: (**a**) inside. (**b**) surface.

Since the composition of shielding gas was consistent, the influence of the Marangoni flow can be excluded. And for the cold part of the weld pool lie on the further from the center of the arc, the influence of plasma flow force and electromagnetic force can be ignored. The difference observed in weld pool flow pattern was attributed to the varied degrees-of-freedom of weld pool.

3.4. Geometry and Microstructure of Weld Bead

The weld geometry often qualifies melting characteristics of base metal, amount of weld deposition, welding heat input, energy distribution, and regulation of the flow of liquid metal to control its shape at various parameters [6]. The energy distribution and the metal transfer in the welding process depend on the waveforms of S-GMAW, which were both the main factors that affects the fluidity of the weld pool, and further affects the weld microstructure.

3.4.1. Geometry of Weld Bead

The change of waveforms had remarkable influence on weld geometry. It is of great significance to study the weld bead geometry of S-GMAW under different wave forms. Transverse sections are shown in Figure 19 for beads on plate deposition which contains partial penetration (the thickness of base material was 4 mm) as well as fully penetration (the thickness of base material was 2 mm), respectively.

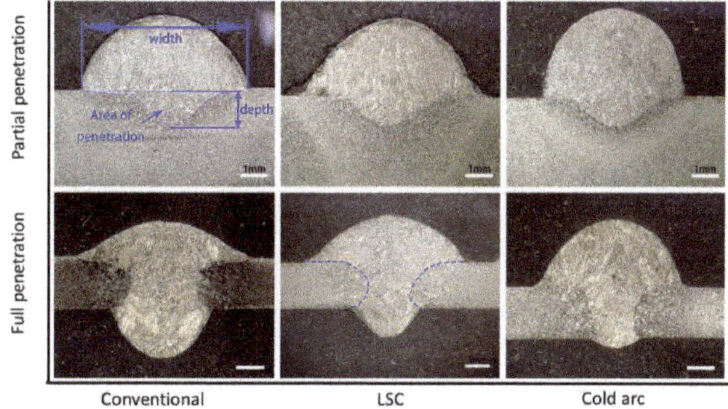

Figure 19. The geometry of the weld beads with different waveforms (wire feed rate: 3 m/min, voltage: 19 V).

As shown in Figure 19, the area of base metal fusion of Cold Arc was smaller than that of the other two waveforms. As mentioned in Section 3.1, the average heating powers of LSC and Cold Arc to the base material were similar. The difference of weld pool behaviors was the main factor caused the difference of the area of bead deposit.

The oscillation intensified the convection of the weld pool, which led to the metal in the hot part to flow to the pool boundary, which intensified the melting of the base material. The impact of the conventional process and LSC on the weld pool was much larger than that of Cold Arc, so the area of base metal fusion for the Cold Arc was the smallest. Due to the oscillation amplitude of the fully penetrated pool was larger than that of the partly penetrated pool, as well as the bad heat dissipation of thin plate, the area of full penetration base metal fusion was much larger than that of the partial penetration.

The influence of electric explosion on the weld pool changed the depth and width of weld pool. Figure 20 listed the variation of weld geometry with increasing of wire feed rate. Due to the narrow range of weld parameters of the full penetration pool, this paper only listed the weld forming parameters of the partial penetration pool with different wire feed rate.

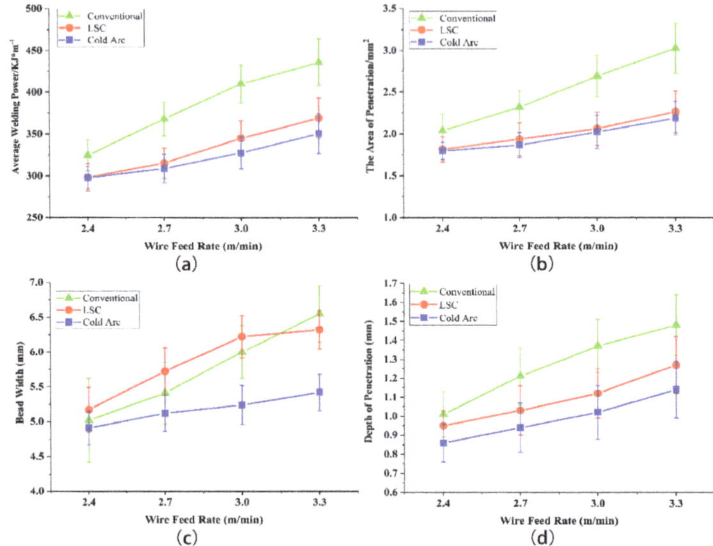

Figure 20. Effect of wire feed rate on geometry of fusion (partial penetration) in bead on plate deposition of S235JR: (**a**) average welding power; (**b**) area of penetration; (**c**) width; and (**d**) depth.

As shown in Figure 20, the main factor that determined weld geometry was welding heat input. The influence of weld pool behavior on weld geometry and penetration area was not obvious. Low wire feeding speed resulted in smaller peak current, lower energy input difference, lower effect of electric explosion impact on the pool, and more consistent weld formation of three waveforms. The difference of weld forming was more obvious with the increase of wire feeding speed.

3.4.2. Microstructure of Weld Metals

The waveform of S-GMAW has great influence on the microstructure of weld bead. It determines the dynamic behavior of the weld pool, which further influences the solidification behavior of the weld pool. In a large part, solidification behavior determines microstructure of weld metal. This microstructure of weld metals is probably due to the complex interactions between weld thermal cycle, cooling rate, and the prior austenite grain size [22]. In this study, the weld metal composition remained almost constant under the conditions that the same base metal, filler metal, and shielding gas were used in all experiments, therefore, the change of microstructure is related to the change of arc heat input and weld pool behavior.

For the same welding condition, the weld pool thermal behaviors between the three waveform processes were completely different, which resulting in obvious and different weld microstructure, as shown in Figures 21 and 22. The proeutectoid ferrite (PF) volume fraction precipitated at the initial austenite grain boundary were related to the welding heat cycle. The lower the cooling rate of weld bead, the greater the amount of PF formation. Poor heat dissipation can aggravate this phenomenon, as the full penetration weld was prepared with 2 mm steel plate, which inhibited the heat transfer and reduced the cooling rate, it caused thicker PF compared with the partly penetrated weld pool.

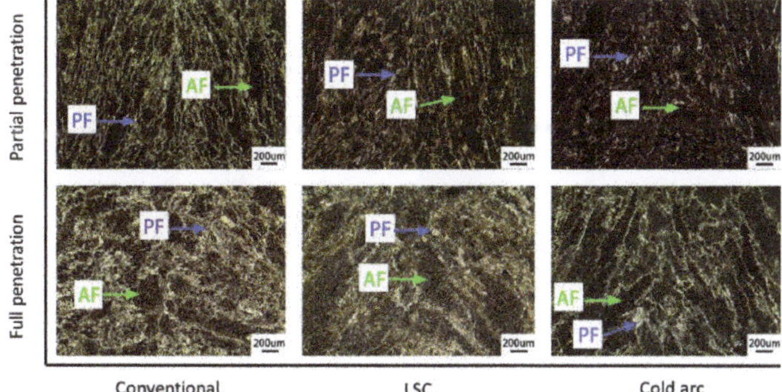

Figure 21. The microstructures of the weld beads with different waveforms (magnification: 50 times).

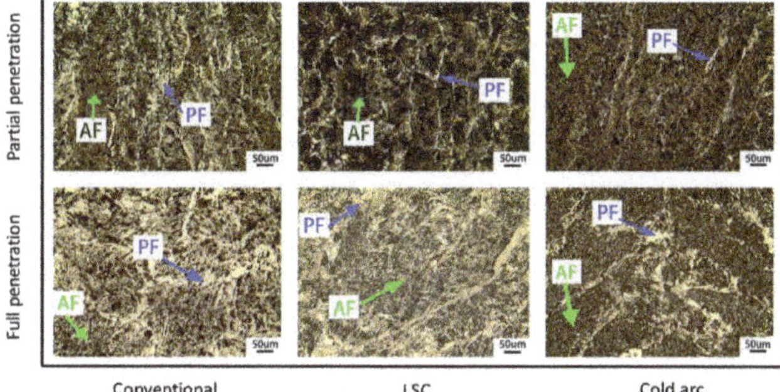

Figure 22. The microstructures of the weld beads with different waveforms (magnification: 200 times).

The heat input of conventional process was larger than the LSC and the Cold Arc, resulting in larger PF microstructure. The heat input of LSC was close to Cold Arc, but the oscillating behavior of the weld pool aggravated the heat exchange between the metal liquid in the high temperature zone and the low temperature zone, resulting in the slow cooling rate and the precipitation of more PF in the initial austenite grain boundary.

PF usually precipitates at the initial austenite grain boundary, and the distribution of PF reflected the initial austenite grain boundary [23]. The directivity of PF structure in partly penetrated weld bead was obvious, which indicated that most of the initial austenite grain were columnar. In the fully penetrated weld bead of conventional process and LSC, the structure of PF microstructure no longer had direction, which indicated that the solidification and growth process of initial austenite grains could be changed by the oscillation of the weld pool.

In the weld microstructure, acicular ferrite (AF) can effectively increase the mechanical properties (particularly toughness) of the weld. Therefore, the volume fraction of acicular ferrite in Figure 21 was measured, image processing software was used to measure the area fraction of different phases in the Figure. The area fraction of AF and PF phases in the metallographic picture were approximately equal to their respective volume fractions, and the results were shown in Figure 23. The results show that the volume fractions of AF were significantly different, due to the change of the weld heat gradient within the different waveforms. The heat gradient of the weld pool of the fully penetrated weld bead was smaller than that of the partial penetration, which resulted in a decrease of the volume fraction of AF. The weld pool oscillated continuously in the traditional process and LSC, the weld pool was agitated and the thermal gradient declined, resulting in the decrease of AF content. The AF content in the weld microstructure of the traditional process and LSC was lower than that of Cold Arc.

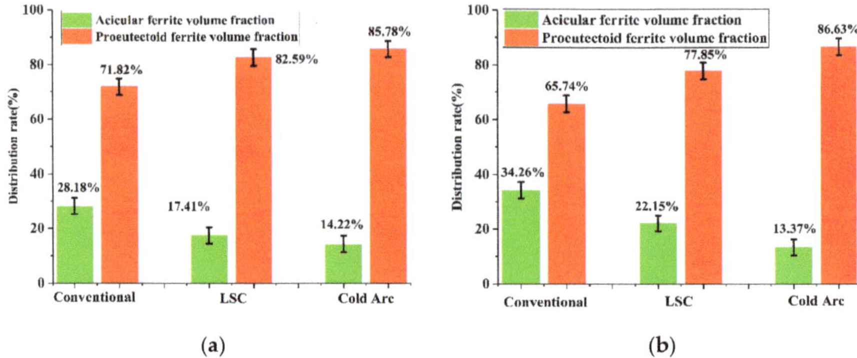

Figure 23. Acicular ferrite and proeutectoid ferrite volume fraction in weld deposits: (**a**) partial penetration; (**b**) full penetration.

The AF microstructures of different bead-on-welds as shown in Figure 24 reveal different grain size. The average grain sizes of AF grain were calculated using intercept method (as per ASTM E112-10). In general, the grain size of weld metals can be typically correlated with the heat input or the cumulative effect of weld parameters. As mentioned in Table 3, the heat input of traditional process was obviously larger than that of LSC and Cold Arc, which caused the grain size of conventional process was larger than the others. The heat input of LSC was approximately equal with the Cold Arc, the size of acicular ferrite precipitated in the weld of LSC was larger than that of Cold Arc. In the case when the heat input was approximately equal, the change in AF size also verified the difference of thermal gradient of welding pools between different waveforms. The decrease of thermal gradient resulted in AF grain grow.

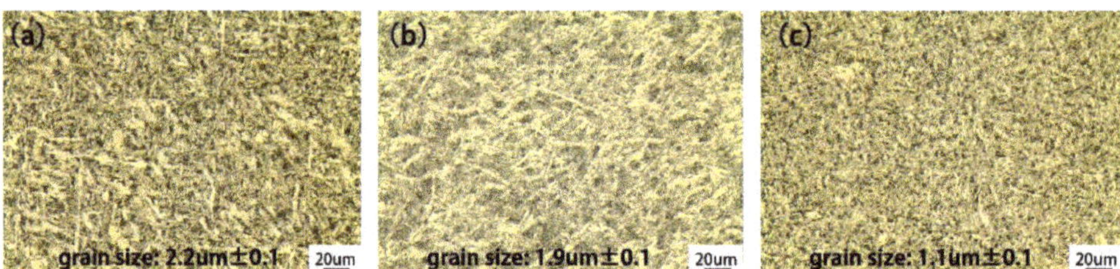

Figure 24. Optical micrographs of acicular ferrite (**a**) conventional, (**b**) LSC, and (**c**) cold Arc.

The weld microstructure of low carbon steel was the result of solidification and solid phase transformation of the weld pool in non-equilibrium state. The columnar-to-equiaxed transition caused by the oscillation of the weld pool had not been observed, but the oscillation homogenized the temperature distribution of the weld pool, reduced the temperature gradient of the weld pool, and resulted in the coarsening of the weld structure in the process of solid phase transformation.

4. Conclusions

In this paper, the behavior characteristics of S-GMAW weld pool were studied, and the differences of the pool behavior and weld microstructure were compared under different waveforms. The conclusions are as follows:

(1) In short-circuit period, the duration of destabilization and break-up of the liquid bridge is mainly related to the surface tension of the liquid metal, not the loop current. However, the rise rate of the loop current can effectively shorten the stability time of the liquid bridge and promote the

formation of the neck of the short-circuit liquid bridge. The liquid bridge explosion is related to the instantaneous power density of liquid bridge metal.

(2) The weld pool oscillation is triggered by the pressure of the electric explosion. The oscillation of the weld pool can be monitored visually by high-speed photography imaging. The oscillation of the weld pool has natural frequencies which decrease with the increase of volume of weld pool. In the case of partial penetration, only one natural oscillation frequency can be detected. In the case of full penetration two different oscillation frequencies can be detected.

(3) The shape of slag on the surface of the weld pool and the flow behavior of the weld pool can reflect the penetration state of the weld pool. The different boundary conditions between the partial and full penetration cause different flow behavior of the weld pool, which leads to the fact that the slag tends to aggregate into large blocks in partial penetration, while the slag in the fully penetrated weld pool cannot aggregate into blocks. Large slag island can be deformed or split apart with different impact strength of electrical explosions.

(4) Compared with the influence of weld heat input on the size of weld pool, the effect of weld pool oscillation is not obvious. The oscillation imparts a negative effect on the weld microstructure, along with the aggravation of the weld pool oscillation, the content and size of proeutectoid ferrite in the weld microstructure increases, the content of acicular ferrite decreases while the grain size increases.

Author Contributions: Methodology, T.C.; Software, T.C.; Investigation, T.C., P.Z., Writing—Original Draft Preparation, T.C.; Writing—Review and Editing, S.X., T.C., B.W.; Supervision, S.X.; Project Administration, S.X., W.L.; Funding Acquisition, S.X.

Funding: This work was funded by the National Natural Science Foundation of China, grant No.51675269 and the Priority Academic Program Development of Jiangsu Higher Education Institutions (PAPD).

Conflicts of Interest: The authors declare no conflict of interest.

References

1. Kah, P.; Suoranta, R.; Martikainen, J. Advanced gas metal arc welding processes. *Int. J. Adv. Manuf. Technol.* **2013**, *65*, 655–674. [CrossRef]
2. Mvola, B.; Kah, P.; Layus, P. Review of current waveform control effects on weld geometry in gas metal arc welding process. *Int. J. Adv. Manuf. Technol.* **2018**, *96*, 4243–4265. [CrossRef]
3. Matusiak, J.; Pfeifer, T.; Martikainen, J. The research of technological and environmental conditions during low-energetic gas-shielded metal arc welding of aluminium alloys. *Weld. Int.* **2013**, *27*, 338–344. [CrossRef]
4. Fronius. TPS/I Steel Edition Optimized for Manual Steel Applications. 2019. Available online: https://www.fronius.com/en/welding-technology/world-of-welding/tpsi-steel-edition (accessed on 22 September 2019).
5. Liu, A.; Tang, X.; Lu, F. Weld pool profile characteristics of Al alloy in double-pulsed GMAW. *Int. J. Adv. Manuf. Technol.* **2013**, *68*, 2015–2023. [CrossRef]
6. Zhang, Z.; Xue, J. Profile Map of Weld Beads and Its Formation Mechanism in Gas Metal Arc Welding. *Metals* **2019**, *9*, 146. [CrossRef]
7. Yao, P.; Zhou, K.; Huang, S. Process and Parameter Optimization of the Double-Pulsed GMAW Process. *Metals* **2019**, *9*, 1009. [CrossRef]
8. Li, C.; Shi, Y.; Gu, Y.; Yuan, P. Monitoring weld pool oscillation using reflected laser pattern in gas tungsten arc welding. *J Mater. Process. Tech.* **2018**, *255*, 876–885. [CrossRef]
9. Guo, J.; Zhou, Y.; Liu, C.; Wu, Q.; Chen, X.; Lu, J. Wire Arc Additive Manufacturing of AZ31 Magnesium Alloy: Grain Refinement by Adjusting Pulse Frequency. *Materials* **2016**, *9*, 823. [CrossRef]
10. Yudodibroto, B.Y.B.; Hermans, M.J.M.; Richardson, I. Observations on Droplet and Arc Behaviour during Pulsed GMAW. *Weld. World* **2009**, *53*, R171–R180. [CrossRef]
11. Liu, A.; Tang, X.; Lu, F. Study on welding process and prosperities of AA5754 Al-alloy welded by double pulsed gas metal arc welding. *Mater. Des.* **2013**, *50*, 149–155. [CrossRef]
12. Hälsig, A.; Pehle, S.; Kusch, M.; Mayr, P. Reducing potential errors in the calculation of cooling rates for typical arc welding processes. *Weld. World* **2017**, *61*, 745–754. [CrossRef]

13. Hälsig, A.; Kusch, M.; Mayr, P. New Findings on The Efficiency of Gas Shielded Arc Welding. *Weld. World* **2012**, *56*, 98–104. [CrossRef]
14. Hälsig, A.; Mayr, P. Energy balance study of gas-shielded arc welding processes. *Weld. World* **2013**, *57*, 727–734. [CrossRef]
15. Yudodibroto, B.Y.B. Liquid Metal Oscillation and Arc Behaviour during Welding. Ph.D. Thesis, Delft University of Technology, Delft, The Netherlands, 2010.
16. Hermans, M.J.M.; Ouden, G.D. Process behavior and stability in short circuit gas metal arc welding. *Weld. J.* **1999**, *78*, 137s–141s.
17. Zacksenhouse, M.; Hardt, D.E. Weld Pool Impedance Identification for Size Measurement and Control. *J. Dyn. Syst. Meas. Control* **1983**, *105*, 179–184. [CrossRef]
18. Wang, B.; Xue, S.; Ma, C.; Wang, J.; Lin, Z. Study in Wire Feedability-Related Properties of Al-5Mg Solid Wire Electrodes Bearing Zr for High-Speed Train. *Metals* **2017**, *7*, 520. [CrossRef]
19. Ahsan, M.R.U.; Cheepu, M.; Kim, T.H. Mechanisms of weld pool flow and slag formation location in cold metal transfer (CMT) gas metal arc welding (GMAW). *Weld. World* **2017**, *61*, 1275–1285. [CrossRef]
20. Yi, L.; Yang, Z.; Xiaojian, X. Energy Characterization of Short-Circuiting Transfer of Metal Droplet in Gas Metal Arc Welding. *Metall. Mater. Trans. B* **2015**, *46*, 1924–1934. [CrossRef]
21. Ronda, J.; Estrin, Y.; Oliver, G.J. Modelling of welding. A comparison of a thermo-mechano-metallurgical constitutive model with a thermo-viscoplastic material model. *J. Mater. Process. Tech.* **1996**, *60*, 629–636. [CrossRef]
22. Sen, M.; Mukherjee, M.; Singh, S.K. Effect of double-pulsed gas metal arc welding (DP-GMAW) process variables on microstructural constituents and hardness of low carbon steel weld deposits. *J. Manuf. Process.* **2018**, *31*, 424–439. [CrossRef]
23. Hunt, A.C.; Kluken, A.O.; Edwards, G.R. Heat input and dilution effects in microalloyed steel weld metals. *Weld. J.* **1994**, *731*, S9–S15.

Article

Effect of Heat Input on Weld Formation and Tensile Properties in Keyhole Mode TIG Welding Process

Zhenyu Fei [1,2], Zengxi Pan [1,2], Dominic Cuiuri [1,2], Huijun Li [1,2], Bintao Wu [1], Donghong Ding [3,*] and Lihong Su [1,2]

[1] School of Mechanical, Materials, Mechatronic and Biomedical Engineering, University of Wollongong, Northfield Avenue, Wollongong, NSW 2522, Australia; zf996@uowmail.edu.au (Z.F.); zengxi@uow.edu.au (Z.P.); dominic@uow.edu.au (D.C.); huijun@uow.edu.au (H.L.); bw677@uowmail.edu.au (B.W.); lihongsu@uow.edu.au (L.S.)

[2] Defence Materials Technology Centre, 24 Wakefield Street, Hawthorn, VIC 3122, Australia

[3] School of Mechatronic Engineering, Foshan University, 33 Guangyun Road, Foshan 528225, China

* Correspondence: dd443@uowmail.edu.au; Tel.: +86-13054466797

Received: 20 October 2019; Accepted: 4 December 2019; Published: 7 December 2019

Abstract: Keyhole mode Tungsten Inert Gas (K-TIG) welding is a novel advanced deep penetration welding technology which provides an alternative to high power density welding in terms of achieving keyhole mode welding. In order to facilitate welding procedure optimisation in this newly developed welding technology, the relationship among welding parameters, weld formation and tensile properties during the K-TIG welding was investigated in detail. Results show that except for travel speed, the heat input level also plays an important role in forming undercut defect by changing the plasma jet trajectory inside keyhole channel, leading to the formation of hump in the weld centre and exacerbation of undercut formation. Both undercut defect and root side fusion boundary can act as a stress concentration point, which affects the fracture mode and tensile properties considerably. The research results provide a practical guidance of process parameter optimisation and quality assurance for the K-TIG welding process.

Keywords: K-TIG; heat input; weld formation; tensile properties; welding procedure optimisation

---

1. Introduction

Tungsten Inert Gas welding, also known as TIG or GTAW, is a clean and widely used welding technology for metals. Nevertheless, a main disadvantage of this welding process is the limited penetration ability. It may require multipasses to complete the weld for medium thickness materials, which results in very low productivity. In order to overcome this drawback, K-TIG was developed as a new TIG variant to increase the penetration capability and production efficiency. In K-TIG welding, a free burning arc is applied as the heat source to melt the weldment by forming a cylindrical keyhole [1], which provides a nice alternative to high power density welding. This type of keyhole welding is characterized by its dependence on arc force, as opposed to ablation pressure in laser-based and electron beam-based welding, and a combination of stagnation and recoil pressure in plasma arc welding (PAW), as reported by Jarvis and Ahmed [2]. Compared with standard melt-in mode TIG welding, K-TIG process possesses higher energy density, process efficiency and greater penetration ability. Although heat density of K-TIG is far lower compared to high power density welding, it is easier to operate and more cost-effective, as stated by Liu et al. [3]. In addition, Liu et al. [4] reported that K-TIG was able to achieve continuous open keyhole and was more stable than PAW. It was also demonstrated by Liu et al. [5] that K-TIG had much wider operating window compared with PAW owing to the fact that it has much smaller arc pressure/arc current ratio.

To date, K-TIG has found applications in industries for joining medium thickness materials, say titanium [6], zirconium [7], stainless steel [8], low carbon steel [9] and dissimilar metal welding [10]. It was shown in these studies that K-TIG can complete the medium thickness weld in a single pass with a V-shaped morphology. It was also shown that K-TIG produced a welded joint with smaller fusion zone and larger heat-affected zone compared with conventional TIG welding process. In addition, it has been demonstrated that K-TIG has the potential to increase the productivity while maintaining the mechanical properties at a level similar to the joints produced by conventional TIG welding. Recently, it was reported by Cui et al. [11] that K-TIG can even be used in underwater condition and can produce a duplex stainless-steel weld that meets underwater welding standards. In addition, because of the fact that the material properties of some welded joints produced by K-TIG are not satisfactory, several optimisation technologies have been introduced into the K-TIG welding process. For example, Fei et al. [12] introduced filler materials into K-TIG to improve the weld microstructure and mechanical properties of armour steel weld. Fei et al. [13] also applied specially designed interlayer to tailor the microstructure and solidification sequence of K-TIG-welded armour steel joint. Post-weld heat treatment and ultrahigh frequency were utilised by Xie et al. [14] to improve both the mechanical properties and corrosion resistance of 430 ferritic stainless steel welded joint. On the other hand, since K-TIG has a narrower operating window for carbon steel because of its higher thermal conductivity, several studies have focused on the keyhole behaviour and stability improvement, such as the application of back purging unit [15], one pulse one keyhole technique [16] and high frequency pulse current waveform [17]. Furthermore, in order to develop on-line control strategy for the K-TIG welding process, several penetration monitoring systems have been developed, such as arc voltage-based oscillation frequency sensing [18] and combination of acoustics and vision-based sensing [19].

Overall, the current research directions of the K-TIG welding technology can be divided into four parts, namely metallurgical qualification of K-TIG welded joints, optimisation technology for joint microstructure and properties, process dynamics and stability improvement as well as development of on-line monitoring system. Up to date, there has been little research on the relationship among process parameters, weld formation and mechanical properties in the K-TIG welding process. More importantly, unlike conventional TIG welding for which the weld formation can be partially controlled or compensated by the operators, the K-TIG operates in a completely automated mode, with the weld formation being entirely dependent on the welding parameter combination. Although in some studies, small amount of welding trials were conducted to test the influence of welding parameters in the K-TIG welding, the main purpose was to find a parameter combination to achieve full penetration, as did by Cui et al. [8] and Feng et al. [20]. The mechanism for defect formation and the correlation between weld formation and mechanical properties in the K-TIG welding process have not been addressed. To a practical view, the mechanical properties, especially tensile properties, are easily affected by weld formation which is a function of process parameters used during welding. Figuring out the relationship among welding parameters, weld formation, and mechanical properties is helpful to the process optimisation and is crucial for the avoidance of defect formation and premature failure during service.

In this study, K-TIG welding was performed on 6.2 mm thickness high hardness armour (HHA) plates using different welding parameter combination, aiming at evaluating the effect of heat input on weld formation and tensile properties. The mechanism by which various behaviours occurred has been discussed. The results would provide a practical guidance of parameter optimisation and quality assurance for the K-TIG welding process.

2. Materials and Methods

The material used for investigation is 500 grade armour steel, also known as HHA. Detailed chemical composition and mechanical properties of the base plate are listed in Tables 1 and 2, respectively. The chemical composition was measured by atomic emission spectroscopy. The base metal shows tempered martensite microstructure, as shown in Figure 1. It is worth noting here that as this paper is intended to analyse the relationship among welding parameters, weld formation and tensile properties, the material-dependent aspect will not be discussed in detail.

Table 1. Chemical composition of HHA (wt%).

C	Si	Mn	P	S	Ni	Cr	Mo	B	Fe
0.27	0.3	0.3	0.014	0.0025	0.19	1.05	0.25	0.0012	Bal.

Table 2. Mechanical properties of HHA.

Materials	Ultimate Tensile Strength (MPa)	Elongation (%)	Hardness (HV)
HHA	1775	14	495

Figure 1. Microstructure of the HHA base metal.

As depicted in Figure 2, the experimental equipment designed to conduct the welding trials was composed of a K-TIG power supply, a data acquisition device and a high-speed camera. The welding output was provided by a special power source with capability up to 1000 A. A high-speed camera in combination with a 10% transparency neutral filter was placed beside the worktable in order to record the arc behaviour. An inert gas back purge unit was designed to provide shielding from the atmospheric contamination on the root side of the weld pool. The argon shielding gas was transferred into the back purging unit from the left-hand side hose hole and allowed to be ejected through the switch on the right-hand side to avoid accumulation of pressure inside the unit. The image capturing was conducted after the process reached stable. During welding, the torch and high-speed camera were kept stationary while the worktable moved at the pre-set speed.

To test the influence of process parameters on weld formation and tensile properties, seven butt-joint welding experiments were performed on 6.2 mm thickness HHA plates by using control variable method. Process window identification tests were conducted to find the appropriate parameter combination leading to full penetration with the absence of both incomplete penetration because of insufficient heat input and weld pool collapse because of excessive heat input. Detailed parameter combination is depicted in Table 3. The heat input was calculated by the product of welding current and voltage divided by the travel speed plus a thermal coefficient of 0.6. The arc length was measured by the stacked thickness measurement device before welding. The dimension of the plates used in this study is 250 mm × 75 mm × 6.2 mm. The welding current and travel speed were chosen as variables. The fixed parameters for all experimental groups are depicted in Table 4. After welding, both transverse and longitudinal cross sections of the welds were cut from the weldments. The longitudinal cross section was cut in the arc extinguishment area in a bid to capture the in-situ profile during welding. The surface morphology and cross section macrographs were captured by Nikon digital camera and

Leica M205A stereomicroscope, respectively. The mounted samples were etched with 2 wt% nital for both macro and microstructure observation. Microhardness test was undertaken using Struers DuraScan-70 automatic hardness tester 2 mm below the surface with 0.5 mm interval under a load of 1 kg for 10 s dwell time. The preparation of tensile samples and the execution of tensile tests were in accordance with ASTM E8/E8M-16a guideline. The tensile samples were cut perpendicular to the welding direction using Wire Electrical Discharge Machining. Tensile tests were performed in as-cut condition without any finishing on the surface of tensile samples with 1 mm/min crosshead speed and 25 mm gauge length (Figure 3). The thickness of all tensile test specimens is 6.2 mm. A CCD camera was used to measure the total elongation by capturing the movement of two white spots with 25-mm interval on the tensile specimens. In order to ensure the reproducibility of tensile results, three samples were tested and the average value was considered.

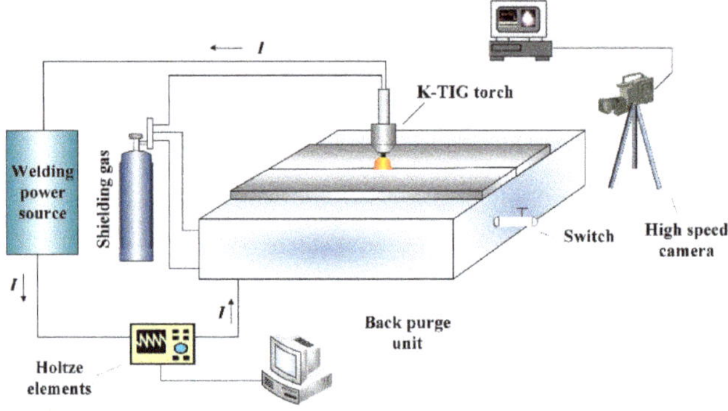

Figure 2. Schematic of experimental system.

Table 3. Welding parameter combination.

Test No.	Welding Current (A)	Arc Voltage (V)	Travel Speed (cm/min)	Heat Input (kJ/cm)
1	450	16.41	34.2	7.78
2	465	16.52	34.2	8.1
3	480	16.69	34.2	8.42
4	510	17.11	34.2	9.18
5	450	16.59	30	8.96
6	450	16.44	26	10.22
7	450	16.57	24	11.16

Table 4. Fixed welding parameters (all experiments).

Process Parameters	Details
Electrode material	Lanthanated tungsten
Electrode diameter	6.4 mm
Electrode tip angle	45 degree
Shielding gas	99.95% Ar
Shielding gas flow rate	20 l/min
Back purging gas	99.95% Ar
Back purging gas flow rate	10 l/min
Post flow shielding time	10 s
Arc length	3 mm

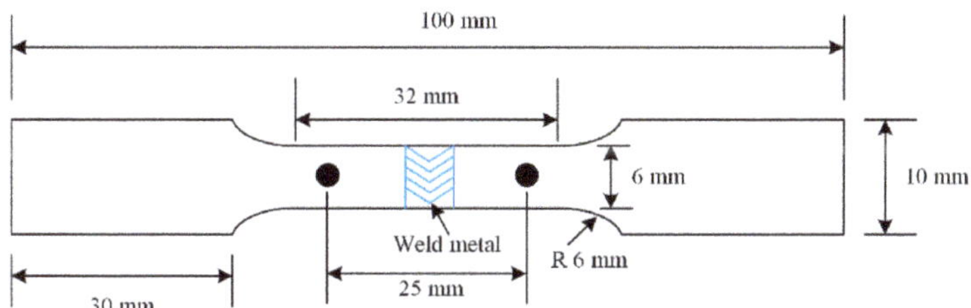

Figure 3. Dimension of tensile specimen.

3. Results and Discussion

3.1. Surface Formation

Tables 5 and 6 illustrate both the face side and root side surface formation for the weldments with different parameters. It can be observed that when the travel speed was fixed at 34.2 cm/min, undercut was experienced for all the weldments, as marked by red arrows and shown in Table 5. When the travel speed was reduced to 30 cm/min, the undercut defect disappeared, together with smooth surface formation on both sides, as shown in Table 6. A further decrease in travel speed led to wavy ripples along the face side weld edges, as highlighted by red circles, although undercut was not experienced. Also, even irregular bead face was produced on the root side when the travel speed was reduced to 24 cm/min in Test 7 sample.

Table 5. Weld surface formation with different arc current.

Test No.	Defect Type	**Fixed parameters:** Speed = 34.2 cm/min, Arc length = 3 mm		Variable
		Face side	**Root side**	**Current**
1	Undercut			450 A
2	Undercut			465 A
3	Undercut			480 A
4	Undercut			510 A

Table 6. Weld surface formation with different travel speed.

Test No.	Defect Type	Fixed parameters: Arc length = 3 mm, Welding current = 450 A		Variable
		Face side	Root side	Speed
5	N/A	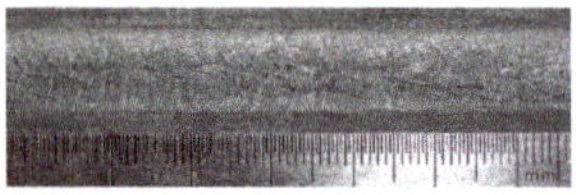	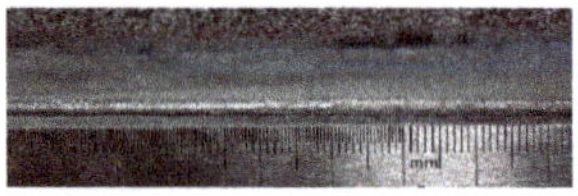	30cm/min
6	Ripple	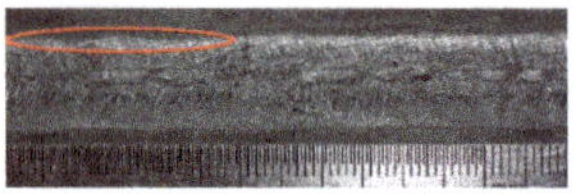	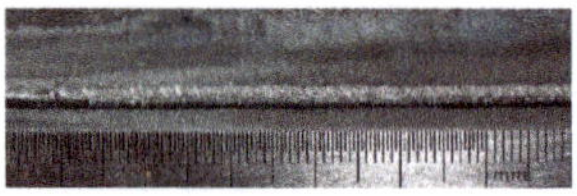	26 cm/min
7	Ripple	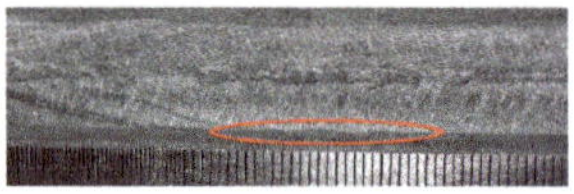	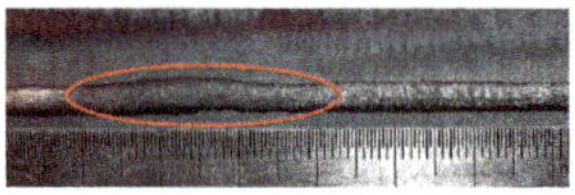	24 mm/min

It is known that high welding speed is easy to cause undercut formation. This is because the molten metal experiences much stronger backwards momentum at higher travel speed. The mass flow rate is the highest along the centreline of the weld, which tends to draw molten metal towards the centreline of the flow and induces lack of melt on the side walls on the face side [21]. Thus, undercut was consistently observed when high travel speed was used. Considering that the Tests 4 and 5 samples possess similar heat input and different travel speed, the presence of undercut in Test 4 may further indicate that travel speed plays a part in the formation of undercut. In addition, as reported by Tomsic and Jackson [22], high heat input makes the keyhole widen, more molten metal displaced by the keyhole flows back to the trailing region for a longer distance and a uniform bead is hard to obtain. This could be the reason for the formation of wavy ripples on the face side and irregular bead face on the root side. As the mitigation of undercut defect was observed with increasing arc current, it is reasonable to believe that the heat input also plays certain role in the formation of undercut in the K-TIG welding process. The effect of heat input on the undercut formation is discussed in following section.

3.2. *Weld Longitudinal Cross-Section*

The weld longitudinal cross-section and arc behaviour are shown in Figure 4. After arc was extinguished, the weld pool was suddenly frozen. The profile inside the keyhole channel was retained. When lower heat input was used, as in Test 1 sample, strong arc plasma was experienced on the face side, together with very high deflection angle relative to the horizontal line, as shown in Figure 4a. As the heat input increased, the amount of arc plasma on the face side decreased significantly, along with the increase in the slope of keyhole leading wall, as shown in Figure 4b. What is more, the deflection angle of plasma cloud relative to the horizontal line decreased with increasing heat input.

As can be seen from the schematic in Figure 4, when lower heat input was used, there was insufficient heat deposited on the lower half of the keyhole leading wall. Thus, the keyhole leading wall exhibited a severely curved shape. Only small amount of the arc plasma can be ejected via the keyhole exit, leaving majority of them guided towards the face side. As the heat input was increased, more heat was deposited on the keyhole bottom region. The bottom region (A zone) of the keyhole leading wall in Figure 4a was melted away into liquid and displaced to the rear part to form the weld, which makes the keyhole exit deviate less from the welding torch. Therefore, larger amount of the arc plasma could be easily ejected through the keyhole exit, leaving very small amount of them guided towards the face side. In addition, the increase in heat input further decreased the slope of keyhole rear edge, which allowed the plasma cloud to be deflected at a very small angle.

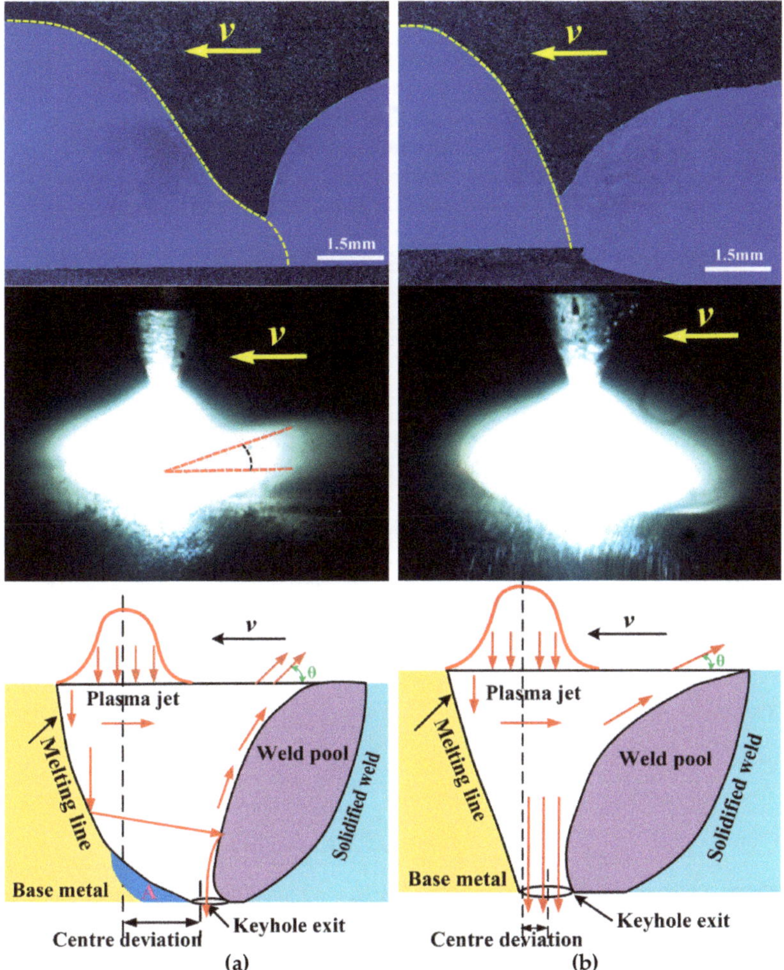

Figure 4. Weld longitudinal cross section and arc shape; (**a**) Test 1; (**b**) Test 7.

3.3. Weld Transverse Cross-Section

The weld macrographs are depicted in Figure 5. It can be seen that both the face side and root side width increased with increasing heat input. More importantly, there existed slight hump in Tests 1 and 2 samples, as shown in Figure 5a,b. The hump decreased and even disappeared with increasing heat input, as shown in Figure 5c,d. This is consistent with the mitigation or elimination of undercut defect.

As mentioned above, the heat input is also believed to be a controlling factor in the formation of undercut defect. Also, the variation trend of heat input, hump and undercut formation are consistent. Thus, it is necessary to consider the interaction among them. As mentioned above, when a high heat input was used, the keyhole front wall was less curved, majority of the plasma gas jet spurted out via the keyhole exit and very small amount of them was reflected by the keyhole front wall towards the rear keyhole wall, as indicated in Figure 6a. Therefore, the face side weld formation was smooth with the absence of humping and evident undercut. While if the heat input was reduced, the keyhole wall on the front side bent against the travel direction and the distance between the welding torch and the keyhole exit would increase, as demonstrated previously in Figure 4a. In this case, the plasma gas flow field changed dramatically as depicted in Figure 6b. Almost all of the plasma jet was guided towards the lower region of the keyhole wall on the backside. As the welding current used in K-TIG welding process was relatively high, the pressure associated with the plasma jet flow could overwhelm the surface tension coming from the weld pool on the backside. Therefore, the trailing weld pool was severely deformed. Certain amount of the weld pool was pushed towards the face side, and hump appeared in the weld centre. This kind of hump would reduce the amount of liquid metal on the edges of the weld pool on the face side, which may facilitate the formation of undercut. This implies that the heat input level is also an important parameter in determining the undercut formation.

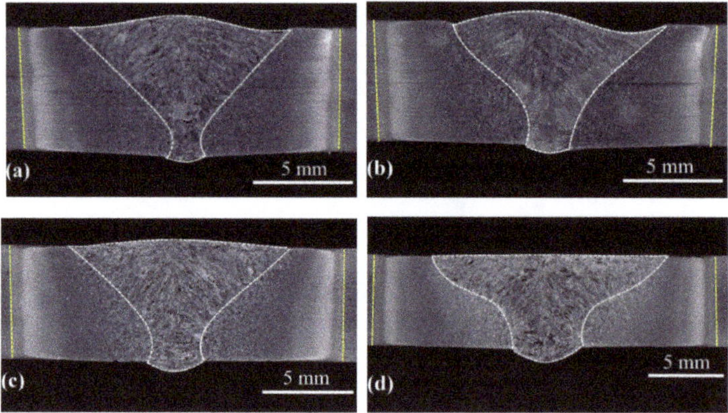

Figure 5. Macrographs of four typical welds with different heat input; (**a**) Test 1 (7.78 kJ/cm); (**b**) Test 2 (8.1 kJ/cm); (**c**) Test 4 (9.18 kJ/cm); (**d**) Test 7 (11.16 kJ/cm).

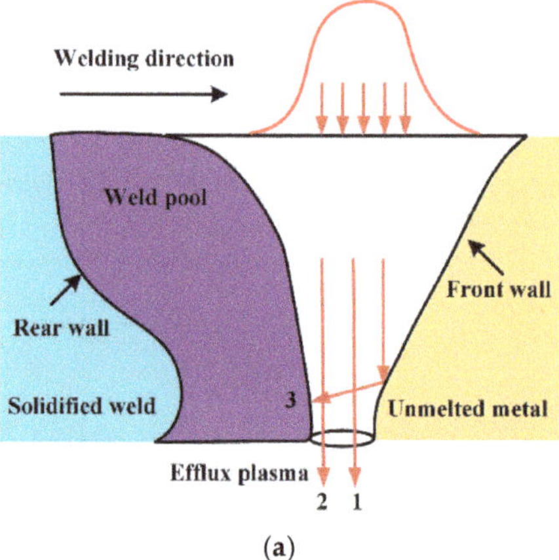

(a)

Figure 6. *Cont.*

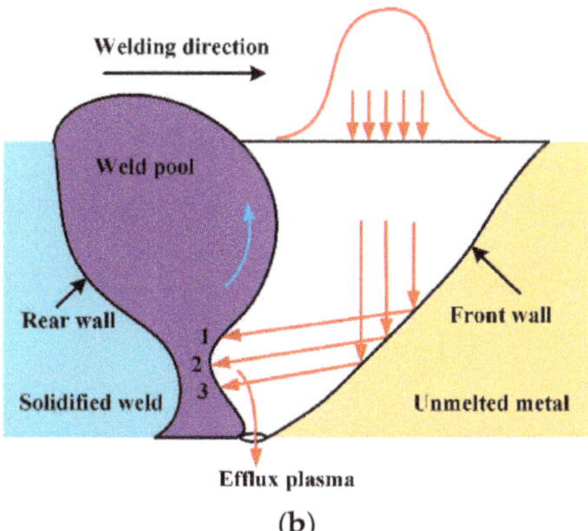

(b)

Figure 6. Schematic of plasma gas flow inside keyhole; (**a**) high heat input; (**b**) low heat input.

3.4. Weld Microstructure

The weld microstructure with different heat input is depicted in Figure 7. The weld metal in Test 1 consists of martensite and small amount of bainite, whereas only bainite with larger grain size is observed in the weld metal of Test 7. The same is true for the microstructure in the coarse grain heat-affected zone (CGHAZ) where a combination of martensite with bainite is present in Test 1 and only bainitic microstructure is found in Test 7. When it comes to fine grain heat-affected zone (FGHAZ), the microstructure consists predominantly of martensite. However, it is evident that the fraction of bainite in FGHAZ of Test 7 is higher than that of Test 1.

It is known that microstructure is closely linked to welding thermal cycle and hardenability. During welding thermal cycle, CGHAZ and FGHAZ were heated to austenite phase region. Upon cooling, the reformed austenite would transform to martensite, bainite or ferrite, depending on the cooling rate and hardenability of the base metal. Since the base metal contains large amount of carbon and chromium, the hardenability is relatively high, which favours the formation of martensite and/or bainite instead of ferrite. As the heat input increases, the cooling rate decreases, which favours the formation of bainite rather than martensite. The same is true for the weld metal as it still experiences phase transformation from austenite to various transformation products upon cooling, although it was melted during welding thermal cycle. That is why higher fraction of martensite was formed in the weld and HAZ with lower heat input.

3.5. Hardness

The hardness distribution of various welds is depicted in Figure 8. Lower hardness is observed in the weld metal region. The hardness increases when moving from the weld metal to the CGHAZ and FGHAZ. After that, the hardness decreases again to the lowest point when it comes to the over-tempered region. Finally, the hardness would increase continuously until it reaches the hardness value of the base metal (495 Hv). The hardness variation trend is very similar to that found in the armour steel weld produced with conventional fusion welding, as shown by Reddy et al. [23]. In addition, the overall hardness decreases with increasing heat input. Once the heat input reaches 10.22 kJ/cm and beyond, the overall hardness decrease becomes evident.

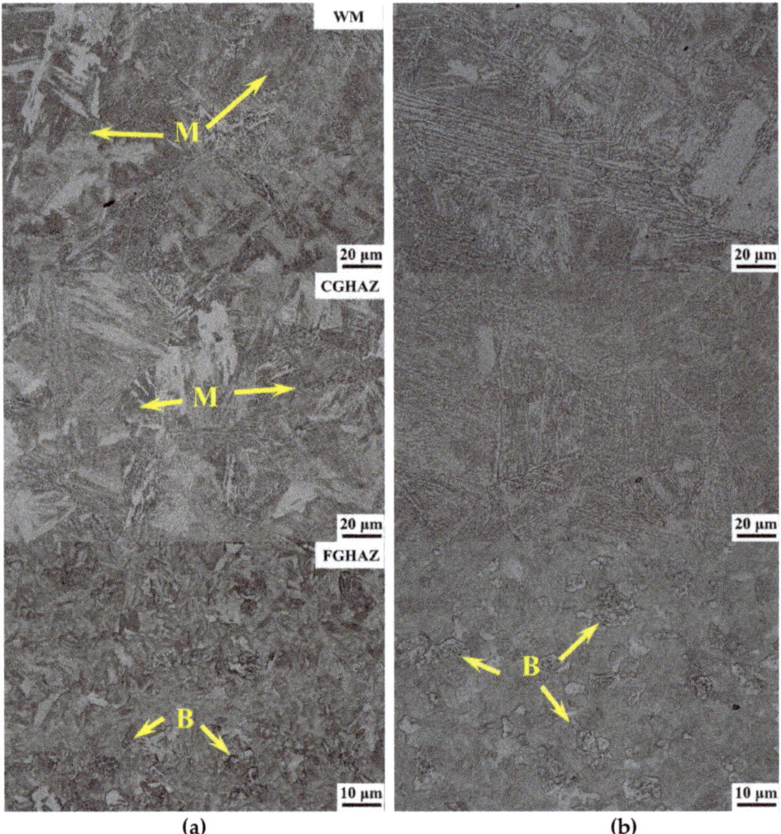

Figure 7. Microstructure in weld metal and heat-affected zone (HAZ); (**a**) Test 1 (HI = 7.78 kJ/cm); (**b**) Test 7 (HI = 11.16 kJ/cm) Note: M and B represent martensite and bainite respectively.

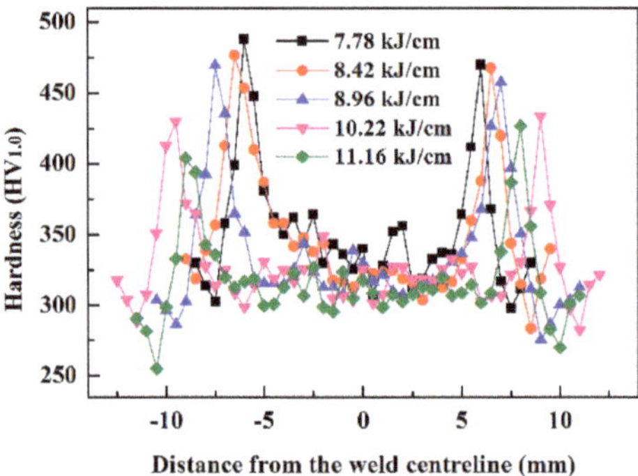

Figure 8. Hardness distribution across weld with varying heat input.

The variation trend of hardness across the weld is in good agreement with the microstructure change. As martensite and smaller grain size favours higher hardness, the increase in hardness from weld metal to FGHAZ is unexpected because of the increased fraction of martensite and decreased grain size from weld metal to FGHAZ, as shown in Figure 7. In the over-tempered region, the martensite was heavily tempered, which led to the reduction in dislocation density, precipitation of carbides and reduction in solid solution strengthening in the matrix. Thus, the hardness in the over-tempered region decreased dramatically compared with the HHA base metal (495Hv).

When the heat input was increased, the fraction of martensite decreased in both the weld metal and HAZ, as shown in Figure 7, because of the slower cooling rate induced from higher heat input. Furthermore, the softening behaviour in the over-tempered region became severer because of prolonged holding time in this region with increasing heat input, leading to more significant reduction in dislocation density, precipitation of carbides and reduction in solid solution strengthening. That is why the overall hardness of the welded joints decreased with increasing heat input.

3.6. Tensile Properties

The tensile results of the welded joints with different heat input are shown in Figure 9, while the fracture location of each welded joint is shown in Figure 10. Two tensile samples for each setting were used to present the fracture location on both face and root side, with the dotted yellow lines representing the weld metal region on both face and root sides. Figure 9b shows that the elongation of Test 1 (6.47%) and Test 3 (7.65%) is much lower than the other four samples although the ultimate tensile strength (UTS) remains at a high level. The fracture behaviour of these two samples is depicted in Figure 10a,b respectively. They both fractured from the face side undercut. Once the heat input was increased to more than 8.96 kJ/cm, as is the case in Tests 4–7 samples, the elongation reaches more than 10%. Although the Tests 4–7 tensile samples all fractured in the weld metal, the Tests 4 and 5 samples fractured in the weld metal from the root side fusion boundary (Figure 10c,d), while the fracture location of Tests 6 and 7 samples were randomly located in the weld metal (Figure 10e,f). It is worth noting here that the wavy ripples seem to have no evident effect on the tensile properties in Tests 6 and 7 samples.

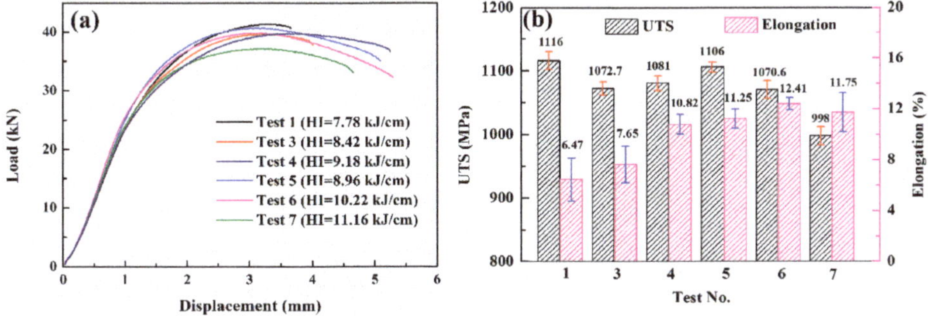

Figure 9. Tensile properties; (**a**) tensile curves for welded joints with different heat input; (**b**) tensile results for welded joints with different heat input.

It is well accepted that the tensile strength is closely linked to the overall hardness of the welded joint. With the increase in heat input from 8.96 kJ/cm to 11.16 kJ/cm, the overall hardness decreases, as shown in Figure 8, which is probably a result of the change in cooling rate and microstructure. In addition, higher heat input also tends to produce wider weld metal and HAZ, which could increase the width of reduced hardness region, as is the case in this study. Both these two factors can contribute to the decrease in tensile strength with increasing heat input. The significant reduction in elongation in Tests 1 and 3 is a result of undercut defect, which is a well-known stress concentration point. This can also be confirmed by fractography, as shown in Figure 11a,c and e. Note that the left-hand side on the fracture surface is the face side of the tensile sample, whereas the right-hand side is the root side of the tensile sample. The fracture surface in Test 1 sample consists predominantly of delamination both across two planes and on a single plane, as indicated by yellow and red arrows, respectively in

Figure 11a, as well as small amount of radial zone on the left. This means that after quick fracture initiation and propagation, the tensile sample experienced delamination fracture mode along the vertical direction down to the root side, as shown in Figure 11b. The recovery of elongation in Tests 4–7 samples indicates that the mitigation or elimination of undercut effect is an effective way to improve the tensile properties by reducing or eliminating stress concentration. However, it is worth mentioning that although the tensile properties of Test 4 were not affected by slight undercut in this study, it is still highly desirable to eliminate it in real fabrication. Even in the absence of undercut, the fracture still presents two different routes. As can be seen in Figure 11c, after crack initiated from the root side fusion boundary (dark grey area on the right-hand side), quick crack propagation occurred until reaching the face side, as evidenced by the completely radial zone in the fracture surface. The schematic is shown in Figure 11d. This indicates that the fusion boundary on the root side can also act as a stress concentration point in the absence of appreciable backside reinforcement, although it is less harmful than undercut. Thus, it is recommended that post-weld surface treatment be carried out for the K-TIG welded joint if the workpiece root side is accessible. With the formation of appreciable reinforcement on the root side, as is the case in Test 6, evident fibrous zone is present in the centre region, as indicated by the dotted yellow circle in Figure 11e. This is a result of plastic deformation occurring before crack initiated. In addition, there is larger amount of shear lip formed on the fracture surface, indicating that this sample followed a regular fracture route in which the stress condition changed to shear from tension and resulted in a fracture around 45 degree relative to the maximum stress direction after crack reached the edge. The schematic of fracture path is shown in Figure 11f.

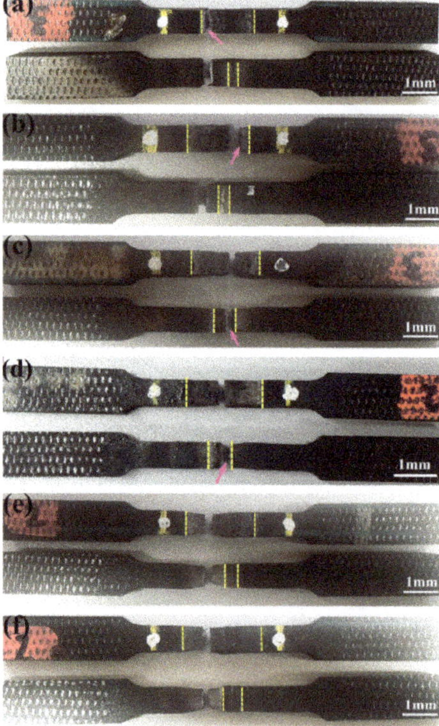

Figure 10. Fracture behaviour of welded joints; (**a**) Test 1 (HI = 7.78 kJ/cm); (**b**) Test 3 (HI = 8.42 kJ/cm); (**c**) Test 5 (HI = 8.96 kJ/cm); (**d**) Test 4 (HI = 9.18 kJ/cm); (**e**) Test 6 (HI = 10.22 kJ/cm); (**f**) Test 7 (HI = 11.16 kJ/cm).

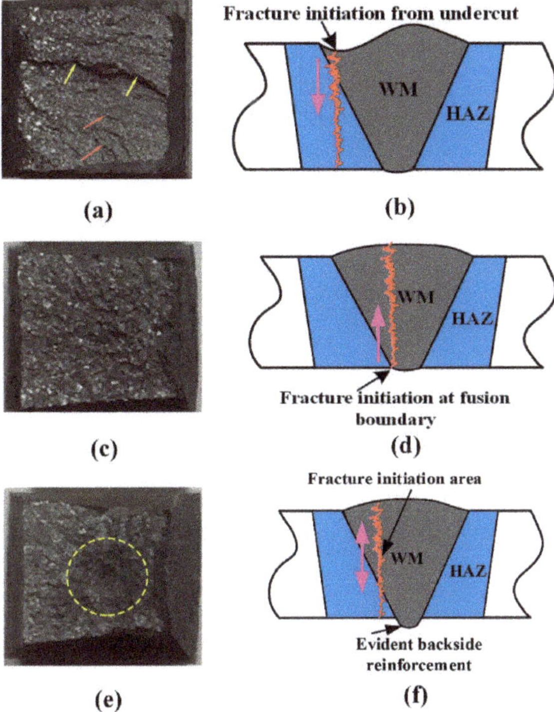

Figure 11. Schematic of fracture mechanism; (**a**,**c**,**e**): Fractography of Test 1 (HI = 7.78 kJ/cm), Test 5 (HI = 8.96 kJ/cm) and Test 6 (HI = 10.22 kJ/cm) respectively. (**b**,**d**,**f**) The corresponding schematic of fracture mechanism for these three samples.

It is worth noting that the much-reduced elongation would make the prediction of service life ineffective and lead to premature failure. In addition, as the stress concentration in the root side fusion boundary may also lead to premature failure in highly stressed or fatigue conditions, it is highly desirable and necessary to eliminate any stress concentration and avoid the formation of undercut defect in the K-TIG welded joint. It is suggested that the heat input level be carefully monitored during the welding process through either arc behaviour or other feedback signals described in the literature, together with appropriate use of welding speed. The results presented in this study can be useful in the real fabrication process and could be used as a guidance of parameter optimisation and quality assurance for the K-TIG welding process.

4. Conclusions

1. Undercuts were easily produced when the travel speed was fixed at 34.2 cm/min, while wavy ripples were experienced when the heat input was too high. The formation of hump in the weld centre is a result of low heat input, which changes the profile of weld longitudinal cross section and plasma trajectory inside the keyhole channel and leads to the exacerbation of undercut formation. This indicates that both travel speed and heat input are the contributing factors for undercut formation.
2. The tensile properties of the K-TIG welded joint were closely correlated to the weld formation. The undercut defect decreased the elongation considerably by imposing stress concentration and inducing delamination fracture mode. In addition, although the tensile properties were not appreciably affected by the height of root side reinforcement, stress concentration tended to occur

near the root side fusion boundary in the absence of appreciable reinforcement, which would also affect the regular fracture route. The wavy ripples produced with high heat input had no evident effect on the tensile properties. It is required that a combination of appropriate heat input level and travel speed be selected, leading to the absence of undercut defect, either appreciable root side reinforcement or a combination of post-weld surface treatment with the absence of appreciable root side reinforcement. It has been demonstrated that the welding parameters should be carefully controlled to avoid defect formation and maintain tensile properties in the K-TIG-welded joint.

Author Contributions: Conceptualization, Z.F., B.W. and D.D.; methodology, Z.F. and L.S.; software, D.D. and L.S.; validation, Z.P. and D.C.; formal analysis, Z.F., B.W. and D.C.; investigation, Z.F. and D.C.; resources, Z.P. and H.L.; data curation, D.D., H.L. and Z.F.; writing—original draft preparation, Z.F.; writing—review and editing, Z.P., H.L. and D.C.; supervision, D.C., Z.P. and H.L.; project administration, L.S. and H.L.; funding acquisition, Z.P and H.L.

Funding: This research has been conducted with the support of the Australian Government Research Training Program Scholarship. This paper includes research that was supported by DMTC Limited (Australia). The authors have prepared this paper in accordance with the intellectual property rights granted to partners from the original DMTC project. This work was partially supported by NSFC under grant No. 51805085.

Conflicts of Interest: The authors declare that they have no conflict of interest.

References

1. Jarvis, B.L. Keyhole Gas Tungsten Arc Welding: A New Process Variant. Ph.D. Thesis, University of Wollongong, Wollongong, Australia, 2001.
2. Jarvis, B.L.; Ahmed, N.U. Development of keyhole mode gas tungsten arc welding process. *Sci. Technol. Weld. Join.* **2000**, *5*, 1–7. [CrossRef]
3. Liu, Z.M.; Fang, Y.X.; Cui, S.L.; Yi, S.; Qiu, J.Y.; Jiang, Q.; Liu, W.D.; Luo, Z. Sustaining the open keyhole in slow-falling current edge during K-TIG process: Principle and parameters. *Int. J. Heat. Mass. Trans.* **2017**, *112*, 255–266. [CrossRef]
4. Liu, Z.M.; Fang, Y.X.; Cui, S.L.; Luo, Z.; Liu, W.D.; Liu, Z.Y.; Jiang, Q.; Yi, S. Stable keyhole welding process with K-TIG. *J. Mater. Process. Technol.* **2016**, *238*, 65–72. [CrossRef]
5. Liu, Z.M.; Fang, Y.X.; Cui, S.L.; Yi, S.; Qiu, J.Y.; Jiang, Q.; Liu, W.D.; Luo, Z. Keyhole thermal behavior in GTAW welding process. *Int. J. Therm. Sci.* **2017**, *114*, 352–362. [CrossRef]
6. Lathabai, S.; Jarvis, B.L.; Barton, K.J. Comparison of keyhole and conventional gas tungsten arc welds in commercially pure titanium. *Mater. Sci. Eng. A* **2001**, *299*, 81–93. [CrossRef]
7. Lathabai, S.; Jarvis, B.L.; Barton, K.J. Keyhole gas tungsten arc welding of commercially pure zirconium. *Sci. Technol. Weld. Join.* **2008**, *13*, 573–581. [CrossRef]
8. Cui, S.W.; Shi, Y.H.; Sun, K.; Gu, S.Y. Microstructure evolution and mechanical properties of keyhole deep penetration TIG welds of S32101 duplex stainless steel. *Mater. Sci. Eng. A* **2018**, *709*, 214–222. [CrossRef]
9. Fei, Z.Y.; Pan, Z.X.; Cuiuri, D.; Li, H.J.; Wu, B.T.; Ding, D.H.; Su, L.H.; Gazder, A.A. Investigation into the viability of K-TIG for joining armour grade quenched and tempered steel. *J. Manuf. Process.* **2018**, *32*, 482–493. [CrossRef]
10. Fei, Z.Y.; Pan, Z.X.; Cuiuri, D.; Li, H.J.; Van Duin, S.; Yu, Z.P. Microstructural characterization and mechanical properties of K-TIG welded SAF2205/AISI316L dissimilar joint. *J. Manuf. Process.* **2019**, *45*, 340–355. [CrossRef]
11. Cui, S.W.; Xian, Z.Y.; Shi, Y.H.; Liao, B.Y.; Zhu, T. Microstructure and Impact Toughness of Local-Dry Keyhole Tungsten Inert Gas Welded Joints. *Materials* **2019**, *12*, 1638. [CrossRef]
12. Fei, Z.Y.; Pan, Z.X.; Cuiuri, D.; Li, H.J.; Wu, B.T.; Su, L.H. Improving the weld microstructure and material properties of K-TIG welded armour steel joint using filler material. *Int. J. Adv. Manuf. Technol.* **2019**, *100*, 1931–1944. [CrossRef]
13. Fei, Z.Y.; Pan, Z.X.; Cuiuri, D.; Li, H.J.; Gazder, A.A. A Combination of Keyhole GTAW with a Trapezoidal Interlayer: A New Insight into Armour Steel Welding. *Materials* **2019**, *12*, 3571. [CrossRef]
14. Xie, Y.; Cai, Y.C.; Zhang, X.; Luo, Z. Characterization of keyhole gas tungsten arc welded AISI 430 steel and joint performance optimization. *Int. J. Adv. Manuf. Technol.* **2018**, *99*, 347–361. [CrossRef]

15. Liu, Z.M.; Fang, Y.X.; Qiu, J.Y.; Feng, M.N.; Luo, Z.; Yuan, J.R. Stabilization of weld pool through jet flow argon gas backing in C-Mn steel keyhole TIG welding. *J. Mater. Process. Technol.* **2017**, *250*, 132–143. [CrossRef]
16. Cui, S.L.; Liu, Z.M.; Fang, Y.Q.; Luo, Z.; Manladan, S.M.; Yi, S. Keyhole process in K-TIG welding on 4 mm thick 304 stainless steel. *J. Mater. Process. Technol.* **2017**, *243*, 217–228. [CrossRef]
17. Fang, Y.X.; Liu, Z.M.; Cui, S.L.; Zhang, Y.; Qiu, J.Y.; Luo, Z. Improving Q345 weld microstructure and mechanical properties with high frequency current arc in keyhole mode TIG welding. *J. Mater. Process. Technol.* **2017**, *250*, 280–288. [CrossRef]
18. Cui, Y.X.; Shi, Y.H.; Hong, X.B. Analysis of the frequency features of arc voltage and its application to the recognition of welding penetration in K-TIG welding. *J. Manuf. Process.* **2019**, *46*, 225–233. [CrossRef]
19. Zhu, T.; Shi, Y.H.; Cui, S.W.; Cui, Y.X. Recognition of Weld Penetration During K-TIG Welding Based on Acoustic and Visual Sensing. *Sens. Imaging* **2019**, *20*, 3. [CrossRef]
20. Feng, Y.Q.; Luo, Z.; Liu, Z.M.; Li, Y.; Luo, Y.C.; Huang, Y.X. Keyhole gas tungsten arc welding of AISI 316L stainless steel. *Mater. Des.* **2015**, *85*, 24–33. [CrossRef]
21. Eriksson, I.; Powell, J.; Kaplan, A.F.H. Measurements of fluid flow on keyhole front during laser welding. *Sci. Technol. Weld. Join.* **2011**, *16*, 636–641. [CrossRef]
22. Tomsic, M.J.; Jackson, C.E. Energy distribution in keyhole mode plasma arc welds. *Weld. J.* **1973**, 110s–115s.
23. Reddy, G.M.; Mohandas, T.; Papukutty, K.K. Effect of welding process on the ballistic performance of high-strength low-alloy steel weldments. *J. Mater. Process. Technol.* **1998**, *74*, 27–35. [CrossRef]

Review

Review on Friction Stir Processed TIG and Friction Stir Welded Dissimilar Alloy Joints

Sipokazi Mabuwa * and Velaphi Msomi

Faculty of Engineering and the Built Environment, Cape Peninsula University of Technology, P.O. Box 1906, Bellville 7535, South Africa; msomiv@gmail.com

* Correspondence: sipokazimabuwa@gmail.com; Tel.: +27-219538778

Received: 18 October 2019; Accepted: 10 December 2019; Published: 17 January 2020

Abstract: There is an increase in reducing the weight of structures through the use of aluminium alloys in different industries like aerospace, automotive, etc. This growing interest will lead towards using dissimilar aluminium alloys which will require welding. Currently, tungsten inert gas welding and friction stir welding are the well-known techniques suitable for joining dissimilar aluminium alloys. The welding of dissimilar alloys has its own dynamics which impact on the quality of the weld. This then suggests that there should be a process which can be used to improve the welds of dissimilar alloys post their production. Friction stir processing is viewed as one of the techniques that could be used to improve the mechanical properties of a material. This paper reports on the status and the advancement of friction stir welding, tungsten inert gas welding and the friction stir processing technique. It further looks at the variation use of friction stir processing on tungsten inert gas and friction stir welded joints with the purpose of identifying the knowledge gap.

Keywords: friction stir welding; tungsten inert gas welding; friction stir processing; dissimilar aluminium alloys joints; dissimilar metal joints

---

1. Introduction

Aluminium alloys are known to be good candidates for different applications in various fields like aerospace, food packaging, automotive industries, etc. Their good candidacy came from the fact that these metals are light in weight, have good mechanical properties, good corrosion resistance, etc. Various aluminium alloys possess different mechanical and thermal properties and these differences are influenced by the alloying elements used in producing each alloy [1,2]. Most industries are opting towards using dissimilar alloys in producing various components. This option is meant to reduce the costs that are involved in using similar alloys [3]. In as much as this approach is a cost-saving measure, however, there are also challenges associated with it. This includes the welding technique suitable at welding dissimilar alloys. The most used welding techniques involve tungsten inert gas (TIG) welding and friction stir welding (FSW). The TIG welding technique has been dominant in joining similar and dissimilar aluminium alloys until the birth of FSW.

There have been some challenges that were involved in joining dissimilar alloys through TIG technique. Those challenges include porosity, solidification cracking, thermal residual stresses, etc. These challenges have led to the discovering of the post-processing technique called friction stir processing. Friction stir processing (FSP) is a technique used to modify the microstructure of a metal through the use of a non-consumable rotating tool. FSP originated from friction stir welding which was initially established by The Welding Institute. FSP uses the same principle as FSW but does not join metals rather modifies the local microstructure in the near-surface layer of metals [4]. Figure 1 shows the schematic diagram of FSP and FSW techniques.

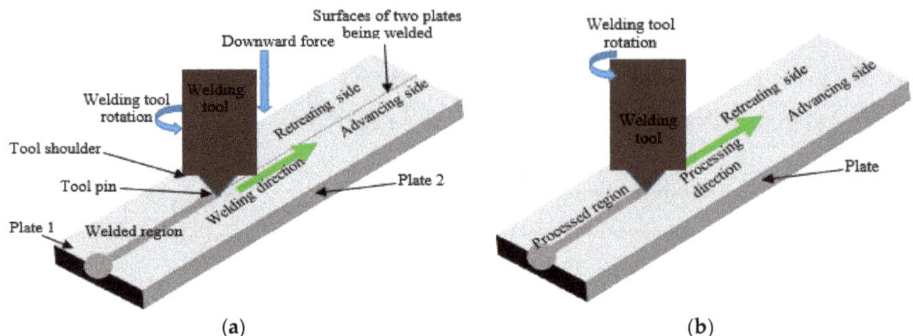

Figure 1. (**a**) FSW (friction stir welding) and (**b**) FSP (Friction stir processing) technique.

FSP works by plunging a specific cylindrical non-consumable tool into the plate and kept stationary for a few seconds. This is done so as to allow the stabilization in an input temperature required for processing. The rotating tool gets released so that it travels along the surface of the metal being processed. The tool travels from the start to the end of the plates resulting in the attainment of the processing. When the processing is finished, the tool is then unplunged, leaving a small hole, or rather travels to an offset distance to avoid leaving a hole. The side in which the tangential velocity of the tool surface is parallel to the traverse direction is called the advancing side, and the non-parallel side is called the retreating side [5,6].

This paper is aiming at reviewing the works that deals with the processing of similar and dissimilar joints produced by the TIG and FSW welding techniques.

2. Review on Friction Stir Welding, TIG Welding and Friction Stir Processing

2.1. Friction Stir Welding of Similar and Dissimilar Alloys

Recent studies have revealed that the material positioning during FSW of dissimilar plays an important role towards the strength of the weld. The good weld is produced when the hardest material is positioned on the retreating side while keeping the softer one on the advancing side during welding [7,8]. In an attempt towards analyzing the impact of material positioning during FSW dissimilar alloys, various studies have been performed in this regard. Dilip et al. [9] have performed the FSW of AA2219-T87 and AA5083-321 with the aim of performing the microstructural analysis of the joint. The weaker material (AA2219-T87) was positioned on the advancing side while the stronger material (AA5083-H321) was kept on the retreating side. The microstructural analysis revealed that the joint and the retreating side were dominated by the material which was placed on the advancing side (AA2219-T87). The microhardness value corresponding to the weaker material was observed on the retreating side where most tensile failure occurred.

Friction stir welding of the 3 mm thick AZ31B magnesium alloy and AA5052-H32 was performed by Taiki et al. [10]. The aluminium plate was positioned on the advancing side and magnesium (Mg) plate on the retreating side during welding. There was a variation in welding speed and tool speed. The microstructure analysis revealed that the joint was dominated by the AA5052-H32. It was also noted from the microstructural analysis that the dominating AA5052 had refined grains compared to parent material although the hardness value dropped compared to AA5052-H32 base metal. Hardness distributions of the cross-section revealed that the intermetallic compounds (IMCs) partly existed in the stir zone (SZ). All the samples failed at the center of the joint during tensile tests analysis. This failure location showed that the joint was dominated by the material that was positioned on the advancing side during welding.

Cavaliere & Panella, [11] conducted a study on the effect of tool position on fatigue properties of dissimilar 4 mm thick AA2024 and AA7075 plates joined by FSW. The AA2024 was positioned on the advancing side while AA7075 was kept on the retreating side. The joint attained when the tool was positioned 1 mm off the center (towards AA7075) had a higher hardness value compared to the joint attained when tool was 1.5 mm off the center of the weld. The maximum tensile properties of both joints were lower than the parent materials. Both joints revealed a ductile failure mode characterized by the presence of very fine dimples. The strong effect on fatigue crack growth was attributed to the positive fracture resistance (Kr) value measured on the cross-section of the different welds.

Peng et al. [12] performed friction stir welded on the 5 mm AA5A06-H112 and AA6061-T651 plates. This welding was performed under controlled cooling conditions i.e., forced air cooling (FAC) and natural cooling (NC) conditions. The AA5A06-H112 was positioned on the advancing side while AA6061-T651 was kept on the retreating side. The 0.5 MPa pressure was used to blow the air towards the welding direction which then intersected the surface of materials at angle of 30°. The microstructural analysis and microhardness test results for the joint produced under FAC were found to be higher compared to those that of the joint produced under NC condition. The tensile results for the joint produced under FAC condition were 10% higher than those produced at NC condition. The joint produced under NC condition had coarser grains compared to the joint produced under FAC condition. Both joints had ductile failure mode but the dimple size for the joint produced under FAC were higher than the joint produced under NC condition.

Shah et al. [13] investigated the influence of the tool eccentricity towards the friction stir welded dissimilar metals joint quality. They discovered that placing the stronger material on the advancing side improves the tensile strength and the percentage elongation of the joint. Their metallurgical analysis revealed that the tool eccentricity also plays a vital role towards the material flow however, there are some limitations when it comes to material mixing. The analysis of the joint formed when two dissimilar alloys are used in friction stir welding normally focuses on the mechanical properties. However, Giraud et al. [14] have gone to the extent of analyzing the compounds that are being formed during the FSW of dissimilar alloys. They have discovered that there are intermetallic compounds (IMCs) that are formed during the FSW of dissimilar alloys. These IMCs have a brittle nature which could lead to greater mechanical weakness.

Khodir and Shibayanagi [15] assessed the joint formed when AA2024-T3 was friction stir welded with AZ31 magnesium alloy. Their study involved the variation of welding speed at a constant rotational speed. The AA2024-T3 was located on the advancing side for all the welding. The microstructural analysis revealed that the increase in welding speed impacted the phase redistribution in the stir zone. The AA2024-T3 was distributed towards the lower regions of the stir zone while the AZ31 dominated the upper regions below the tool shoulder of the stir zone. The microstructural analysis also revealed a consistent formation of laminates structures in the SZ near the advancing side boundary between SZ and thermal affected zone (TMAZ) which were independent from welding speed variation. There were also intermetallic compounds that were formed in the SZ which contributed towards the fluctuation of the hardness distribution.

Rodriguez et al. [16] have friction stir welded AA6061-T6 and AA7050-T7451 with the purpose of assessing the microstructure and mechanical properties of the dissimilar welded joint. Their study involved the variation of rotational speed while keeping the welding speed constant. The AA7050-T7451 was positioned on the advancing side while AA6061–T6 was kept on the retreating side during welding. Tensile analysis revealed that the joints produced at lower speed were weaker than the base metals hence the fracture occurred at the SZ. The joints that were produced at higher rotational speed were stronger than AA6061-T6 base metal hence the fracture occurred consistently towards the AA6061-T6. The variation in fracture location was found to be directly linked with the material mixing at the SZ. The microstructural analysis revealed the ductile mode of failure. Moreover, the energy dispersive X-ray spectroscopy (EDS) results reveal the existence of three distinct layers where layer 1 had a nominal composition of AA6061, layer 2 had a composition of AA7050 and layer 3 had the combination of the two. Similar results were reported by Gou et al. [17] when they performed FSW on dissimilar AA6061-AA7075.

Mofid et al. [18] performed a study on the friction stir welding of the 3-mm thick AZ31C-O magnesium alloy to AA5083 in air and under nitrogen liquid. Their study involved the tracking of the temperature profile during welding and they attained this through the installation of thermocouples. There was a notable decrease in IMCs formation for the joints produced under liquid nitrogen compared to joints produced through air. The X-ray diffraction (XRD) analysis results exhibited the intermetallic phases of Al_3Mg_2, $Al_{12}Mg_{17}$ and Al_2Mg_3. The stir zone of the welds produced under nitrogen atmosphere showed a smoother interface compared to welds produced through air atmosphere. The attained maximum temperature during the welding was 676 K and 651 K respectively during air weld and under water weld.

Friction stir welding of AA2024-T365 and AA5083-H111 was performed by El-Hafez and El-Megharbel [19]. Variation in process parameters and pin profiles were employed with the purpose of analyzing their influence to the microstructure and tensile properties. The stronger material (AA2024-T365) was positioned on the advancing side throughout the welding. The combination of the highest speeds of 1120 rpm and 1400 rpm with 80 mm/min achieved the best strength and joint efficiency of 90% and this was due to sufficient heat being generated. Square pin profile produced higher strength joints compared to triangular and stepped profiles. Locating AA2024 on the advancing side (AS) played a significant role towards joint strength improvement. Cole et al. [20] also reported that the material placed on the advancing side dominates a major portion of the weld zone.

Vivekanandan et al. [21] used vertical milling machine for the friction stir welding of AA6035 and AA8011 with the aim of evaluating the mechanical properties of the dissimilar weld joint. The varying welding speed at a constant rotational speed was employed throughout the welding. The welds produced at the welding speed of 60 mm/min were found to be the best results compared to other speed combinations. This parameter combination produced fine grains at the center of the weld which contributed to the increase in hardness value. The dissimilar friction stir welding of undiluted copper and AA1350 sheet with a thickness of 3 mm was investigated by Li et al. [22]. The AA1350 was placed on the advancing side throughout the welding performance. The microstructural results in the nugget zone showed the vortex-like pattern and lamella structure. There was no formation of IMCs in the nugget zone. The hardness dispersion revealed that the hardness on the copper side was higher than that on the AA1350 side and the hardness at the bottom of the nugget was generally higher than those previously mentioned. The tensile properties of the dissimilar welds were all lower than those of the base metals. A ductile-brittle mixed fracture surface was observed on the dissimilar joints of the tensile tested specimens.

Friction stir welding was applied on the 1.3 mm thick stiffened AA2024-T3 panels with the aim to analyze the crack growth behaviour [23]. The experimental tests were correlated to linear elastic finite element method and dual boundary element method (DBEM). It was found that the DBEM showed better results and accurate as the stress level increased as the crack was approaching the stiffener. A Similar study was conducted by Citarella et al. [24] using a hybrid technique to assess the fatigue performance of multiple cracked friction stir welded AA2024-T3 joints. The crack propagation experimental tests were evaluated using the contour method in order to analyze the distribution of the residual. The metallographic analysis results showed a visible initial defect which led to initial crack for the simulation. The experimental fracture surface confirmed the crack propagation. The numerical crack was comparable to fatigue area shown by the post- mortem fractography.

Sheng et al. [25] used friction stir welding technique to join the AA6005—T4 plates with the purpose of investigating the weldability, microstructure and mechanical properties of the said alloy. The microstructural analysis results showed recrystallized grains in the nugget zone with equiaxed grain sizes of about 2.2 μm. The maximum ultimate tensile strength of about 174 MPa equivalent to 83.8% of the base material was obtained. A microhardness was reduced to 58 HV0.2 by the dissolution of phase β. In another study, an impact of using a bobbin type tool in friction stir welding of AA6082-T6 plates at different rotational speeds was investigated [26]. This investigation involved the variation of tool rotational speed. The tensile strength was found to be increasing linearly with rotational speed. However, it was discovered that the tensile strength reached its maximum when the rotational speed of 800 rpm was employed. The tensile strain of 7.9% was achieved at the same rotational speed. However, the strength and hardness were found to be having an inverse relationship with the increment of heat input at the speed beyond 800 rpm. The fractographic results showed a dimple fracture with white second phase particles of AlFeMnSi.

The 6 mm thick sheets of AA6061 and AA5086 were friction stir welded together to analyze the evolution of microstructure in the stir zone and its influence on tensile properties of the joints [27]. The welding parameters used were the rotational speed of 500 rpm, traverse speed of 35 mm/min and the axial force of 4.9 kN. The tensile properties of the joints correlated with microstructural features and microhardness values. The dissimilar joint exhibited a maximum hardness of 115 HV and a joint efficiency of 56% which was higher than the hardness of the base metals. This was attributed to the defect-free stir zone formation and grain size strengthening. Table 1 below give a tabulated review of the above literature. The idea behind the incorporation of this table is to show the typical positioning of the materials during welding of dissimilar materials and alloys. Table 1 also shows the mostly used tool material and tool profile in performing welding of dissimilar alloys/materials.

Table 1. Friction stir welding of dissimilar materials/alloys (RS—Retreating side, AS—Advancing side, SZ—Stir zone, TRS—Tool rotational speed, WS—Welding speed, El—Elongation, YS—Yield strength, JE—Joint efficiency, NS—Not specified.).

Material Used	FSW Tool	Welding Parameters	Material Positioning	Comments	Reference
AA2219-T87 and AA5083-H321	Material—M2 grade tool steel, straight cylindrical shape.	TRS—650 rpm, WS—55 mm/min, Axial load—9.8 kN.	AS—AA2219-T87, RS—AA5083-H321	Maximum UTS: 265 MPa, YS: 228 MPa, El 13, JE—61%	[9]
AA5052H and AZ31B	JIS SKD61 tool steel	Optimum: TRS—1000 rpm, WS—200 mm/min, Tool tilt 3°	AS—AA5052, RS—AZ31	The maximum UTS—147 MPa, YS—64 MPa, El—3.4%, JE—61%, microhardness—60 HV.	[10]
AA5052-H32 and AA6061-T651	Material—H13 steel, M6-threaded tool with tri-flats and an 8° taper.	TRS—1120 rpm, WS—90 mm/min, Tool tilt 2.5°, Dwell time 10 s	AS—AA6061 and RS—AA5052, AS—AA5052 and RS—AA6061	Better results with AA6061 on the AS. Maximum UTS—215.2 MPa, YS—141 MPa and El of 7.6%, microhardness SZ—~80 HV.	[13]
AA2024-T3 and AZ31	SKD61 Tool steel (threaded).	TRS—2500 rev/min, WS—200, 300, 400 and 550 mm/min,	AS—AA2024-T3 and RS—AZ31	In the HAZ AND TMAZ of AA2040A hardness distribution was significantly affected by increasing welding speed HAZ and TMAZ of 2024 A	[15]
AA5086-O and AA6061-T6	High speed steel, straight cylindrical, threaded cylindrical and tapered cylindrical	AA6061-AA6061: TRS—1300 rpm to 1200 rpm and WS—35 mm/min, axial force—6 kN, AA5086-AA5086: TRS—500 rpm, WS—5 mm/min, axial force—4.6 kN	AS—AA5086, RS—AA6061	Dissimilar joint at TRS of 500 rpm and WS of 10 mm/min. cylindrical plain tool resulted in Maximum UTS—140 MPa, YS—120 MPa, El—5.5%, JE—56%.	[27]

2.2. TIG Welding of Similar and Dissimilar Alloys and Metals

One of the most critical factor to consider for TIG welding is the filler metal, which mainly depends on the alloys to be welded. Ishak et al. [28] investigated the welding of dissimilar AA6061 and AA7075 using different filler metals, i.e., ER4043 (Si-reach) and ER5356 (Mg–reach). The depth analysis revealed that ER5356 penetrated deeper compared to the ER4043 and the depth of penetration plays an important role towards the strength of the weld or joint. The microstructural analysis revealed the existence of fusion zones that are normally identified on dissimilar TIG joints. The fusion zone (FZ) filled with ER5356 had finer grain sizes compared to the FZ filled with ER4043. Average hardness values for ER5356 filler specimens were higher compared to ER4043 filler specimens. TIG welding using the ER5356 filler yielded better joint compared to ER4043.

Borrisutthekul et al. [29] have evaluated the feasibility of using TIG welding technique in joining the dissimilar materials, i.e., steel plate and aluminium alloy plate. The microstructural analysis reveal that there was an existence of fusion zone (FZ) and heat affected zone (HAZ) which are the characteristics of the TIG welding. All the specimens were fractured on the same location during the tensile analysis, i.e., HAZ of aluminium alloy side. This type of behaviour was due to the growth of grain sizes that were observed through microstructural analysis. The existence of intermetallic reaction layers was also observed during the microstructural analysis.

Most studies that are studying different aspects of welding seem to be dominated by mostly two dissimilar aluminium alloys, i.e., AA5083 and AA6061 [30–35]. Waleed and Subbaiah [30] have evaluated the effect of using ER4047 filler rod in welding aluminium alloys AA5083-H111 and AA6061-T6. Similar analysis was also performed by other researchers with the focus on different aspects and different welding parameters on different grades of AA5083 [31–35]. Waleed and Subbaiah focused on analyzing the mechanical behaviour of the joint formed through the use of ER4047 filler rod. The tensile strength of the joint was lower than that of the base metals. The hardness value of the joint was varying in each side of the joint. This variation was caused by the formation of the magnesium-silicon (Mg_2Si) precipitates on the AA6061 side. The microstructural analysis showed the elongation of grains towards the rolling direction. There was also an existence of cavities and micro-pores at the intersection point of the weld. There was a notable decrease in ductility and this decrease was caused by the presence of columnar grain.

In another study, the AA2195 was joined using the ultrasonic assisted TIG welding with the aim of analyzing the weld characteristics in terms of size and porosity [36]. It was found that the pores existed at the weld adjacent to the surface. The size of the porosity was found to increase with the decrease in welding speed. Additionally, the increase in ultrasonic power resulted to the decrease in weld porosity. Wang et al. [37] used the TIG welding technique to fabricate the AA7A05-T6/AA5A06-O dissimilar joint in order to study the mechanical properties of the said joint. The results revealed that the dissimilar joint had a tensile strength which was 78.8% and yield strength of 97.24% of the base metal (AA5A06-O). The elongation was about 84.29%to that of the base metal AA7A05-T6. The microstructural analysis results showed a coarse grain sizes due to high heat input which resulted in hardness and strength drop.

Narayanan et al. [38] have evaluated the impact of TIG welding parameter variation on the AA5083 joints. The welding current and the shielding gas flow rate were the two parameters that were being varied for the duration of the study. The tensile results and the hardness value for the joint was lower than that of the commercial base metal. The microstructural analysis showed that the grains in the HAZ region were coarser compared to the base metal hence the brittle failure. The welding quality improvement of AA6031 plates using an automated TIG welding system was performed by Mohan [39]. The mechanical analysis showed that the joint performance was found to be way lower than that of the base metal. There was an inverse proportionality observed between the welding speed and the tensile strength of the joint. There was also a variation of hardness value across the weld.

Automated pulse TIG welding using AA5083 and AA6061 dissimilar plates was performed by Baghel & Nagesh, [40]. The main purpose was to evaluate the mechanical behaviour of the joint formed through this technique. The radiographical analysis revealed the presence of porosities which were caused by the lack of proper penetration. The tensile results for the joint were lower than the base metal. There was a variation of hardness value which was caused by non-uniformity of the grain sizes across the weld. The surface fracture exhibited the ductile failure mode. The impact of the welding speed variation towards the quality of the AA5083 TIG welded joints was analyzed by KumarSingh et al. [41]. All the other parameters were kept constant but only the welding speed that was varying. The tensile results showed a linear relationship with the welding speed until 100 mm/min. The notable decrease in tensile results was observed at the welding speed beyond 100 mm/min. The microstructure of the weld pool showed a refined grain size in comparison to the base metal.

TIG welding of dissimilar AA2014 and AA5083 was investigated by Sayer et al. [42]. One-sided TIG welding was applied with two passes. The microstructural analysis results in the weld region showed nonhomogeneous less equiaxed grain distribution with bigger diameters when compared to AA2014 and AA5083-O base metals. The grain size increase was said to be due to severe heat input. The tensile test results were lower than those of base metal. The tensile test specimens fractured in the welded region revealing brittle mode of failure. There was a variation in hardness across the weld with a sharp decrease at the center. This sharp decrease at the center was caused by the high silicon content in the filler material which dominated the center of the weld. Singh et al. [43] reported the mechanical properties of TIG welding at different parameters with and without the use of flux. The welding parameters used were all varied with the purpose of determining the optimal welding parameter combination. The variation of current effected the decrease in hardness value of the joint. The hardness values for joint formed with flux were higher compared to those formed without flux.

Dissimilar AA7075-T651 and AA6061-T6 plates were TIG welded with the aim of investigating the hardness of the center of the weld joint [44]. The Al-Si alloy filler wire type was used in performing all the welding. The maximum hardness value for the joint was found to be lower than that of the base metals. There were voids that were observed through microstructural analysis. Kumar et al. [45] also discovered that the use of pulsed current during TIG welding improves the mechanical properties of the welded joint in comparison to continuous current welded ones. This was found to be caused by the microstructural grain refinement which occurs in the fusion zone. Table 2 summarizes the mostly studied welding parameters during welding similar and dissimilar alloys/materials. It also shows the typical plates profile and filer wire used for different analysis.

Table 2. TIG welding of dissimilar materials/alloys. (WC—welding current, WS—welding speed, *Q*—gas flow rate, FZ—fusion zone, YS—yield strength, *V*—voltage, *F*—frequency, FW—filler wire, BM—base metal, UP—ultrasonic power, NS—not specified).

Material Used	Thickness	Welding Parameters	Comments	Reference
AA2195	2 mm	Ultrasonic TIG welding and ordinary TIG, *Q*—15 L/min, WC—70 A, WS—100 to 200 mm/min, UP—0 to 40 W, *F*—35 kHz	Porosity decreases with a decrease in welding speed for normal TIG. Porosity and pore size in UP TIG first decrease and increase with an increase in the UP.	[36]
AA5A06-O and AA7A05-T6	NS	Butt joint TIG, FW—ER5356, WC—260 A, *V*—25 V, WS—200 mm/min, *Q*—24 L/min, SG—99.99% argon.	The maximum UTS—78.87% BM, YS—97.24% BM, El—84.29% of BM, JE—61%, FZ had coarse grains resulted to 120 HV microhardness.	[37]
AA2014 and AA5083	5 mm	V joint, *Q*—10 L/min, SG—argon, *V*—14 V, WC—140 to 150 A, FW—TAL 4043	Maximum UTS—175 MPa, YS—128 MPa and El of 2.6%, microhardness SZ—~125 HV.	[42]
AA5083-O and AA6061-T651	6.35 mm	60° groove weld joint, FW—ER 5356, SG—pure argon, *V*—13.2 V, *Q*—6 L/min, WS—155 mm/min, WC—105 to 175 A, *F*—2 Hz.	FZ microhardness increased with higher cooling rate, finer dimples, higher YS and UTS.	[40]
AA5083	3 mm	Pulsed TIG, WC—118 to 134 A, WS—90 to 105 mm/min, *Q*—of 6 to 7 L/min	Maximum UTS at WC—134 A, with *Q*—7 L/min and WS—98 mm/min. Fused welding seams.	[41]

2.3. Friction Stir Processing of Similar/Dissimilar Alloys/Metals

Friction stir processing is a fairly new material processing technique. This then suggests that there are many works that are still in progress focusing in different aspects of this new technique. This involves the processing of plates and welded joints. The evaluation of the mechanical properties of the friction stir processed dissimilar AA2024 and AA6061 welded joint was performed by Hameed et al. [46]. The friction stir processed joints used were formed through the use of FSW technique. The authors did the mechanical analysis of the processed joint in comparison with the unprocessed joints. The parameters used in performing the processing were similar to those used to perform FSW. The tensile properties of the processed joint were higher than the unprocessed joint. The hardness value for the processed joint was higher than the unprocessed joint. The microstructural analysis for processed joint reveal finer grain sizes compared to the unprocessed one.

Karthikeyan and Kumar, [47] studied the relationship between process parameters and mechanical properties of a single pass friction stir processed AA6063-T6 plate. Processing was performed at different axial forces, traverse speeds and tool rotational speeds. The tensile results revealed a linear relationship with the axial force than any other parameter used. The improvement in ductility was found to be linearly depending on the axial force than on the other parameters. The impact of applying the FSP on 6 mm AA6056-T4 plates was investigated by Hannard et al. [48]. It is well known that the ductility of each material plays a very crucial role towards the formability of the material and the chosen material is mostly used in forming different components and structures. Hannard et al. discovered that the ductility of the plate increased with the increase in number of processing pulses. The existence of pores from the base metal was suppressed completely by the increase in multi-pass FSP. The multi-pass FSP was found to be the method in breaking the intermetallic particles and to redistribute them homogenously. Hannard et al. work have proven that the proper processing of the material occurs when the multi-pass FSP is used.

Mazaheri et al. [49] have shown the capabilities of FSP in producing surface composites. This capability was tested when they used the FSP technique to produce the $A356/A_2O_3$ surface composites. The microstructural analysis results of the $A356/A_2O_3$ indicated that A_2O_3 particles were well distributed in the aluminium matrix, and good bonding was also observed. The nanoindentation technique revealed that the microhardness for $A356/A_2O_3$ and $A356\text{–}nA_2O_3$ surface composites was higher than the samples processed without A_2O_3 particles and the as-received A356 material. A similar study was performed by Kalashnikova and Chumaevskii [50] where they used the FSP technique to develop surface composite between titanium carbide (TiC) and AA6082. The microhardness of the composite was found to be higher compared to the AA6082 metal. The tensile properties of the composite were found to be matching those of AA6082.

The effect of FSP on AA2024-T3 was studied by Hashim et al. [51]. The performance of the FSP was based on the pin-less cylindrical shoulder. The hardness results revealed that the application of FSP increased the hardness of the processed sample compared to the base material. There was also a notable increase in tensile properties of the processed sample compared to the base metal. The microstructural grain size was also refined compared to the base metal and this was found to be in correlation with tensile results.

The impact of FSP technique on the mechanical properties of cast Al-Si base alloy was analyzed by Tsai and Kao [52]. The tensile properties of cast AC8A alloy were improved after FSP, particularly the tensile elongation, which increased from <1% to 15.4%. FSP resulted in improvement of the tensile strength as the result of a combination of dissolution, coarsening and strengthening precipitates, which were attained by the FSP parameters. Jana et al. [53] investigated the FSP effect on fatigue behaviour of the cast Al–7Si–0.6 Mg alloy. The results showed five times improvement in fatigue life for a hypoeutectic Al–Si–Mg cast alloy. FSP eliminated the porosities and refined the silicon (Si) particles resulting in a decrease of the crack growth rate. In addition, FSP resulted in both break-ups of the dendritic microstructure and complex material mixing.

Kurt et al. [54] performed FSP on the aluminium alloy AA1050 to improve respective mechanical properties. Samples were subjected to the various tool rotating and traverse rates with and without silicon carbide (SiC) powders. The optimum processing parameters that were found to give better results were the rotational speed of 1000 rpm and traverse speed of 20 mm/min. The results revealed that FSP reduced the AA1050 grain size, which subsequently increased its hardness. A good dispersion of SiC was obtained and a good formation of composite layer. The hardness of the formed composite surfaces was improved significantly compared to that of base metal. Bending strength of the produced metal matrix composite was significantly higher than that of the processed specimen and untreated base metal.

The impact of using various chromium molybdenum (Cr-Mo) steel tool profiles in performing the FSP on AA2014 was studied by John et al. [55]. The hexagonal profile was found to be the best profile in producing good mechanical properties of the processed sample. The highest value for the hardness was also achieved through the hexagonal pin profile. The hexagonal pin profile was found to be suitable in producing highly refined grains compared to other profiles tested in the study. The influence of FSP on the microstructure and mechanical properties in terms of hardness for AA6061 sheet was investigated by Prakash and Sasikumar [56]. The cylindrical shaped high steel tool was employed in performing the multi-pass FSP. The microstructural evaluation reveal that the grain size of the processed specimens was 70% smaller than the base metal. There was also a linear relationship between the hardness value and the number of pulses used. The tensile properties were found to be linearly depending on the number of pulses used during FSP.

Sinhmar et al. [57] have analyzed comparatively the mechanical properties of the processed and unprocessed AA7039. The modified surfaces were characterized in respect to macrostructure, microstructure, hardness and tensile properties. The results showed an increase in ductility from about 13.5% to 23.6% while the ultimate and yield strength were adversely affected. The results showed higher ductility on the longitudinal direction than in traverse direction. The multi-pass friction stir processing produced higher hardness than the single pass one. Santella et al. [58] showed that FSP created a uniform distribution of broken second-phase particles of A319 and A356 and eliminated the coarse and heterogeneous structure of the alloys. The study was performed to assess the mechanical properties and reported that the tensile and fatigue behaviour of the material were improved with friction stir technique. The transmission electron microscopy (TEM) observations revealed the generation of a fine-grained structure of 5–8 μm for FSP A356. Furthermore, TEM examinations revealed that the coarse Mg_2Si precipitates in the as-cast A356 sample disappeared after FSP, indicating the dissolution of most of the Mg_2Si precipitates during FSP. FSP was found to be generally beneficial for dissolution of precipitates and structure homogenization [59].

Wrought AA5059 was friction stir processed by Izadi et al. [60] with the purpose of finding the best tool profile suitable for such class of aluminium alloy. Amongst the profile tested, the 3-flat threaded pin profile outshine the profile in all aspects. The microstructural analysis revealed that the average grain sizes were about 1.24 μm and this size was far less than the grains of the base metal. This grain size contributed towards the improvement of the microhardness. The yield strength and the ultimate tensile stress were also found to be higher than the base metal. The percentage elongation was also found to be higher than 25%. Ni et al. [61] have used FSP to modify the surface of cast Mg-9Al-1Zn alloy. The processed specimens were found to be defect free with fine-grain microstructure dominated by fine β-Mg17Al12 particles. The fatigue properties of the processed specimens were found to be higher than the base metal. The employment of FSP resulted to the transformation of quasi-cleavage fracture to dimple fracture. It was also found that the employment of FSP brought about the suppression of porosities and coarse β particles.

Sakurada et al. [62] were the first to perform a study on underwater FSW on AA6061. Their results showed that it was possible to generate enough friction for processing even though the workpieces were underwater. The stirred region of the underwater weld joint showed a finer microstructure in comparison to the one exposed to room temperature air conditions. The hardness of the underwater specimens was found to be relatively higher than those of the room temperature based specimens. Hofmann & Vecchio [63] studied the effect of submerging FSP on the grain size of AA6061-T6 compared to in air FSP. Their results showed that more grain refinement was attained under submerged conditions due to a faster cooling rate. They also demonstrated the feasibility of predicting the grain size of the processed specimens through the use of boundary migration model.

Zhang et al. [64] performed the joint analysis produced through the processing that was performed under water. They used variation in rotational speed in assessing the joint quality. They discovered that the fracture of the underwater joints was mostly dependent on the tool rotational speed. Their tensile analysis results showed a linear relationship with the rotational speed. Darras & Kishta [65] investigated the friction stir processing of AZ31 magnesium alloy in normal and submerged conditions. There were three condition used in performing the analysis of the joints i.e., air, hot underwater and cold underwater. The grain size for the cold underwater specimens was relatively smaller than the specimens produced at other different conditions. The thermal results revealed that the highest peak temperature for weld was in air-based processing compared to the other conditions. The longest processing duration was found when the processing was performed on air. Sabari et al. [66] have performed similar study on different material and processing parameters. They reported that the higher temperature gradient (along transverse and longitudinal weld axis) and higher cooling rate in underwater friction stir welds were a result of uniform heat absorption capacity of water when compared to the air-cooled welds.

El-Danaf et al. [67] have used commercial AA5083 rolled plates in analyzing the impact of FSP towards the ductility and the grain size of the processed specimens. The microstructure analysis showed a fine grain and an average disorientation angle of about 24°. Ductility was enhanced with a factor ranging between 2.6 and 5 when compared to the base metal. The strain rate sensitivity of the processed material was 0.33 while for the base metal was 0.018. Akinlabi et al. [68] investigated the effect of the tool rotational and traverse speeds as well as the number of passes on tribological characteristics of the modified surfaces. The FSPed samples exhibited lower wear rates than the as-cast A390 hypereutectic Al–Si alloy. The wear rates were found to decrease by reducing the tool rotational speed while increasing the tool traverse speed. There was a notable inverse correlation between the wear rate and the number of FSP passes.

Toma et al. [69] investigated the effect of FSP tool cutting depth on the mechanical properties of AA6061-T6. The cylindrical tool without the pin was employed in performing this analysis. The hardness was found to be increasing with the increase in cutting depth. The engineering flaws granules became smaller and the size of these granules increased with cutting depth. The tensile properties of the processed specimens were found to be improving with the increase in cutting depth. Abrahams et al. [70] investigated the properties and microstructure of friction stir processed 7075-T651 using various tool designs. Trials were conducted on AA5005-H34 with the aim of determining the most suitable FSP tool design out of all the considered pin profiles. Fully recrystallized fine microstructure and a defect free processed zone were achieved through the use of some of the FSP pin profiles. The grain sizes were reduced from the initial 192 μm pancake-like microstructure for AA5005-H34 base material to the range between 10 and 20 μm in the processed regions. The similar behaviour was also observed on the case of AA7075-T651. The traverse speed had a greater influence on the microhardness and mechanical properties compared to the tool rotational speed. It was also discovered the traverse speed suppressed the precipitates free zones which have negative impact towards the mechanical properties of the material.

The effect of the processing parameters of friction stir processing on the microstructure and mechanical properties of AA6063 was performed by Zhao et al. [71]. Post FSP produced fine equiaxed α-Al grains formed in the weld nugget of AA6063. The size of those α-Al grains was increasing with the increase in rotational speed. Tunnel defects were observed in the TMAZ region for a low tool rotation speed. When the rotational speed exceeded 700 rpm, a good combination interface was formed between the weld nugget (WN) and the TMAZ. Electron backscatter diffraction results showed that the fraction of the high-angle grain boundary was increased after FSP in the WN. The TEM analysis results showed that the densities of needle-shaped precipitates were reduced in the WN. There was an observed linear relationship between the ultimate tensile strength (UTS) and the tool rotational speed.

Rouzebehani et al. [72] have used AA7075 plate to perform FSP underwater and room temperature with the purpose of analyzing the metallurgical and mechanical properties. The variable process parameters were used. The temperature during FSP was monitored and recorded using the K-type thermocouple placed underneath the plate close to the abutting line of the workpiece. The average grain and precipitate sizes of the weld nugget zones were significantly reduced by the submerged conditions. The best metallurgical and mechanical properties were achieved when the processing was performed under water. There are numerous attempts that are being reported where the FSP technique is being utilized to produce surface composite. These attempts look into different alloys of aluminium and different dopants. Singh et al. [73] have produced surface composite through the use of FSP technique. The approach used by Singh et al. was to deposit SiC particles inside the holes drilled on the surface of AA6063 plate. The microhardness of the fabricated composite was relatively high compared to the one for the base metal. It was discovered that the increase in microhardness was due to the pining effect of hard SiC particles. The good bonding between the SiC particles and AA6063 results to the improvement of tensile strength of the composite when compared with base metal.

The microstructural modification of AA206 through the use of FSP was also reported by Sun et al. [74]. This modification was performed so as to comparatively evaluate the mechanical properties of processed and unprocessed AA206 material. A 6.26 mm and 16 mm thick plates were used for tensile and fatigue test respectively. The two key processing parameters were tool rotation speed and tool traverse speed. The results showed an increase in both yield strength and UTS after FSP when compared to those of the base metal. There was a notable improvement in yield strength and UTS on the processed plates compared to the base material. The percentage of elongation and fatigue strength also increased compared to the unprocessed ones.

Thakral et al. [75] used FSP to enhance the tensile properties and hardness of the TIG welded AA6061-T6 joint. Tensile results showed that the average UTS value for the base metal was 299 MPa, 85 MPa for the TIG welded joint and 125 MPa for the friction stir processed (FSPed) TIG welded joint. An increase of 48% was reported on the UTS on the TIG welded joint. The hardness values of FSP TIG specimen ranged from 72–74 HV which was almost similar to that of base metal which was 74 HV whereas in TIG specimen hardness value ranged from 66–68 HV. Microstructural analysis was performed on the weld zone to evaluate the effect of welding parameters on welding quality and grain structure. The microstructure of FSPed TIG joint showed very fine equiaxed recrystallized grains compared to the microstructure of TIG joint.

The effect of a single pass FSP on the mechanical properties and microstructure of the commercially pure aluminium was investigated by Yadav and Bauri [76]. The grain size of the FSPed specimens were way smaller than those of the base metal. The TEM results showed fine grains with well-defined boundaries. The tensile results showed UTS increase of about 25% while the ductility decreased by about 10%. The impurity particles observed in TEM resulted in the yield strength decrease. The hardness also improved substantially compared to the base metal. Feng et al. [77] investigated the effect of SFSP on the microstructure of the AA2219 sheet. The grain size on the stir zone was less than that of the base metal (BM). The area fraction of the ultra-fine grains in the stir zone increased as heat input decreased. The results showed a decrease in microhardness of the SFSP stir zone compared to that of the unprocessed BM. The processed zone exhibited microhardness that was higher than that of the base metal.

The 6-mm thick aluminium alloy AA6082 was subjected to underwater FSP to test the changes in the UTS [78]. The high carbon high chromium steel rod of diameter 20 mm material was used as the processing tool for this investigation. The result revealed that the maximum tensile strength of the underwater joints was higher than that of the normal air. The effect of SFSP on the mechanical and microstructural properties of 10 mm thick AA7075 was investigated by Nourbakhsk and Atrian [79].

A thermocouple was used to record the temperature of water during the processing. The single pass FSP was used in carrying out the analysis. The results obtained from the submerged processing were similar to those obtained by other researchers [64,65,72]. Mabuwa and Msomi [80] used a single pass FSP to improve the mechanical properties of the TIG and friction stir welded AA5083-H111 joints. The processing was performed under normal room conditions. The FSPed joints showed better mechanical properties compared to the unprocessed ones. The DRX that happened during FSP resulted in ultra-fine grain refinement of the FSPed joints. Table 3 presents typical summary of the friction stir processed literature with the purpose of showing the mostly affected material property post friction stir processing. Table 4 shows different types of tool that are used on FSP extracted from cited literature. Figure 2 shows typical stress strain curves for the unprocessed and friction stir processed joints.

Table 3. Typical results of friction stir processed plate and joints. (TS—traverse speed, RTS—tool rotational speed, UTS—ultimate tensile strength, SZ—stir zone, NS—not specified.).

Material Used	FSP Tool	Condition	Surface/Joint	Processing Parameters	Enhanced Property	Reference
AA6063-T6	NS	Normal	surface	Optimum: TRS—1200 and 1400 rpm, TS—40.2 mm/min, axial force—10 kN	UTS, ductility, microhardness	[47]
AA6061	EN 31 steel	Normal	TIG welded joint	TRS—1200 rpm, TS—75 mm/min	UTS, microhardness	[75]
Pure Al (99.2%)	M2 steel	Normal (air)	surface	TRS—640 rpm, TS—150 mm/min	significant improvement in UTS and hardness. Marginal increase in ductility	[76]
AA2219-T6	standard tool steel	Submerged	Surface	TRS—600 to 800 rpm, TS—200 mm/min, Tool tilt 2.5°	As the TRS increase the SZ hardness decreases; Refinement of grains.	[77]
AA7075	NS	Submerged and normal (air)	Surface	TRS 800 and 1250 rpm, TS and 40 and 63 mm/min	Ductility; Tensile strength; Uniform grain sizes	[79]
AA5083-H111	high-carbon steel	Unsubmerged	TIG joint, FSW joint	TRS—1000 rpm and TS—30 mm/min Tool tilt 2°	Ductility; Grain sizes refined and UTS	[80]

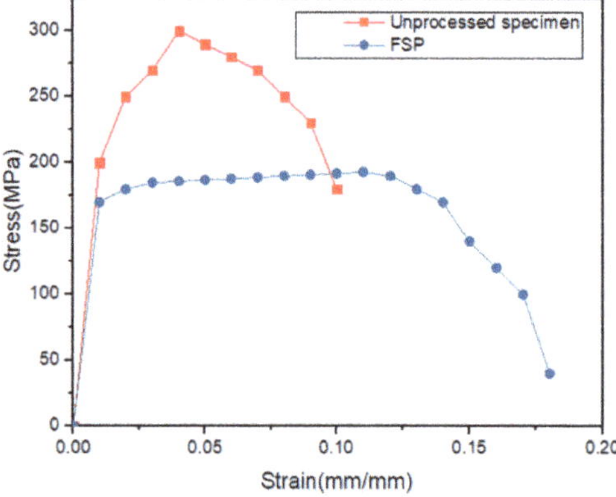

Figure 2. Tensile stress and strain curves for processed and unprocessed surface.

Table 4. Typical FSP tools. (PL—Pin length, PD—Pin diameter, SD—Shoulder diameter, SL—Shoulder length, PH—Pin height, NS—Not specified, SBPL—Square base pin length, CHL—Conical head length).

Material Processed	Tool Material	Tool Profile	Tool Dimensions	Reference
A356 Aluminium	H13 steel	Threaded	PL: 4 mm, PD: 3.6 mm, SD: 18 mm, SL: 10 mm.	[49]
AC8A alloy	NS	NS	SD—18 mm, PD—5.2 mm, PH—2.7 mm,	[52]
Cast F357	NS	Conical pin with a stepped spiral feature	PD—4 mm, PH—2 mm, SD—12 mm,	[53]
AA6061-T6	Alloy steel	Concave shoulder	SD—15 mm,	[69]
AA6063-T5	H13 steel	Conical pin	SD—18 mm, PL—5.7 mm, end PD—4 mm, Root PD—6 mm,	[71]
AA7075-T651 and AA5005-H34	NS	Square base pin with conical head	PL—3.2 mm, SD—14 mm, PD—3.6 mm, SBPL—2.3 mm, CHL—3 mm.	[70]
AA7075-T651 and AA5005-H34	NS	Pyramid shape	PL—3.2 mm, PD—3 mm,	[70]
	NS	Conical threaded	PL—3.2 mm, PD—3 mm.	

2.4. *Advantages of Friction Stir Welding, TIG Welding and Friction Stir Processing*

Table 5 presents the advantages of the friction stir welding, TIG welding and friction stir processing techniques.

Table 5. Advantages of Friction stir welding, TIG welding and friction stir processing (IMC—intermetallic compounds, BM—base materials, UFG—ultra-fine grains, DR—dynamic recrystallization, El—elongation, YS—yield strength, UTS—ultimate tensile strength).

Feature	Friction Stir Welding	TIG Welding	Friction Stir Processing
Environment	Considered as a green technology technique, due to its clean and environment-friendly. No toxic gas emission nor radiation involved [81,82].	Uses gas (helium/argon) [37,40,42].	Also a green technology since same principle as from FSW.
Surface finish	Good surface finish and no welds finishing costs [83–85]	Weld joint with pores and cracks [30,36,80]. Sometimes requires grinding to get good surface.	Good surface finishing especially with pin-less tool [86]. No pores or cracks on FSPed welded joint [80,87].
Microstructure evolution	uniform arrangement with fine grains [13–15,17,80].	Nonhomogeneous, coarse less equiaxed grain distribution with bigger diameters [30,37,42,80].	Thermal mechanical effects and DRX during FSP results in coarse grains transformation to UFG. [52,54,60,70,87–91].
Microhardness	Microhardness of the joint increased when compared to base BM ones [11,27]. But remains circumstantially.	Microhardness decreases [37–39].	FSP resulted in an increase in hardness when compared to the unprocessed surface [89,90,92,93].
Tensile properties (UTS, YS, El)	Tensile properties are mostly lower when welding dissimilar alloys due to the formation of IMC (circumstantial) compared to the BM ones [10–13,15,21,94].	Tensile properties decrease compared to BM ones [38,40,41].	FSP also improves the tensile properties of a material [54,56,68,87]. Improved ductility of the processed surface in comparison to the unprocessed one [80,95].

2.5. Mostly Processed Welded Structures

Based on the literature cited the most friction stir processing has been performed to enhance the properties of the base metal rather than welded joints. There are limited works [46,60] reporting on processing the friction stir welded joints with only [76,80] reported on using FSP as a post processing of the TIG and FSWed welded joints. No other searchable literature was obtained on post processing for both FSW and TIG welded joints. This then suggest that there are more opportunities of exploring the impact of employing FSP technique on TIG welded structures.

2.6. Mostly Processed Grades

The most processed aluminium alloy series based on the presented literature was 6xxx (AA6061, AA6082, AA6063, AA6056), with AA6061 taking a lead. Following the 6xxx was the 7xxx (AA7075, AA7039), 2xxx (AA2024, AA2219, AA2014) and the 5xxx (AA5005, AA5083, AA5059).

3. Concluding Remarks

In all the work that has been performed thus far, it has been noted that all the focus has been on FSP as an enhancement technique on aluminium alloys, magnesium, and other alloys. It is also noticed that the common mechanical properties analyzed include the tensile test, fatigue and microhardness. These properties are studied correlatively with the microstructure. Very few works thus far, which considers friction stir processing as a post weld processing technique for tungsten inert gas dissimilar alloy welded joint. There is minimal to no trace of any literature on submerged friction stir processing of tungsten inert gas and friction stir welded dissimilar alloy joints. Therefore, this opens an opportunity for the exploration on the impact of friction stir processing and submerged friction stir processing towards the properties of the friction stir welded and tungsten inert gas welded dissimilar joints.

Author Contributions: Both authors have fully contributed equally to all the work produced. All authors have read and agreed to the published version of the manuscript.

Funding: This research received no external funding.

Acknowledgments: The authors would like to thank the Cape Peninsula University of Technology for allowing this study to be performed.

Conflicts of Interest: The authors declare no conflict of interest. The authors declare there is no funding involved in this study.

References

1. Cam, G.; Kocak, M. Progress in joining of advanced materials. *Int. Mater. Rev.* **1998**, *43*, 1–44. [CrossRef]
2. Nicholas, E.D. Developments in the friction stir welding of metals. In Proceedings of the 6th International Conference on Aluminium Alloys, Toyohashi, Japan, 5–10 July 1998; pp. 139–151.
3. Patel, V.; Li, W.; Wang, G.; Wang, F.; Vairis, A.; Niu, P. Friction stir welding of dissimilar aluminum alloy combinations: State-of-the-art. *Metals* **2019**, *9*, 270. [CrossRef]
4. Sun, N. Friction Stir Processing of Aluminium Alloys. Master's Thesis, University of Kentucky, Lexington, KY, USA, 2009.
5. Marczyk, J.; Nosal, P.; Hebda, M. Effect of Friction Stir Processing on Microstructure and Microhardness of Al-TiC Composites. In Proceedings of the Student's Conference, Freiberg, Germany, 8–9 November 2018.
6. Węglowski, M.S. Friction stir processing—State of the art. *Arch. Civ. Mech. Eng.* **2018**, *18*, 114–129. [CrossRef]
7. Chaudhari, R.; Parekh, R.; Ingle, A. Reliability of Dissimilar Metal Joints Using Fusion Welding: A Review. In Proceedings of the International Conference on Machine Learning, Electrical and Mechanical Engineering (ICMLEME'2014), Dubai, UAE, 8–9 January 2014.
8. Simar, A.; Joncheere, C.; Deplus, K.; Pardoen, T.; de Meester, B. Comparing similar and dissimilar friction stir welds of 2017–6005A aluminium alloys. *Sci. Technol. Weld. Join.* **2013**, *15*, 254–259. [CrossRef]

9. Dilip, J.J.S.; Koilraj, M.; Sundareswaran, V.; Janaki Ram, G.D.; Koteswara Rao, S.R. Microstructural characterization of dissimilar friction stir welds between AA2219 and AA5083. *Trans. Indian Inst. Met.* **2010**, *63*, 757–764. [CrossRef]
10. Morishige, T.; Kawaguchi, A.; Tsujikawa, M.; Hino, M.; Hirata, T.; Higashi, K. Dissimilar welding of Al and Mg alloys by FSW. *Mater. Trans.* **2008**, *49*, 1129–1131. [CrossRef]
11. Cavaliere, P.; Panella, F. Effect of tool position on the fatigue properties of dissimilar 2024-7075 sheets joined by friction stir welding. *J. Mater. Process. Technol.* **2008**, *206*, 249–255. [CrossRef]
12. Peng, G.; Yan, Q.; Hu, J.; Chen, P.; Chen, Z.; Zhang, T. Effect of forced air cooling on the microstructures, tensile strength, and hardness distribution of dissimilar friction stir welded AA5A06-AA6061 joints. *Metals* **2019**, *9*, 304. [CrossRef]
13. Shah, L.H.A.; Sonbolestan, S.; Midawi, A.R.H.; Walbridge, S.; Gerlich, A. Dissimilar friction stir welding of thick plate AA5052-AA6061 aluminum alloys: Effects of material positioning and tool eccentricity. *Int. J. Adv. Manuf. Technol.* **2019**, *105*, 889–904. [CrossRef]
14. Giraud, L.; Robea, H.; Claudina, C.; Desrayaud, C.; Bocherd, P.; Feulvarcha, E. Investigation into the dissimilar friction stir welding of AA7020-T651 and AA6060-T6. *J. Mater. Process. Technol.* **2016**, *235*, 220–230. [CrossRef]
15. Khodir, S.A.; Shibayanagi, T. Dissimilar friction stir welded joints between 2024-T3 aluminum alloy and AZ31 magnesium alloy. *Mater. Trans.* **2007**, *48*, 2501–2505. [CrossRef]
16. Rodriguez, R.I.; Jordon, J.B.; Allison, P.G.; Rushing, T.; Garcia, L. Microstructure and mechanical properties of dissimilar friction stir welding of 6061-To-7050 aluminum alloys. *Mater. Des.* **2015**, *83*, 60–65. [CrossRef]
17. Guo, J.F.; Chen, H.C.; Sun, C.N.; Bi, G.; Sun, Z.; Wei, J. Friction stir welding of dissimilar materials between AA6061 and AA7075 Al alloys effects of process parameters. *Mater. Des.* **2014**, *56*, 185–192. [CrossRef]
18. Mofid, M.A.; Abdollah-Zadeh, A.; Gürza, C.H. Investigating the formation of intermetallic compounds during friction stir welding of magnesium alloy to aluminum alloy in air and under liquid nitrogen. *Int. J. Adv. Manuf. Technol.* **2014**, *71*, 1493–1499. [CrossRef]
19. El-Hafez, H.A.; El-Megharbel, A. Friction stir welding: Dissimilar aluminum alloys. *World J. Eng. Technol.* **2018**, *6*, 408–419. [CrossRef]
20. Cole, E.G.; Fehrenbacher, A.; Duffie, N.A.; Zinn, M.R.; Pfefferkorn, F.E.; Ferrier, N.J. Weld temperature effects during friction stir welding of dissimilar aluminum alloys 6061-T6 and 7075-T6. *Int. J. Adv. Manuf. Technol.* **2014**, *71*, 643–652. [CrossRef]
21. Vivekanandan, P.; Arunachalam, V.P.; Prakash, T.; Savadamuthu, L. The experimental analysis of friction stir welding on aluminium composites. *Int. J. Metall. Eng.* **2012**, *1*, 1–6.
22. Li, X.; Zhang, D.; Qiu, C.; Zhang, W. Microstructure and mechanical properties of dissimilar pure copper/1350 aluminum alloy butt joints by friction stir welding. *Trans. Nonferrous Met. Soc. China* **2012**, *22*, 1298–1306. [CrossRef]
23. Sepe, R.; Armentani, E.; Di Lascio, P.; Citarella, R. Crack growth behavior of welded stiffened panel. *Proc. Eng.* **2015**, *109*, 473–483. [CrossRef]
24. Citarella, R.; Carlone, P.; Lepore, M.; Sepe, R. Hybrid technique to assess the fatigue performance of multiple cracked FSW joints. *Eng. Fract. Mech.* **2016**, *62*, 38–50. [CrossRef]
25. Sheng, X.; Li, K.; Wu, W.; Yang, W.; Liu, Y.; Zhao, Y.; He, G. Microstructure and mechanical properties of friction stir welded joint of an aluminum alloy sheet 6005A-T4. *Metals* **2019**, *9*, 1152. [CrossRef]
26. Li, Y.; Gong, W.; Sun, D. Effect of tool rotational speed on the microstructure and mechanical properties of bobbin tool friction stir welded 6082-T6 aluminum alloy. *Metals* **2019**, *9*, 894. [CrossRef]
27. Ilangovan, M.; Boopathy, S.R.; Balasubramanian, V. Microstructure and tensile properties of friction stir welded dissimilar AA6061-AA5086 aluminium alloy joints. *Trans. Nonferrous Met. Soc. China* **2015**, *25*, 1080–1090. [CrossRef]
28. Ishak, M.; Noordin, M.N.F.; Shah, L.H.A. Feasibility study on joining dissimilar aluminum alloys AA6061 and AA7075 by tungsten inert gas (TIG). *J. Teknol.* **2015**, *75*, 79–84. [CrossRef]
29. Borrisutthekul, R.; Mitsomwang, P.; Rattanachan, S.; Mutoh, Y. Feasibility of using TIG welding in dissimilar metals between steel/aluminum alloy. *Energy Res. J.* **2010**, *1*, 82–86.
30. Waleed, W.A.; Subbaiah, K. Effect of ER4047 filler rod on tungsten inert gas welding of AA5083-H111 and AA6061-T6 aluminium alloys. *JCHPS* **2017**, *7*, 210–213.

31. Sefika, K. Multi-response optimization using the Taguchi based grey relational analysis: A case study for dissimilar friction stir butt welding of AA6082-T6/AA5754-H111. *Int. J. Adv. Manuf. Technol.* **2013**, *68*, 795–804.
32. Subbaiah, K.; Geetha, M.; Sridhar, N.; Koteswara Rao, S.R. Comparison of Tungsten Inert Gas and Friction Stir Welding of AA 5083-H321 Aluminium Alloy Plates. Trends in Welding Research. In Proceedings of the 9th International Conference, ASM International, Chicago, IL, USA, 4–8 June 2012.
33. Leitao, C.; Louro, R.; Rodrigues, D.M. Analysis of high temperature plastic behavior and its relation with weld ability in friction stir welding for aluminum alloys AA5083-H111 and AA6082-T6. *Mater. Des.* **2012**, *37*, 402–409. [CrossRef]
34. Menzemer, C.C.; Lam, P.C.; Wittell, C.F.; Srivarsan, T.S. A study of fusion zone microstructures of arc-welded joints made from dissimilar aluminum alloys. *J. Mater. Eng. Perform.* **2001**, *10*, 173–178. [CrossRef]
35. Palanivel, R.; Koshy Mathews, P.; Murugan, N. Optimization of process parameters to maximize ultimate tensile strength of friction stir welded dissimilar aluminum alloys using response surface methodology. *J. Cent. South Univ.* **2013**, *20*, 2929–2938. [CrossRef]
36. Chen, Q.; Ge, H.; Yang, C.; Lin, S.; Fan, C. Study on pores in ultrasonic-assisted TIG weld of aluminum alloy. *Metals* **2017**, *7*, 53. [CrossRef]
37. Wang, W.; Cao, Z.; Liu, K.; Zhang, X.; Zhou, K.; Ou, P. Fabrication and mechanical properties of tungsten inert gas welding ring welded joint of 7A05-T6/5A06-O dissimilar aluminum alloy. *Materials* **2018**, *11*, 1156. [CrossRef] [PubMed]
38. Narayanan, A.; Mathew, C.; Baby, V.Y.; Joseph, J. Influence of gas tungsten arc welding parameters in aluminium 5083 alloy. *IJESIT* **2013**, *2*, 269–277.
39. Mohan, P. Study the Effects of Welding Parameters on TIG Welding of Aluminum Plate. Master's Thesis, National Institute of Technology, Rourkela, India, 2014.
40. Baghel, P.K.; Nagesh, D.S. Mechanical properties and microstructural characterization of automated pulse TIG welding of dissimilar aluminum alloy. *IJEMS* **2018**, *25*, 147–154.
41. KumarSingh, S.; Tiwari, R.M.; Kumar, A.; Kumar, S.; Murtaza, Q.; Kumar, S. Mechanical Properties and Microstructure of Al-5083 by TIG. In Proceedings of the International Conference on Processing of Materials, Minerals and Energy, Ongole, India, 29–30 July 2018.
42. Sayer, S.; Yeni, C.; Ertugrul, O. Comparison of mechanical and microstructural behaviors of tungsten inert gas welded and friction stir welded dissimilar aluminum alloys AA 2014 and AA 5083. *Kovove Mater.* **2011**, *49*, 155–162. [CrossRef]
43. Singh, G.; Singh, F.; Singh, H. A Study of mechanical properties on TIG welding at different parameters with and without use of flux. *IJTIR* **2015**, *16*.
44. Patil, C.; Patil, H.; Patil, H. Experimental investigation of hardness of FSW and TIG joints of aluminium alloys of AA7075 and AA6061. *Frat. Integrità Strutt.* **2016**, *10*, 325–332. [CrossRef]
45. Kumara, S.T.; Balasubramanian, V.; Sanavullah, M.Y. Influences of pulsed current tungsten inert gas welding parameters on the tensile properties of AA 6061 aluminum alloy. *Mater. Des.* **2007**, *28*, 2080–2092. [CrossRef]
46. Hameed, A.M.; Resan, K.K.; Eweed, K.M. Effect of Friction Stir Processing Parameters on the Dissimilar Aluminum Alloys. In Proceedings of the ASME International Mechanical Engineering Congress and Exposition, Houston, TX, USA, 13–19 November 2015.
47. Karthikeyan, L.; Kumar, V.S. Relationship between process parameters and mechanical properties of friction stir processed AA6063-T6 aluminum alloy. *Mater. Des.* **2011**, *32*, 3085–3091. [CrossRef]
48. Hannard, F.; Castin, E.; Pardoen, T.; Mokso, T.; Maire, E.; Simar, A. Ductilization of aluminium alloy 6056 by friction stir processing. *Acta Mater.* **2017**, *130*, 121–136. [CrossRef]
49. Mazaheri, Y.; Karimzadeh, F.; Enayati, M.H. A novel technique for development of A356/Al2O3 surface nanocomposite by friction stir processing. *J. Mater. Process. Technol.* **2011**, *211*, 1614–1619. [CrossRef]
50. Kalashnikova, T.A.; Chumaevskii, A.V.; Rubtsov, V.E.; Ivanov, A.N.; Alibatyro, A.A.; Kalashnikov, K.N. Structural Evolution of Multiple Friction Stir Processed AA2024. In Proceedings of the International Conference on Advanced Materials with Hierarchical Structure for New Technologies and Reliable Structures (AMHS'17), Tomsk, Russia, 9–13 October 2017.
51. Hashim, F.A.; Salim, R.K.; Khudair, B.H. Effect of friction stir processing on (2024-T3) aluminum alloy. *IJIRSET* **2015**, *4*, 1822–1829.

52. Tsai, F.; Kao, P. Improvement of mechanical properties of a cast Al-Si base alloy by friction stir processing. *Mater. Lett.* **2012**, *80*, 40–42. [CrossRef]
53. Jana, S.; Mishra, R.; Baumann, J.; Grant, G. Effect of friction stir processing on fatigue behavior of an investment cast Al–7Si–0.6 Mg alloy. *Acta Mater.* **2010**, *58*, 989–1003. [CrossRef]
54. Kurt, A.; Uygur, I.; Cete, E. Surface modification of aluminium by friction stir processing. *J. Mater. Process. Technol.* **2010**, *211*, 313–317. [CrossRef]
55. John, J.; Shanmuganatan, S.P.; Kiran, M.B.; Senthil Kumar, V.S.; Krishnamurthy, R. Investigation of friction stir processing effect on AA 2014-T6. *Mater. Manuf. Process.* **2019**, *34*, 159–176. [CrossRef]
56. Prakash, T.; Sasikumar, P. The Influences of the friction stir processing on the microstructure and hardness of AA6061 aluminium sheet metal. *JMET* **2013**, *1*, 66–72.
57. Sinhmar, S.; Dwivedi, D.K.; Pancholi, V. Friction Stir Processing of AA 7039 Alloy. In Proceedings of the International Conference on Production and Mechanical Engineering, Bangkok, Thailand, 30–31 December 2014.
58. Santella, M.L.; Engstrom, T.; Storjohann, D.; Pan, T.Y. Effects of friction stir processing on mechanical properties of the cast aluminum alloys A319 and A356. *Scr. Mater.* **2005**, *53*, 201–206. [CrossRef]
59. Kuncická, L.; Král, P.; Dvorák, J.; Kocich, R. Texture evolution in biocompatible Mg-Y-Re alloy after friction stir processing. *Metals* **2019**, *9*, 1181. [CrossRef]
60. Izadi, H.; Nolting, A.; Munro, C.; Gerlich, A.P. Effect of Friction Stir Processing Parameters on Microstructure and Mechanical Properties of AL 5059. In Proceedings of the 9th International Conference on Trends in Welding Research, Chicago, IL, USA, 4–8 June 2012.
61. Ni, D.R.; Wang, D.; Feng, A.H.; Yao, G.; Ma, Z. Enhancing the high-cycle fatigue strength of Mg–9Al–1Zn casting by friction stir processing. *Scr. Mater.* **2009**, *61*, 568–571. [CrossRef]
62. Sakurada, D.; Katoh, K.; Tokisue, H. Underwater Friction welding of 6061 aluminum alloy. *JILM* **2002**, *52*, 2–6. [CrossRef]
63. Hofmann, D.C.; Vecchio, K.S. Thermal history analysis of friction stir processed and submerged friction stir processed aluminum. *MSEA* **2007**, *465*, 165–175. [CrossRef]
64. Zhang, H.J.; Liu, H.J.; Yu, L. Microstructure and mechanical properties as a function of rotation speed in underwater friction stir welded aluminum alloy joints. *Mater. Des.* **2011**, *32*, 4402–4407. [CrossRef]
65. Darras, B.; Kishta, E. Submerged friction stir processing of AZ31 Magnesium alloy. *Mater. Des.* **2013**, *47*, 133–137. [CrossRef]
66. Sabari, S.S. Evaluation of Performance of Friction Stir Welded AA2519-T87 Aluminium Alloy Joints. Ph.D. Thesis, Annamalai University, Tamil Nadu, India, 1 September 2016.
67. El-Danaf, E.A.; El-Rayes, M.M.; Soliman, M.S. Friction stir processing: An effective technique to refine grain structure and enhance ductility. *Mater. Des.* **2010**, *31*, 1231–1236. [CrossRef]
68. Akinlabi, E.T.; Mahamood, R.M.; Akinlabi, S.A.; Ogunmuyiwa, E. Processing parameters influence on wear resistance behaviour of friction stir processed Al-TiC composites. *Adv. Mater. Sci. Eng.* **2014**, *2014*, 724590. [CrossRef]
69. Toma, E.; Karash, B.Y.; Saeed, R.; Taqi, M.; Qasim, E. The effect of the cutting depth of the tool friction stir process on the mechanical properties and microstructures of aluminium alloy 6061-T6. *AJMA* **2015**, *3*, 33–41.
70. Abrahams, R.; Mikhail, J.; Fasihi, P. Effect of friction stir process parameters on the mechanical properties of 5005-H34 and 7075-T651 aluminium alloys. *Mater. Sci. Eng.* **2019**, *751*, 363–373. [CrossRef]
71. Zhao, H.; Pan, Q.; Qin, Q.; Wu, Y.; Su, X. Effect of the processing parameters of friction stir processing on the microstructure and mechanical properties of 6063 aluminum alloy. *Mater. Sci. Eng.* **2019**, *751*, 70–79. [CrossRef]
72. Rouzbehani, R.; Kokabi, A.H.; Sabet, H.; Paidar, M.; Ojo, O.O. Metallurgical and mechanical properties of underwater friction stir welds of Al7075 aluminum alloy. *J. Mater. Process. Technol.* **2018**, *262*, 239–256. [CrossRef]
73. Singh, I.; Singh, T.; Singh, R.; Singh, S.G. Fabrication of AA-6063/Sic composite material by using friction stir processing. *IJAR* **2017**, *5*, 1652–1656. [CrossRef]
74. Sun, N.; Jones, W.J.; Apelian, D. Friction stir processing of aluminum alloy A206: Part II—Tensile and fatigue properties. *Int. J. Met.* **2019**, *13*, 244–254. [CrossRef]
75. Thakral, R.; Sanjeev, S.; Taljeet, S. Experimental analysis of friction stir processing of TIG welded aluminium alloy 6061. *IJIRST* **2018**, *4*, 51–57.

76. Yadav, D.; Bauri, R. Effect of friction stir processing on microstructure and mechanical properties of aluminium. *MSEA* **2012**, *539*, 85–92. [CrossRef]
77. Feng, X.; Liu, H.; Lippold, J.C. Microstructure characterization of the stir zone of submerged friction stir processed aluminum alloy 2219. *Mater. Charact.* **2013**, *82*, 97–102. [CrossRef]
78. Singh, H.; Kumar, P.; Singh, B. Effect of under surface cooling on tensile strength of friction stir processed aluminium alloy 6082. *AJEAT* **2016**, *5*, 40–44.
79. Nourbakhsh, S.H.; Atrian, A. Effect of submerged multi-pass friction stir process on the mechanical and microstructural properties of Al7075. *J. Stress Anal.* **2017**, *2*, 51–56.
80. Mabuwa, S.; Msomi, V. Effect of friction stir processing on gas tungsten arc welded and friction stir welded 5083-H111 aluminium alloy joints. *Adv. Mater. Sci. Eng.* **2019**, *2019*, 3510236. [CrossRef]
81. Dawood, H.I.; Mohammed, K.S.; Raja, M.Y. Advantages of the Green Solid State FSW over the Conventional GMAW Process. *Adv. Mater. Sci. Eng.* **2014**, *2014*, 105713. [CrossRef]
82. Bevilacqua, M.; Ciarapica, P.E.; D'Orazio, A.; Forcellese, A.; Simoncini, M. Sustainability analysis of friction stir welding of AA5754 sheets. *Procedia CIRP* **2017**, *62*, 529–534. [CrossRef]
83. Nandan, R.; DebRoy, T.; Bhadeshia, H.K.D.H. Recent advances in friction-stir welding—Process, weldment structure and properties. *Prog. Mater. Sci.* **2008**, *53*, 980–1023. [CrossRef]
84. Shukla, R.K.; Shah, P.K. Comparative study of friction stir welding and tungsten inert gas welding process. *IJST* **2010**, *3*, 667–671.
85. Nicholas, E.D.; Kallee, S.W. Friction stir welding-a decade on. In Proceedings of the IIW Asian Pacific International Congress, Sydney, Australia, 29 October–2 November 2000.
86. Costa, M.I.; Verdera, D.; Vieira, M.T.; Rodrigues, D.M. Surface enhancement of cold work tool steels by friction stir processing with a pinless tool. *Appl. Surf. Sci.* **2014**, *296*, 214–220. [CrossRef]
87. Li, K.; Liu, X.; Zhao, Y. Research status and prospect of friction stir processing technology. *Coatings* **2019**, *9*, 129. [CrossRef]
88. Wang, L.Q.; Xie, L.C.; Lv, Y.T.; Zhang, L.C.; Chen, L.Y.; Meng, Q.; Qu, J.; Zhang, D.; Lu, W.J. Microstructure evolution and superelastic behavior in Ti-35Nb-2Ta-3Zr alloy processed by friction stir processing. *Acta Mater.* **2017**, *131*, 499–510. [CrossRef]
89. Sun, P.; Wang, K.; Wang, W.; Zhang, X. Influence of process parameter on microstructure of AZ31 magnesium alloy in friction stir processing. *Hot Work. Technol.* **2008**, *37*, 99.
90. Thompson, B.; Doherty, K.; Su, J.; Mishra, R. Nano-sized grain refinement using friction stir processing. In *Friction Stir Welding and Processing VII*; Springer: Cham, Switzerland, 2013; pp. 9–19.
91. Xin, R.L.; Zheng, X.; Liu, Z.; Liu, D.; Qiu, R.S.; Li, Z.Y.; Liu, Q. Microstructure and texture evolution of an Mg-Gd-Y-Nd-Zr alloy during friction stir processing. *J. Alloy. Compd.* **2016**, *659*, 51–59. [CrossRef]
92. Han, J.Y.; Chen, J.; Peng, L.M.; Zheng, F.Y.; Rong, W.; Wu, Y.J.; Ding, W.J. Influence of processing parameters on thermal field in Mg-Nd-Zn-Zr alloy during friction stir processing. *Mater. Des.* **2016**, *94*, 186–194. [CrossRef]
93. Khodabakhshi, F.; Gerlich, A.P.; Svec, P. Fabrication of a high strength ultra-fine grained Al-Mg-SiC nanocomposite by multi-step friction-stir processing. *Mater. Sci. Eng. A* **2017**, *698*, 313–325. [CrossRef]
94. Msomi, V.; Mbana, N.; Mabuwa, S. Microstructural analysis of the friction stir welded 1050-H14 and 5083-H111 aluminium alloys. *Mater. Today Proc.* **2019**. [CrossRef]
95. Raj, K.H.; Sharma, R.H.; Singh, P.; Dayal, A. Study of friction stir processing (FSP) and high pressure torsion (HPT) and their effect on mechanical properties. *Proc. Eng.* **2011**, *10*, 2904–2910. [CrossRef]

Review

A Short Review on Welding and Joining of High Entropy Alloys

João G. Lopes and João Pedro Oliveira *

UNIDEMI, Department of Mechanical and Industrial Engineering, NOVA School of Science and Technology, Universidade NOVA de Lisboa, 2829-516 Caparica, Portugal; jcg.lopes@campus.fct.unl.pt

* Correspondence: jp.oliveira@fct.unl.pt; Tel.: +351-21-294-96-18

Received: 27 December 2019; Accepted: 31 January 2020; Published: 2 February 2020

Abstract: High entropy alloys are one of the most exciting developments conceived in the materials science field in the last years. These novel advanced engineering alloys exhibit a unique set of properties, which include, among others, good mechanical performance under severe conditions in a wide temperature range and high microstructural stability over long time periods. Owing to the remarkable properties of these alloys, they can become expedite solutions for multiple structural and functional applications. Nevertheless, like any other key engineering alloy, their capacity to be welded, and thus become a permanent feature of a component or structure, is a fundamental issue that needs to be addressed to further expand these alloys' potential applications. In fact, welding of high entropy alloys has attracted some interest recently. Therefore, it is important to compile the available knowledge on the current state of the art on this topic in order to establish a starting point for the further development of these alloys. In this article, an effort is made to acquire a comprehensive knowledge on the overall progress on welding of different high entropy alloy systems through a systematic review of both fusion-based and solid-state welding techniques. From the current literature review, it can be perceived that welding of high entropy alloys is currently gaining more interest. Several high entropy alloy systems have already been successfully welded. However, most research works focus on the well-known CoCrFeMnNi. For this specific system, both fusion and solid-state welding have been used, with no significant degradation of the joints' mechanical properties. Among the different welding techniques already employed, laser welding is predominant, potentially due to the small size of its heat source. Overall, welding of high entropy alloys is still in its infancy, though good perspectives are foreseen for the use of welded joints based on these materials in structural applications.

Keywords: welding; high entropy alloys; laser welding; friction stir welding; microstructure; review

---

1. Introduction

Throughout recent years, high entropy alloys (HEAs) have arisen as a new reality in the advanced materials domain. The motivation behind this innovative class of engineering materials was based on the purpose of broadening the boundaries of knowledge in what concerns alloys with more than one main constituent. This concept can be traced back to 2004, when the first papers concerning this subject were published [1,2], whereas, at the present time, several other works [3–12] have also been presented to extend the available information regarding HEAs.

Currently, a common definition for these alloys was introduced by one of the precursor works, where high entropy alloys are described as metallic compounds with at least five principal elements, with the percentage of each component varying between 5 and 35 at.% [2]. However, other definitions include a wider range of materials, with different component amounts, which can also be considered high entropy alloys [3]. Several review papers address and discuss the different nomenclatures regarding high entropy alloys [3,13–17], and as such, we refrain from addressing this topic in this paper.

As established by Yeh [18], these innovative multi-element compositions confer to these materials four distinct core effects:

The "high entropy effect", associated to the high configurational entropy that stabilizes the formation of simple solid-state phases such as body-centered-cubic (BCC) or face-centered-cubic (FCC), while inhibiting the development of brittle intermetallic compounds. However, it must be noted that high entropy alloys can still exhibit brittle-like behavior, as evidenced for the AlCoCrCuFeNi system [19].

The "lattice distortion effect", that occurs due to the lack of a dominant element in the composition of the alloy, resulting in different atoms of different sizes occupying the lattice positions of the crystal structure, promoting its distortion and affecting the physical and mechanical properties of the alloy. Depending on the selected atomic elements and their concentration, distinct phases, with potentially different mechanical properties can form as evidenced in the work of Wu et al. [20]. As such, the lattice distortion effect can be used to promote a phase over the other for specific alloy systems. Recent work by He et al. [21] has shown that not only the atomic ratio of the element that compose the high entropy alloy affect the lattice distortion, but the alloy Poisson ration must also be considered.

The "sluggish diffusion effect", that constrains the atomic diffusion of the elements and inhibits the phase transformations that require such phenomenon to occur. As a result, higher recrystallization temperatures can be achieved, and formation of nano-precipitates and amorphous structures are susceptible to occur, as such, second-phase precipitation only occurs after extremely long periods [22]. Bhattacharjee et al. [23] showed that heat treatments with the duration of 1 h at or below 800 °C in severely deformed CoCrFeNiMn high entropy alloys would not result in any significant grain growth. However, when the temperature was of 900 or 1000 °C, massive grain growth was observed. Regarding the extremely long times and temperatures required for second-phase precipitation, Pickering et al. [22] showed that annealing at 700 °C for times above 500 h would lead to the formation of $M_{23}C_6$ and σ phase.

The "cocktail effect", which refers to the enhancement of the established properties of the alloy, which cannot be attributed independently to any of the elements that compose the material [24].

Considering the above-mentioned features and the prospect of customizing the composition of high entropy alloys, a new path for a wide range of applications can be expected. As such, owing to these properties, the outstanding performance of HEAs to operate under extreme conditions is subjected to intensive research. Depending on the composition of the alloy, the corrosion resistance [25–28], the ability to sustain high cyclical loading [29,30], wear resistance [31–33], and the good performance at both high [34,35] and cryogenic temperatures [36–38] are some of the key features that these alloys exhibit towards being novel solutions for structural and functional applications [39]. Miracle et al. [40] suggest that high entropy alloys can be used as structural materials in transportation and energy sectors. The same author [41] proposes that some high entropy alloys can be used in functional applications that require resistance to radiation damage or when in need of diffusion barriers, as in the microelectronics sector. Replacement of conventional materials by entropy alloys is suggested to occur in the future, and materials to be replaced include stainless steels, Al-, Ti-, and Ni-based alloys. This is related to the fact that high entropy alloys can be fine-tuned to simultaneously present, if desired, low density and high mechanical strength or other combination of properties. These properties will be dictated by the elements that compose the high entropy alloy system. Table 1 summarizes the mechanical properties (yield stress, ultimate tensile stress, and elongation) for multiple high entropy alloy systems at different temperatures.

Table 1. Mechanical properties of various high entropy alloys (HEAs) under different temperatures.

Alloy System	Mechanical Properties	Temperature (K)									Refs.
		77	296	673	873	1073	1273	1473	1673	1873	
CoCrFeNiMn	σ_{ys} (MPa)	759	410	-	-	-	-	-	-	-	[42]
	σ_{uts} (MPa)	763	1280	-	-	-	-	-	-	-	
	Elongation (%)	71	57	-	-	-	-	-	-	-	
CoCrFeMn	σ_{ys} (MPa)	481	272	-	-	-	-	-	-	-	[38]
	σ_{uts} (MPa)	1003	567	-	-	-	-	-	-	-	
	Elongation (%)	65	47	-	-	-	-	-	-	-	
CoCrFeNiAl	σ_{ys} (MPa)	-	250	155	150	-	-	-	-	-	[43]
	σ_{uts} (MPa)	-	-	-	-	-	-	-	-	-	
	Elongation (%)	-	>50	>50	>50	-	-	-	-	-	
CrFeNiMnAl	σ_{ys} (MPa)	-	910	755	325	-	-	-	-	-	[43]
	σ_{uts} (MPa)	-				-	-	-	-	-	
	Elongation (%)	-	>50	>50	>50	-	-	-	-	-	
CoCrFeNiAlTi	σ_{ys} (MPa)	-	1420	1285	795	285	-	-	-	-	[43]
	σ_{uts} (MPa)	-	-	-	-	-	-	-	-	-	
	Elongation (%)	-	18	24	>50	>50	-	-	-	-	
CrFeNiMnAlTi	σ_{ys} (MPa)	-	1280	1100	355	-	-	-	-	-	[43]
	σ_{uts} (MPa)	-				-	-	-	-	-	
	Elongation (%)	-	31	>50	>50	-	-	-	-	-	
CoCrFeNiV	σ_{ys} (MPa)	477	470	-	-	-	-	-	-	-	[37]
	σ_{uts} (MPa)	1000	626	-	-	-	-	-	-	-	
	Elongation (%)	62	36	-	-	-	-	-	-	-	
HfNbTaTiZrW	σ_{ys} (MPa)	-	1550	-	-	577	409	345	-	-	[44]
	σ_{uts} (MPa)	-		-	-				-	-	
	Elongation (%)	-	26.3	-	-	>35	>35	>35	-	-	
HfNbTaTiZrMoW	σ_{ys} (MPa)	-	1637	-	-	1065	736	703	-	-	[44]
	σ_{uts} (MPa)	-		-	-				-	-	
	Elongation (%)	-	15.5	-	-	>35	>35	>35	-	-	
HfNbTaTiZr	σ_{ys} (MPa)	-	929	790	675	535	295	92	-	-	[45]
	σ_{uts} (MPa)	-	-	-	-	-	-	-	-	-	
	Elongation (%)	-	-	-	-	-	-	-	-	-	
VNbMoTaW	σ_{ys} (MPa)	-	1246	-	862	846	842	735	656	477	[46]
	σ_{uts} (MPa)	-	1270	-	1597	1536	1454	943	707	479	
	Elongation (%)	-	1.7	-	13	17	19	7.5			
NbMoTaW	σ_{ys} (MPa)	-	1058	-	561	552	548	506	421	405	[46]
	σ_{uts} (MPa)	-	1211	-	-	-	1008	803	467	600	
	Elongation (%)	-	-	-	-	-	-	-	-	-	

To ensure the viability of these alloys to be used in complex shaped structures, their weldability is an important issue that needs to be addressed. Any advanced engineering alloy will require welding to either obtain complex shape structures or to couple its properties to those from another material. As such, evaluating the weldability of novel alloys is fundamental to further expand its potential applications. Additionally, because these alloys are now being more studied, it is possible to adjust their chemistry or microstructure to avoid weldability issues, such as liquation cracking. The sooner these potential issues are found, the easier and more cost-effective it is to find a solution.

This paper analyses the overall progress achieved using well-known welding techniques and weldability studies on HEAs. First, focus is given to fusion-based techniques, in a second instance, a review of solid-state ones is provided.

2. Current Progress on Welding HEAs

Among several methods for joining materials, welding comprises a broad range of techniques, and it is an expedite and often a reliable way to produce permanent and continuous joints. When proper design specifications and process optimization are used, welding technologies become unquestionably competitive for several industrial sectors, such as infrastructure construction and transportation. This category of processes is known to be capable of achieving particularly strong and resistant bonds when exposed to static and dynamic forms of loading [47,48]. As such, welding technologies can be divided according to the diverse mechanisms and processes required to achieve such joints.

Fusion-based welding processes are based on melting and solidification of metal and are currently the most established in industry. The resultant joints are typically characterized by exhibiting three distinct regions. The fusion zone (FZ), which is the region that undergoes melting and subsequent solidification; the heat affected zone (HAZ), that is characterized by experiencing temperatures that can promote solid-state transformation and never changes its state (i.e., the material never becomes fully liquid during the process); and, lastly, the base material (BM), which is the region that remains unchanged throughout the process.

Often, fusion-based welding methods are known to impair the mechanical properties of the joints due to metallurgical incompatibility between the materials to be joined, development of high residual stresses, mismatch of thermomechanical properties of the BMs, or due to the sensitivity of the material to the weld thermal cycle. In those cases, solid-state techniques are a viable option capable of solving some of the above-mentioned problems. One well-known example is dissimilar welding of aluminium to steels [49]: While fusion welding can be used to join this dissimilar combination, the resulting mechanical properties are often poor due to the formation of brittle intermetallic compounds upon mixing of the liquid phases of the two BMs. However, solid-state methods can effectively join these materials allowing their use in structural applications.

Solid-state welding is based on intense friction, plastic deformation, and diffusion mechanisms that aid in the formation of a joint between the BMs [50]. The versatility of such approach is proven by the ability to connect materials that are difficult or impossible to join through fusion-based processes, avoiding the development of undesired phases, distortions, and high residual stresses that may occur during the liquid-solid state interchange. These processes where plastic deformation occurs, exhibit a thermomechanically affected zone (TMAZ), as well as an HAZ. The microstructural evolution of the joints will depend on the material susceptibility to the combined effect of temperature and deformation in the TMAZ, and of temperature in the HAZ.

Another possibility for materials joining is through brazing and soldering. These are mainly characterized by the introduction of a filler metal into the joint region. In such processes, the metallic filler has a melting point lower than that of the materials that are to be welded, and the welds are established by diffusion between the filler and BMs [47].

Regarding the overall developments of welding of HEAs, a survey on the contemporary literature shows that an effort is being made towards the understanding of the microstructural evolution and optimization of these joining processes [51]. Until now, the number of publications regarding welding of HEAs is increasing throughout the years, as the developments on these alloys become more evident, as it can be observed in Figure 1.

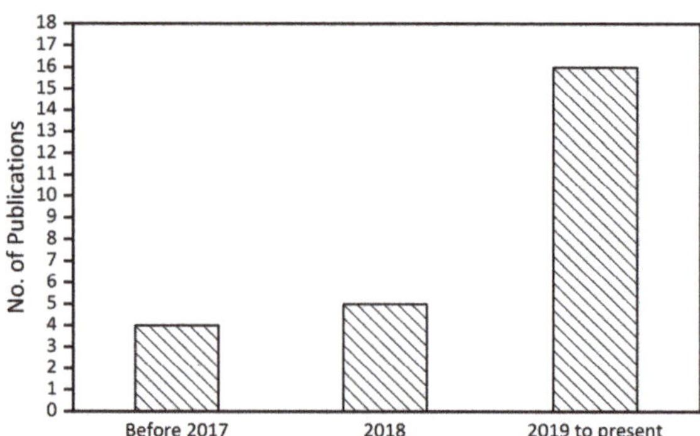

Figure 1. Number of publications on welding of high entropy alloys over time.

However, to the best of the authors' knowledge, most studies are focused on fusion-based welding processes. Still, it is also possible to perceive that some efforts are also being dedicated to the study of other welding techniques on HEAs, especially in what concerns solid-state welding processes. Considering this, the next topics that are presented in this paper mainly concern the overall progress on fusion-based and solid-state welding processes on HEAs. Nevertheless, information about welding HEAs through brazing is also available, as presented by Lin et al. [52], where dissimilar joints between CoCrFeMnNi/CoCrFeNi alloy and CoCrFeMnNi/316 stainless steel were investigated.

2.1. Fusion-based Welding of HEAs

Several studies [53–69] have already been reported on fusion-based welding of HEAs. To date, these works focus mainly on laser-based techniques, although some information is also available on Electron Beam and Gas Tungsten Arc Welding (GTAW) techniques.

In terms of the materials, the most used HEA is the CoCrFeNiMn one. This is probably related to the fact that this alloy system is the most studied within the field of HEAs [22,25,34,70–72]. For this reason, in this section of the paper (2.1.1), we first address the current status on fusion-welding of the CoCrFeNiMn HEA system, and the following one (2.1.2) focuses on the remaining materials that were already welded.

2.1.1. CoCrFeNiMn HEA System

Concerning the investigation of the welded joints behavior on HEAs obtained through laser-based techniques, Kashaev et al. [53] reported a study on the CoCrFeNiMn system, where the base material was fabricated via self-propagating high temperature synthesis, a process where reagents are ignited and then, due to an exothermic reaction, a given product is formed [73]. The as-produced HEA exhibited columnar FCC grains with MnS precipitates and Cr-rich carbides on its microstructure, and, as a consequence of the BM manufacturing method, its matrix composition revealed a reduced content of Mn and the existence of several impurities. These impurities did not lead to any strengthening effect, most likely due to their large grain size and low volume fraction. After welding, using a laser power of 2 kW and a welding speed of 5 m/min, several changes in texture and microstructure of the FCC matrix occurred. The precipitation of nanoscale intermetallic B2 phase compounds in the welded region promoted an increase of the microhardness in the fusion zone, as evidenced in Figure 2. The nanoscale B2 particles were seen to be mainly composed by Ni and Al, with the latter element being an impurity of the starting powders. The formation of the B2 phase was predicted by thermodynamic calculations, showing the interest of such approach to predict and explain the developed microstructures in the fusion zone of the joint. During fusion welding, there is an intense mixing of elements within the molten pool. As such, it is possible that the B2 particles are formed due to the local mixing of Ni and Al favoring the formation of this phase in the fusion zone of the joint.

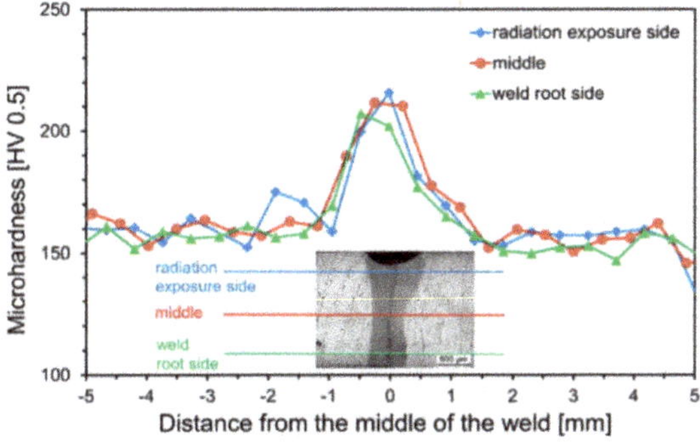

Figure 2. Microhardness profile of a laser beam welded joint (Reproduced from [53], with permission from Elsevier, 2018).

The measured increase of hardness on the welds was suggested by the authors to be an asset for structural applications of this class of HEA. However, no assessement of the tensile properties of the welded joints was performed, thus it was not possible to state the suitabilty of these joints to be employed as structural parts. Though this was not studied in the abovementioned paper, it can be hypothesized that it may be possible to slightly change the microstructure (and resulting properties) in the fusion zone, by controlling the heat input. For example, lower heat input leads to higher cooling rates, which restricts grain growth in the fusion zone of the joint [74].

In a follow-up work, Kashaev et al. [54] addressed the impact of laser welding on the mechanical performance of the welded joints. Due to the coarse grain structure of the base material, with a grain size ranging from 250–500 μm, a refined structure in the fusion zone was observed (100–300 μm). Of special relevance in this work is the fact that the welding process did not impair the mechanical properties of the HEA joints, and the tensile and fatigue behavior of both base material and welded joints were similar (refer to Figure 3). Fracture of the welded joints occurred in the BM, which can be explained by the lower hardness of this region, which promoted strain accumulation.

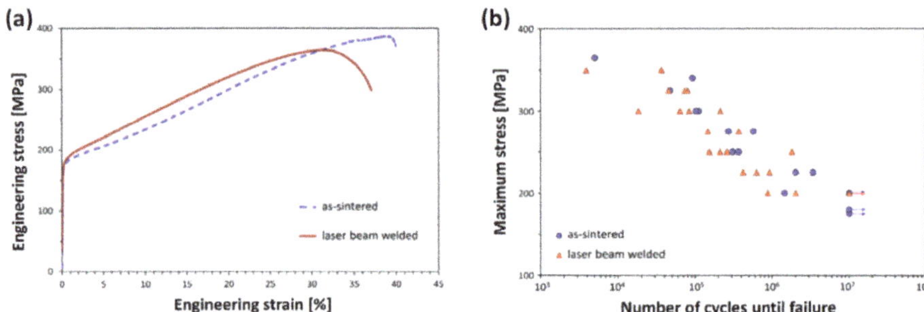

Figure 3. Laser beam welds characterization: (**a**) Tensile testing; (**b**) fatigue testing (Adapted from [54], with permission from Elsevier, 2019).

Other researchers have also used laser welding for similar joining of CoCrFeNiMn HEAs. Jo et al. [55] also found that the hardness of the fusion zone of the joint was higher than in the base materials and also that the FCC structure was preserved. These results show the good reproducibility in terms of mechanical properties in laser welded CoCrFeMnNi HEAs obtained by different research groups. The higher hardness of the fusion zone was attributed to fine dendritic arm spacing and composition inhomogeneity. However, it must be noticed that the BM hardness was that of an as-annealed CoCrFeNiMn HEA [75]. It is known that CoCrFeNiMn alloys can exhibit higher hardness under appropriate heat treatment conditions, which would lead to a lower hardness region in the FZ if the BM was heat treated prior to welding. Obviously, the condition of the BM will impact the microstructural evolution of the welded joints, especially in what concerns strain accumulation upon mechanical testing.

A study on laser welding of as-cast and as-rolled CoCrFeNiMn HEA, was a performed by Nam et al. [56,57] in order to assess their viability for cryogenic applications. On a similar laser welding case, increasing the welding speed, which varied between 6, 8, and 10 m/min, and corresponds to a decrease in the heat input, revealed that the density of shrinkage voids, primary dendrite arm spacing, and dendrite packet size decreased. It should be noted that shrinkage avoids can be avoided/minimized upon careful optimization of the process parameters [76]. The microhardness profiles showed that, on the casted samples, the FZ presented higher values than the BM, which was attributed to the differences in grain size between both regions. Nevertheless, on the as-rolled specimens, no pronounced variation in hardness was observed, due to the existence of similar grain size in both the BM and FZ.

As depicted in Figure 4, the tensile properties of the as-cast specimens were similar to those of the BMs. However, the same did not occur in the as-rolled condition, where the welded region showed lower tensile strength values when compared to the BM, which resulted from the larger grain size on the FZ than in the BM. For the as-cast samples, at a testing temperature of 298 K, fracture occurred near the HAZ/BM interface. The same did not occur on the rolled samples, where fracture occurred in the FZ, which was attributed to higher grain size of this region. Nevertheless, in both cases, the tensile properties of the samples tested at 77 K were superior to those observed as 298 K, which was attributed to the existence of deformation twinning that tends to occur at cryogenic temperatures.

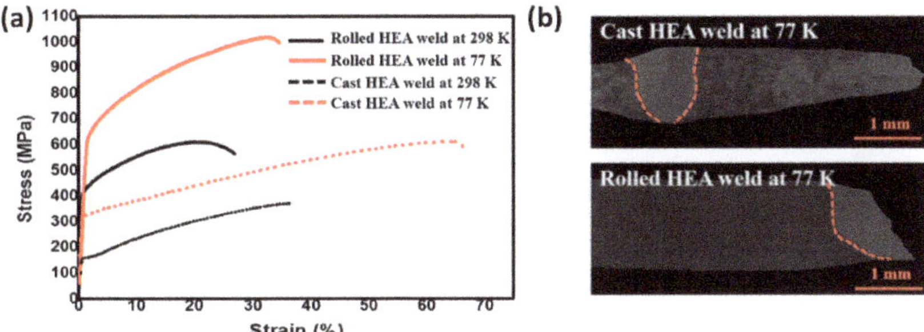

Figure 4. Tensile testing results: (**a**) Comparison between the tensile properties of the weld at different temperatures; (**b**) fracture region (Adapted from [56], with permission from Elsevier, 2019).

In [57], an as-cast and an as-rolled CoCrFeNiMn HEA were welded together, and evident differences of the microstructure were observed between both sides of the joint, as depicted in Figure 5. This phenomenon was attributed to the distinct epitaxial dendritic growth that initiated from the different grain sizes and morphologies in the BMs.

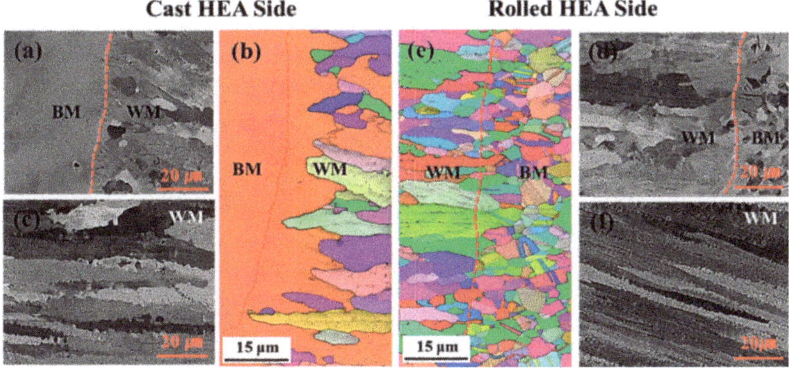

Figure 5. Microstructure of the dissimilar welds: (**a**,**b**) Dendritic growth nucleated from the fusion boundary on the cast HEA (high entropy alloy) side; (**c**) dendrites from near the centerline on the cast WM side; (**d**,**e**) dendritic growth nucleated from the fusion boundary on the rolled HEA side; (**f**) Dendrites from near the centerline on the rolled WM side. BM—Base material; WM—Weld metal (Reproduced from [57], with permission from Taylor & Francis, 2019).

The microhardness increased on the as-cast BM/HAZ interface, but still, no significant change could be observed in the weld/rolled BM interface (refer to Figure 6). Overall, the tensile properties were enhanced when the material was tested in cryogenic conditions due to deformation twinning. During tensile testing, the dissimilar welds fractured on the as-cast BM, exhibiting a comparable behavior to that of the as-cast alloy. In both cases, the viability of the laser welded CoCrFeNiMn HEAs joints for applications in cryogenic environments was evidenced.

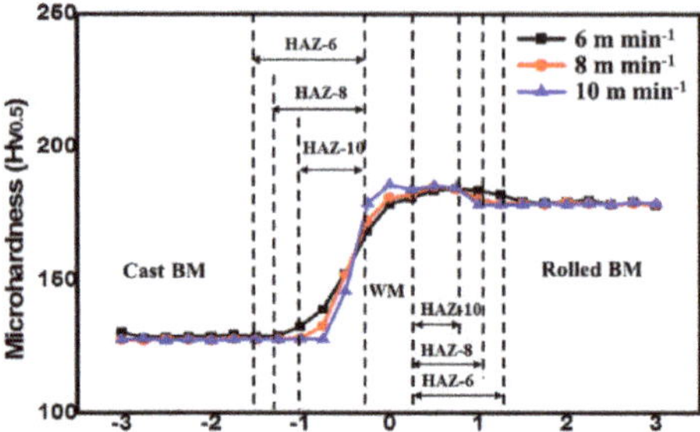

Figure 6. Microhardness profile of the dissimilar weld (Reproduced from [57], with permission from Taylor & Francis, 2019).

In another work on welding of the CoCrFeNiMn system, Chen et al. [58] showed that their laser welded samples had superior properties than that of the BM, since fracture of the joints occurred in that region rather than in the FZ. This could be attributed to the higher hardness of the fusion zone (≈193 HV) when compared to the base material (≈177 HV). The microstructural analysis of the weld region and BM revealed a single FCC phase with several Cr-Mn rich precipitates. In the fusion zone, the average size of the existing precipitates ranged from 0.41 to 0.49 μm, dispersed within the grains and at the grain boundaries. These, owing to their small size, have a pinning effect on dislocations [77], granting the welded region a superior mechanical performance than the BM.

Often, the resultant microstructures in fusion-based welded joints must be modified in order to improve the part mechanical properties. These microstructural modifications are often performed by post-weld heat treatments, which can induce dissolution or formation of new phases/precipitates or promote stress relieving [78–81].

The influence of post-weld heat treatments on laser welded CoCrFeNiMn HEAs was also reported in the literature. Nam et al. [59] studied the effect of post-weld heat treatments on a temperature range of 800 and 1000°C for one hour, on laser beam welds of a cold-rolled CoCrFeNiMn HEA. Before the heat-treatment, the welded region exhibited a larger grain size and inferior tensile strength and hardness than the BM. After being heat-treated, the welds showed superior hardness than the BM, with the FZ preserving the original BM FCC crystal structure and a decrease in the size and fraction of Cr-Mn oxide inclusions. With the increase of temperature of the heat treatment to 1000 °C, the variation in grain size between the weld metal and heat-affected zone decreased, which resulted in approximately the same tensile properties between the welded joint and the BM. This initial work shows the fundamental role of critically selecting the base material initial condition and subsequent post-weld heat treatments. These are fundamental to improve the joint microstructure and consequently its mechanical properties.

Concerning other fusion-based welding techniques, Wu et al. [60,61] investigated Electron Beam Welding and Gas Tungsten Arc Welding (GTAW) on a CoCrFeMnNi HEA. For this purpose, ingots were produced via arc-melting and then thermomechanically processed to achieve a homogeneous equiaxed microstructure. After welding, microstructure characterization evidenced that no major defects existed in the joints, and that the microstructure was mainly composed of dendrites and large columnar grains. Overall, the yield strength of both welded joints was higher than that of the BM. Differences between both welding methods resided on the dendrite arm spacing and on the amount of elemental segregation, which were less evident on the electron beam welds. This can be attributed to the fast cooling rate of the process, which can decrease grain growth and elemental segregation when compared to arc-based techniques [47,48]. The tensile strength of the GTAW samples exhibited, approximately, 80% of the tensile strength and 50% of the ductility of the BM, while the electron beam-welded samples presented a similar behavior to that of the BM. Though sound joints were achieved in both cases, the ability of electron beam welding to localize the heat in a restrict region can be considered an advantage and a potential justification on the superior mechanical properties when compared to arc-based welding of the same alloy.

More recently, Oliveira et al. [62] performed a comprehensive study on GTAW of as-rolled CoCrFeNiMn HEA. Using synchrotron X-ray diffraction analysis, the authors observed that the extension of the HAZ was larger than that determined based on electron microscopy and hardness measurements techniques. In fact, due to the large deformation imposed by cold rolling of the starting BM, recovery phenomenon was seen to occur far away from the weld centerline. This phenomenon translated into a decrease of the residual stresses of the material in that region, though no variations in grain size and hardness were observed.

The use of filler materials during fusion-based welding is often used to control and adjust the chemical composition and resulting microstructures [82,83]. Moreover, in hard-to-join dissimilar pairs, careful selection of the filler material can aid in the inhibition of solidification cracking or other defects than can occur upon solidification [84,85].

Nam et al. [63] evaluated the use of two different filler materials during similar GTAW of an CoCrFeMnNi HEA. The selected filler materials were a 308 L stainless steel, while the other had the same composition as the BM. The results evidenced that both types of welds exhibited a single FCC phase, nevertheless the elemental percentage of Fe increased with the proximity to the weld centerline when the stainless steel filler was used. Regarding the mechanical properties, both welds exhibited superior values than the original BM. The joint obtained using the CoCrFeMnNi filler presented the higher values for the microhardness (≈165 ± 1 HV), as depicted in Figure 7a. The tensile properties were also assessed at room and cryogenic temperatures, exhibiting a comparable behavior to that of the cast BM, as presented in Figure 7b. Overall, the tensile testing provided an insight for the possibility to use stainless steel 308 L as a filler metal to guarantee the applicability of these HEAs in cryogenic environments.

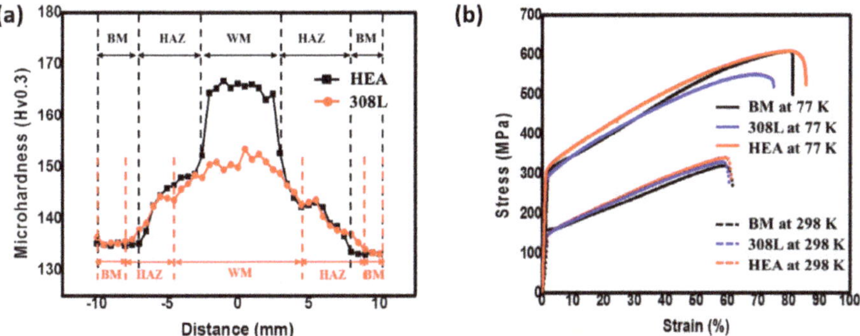

Figure 7. Mechanical behavior of the welded joints: (**a**) Tensile testing; (**b**) microhardness distribution (Adapted from [63], with permission from Elsevier, 2020).

Due to the possibility to control the composition, and therefore the resulting microstructures using filler materials, we hypothesize that in the future, dedicated filler materials can be developed to control and tune the HEA joints properties.

2.1.2. Other HEA Systems

Most of the works on welding of HEAs currently focus on the CoCrFeNiMn system, as exemplified above. However, some researchers started to pay attention to the weldability of other HEA compositions.

The use of GTAW for welding of an $Al_{0.5}$CoCrFeNi HEA was attempted by Sokkalingam et al. [66]. The BM was composed by a near-equiaxed microstructure and, after welding, the HAZ exhibited a microstructure with a size, approximately, double of that of the BM. This is explained by the weld thermal cycle, that is known to induce grain growth in the HAZ, especially closer to the fusion boundary. Additionally, the FZ was characterized by dendritic growth, and near the weld centerline, a fine equiaxed microstructure could be observed. Both the BM and the FZ exhibit a mixed FCC and BCC structure. However, the volumetric fraction of the BCC phase was reduced due to the thermal history experienced by the material during the welding procedure. Mechanical testing showed that the weld region exhibited an inferior microhardness, and the tensile properties experienced a reduction of 6.4% in strength and 16.5% in ductility, when compared to the BM.

The same authors also performed dissimilar welding of $Al_{0.1}$CoCrFeNi HEA to AISI304 stainless steel by GTAW [65]. As depicted in Figure 8, microstructural characterization revealed that the HEA side of the weld was characterized by epitaxial grain growth, while the AISI304 stainless steel side exhibited non-epitaxial grain growth. Nevertheless, towards the weld centerline, the joint tends to exhibit an equiaxed dendritic grain structure. During tensile testing, fracture occurred in the fusion zone, which was attributed to the heterogeneous distribution of the microstructure and lower hardness of the weld. Overall, the applicability of these dissimilar joints for structural applications was confirmed by means of mechanical testing, where the dissimilar joint exhibited superior values for the yield strength and ultimate tensile strength (≈265 and ≈590 MPa, respectively) than that of the HEA side of the BM (≈148 and ≈327 MPa, respectively).

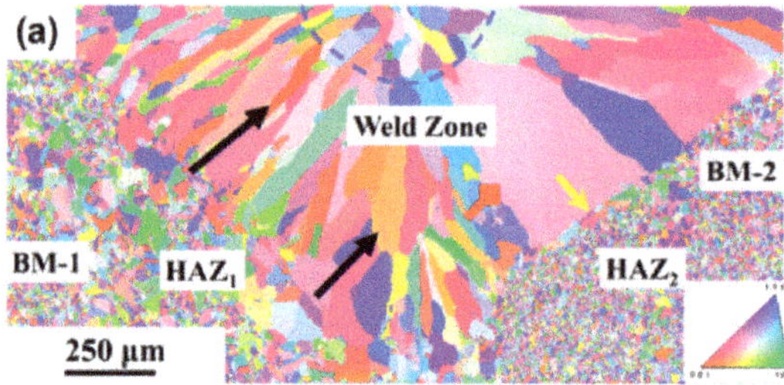

Figure 8. Microstructural characterization by means of EBSD (Electron Backscatter Diffraction) inverse pole figure analysis of the dissimilar welds. BM-1: $Al_{0.1}$CoCrFeNi HEA; BM-2: AISI304 stainless steel (Reproduced from [65] with permission from Cambridge University Press, 2019).

The corrosion behavior of laser welded $Al_{0.5}$CoCrFeNi HEA was studied by Sokkalingam et al. [64], by evaluating the corrosion potential and corrosion current density obtained by potentiodynamic polarization tests. The BM was composed by an equiaxed microstructure composed by Cr-Fe and Al-Ni rich phases and Al-rich particles. During welding, the dissolution of the Al-Ni rich and Al-rich compounds into the CoCrFeNi matrix occurred, resulting in a microstructure exhibiting Cr-Fe rich columnar dendrites with an Al-Ni rich interdendritic region. Overall, the results from the corrosion resistance tests evidenced that the welded joints exhibited a higher corrosion resistance than the BM alone, when exposed to aqueous corrosion environments. This was attributed to the solubility of the Al in the alloy matrix during welding that causes an increase of its corrosion potential, resulting on the reduction of the galvanic circuit in the joint.

More recently, weldability studies on Al_xCoCrCu_yFeNi HEAs have been performed by Martin et al. [67,68]. Cu segregation was seen to promote solidification cracking in the fusion zone of the GTAW joints. By changing the alloy composition, it was possible to mitigate the cracking susceptibility of this HEA class. These works, which were supported by thermodynamic calculations to predict the existing phases on the FZ as function of the alloy composition, show the importance of optimizing the BM starting composition when cracking phenomena, such as hot cracking and liquation cracking, are prone to occur in the materials to be welded. Though this was only observed in the AlCoCrCuFeNi HEA system, it is likely that other HEA compositions may exhibit the problems. If that is the case, the addition of filler materials [48] can also be a potential solution to adjust and improve the chemical composition of the fusion zone.

Panina et al. [69] reported the effects of pre-heating temperature (400, 600, and 800 °C) on the laser weldability of a $Ti_{1.89}NbCrV_{0.56}$ refractory high entropy alloy. Initially, the BM microstructure was mainly composed by BCC grains presenting also small C15 Laves phase particles. Hot cracking occurred when welding was performed with the BM at room temperature and at 400 °C, which was attributed to the low ductility of the alloy. Since during welding thermal stresses are generated, materials with poor ductility can suffer cracking if those stresses are not relieved. Using pre-heating temperatures of 600 °C and 800 °C, however, resulted in defect-free joints. The microstructure of the welds was characterized by columnar grains, where the grain size tended to increase with the increase of pre-heating temperatures. This can be explained based on the effect of changing the pre-heating temperature before welding: Higher pre-heating temperature leads to a slower cooling rate, which promotes more significant grain growth. Due to these slower cooling rates, the grain size was larger in the different FZ. As such, the microhardness of the FZ tended to decrease with the increase of the selected pre-heating temperatures. The mechanical performance of the welds was also assessed through tensile testing at 750 °C, and enhanced tensile properties were observed when pre-heating at 800 °C (ductility of ≈10%, yield strength of 265 MPa and ultimate tensile strength of 285 MPa vs. the 250 MPa maximum stress obtained when fracture occurred in the elastic region on the as-cast specimens). Overall, the results obtained in this study highlight the need for optimizing the welding parameters, such as the pre-heating temperatures, in order to obtain high performing joints.

Currently, it is clear that welding of HEAs is a growing research topic. However, most of the work is focused on the CoCrFeNiMn alloy system using high power beams (laser and electron beam). Though some recent studies have addressed the weldability of other HEA systems, the existence of a significant research gap regarding the weldability of these materials is highly noticeable.

2.2. Solid-State Welding of HEAs

As previously mentioned, welding materials in the solid state can be a reliable and advantageous way to achieve sound joints. The current information regarding welding HEAs using solid-state techniques shows that most studies are focused on friction stir welding (FSW) [55,86–92]. Nevertheless, other possibilities for joining these materials are rotary friction welding [91] and diffusion bonding [92].

Concerning FSW, the literature shows that an effort for the development and comprehension of the microstructural evolution of FSWed HEAs joints is underway. For instance, FSW of a CoCrFeNiMn HEA manufactured by vacuum induction melting, followed by thermomechanical processing was conducted by Jo et al. [55]. After the welding process, the tensile strength and the ductility of the samples exhibited a similar behavior to that of the BM. Ductile fracture occurred in the BM, indicating that the microstructure evolution in the processed region promoted a higher joint strength. The welding process was characterized by inducing dynamic recrystallization, which resulted in grain refinement aided by the temperature increase, coupled with the massive deformation imposed during the process. The microhardness distribution on the welds exhibited higher values than the BM. An EBSD (Electron Backscatter Diffraction) inverse pole analysis on the cross section of the weld exhibited significant grain refinement and a lower proportion of twins at the center of the weld, indicating also that the fraction of low angle grain boundaries tends to decrease with the increase of distance from the weld center. These boundaries have an important role on the mechanical performance of the material acting as barriers to plastic deformation and inhibiting the grain growth mechanisms induced by the increase of temperature.

Typically, it is often preferred that failure of a welded joint occurs in the BM. This is an evidence of the higher resistance of the welded region, which implies that, for structural parts, the main limiting aspect will be the BM mechanical properties. Provided that those properties are ensured to be constant over time, and since FSW is known to be a very reliable process, unlike some arc-based welding processes, the welded joints can be safely used as structural parts in key engineering applications.

A step further into the investigation of the FSW process applied to the CoCrFeNiMn HEA was taken by Xu et al. [86]. In this work, forced cooling was applied to the processed material, aiming at improving the joint properties. The mechanical results showed that it is possible to enhance the mechanical properties of the welds without a ductility loss through the fast cooling of the joint. This enhancement of the material mechanical properties was explained by the inhibition of the static recovery and selected grain growth that can occur during post-annealing.

Zhu et al. [87] performed FSW on cast $CoCrFeNiAl_{0.3}$. Defect-free welds were obtained using two different speeds, 30 and 50 mm/min. In both cases, apart from the original FCC matrix, the results from X-ray diffraction analysis showed that no phase changes occurred. The morphology of the welds exhibited four typical thermomechanically affected areas after FSW: (i) SZ, the stir zone, where refined equiaxed grains resultant from recrystallization were observed; (ii) TMAZ, the thermomechanically affected zone, which exhibited both coarse and fine grains; (iii) HAZ, where columnar grains were observed with an average size of 132 μm; (iv) BM, characterized by columnar grains resultant from the casting process, preferentially oriented in the solidification direction. Because of these microstructural differences, the microhardness was higher at the center of the weld, decreasing with the distance from the center. Additionally, reducing the speed of the tool, which increases the heat input, resulted in slightly a larger grain size on the SZ, which is in good agreement with the effect of FSW process parameters on other materials [93].

In another study, Zhu et al. [88] studied a quaternary HEA composition, $Co_{16}Cr_{28}Fe_{28}Ni_{28}$, in order to study the effects of the reduced Co content on the material mechanical performance. Their work showed that after recrystallization, through thermomechanical processing, the tensile properties were superior to that of common HEAs, as depicted in Figure 9a. Such evidenced the possibility for the enhancement of the alloy mechanical properties through precipitation hardening, given the reduced proportion of Co. By performing FSW on this HEA, varying only the welding speed (ranging between 30 and 50 mm/min), a refined microstructure composed of equiaxed grains was obtained in the SZ, which remained with its original FCC crystal structure. However, at the higher level of welding speed, the formation of a kissing bond [94], which is characterized by the partial penetration of the weld, was inevitable, due to low heat input. In both cases, the formation of a white band was evidenced. Further analysis of this feature revealed the presence of W-rich and Cr-rich particles. The presence of W-rich particles can be explained by the welding tool wear, which is a common occurrence during FSW [95]. The presence of the Cr-rich particles was not explained, requiring further experimental work to justify its presence. Regarding the hardness of the joints, the SZ exhibited a relatively higher hardness than the BM, which was attributed to the distorted crystalline network, high fraction of deformation twins, and refined grain structure (see Figure 9b regarding the hardness profile obtained across the joint).

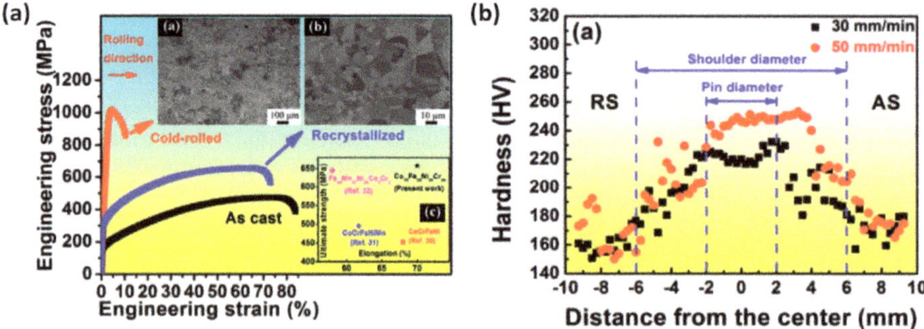

Figure 9. Sample characterization: (**a**) Tensile properties; (**b**) microhardness distribution (Adapted from [88], with permission from Elsevier, 2018).

During FSW, tool wear can occur, and debris can be incorporated in the processed material. The influence of tungsten and chromium carbide particles, caused by the tool deterioration during FSW of a CoCrFeNiMn HEA was accessed by Park et al. [89]. The process parameters comprised a welding speed of 30 mm/min, while the tool rotation varied between 400, 600, 800, and 1000 rpm. The results showed that both the welds and the BM exhibited a single FCC crystal structure. No cracks or voids were found on the welds, although the increase of tool rotation resulted in thinning near the center line, which corresponds to an inferior thickness of the weld when compared to the BM. The formation of a tornado-shaped region on the SZ was also evidenced when performing FSW with rotations speeds higher than 600 rpm (refer to Figure 10a). This tornado-shaped region was characterized by the formation of a secondary phase correspondent to W- and Cr-rich carbides, aided by the tool wear. Overall, superior characteristics regarding the hardness, tensile strength, and joint efficiency were obtained with a rotation speed of 800 rpm, where the grain size was at its lowest value. This was attributed to the different heat inputs that govern the solid-state transformation during the process. As depicted in Figure 10b, a comparison on the carbide size and concentration can be observed between a lower heat input (800 rpm) and a higher heat input sample (1000 rpm). These results show that higher rotation speeds lead to more severe wear of the FSW tool.

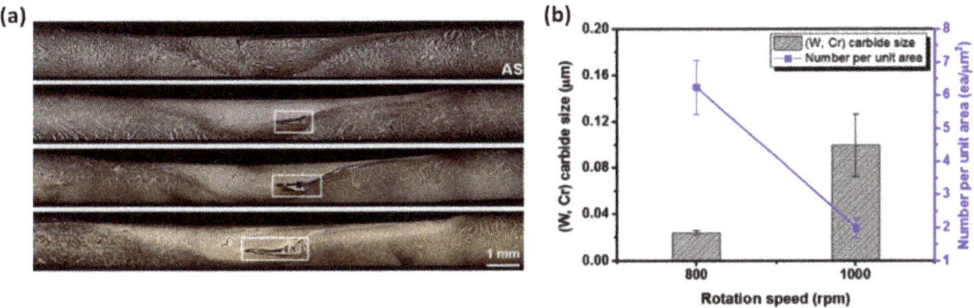

Figure 10. Characterization of friction stir welded (FSWed) CoCrFeNiMn joints: (**a**) Morphology of the welds at different rotation speeds (from top to bottom: 400, 600, 800, 1000 rpm); (**b**) carbide content at 800 rpm and 1000 rpm (Adapted from [89], with permission from Elsevier, 2019).

On another perspective, Shaysultanov et al. [90] performed FSW on a carbon-doped CoCrFeNiMn HEA, with the intent of studying the influence this controlled C addition on the welded joints mechanical performance. After being produced via thermite-type self-propagating high-temperature synthesis, the samples were cold-rolled and annealed at 900 °C for 1 h, to obtain an equiaxed microstructure. The welds resultant from the FSW process were defect-free, while microstructural differences were observed on the grain size on the BM and the SZ, with a change from 9.2 to 4.6 μm, respectively. With this joining process the proportion of $M_{23}C_6$ carbides increased, which was attributed to the rise in temperature triggered by intense plastic deformation, aiding in the precipitation of this phase. Overall, the mechanical performance of the welds exhibited higher values than the BM in both the microhardness (an increase of ≈40 HV) and tensile properties (an increase of ≈80 MPa on the ultimate tensile strength and of ≈200 MPa on the yield strength), which can be attributed to the carbides' precipitation.

As described above, most of the research work on solid-state welding of HEA focuses on FSW. However, other solid-state techniques have also started to be used to join this class of advanced materials.

Rotary friction welding was conducted in a eutectic $AlCoCrFeNi_{2.1}$ HEA by Li et al. [91]. As depicted in Figure 11a–c, similarly shaped joints were obtained with friction pressures of 80 and 120 MPa, while welding at 200 MPa resulted in an increase of the burn-off length and on the size of the resulting flash. Microstructurally, at the center of the weld, on the dynamic recrystallization zone (DRZ), the grains exhibited a refined and equiaxed structure. Additionally, on the TMAZ, the microstructure is composed of bent and elongated grains, while the HAZ exhibited fewer eutectic cells when compared to the BM. An EBSD analysis of the samples produced under the friction pressure of 120 MPa, revealed both an FCC phase, composed mainly by Fe, Co, and Cr, and a B2 phase, composed of AlNi intermetallic compounds and BCC structured-type CrFe precipitates. The tensile properties were superior when the friction pressure was of 200 MPa, where fracture of the joints occurred in the BM. As depicted in Figure 11d–f, the rough fractured surface is the result of the diferent ductility of the hard B2 phase and the soft FCC phase. Nevertheless, the specimens welded with 80 and 120 MPa of friction pressure yielded inferior tensile performance, fracturing at the joint interface, which is due to the the existence of a weld interface and discontinuous distribuition of the B2 phase on the DRZ region.

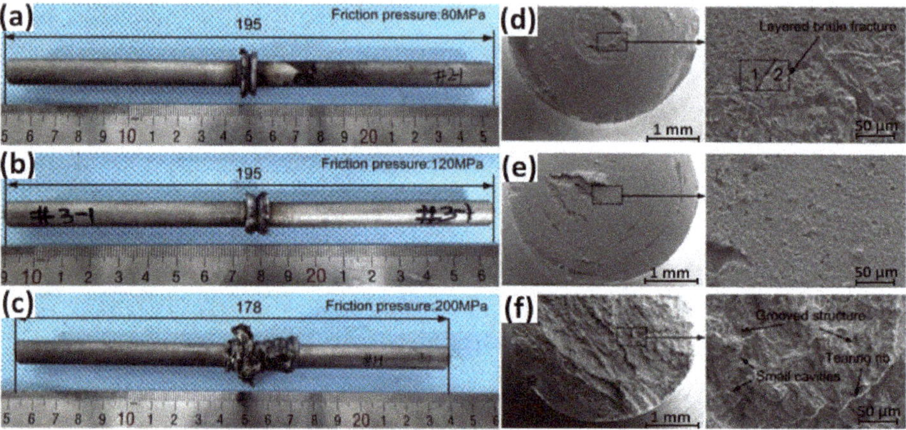

Figure 11. Morphology of the welds and fractured surfaces after tensile testing. The joints were obtained under the friction pressures of: (**a**) and (**d**) 80 MPa; (**b**) and (**e**) 120 MPa; (**c**) and (**f**) 200 MPa (Adapted from [91], with permission from Elsevier, 2020).

Another possibility for welding materials in the solid state is through diffusion bonding. Unlike other solid-state welding processes, this technique proves its purpose when joining materials with a high susceptibility to cracking, as in the case of refractory metals [96–98]. Lei et al. [92] studied vacuum diffusion bonding between the single-phase FCC $Al_{0.85}CoCrFeNi$ HEA and a TiAl alloy. For this purpose, an axial pressure of 30 MPa, a temperature range of 750 to 1050 °C, and a holding time of 30 to 120 min were used. Given the sluggish diffusion effect, characteristic of HEA systems, the atomic diffusion from the TiAl substrate into the HEA substrate was drastically inferior, when compared to the diffusion resultant from the HEA side. As depicted in Figure 12, the obtained bonds were characterized by having three distinct regions, which could be divided according with their microstructural composition: Region I, composed by α_2-Ti_3Al + solid strengthened γ-TiAl; region II, which can be expressed as $Al(Co, Ni)_2Ti$; and region III, characterized by an Cr(Fe, Co) solid solution phase. The formation of voids was noticeable in the interlayer between regions II and III, which was caused by the atomic flux imbalance provided by the process parameters in use. Overall, the optimal results yielded a maximum microhardness of 923 HV in region I, and a maximum shear strength of 71 MPa (at 850 °C and after 90 min of holding time).

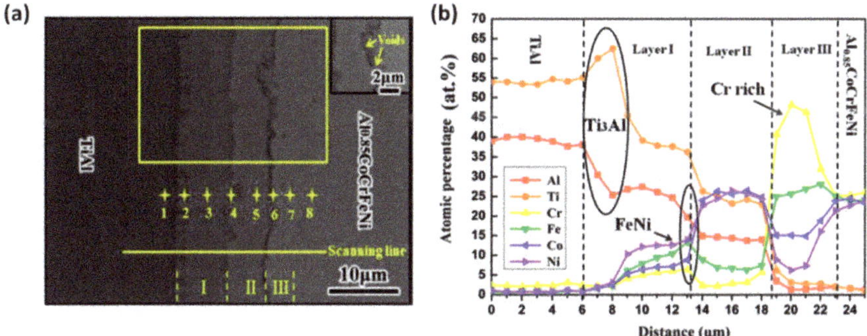

Figure 12. Characterization of the diffusion bonded TiAl/Al0.85CoCrFeNi joints using 950 °C/1 h/30 MPa: (**a**) microstructure under SEM; (**b**) Compositional element distribution taken from the scanning line region (Adapted from [92], with permission from Elsevier, 2020).

As it can be inferred from this work, multiple welding works on HEA currently exist, and increasing attention to these alloys' weldability is emerging. Table 2 compiles the existing studies on welding of different HEA, while Figure 13 details the relative importance of each welding technique already applied to HEAs. As it can be noted, most studies concern the CoCrFeNiMn HEA, showing the need to further extend these research works to other alloys systems.

Table 2. Summary of the welding techniques currently used on HEAs.

Welding Technique	Alloy System	Refs.
Brazing	CoCrFeNiMn CoCrFeNi	[52]
Laser Welding	CoCrFeNiMn TiNbCrV	[53–59] [69]
Gas Tungsten Arc Welding	AlCoCrFeNi AlCoCrCuFeNi CoCrFeNiMn	[64–66] [67,68] [61–63]
Electron Beam Welding	CoCrFeNiMn	[60,61]
Friction Stir Welding	CoCrFeNiMn AlCoCrFeNi CoCrFeNi	[55,86,89,90] [87] [88]
Rotary Friction Welding	AlCoCrFeNi	[91]
Diffusion Bonding	AlCoCrFeNi	[92]

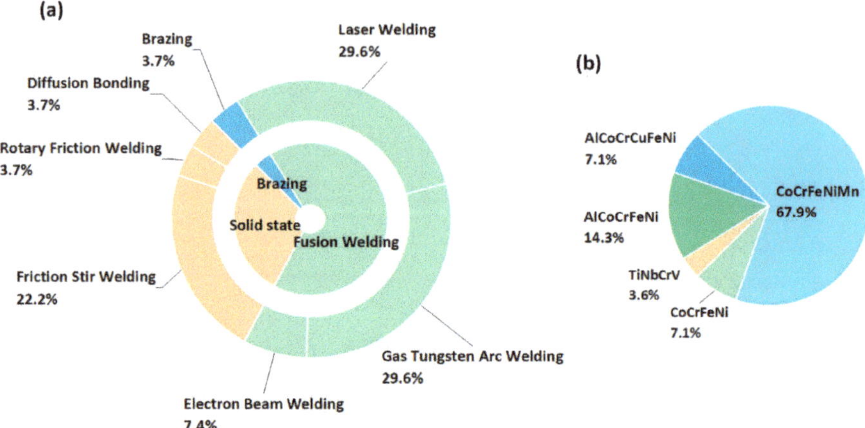

Figure 13. Percentage distribution of the number of papers considered in this study: (**a**) by welding technique; (**b**) by alloy system.

3. Summary and Conclusions

The present article offers an overview on the current status and progress on welding of high entropy alloys. The literature reveals that the main developments on this topic are mainly focused on fusion-based process by means of laser welding, whereas solid-state ones focus on being put on friction stir welding. Nevertheless, it is highly noticeable that the improvement of the welding processes regarding the feasibility of structural and functional applications of HEAs is at its beginning stage.

Regarding fusion-based processes, the current developments show that a step towards the optimization of laser-based methods is being taken. Nevertheless, a need for research on the joining of HEAs through other fusion-based techniques, that are economically more viable than laser-based approaches, is of great interest. Additionally, beyond the welding process, the initial condition of the alloy affects the weld microstructure and its behavior. Such a matter is important to study the four core effects characteristic of HEAs and their impact on the microstructural transformations on the molten pool and mechanical performance of the welds.

Concerning solid-state processes, the literature shows that success on joining CoCrFeNiMn HEAs has been achieved, although there is still a need for the optimization of the process parameters to accomplish high performing joints. Regarding their mechanical behavior, the joints show good mechanical properties, although there is still a need to enhance such solid-state processes into what that makes HEAs competitive, e.g., their outstanding behavior in extreme conditions.

Currently, special emphasis is being put on the most studied CoCrFeNiMn system. However, to further expand the potential applications of HEAs, other alloy systems need to be explored. No major work on dissimilar joints exists yet. This is another key area of interest where the need to couple the properties of different materials is of great interest.

Currently, it is well-known that HEAs can incorporate impurities during their casting. These impurities can lead to the formation on unexpected phases, especially under the non-equilibrium solidification of fusion-based processes. These unexpected phases were already observed for the CoCrFeMnNi system, and were seen to improve the joint mechanical properties. However, it is not necessarily true that these impurities will always be beneficial for the joint microstructure and performance, and as such, efforts should be made to improve the chemical homogeneity of the starting BMs, but also address the potential formation of these phases and propose methods to mitigate them.

All in all, the future of welding of HEA is at its early stages with the potential to further expand the potential applications of these advanced engineering materials.

Author Contributions: Both authors contributed equally to the research of the literature and writing of the review paper. All authors have read and agreed to the published version of the manuscript.

Funding: This research was funded by Fundação para a Ciência e a Tecnologia (FCT - MCTES) via the project UIDB/00667/2020 (UNIDEMI).

Acknowledgments: JGL and JPO acknowledge Fundação para a Ciência e a Tecnologia (FCT - MCTES) for its financial support via the project UIDB/00667/2020 (UNIDEMI).

Conflicts of Interest: The authors declare no conflict of interest.

References

1. Cantor, B.; Chang, I.T.H.; Knight, P.; Vincent, A.J.B. Microstructural development in equiatomic multicomponent alloys. *Mater. Sci. Eng. A* **2004**, *375–377*, 213–218. [CrossRef]
2. Yeh, J.W.; Chen, S.K.; Lin, S.J.; Gan, J.Y.; Chin, T.S.; Shun, T.T.; Tsau, C.H.; Chang, S.Y. Nanostructured high-entropy alloys with multiple principal elements: Novel alloy design concepts and outcomes. *Adv. Eng. Mater.* **2004**, *6*, 299–303. [CrossRef]
3. Miracle, D.B.; Senkov, O.N. A critical review of high entropy alloys and related concepts. *Acta Mater.* **2017**, *122*, 448–511. [CrossRef]
4. Gao, M.C.; Liaw, P.K.; Yeh, J.W.; Zhang, Y. *High.-Entropy Alloys: Fundamentals and Applications*; Springer International Publishing: Cham, Switzerland, 2016; ISBN 9783319270135.
5. Zhang, Y.; Zuo, T.T.; Tang, Z.; Gao, M.C.; Dahmen, K.A.; Liaw, P.K.; Lu, Z.P. Microstructures and properties of high-entropy alloys. *Prog. Mater Sci.* **2014**, *61*, 1–93. [CrossRef]
6. Murty, B.S.; Yeh, J.W.; Ranganathan, S.; Bhattacharjee, P.P. *High-Entropy Alloys*, 2nd ed.; Elsevier: Amsterdam, The Netherlands, 2019; ISBN 9780128160671.
7. Ikeda, Y.; Grabowski, B.; Körmann, F. Ab initio phase stabilities and mechanical properties of multicomponent alloys: A comprehensive review for high entropy alloys and compositionally complex alloys. *Mater. Charact.* **2019**, *147*, 464–511. [CrossRef]
8. Li, W.; Liaw, P.K.; Gao, Y. Fracture resistance of high entropy alloys: A review. *Intermetallics* **2018**, *99*, 69–83. [CrossRef]
9. Li, Z.; Zhao, S.; Ritchie, R.O.; Meyers, M.A. Mechanical properties of high-entropy alloys with emphasis on face-centered cubic alloys. *Prog. Mater Sci.* **2019**, *102*, 296–345. [CrossRef]
10. Chen, J.; Zhou, X.; Wang, W.; Liu, B.; Lv, Y.; Yang, W.; Xu, D.; Liu, Y. A review on fundamental of high entropy alloys with promising high–temperature properties. *J. Alloys Compd.* **2018**, *760*, 15–30. [CrossRef]
11. Couzinié, J.P.; Dirras, G. Body-centered cubic high-entropy alloys: From processing to underlying deformation mechanisms. *Mater. Charact.* **2019**, *147*, 533–544. [CrossRef]
12. George, E.P.; Curtin, W.A.; Tasan, C.C. High entropy alloys: A focused review of mechanical properties and deformation mechanisms. *Acta Mater.* **2019**. [CrossRef]
13. Shi, Y.; Yang, B.; Liaw, P.K. Corrosion-Resistant High-Entropy Alloys: A Review. *Metals* **2017**, *7*, 43. [CrossRef]
14. George, E.P.; Raabe, D.; Ritchie, R.O. High-entropy alloys. *Nat. Rev. Mater.* **2019**, *4*, 515–534. [CrossRef]
15. Tsai, M.-H.; Yeh, J.-W. High-Entropy Alloys: A Critical Review. *Mater. Res. Lett.* **2014**, *2*, 107–123. [CrossRef]
16. Senkov, O.N.; Miracle, D.B.; Chaput, K.J.; Couzinie, J.-P. Development and exploration of refractory high entropy alloys—A review. *J. Mater. Res.* **2018**, *33*, 3092–3128. [CrossRef]
17. Zhang, W.; Liaw, P.K.; Zhang, Y. Science and technology in high-entropy alloys. *Sci. China Mater.* **2018**, *61*, 2–22. [CrossRef]
18. Yeh, J.W. Recent progress in high-entropy alloys. *Ann. Chim. Sci. Mat.* **2006**, *31*, 633–648. [CrossRef]
19. Ng, C.; Guo, S.; Luan, J.; Shi, S.; Liu, C.T. Entropy-driven phase stability and slow diffusion kinetics in an $Al_{0.5}CoCrCuFeNi$ high entropy alloy. *Intermetallics* **2012**, *31*, 165–172. [CrossRef]
20. Wu, Y.D.; Cai, Y.H.; Chen, X.H.; Wang, T.; Si, J.J.; Wang, L.; Wang, Y.D.; Hui, X.D. Phase composition and solid solution strengthening effect in TiZrNbMoV high-entropy alloys. *Mater. Des.* **2015**, *83*, 651–660. [CrossRef]
21. He, Q.; Yang, Y. On Lattice Distortion in High Entropy Alloys. *Front. Mater.* **2018**, *5*. [CrossRef]
22. Pickering, E.J.; Muñoz-Moreno, R.; Stone, H.J.; Jones, N.G. Precipitation in the equiatomic high-entropy alloy CrMnFeCoNi. *Scr. Mater.* **2016**, *113*, 106–109. [CrossRef]

23. Bhattacharjee, P.P.; Sathiaraj, G.D.; Zaid, M.; Gatti, J.R.; Lee, C.; Tsai, C.-W.; Yeh, J.-W. Microstructure and texture evolution during annealing of equiatomic CoCrFeMnNi high-entropy alloy. *J. Alloys Compd.* **2014**, *587*, 544–552. [CrossRef]
24. Dąbrowa, J.; Zajusz, M.; Kucza, W.; Cieślak, G.; Berent, K.; Czeppe, T.; Kulik, T.; Danielewski, M. Demystifying the sluggish diffusion effect in high entropy alloys. *J. Alloys Compd.* **2019**, *783*, 193–207. [CrossRef]
25. Nene, S.S.; Frank, M.; Liu, K.; Sinha, S.; Mishra, R.S.; McWilliams, B.A.; Cho, K.C. Corrosion-resistant high entropy alloy with high strength and ductility. *Scr. Mater.* **2019**, *166*, 168–172. [CrossRef]
26. Pathak, S.; Kumar, N.; Mishra, R.S.; De, P.S. Aqueous Corrosion Behavior of Cast CoCrFeMnNi Alloy. *J. Mater. Eng. Perform.* **2019**, *28*, 5970–5977. [CrossRef]
27. Shang, C.; Axinte, E.; Sun, J.; Li, X.; Li, P.; Du, J.; Qiao, P.; Wang, Y. CoCrFeNi($W_{1-x}Mo_x$) high-entropy alloy coatings with excellent mechanical properties and corrosion resistance prepared by mechanical alloying and hot pressing sintering. *Mater. Des.* **2017**, *117*, 193–202. [CrossRef]
28. Shi, Y.; Yang, B.; Xie, X.; Brechtl, J.; Dahmen, K.A.; Liaw, P.K. Corrosion of Al xCoCrFeNi high-entropy alloys: Al-content and potential scan-rate dependent pitting behavior. *Corros. Sci.* **2017**, *119*, 33–45. [CrossRef]
29. Shukla, S.; Wang, T.; Cotton, S.; Mishra, R.S. Hierarchical microstructure for improved fatigue properties in a eutectic high entropy alloy. *Scr. Mater.* **2018**, *156*, 105–109. [CrossRef]
30. Huo, W.; Fang, F.; Liu, X.; Tan, S.; Xie, Z.; Jiang, J. Fatigue resistance of nanotwinned high-entropy alloy films. *Mater. Sci. Eng. A* **2019**, *739*, 26–30. [CrossRef]
31. Wu, J.M.; Lin, S.J.; Yeh, J.W.; Chen, S.K.; Huang, Y.S.; Chen, H.C. Adhesive wear behavior of Al_xCoCrCuFeNi high-entropy alloys as a function of aluminum content. *Wear* **2006**, *261*, 513–519. [CrossRef]
32. Kong, D.; Guo, J.; Liu, R.; Zhang, X.; Song, Y.; Li, Z.; Guo, F.; Xing, X.; Xu, Y.; Wang, W. Effect of remelting and annealing on the wear resistance of AlCoCrFeNi$Ti_{0.5}$ high entropy alloys. *Intermetallics* **2019**, *114*. [CrossRef]
33. Yang, S.; Liu, Z.; Pi, J. Microstructure and wear behavior of the AlCrFeCoNi high-entropy alloy fabricated by additive manufacturing. *Mater. Lett.* **2019**, 127004. [CrossRef]
34. Joseph, J.; Haghdadi, N.; Shamlaye, K.; Hodgson, P.; Barnett, M.; Fabijanic, D. The sliding wear behaviour of CoCrFeMnNi and Al_xCoCrFeNi high entropy alloys at elevated temperatures. *Wear* **2019**, *428–429*, 32–44. [CrossRef]
35. Fang, Y.; Chen, N.; Du, G.; Zhang, M.; Zhao, X.; Cheng, H.; Wu, J. High-temperature oxidation resistance, mechanical and wear resistance properties of Ti(C,N)-based cermets with Al0.3CoCrFeNi high-entropy alloy as a metal binder. *J. Alloys Compd.* **2020**, *815*, 152486. [CrossRef]
36. Klimova, M.V.; Semenyuk, A.O.; Shaysultanov, D.G.; Salishchev, G.A.; Zherebtsov, S.V.; Stepanov, N.D. Effect of carbon on cryogenic tensile behavior of CoCrFeMnNi-type high entropy alloys. *J. Alloys Compd.* **2019**, *811*, 152000. [CrossRef]
37. Jo, Y.H.; Doh, K.Y.; Kim, D.G.; Lee, K.; Kim, D.W.; Sung, H.; Sohn, S.S.; Lee, D.; Kim, H.S.; Lee, B.J.; et al. Cryogenic-temperature fracture toughness analysis of non-equi-atomic $V_{10}Cr_{10}Fe_{45}Co_{20}Ni_{15}$ high-entropy alloy. *J. Alloys Compd.* **2019**, *809*. [CrossRef]
38. He, Z.F.; Jia, N.; Ma, D.; Yan, H.L.; Li, Z.M.; Raabe, D. Joint contribution of transformation and twinning to the high strength-ductility combination of a FeMnCoCr high entropy alloy at cryogenic temperatures. *Mater. Sci. Eng. A* **2019**, *759*, 437–447. [CrossRef]
39. Yeh, J.W.; Lin, S.J. Breakthrough applications of high-entropy materials. *J. Mater. Res.* **2018**, *33*, 3129–3137. [CrossRef]
40. Miracle, D.B.; Miller, J.D.; Senkov, O.N.; Woodward, C.; Uchic, M.D.; Tiley, J. Exploration and Development of High Entropy Alloys for Structural Applications. *Entropy* **2014**, *16*, 494–525. [CrossRef]
41. Miracle, D.B. Critical Assessment 14: High entropy alloys and their development as structural materials. *Mater. Sci. Technol.* **2015**, *31*, 1142–1147. [CrossRef]
42. Gludovatz, B.; Hohenwarter, A.; Catoor, D.; Chang, E.H.; George, E.P.; Ritchie, R.O. A fracture-resistant high-entropy alloy for cryogenic applications. *Science* **2014**, *345*, 1153–1158. [CrossRef]
43. Stepanov, N.D.; Shaysultanov, D.G.; Tikhonovsky, M.A.; Zherebtsov, S.V. Structure and high temperature mechanical properties of novel non-equiatomic Fe-(Co, Mn)-Cr-Ni-Al-(Ti) high entropy alloys. *Intermetallics* **2018**, *102*, 140–151. [CrossRef]
44. Wang, M.; Ma, Z.; Xu, Z.; Cheng, X. Microstructures and mechanical properties of HfNbTaTiZrW and HfNbTaTiZrMoW refractory high-entropy alloys. *J. Alloys Compd.* **2019**, *803*, 778–785. [CrossRef]

45. Senkov, O.N.; Scott, J.M.; Senkova, S.V.; Meisenkothen, F.; Miracle, D.B.; Woodward, C.F. Microstructure and elevated temperature properties of a refractory TaNbHfZrTi alloy. *J. Mater. Sci.* **2012**, *47*, 4062–4074. [CrossRef]
46. Senkov, O.N.; Wilks, G.B.; Scott, J.M.; Miracle, D.B. Mechanical properties of $Nb_{25}Mo_{25}Ta_{25}W_{25}$ and $V_{20}Nb_{20}Mo_{20}Ta_{20}W_{20}$ refractory high entropy alloys. *Intermetallics* **2011**, *19*, 698–706. [CrossRef]
47. Oliveira, J.P.; Miranda, R.M.; Braz Fernandes, F.M. Welding and Joining of NiTi Shape Memory Alloys: A Review. *Prog. Mater Sci.* **2017**, *88*, 412–466. [CrossRef]
48. Oliveira, J.P.; Santos, T.G.; Miranda, R.M. Revisiting fundamental welding concepts to improve additive manufacturing: From theory to practice. *Prog. Mater Sci.* **2019**, *107*, 100590. [CrossRef]
49. Oliveira, J.P.; Ponder, K.; Brizes, E.; Abke, T.; Ramirez, A.J.; Edwards, P. Combining resistance spot welding and friction element welding for dissimilar joining of aluminum to high strength steels. *J. Mater. Process. Technol.* **2019**, *273*, 116192. [CrossRef]
50. Nandan, R.; Debroy, T.; Bhadeshia, H.K.D.H. Recent advances in friction-stir welding—Process, weldment structure and properties. *Prog. Mater Sci.* **2008**, *53*, 980–1023. [CrossRef]
51. Guo, J.; Tang, C.; Rothwell, G.; Li, L.; Wang, Y.-C.; Yang, Q.; Ren, X. Welding of High Entropy Alloys—A Review. *Entropy* **2019**, *21*, 431. [CrossRef]
52. Lin, C.; Shiue, R.K.; Wu, S.K.; Lin, Y.S. Dissimilar infrared brazing of CoCrFe(Mn)Ni equiatomic high entropy alloys and 316 stainless steel. *Crystals* **2019**, *9*, 518. [CrossRef]
53. Kashaev, N.; Ventzke, V.; Stepanov, N.; Shaysultanov, D.; Sanin, V.; Zherebtsov, S. Laser beam welding of a CoCrFeNiMn-type high entropy alloy produced by self-propagating high-temperature synthesis. *Intermetallics* **2018**, *96*, 63–71. [CrossRef]
54. Kashaev, N.; Ventzke, V.; Petrov, N.; Horstmann, M.; Zherebtsov, S.; Shaysultanov, D.; Sanin, V.; Stepanov, N. Fatigue behaviour of a laser beam welded CoCrFeNiMn-type high entropy alloy. *Mater. Sci. Eng. A* **2019**, *766*, 138358. [CrossRef]
55. Jo, M.G.; Kim, H.J.; Kang, M.; Madakashira, P.P.; Park, E.S.; Suh, J.Y.; Kim, D.I.; Hong, S.T.; Han, H.N. Microstructure and mechanical properties of friction stir welded and laser welded high entropy alloy CrMnFeCoNi. *Met. Mater. Int.* **2018**, *24*, 73–83. [CrossRef]
56. Nam, H.; Park, C.; Moon, J.; Na, Y.; Kim, H.; Kang, N. Laser weldability of cast and rolled high-entropy alloys for cryogenic applications. *Mater. Sci. Eng. A* **2019**, *742*, 224–230. [CrossRef]
57. Nam, H.; Park, S.; Chun, E.; Kim, H. Laser dissimilar weldability of cast and rolled CoCrFeMnNi high-entropy alloys for cryogenic applications. *Sci. Technol. Weld. Joining* **2020**, *25*, 127–134. [CrossRef]
58. Chen, Z.; Wang, B.; Duan, B.; Zhang, X. Mechanical properties and microstructure of laser welded FeCoNiCrMn high-entropy alloy. *Mater. Lett.* **2019**, *262*, 127060. [CrossRef]
59. Nam, H.; Park, C.; Kim, C.; Kim, H.; Kang, N. Effect of post weld heat treatment on weldability of high entropy alloy welds. *Sci. Technol. Weld. Joining* **2018**, *23*, 420–427. [CrossRef]
60. Wu, Z.; David, S.A.A.; Feng, Z.; Bei, H. Weldability of a high entropy CrMnFeCoNi alloy. *Scr. Mater.* **2016**, *124*, 81–85. [CrossRef]
61. Wu, Z.; David, S.A.; Leonard, D.N.; Feng, Z.; Bei, H. Microstructures and mechanical properties of a welded CoCrFeMnNi high-entropy alloy. *Sci. Technol. Weld. Joining* **2018**, *23*, 585–595. [CrossRef]
62. Oliveira, J.P.; Curado, T.M.; Zeng, Z.; Lopes, J.G.; Rossinyol, E.; Park, J.M.; Schell, N.; Fernandes, F.M.B.; Kim, H.S. Gas tungsten arc welding of as-rolled CrMnFeCoNi high entropy alloy. *Mater. Des.* **2020**, *189*, 108505. [CrossRef]
63. Nam, H.; Park, S.; Park, N.; Na, Y.; Kim, H.; Yoo, S.-J.; Moon, Y.-H.; Kang, N. Weldability of cast CoCrFeMnNi high-entropy alloys using various filler metals for cryogenic applications. *J. Alloys Compd.* **2020**, *819*, 153278. [CrossRef]
64. Sokkalingam, R.; Sivaprasad, K.; Duraiselvam, M.; Muthupandi, V.; Prashanth, K.G. Novel welding of Al0.5CoCrFeNi high-entropy alloy: Corrosion behavior. *J. Alloys Compd.* **2020**, *817*, 153163. [CrossRef]
65. Sokkalingam, R.; Muthupandi, V.; Sivaprasad, K.; Prashanth, K.G. Dissimilar welding of $Al_{0.1}$CoCrFeNi high-entropy alloy and AISI304 stainless steel. *J. Mater. Res.* **2019**, 1–12. [CrossRef]
66. Sokkalingam, R.; Mishra, S.; Cheethirala, S.R.; Muthupandi, V.; Sivaprasad, K. Enhanced Relative Slip Distance in Gas-Tungsten-Arc-Welded Al0.5CoCrFeNi High-Entropy Alloy. *Metall. Mater. Trans. A* **2017**, *48*, 3630–3634. [CrossRef]

67. Martin, A.C.; Fink, C. Initial weldability study on $Al_{0.5}CrCoCu_{0.1}FeNi$ high-entropy alloy. *Weld. World* **2019**, *63*, 739–750. [CrossRef]
68. Martin, A.C.; Oliveira, J.P.; Fink, C. Elemental Effects on Weld Cracking Susceptibility in $Al_xCoCrCu_yFeNi$ High-Entropy Alloy. *Metall. Mater. Trans. A* **2019**, *51*, 778–787. [CrossRef]
69. Panina, E.; Yurchenko, N.; Zherebtsov, S.; Stepanov, N.; Salishchev, G.; Ventzke, V.; Dinse, R.; Kashaev, N. Laser Beam Welding of a Low Density Refractory High Entropy Alloy. *Metals* **2019**, *9*, 1351. [CrossRef]
70. Wang, P.; Huang, P.; Ng, F.L.; Sin, W.J.; Lu, S.; Nai, M.L.S.; Dong, Z.L.; Wei, J. Additively manufactured CoCrFeNiMn high-entropy alloy via pre-alloyed powder. *Mater. Des.* **2019**, *168*, 107576. [CrossRef]
71. Wang, B.; Yao, X.; Wang, C.; Zhang, X.; Huang, X. Mechanical properties and microstructure of a NiCrFeCoMn high-entropy alloy deformed at high strain rates. *Entropy* **2018**, *20*, 892. [CrossRef]
72. Ma, D.; Grabowski, B.; Körmann, F.; Neugebauer, J.; Raabe, D. Ab initio thermodynamics of the CoCrFeMnNi high entropy alloy: Importance of entropy contributions beyond the configurational one. *Acta Mater.* **2015**, *100*, 90–97. [CrossRef]
73. Subrahmanyam, J.; Vijayakumar, M. Self-propagating high-temperature synthesis. *J. Mater. Sci* **1992**, *27*, 6249–6273. [CrossRef]
74. Wang, Z.; Oliveira, J.P.; Zeng, Z.; Bu, X.; Peng, B.; Shao, X. Laser beam oscillating welding of 5A06 aluminum alloys: Microstructure, porosity and mechanical properties. *Opt. Laser Technol.* **2019**, *111*, 58–65. [CrossRef]
75. Gu, J.; Ni, S.; Liu, Y.; Song, M. Regulating the strength and ductility of a cold rolled FeCrCoMnNi high-entropy alloy via annealing treatment. *Mater. Sci. Eng. A* **2019**, *755*, 289–294. [CrossRef]
76. Khodabakhshi, F.; Gerlich, A.P. On the stability, microstructure, and mechanical property of powder metallurgy Al–SiC nanocomposites during similar and dissimilar laser welding. *Mater. Sci. Eng. A* **2019**, *759*, 688–702. [CrossRef]
77. Otto, F.; Dlouhý, A.; Somsen, C.; Bei, H.; Eggeler, G.; George, E.P. The influences of temperature and microstructure on the tensile properties of a CoCrFeMnNi high-entropy alloy. *Acta Mater.* **2013**, *61*, 5743–5755. [CrossRef]
78. Montazeri, M.; Ghaini, F.M.; Farnia, A. An investigation into the microstructure and weldability of a tantalum-containing cast cobalt-based superalloy. *Int. J. Mater. Res.* **2011**, *102*, 1446–1451. [CrossRef]
79. Henderson, M.B.; Arrell, D.; Larsson, R.; Heobel, M.; Marchant, G. Nickel based superalloy welding practices for industrial gas turbine applications. *Sci. Technol. Weld. Joining* **2004**, *9*, 13–21. [CrossRef]
80. Elangovan, K.; Balasubramanian, V. Influences of post-weld heat treatment on tensile properties of friction stir-welded AA6061 aluminum alloy joints. *Mater. Charact.* **2008**, *59*, 1168–1177. [CrossRef]
81. Köse, C.; Kaçar, R. The effect of preheat & post weld heat treatment on the laser weldability of AISI 420 martensitic stainless steel. *Mater. Des.* **2014**, *64*, 221–226.
82. Oliveira, J.P.; Panton, B.; Zeng, Z.; Andrei, C.M.; Zhou, Y.; Miranda, R.M.; Fernandes, F.M.B. Laser joining of NiTi to Ti6Al4V using a Niobium interlayer. *Acta Mater.* **2016**, *105*, 9–15. [CrossRef]
83. Miranda, R.M.; Assunção, E.; Silva, R.J.C.; Oliveira, J.P.; Quintino, L. Fiber laser welding of NiTi to Ti-6Al-4V. *In. J. Adv. Manuf. Technol.* **2015**, *81*, 1533–1538. [CrossRef]
84. Young, G.A.; Capobianco, T.E.; Penik, M.A.; Morris, B.W.; McGee, J.J. The mechanism of ductility dip cracking in nickel-chromium alloys. *Weld. J.* **2008**, *87*, 31S–43S.
85. Ramirez, A.J.; Sowards, J.W.; Lippold, J.C. Improving the ductility-dip cracking resistance of Ni-base alloys. *J. Mater. Process. Technol.* **2006**, *179*, 212–218. [CrossRef]
86. Xu, N.; Song, Q.; Bao, Y. Microstructure evolution and mechanical properties of friction stir welded FeCrNiCoMn high-entropy alloy. *Mater. Sci. Technol.* **2019**, *35*, 577–584. [CrossRef]
87. Zhu, Z.G.; Sun, Y.F.; Goh, M.H.; Ng, F.L.; Nguyen, Q.B.; Fujii, H.; Nai, S.M.L.; Wei, J.; Shek, C.H. Friction stir welding of a $CoCrFeNiAl_{0.3}$ high entropy alloy. *Mater. Lett.* **2017**, *205*, 142–144. [CrossRef]
88. Zhu, Z.G.; Sun, Y.F.; Ng, F.L.; Goh, M.H.; Liaw, P.K.; Fujii, H.; Nguyen, Q.B.; Xu, Y.; Shek, C.H.; Nai, S.M.L.; et al. Friction-stir welding of a ductile high entropy alloy: microstructural evolution and weld strength. *Mater. Sci. Eng. A* **2018**, *711*, 524–532. [CrossRef]
89. Park, S.; Park, C.; Na, Y.; Kim, H.S.; Kang, N. Effects of (W, Cr) carbide on grain refinement and mechanical properties for CoCrFeMnNi high entropy alloys. *J. Alloys Compd.* **2019**, *770*, 222–228. [CrossRef]
90. Shaysultanov, D.; Stepanov, N.; Malopheyev, S.; Vysotskiy, I.; Sanin, V.; Mironov, S.; Kaibyshev, R.; Salishchev, G.; Zherebtsov, S. Friction stir welding of a carbon-doped CoCrFeNiMn high-entropy alloy. *Mater. Charact.* **2018**, *145*, 353–361. [CrossRef]

91. Li, P.; Sun, H.; Wang, S.; Hao, X.; Dong, H. Rotary friction welding of $AlCoCrFeNi_{2.1}$ eutectic high entropy alloy. *J. Alloys Compd.* **2020**, *814*. [CrossRef]
92. Lei, Y.; Hu, S.P.; Yang, T.L.; Song, X.G.; Luo, Y.; Wang, G.D. Vacuum diffusion bonding of high-entropy $Al_{0.85}CoCrFeNi$ alloy to TiAl intermetallic. *J. Mater. Process. Technol.* **2020**, *278*, 116455. [CrossRef]
93. Oliveira, J.P.; Duarte, J.F.; Inácio, P.; Schell, N.; Miranda, R.M.; Santos, T.G. Production of Al/NiTi composites by friction stir welding assisted by electrical current. *Mater. Des.* **2017**, *113*, 311–318. [CrossRef]
94. Khan, N.Z.; Siddiquee, A.N.; Khan, Z.A.; Shihab, S.K. Investigations on tunneling and kissing bond defects in FSW joints for dissimilar aluminum alloys. *J. Alloys Compd.* **2015**, *648*, 360–367. [CrossRef]
95. Costa, A.M.S.; Oliveira, J.P.; Pereira, V.F.; Nunes, C.A.; Ramirez, A.J.; Tschiptschin, A.P. Ni-based Mar-M247 superalloy as a friction stir processing tool. *J. Mater. Process. Technol.* **2018**, *262*, 605–614. [CrossRef]
96. Yan, H.; Fan, J.; Han, Y.; Yao, Q.; Liu, T.; Lv, Y.; Zhang, C. Vacuum diffusion bonding W to W-Cu composite: Interfacial microstructure and mechanical properties. *Vacuum* **2019**, *165*, 19–25. [CrossRef]
97. Zhang, P.; Li, Y.; Chen, Z.; Zhang, J.; Shen, B. Oxidation response of a vacuum arc melted NbZrTiCrAl refractory high entropy alloy at 800–1200 °C. *Vacuum* **2019**, *162*, 20–27. [CrossRef]
98. Yao, Q.; Cheng, H.; Fan, J.; Yan, H.; Zhang, C. High strength Mo/Ti6Al4V diffusion bonding joints: Interfacial microstructure and mechanical properties. *Int. J. Refract. Met. Hard Mater* **2019**, *82*, 159–166. [CrossRef]

Article

Mechanical Properties of Friction Stir Welded AA1050-H14 and AA5083-H111 Joint: Sampling Aspect

Velaphi Msomi * and Nontle Mbana

Cape Peninsula University of Technology, Faculty of Engineering and the Built Environment, Mechanical Engineering Department, P.O. Box 1906, Bellville 7535, South Africa; nontle.mbana@gmail.com
* Correspondence: msomiv@gmail.com; Tel.: +27-21-953-8627

Received: 10 December 2019; Accepted: 25 January 2020; Published: 3 February 2020

Abstract: Welding of dissimilar aluminium alloys has been a challenge for a long period until the discovery of the solid-state welding technique called friction stir welding (FSW). The discovery of this technique encouraged different research interests revolving around the optimization of this technique. This involves the welding parameters optimization and this optimization is categorized into two classes, i.e., similar alloys and dissimilar alloys. This paper reports about the mechanical properties of the friction stir welded dissimilar AA1050-H14 and AA5083-H111 joint. The main focus is to compare the mechanical properties of specimens extracted from different locations of the welds, i.e., the beginning, middle, and the end of the weld. The specimen extracted at the beginning of the weld showed low tensile properties compared to specimens extracted from different locations of the weld. There was no certain trend noted through the bending results. All three specimens showed dimpled fracture, which is the characterization of the ductile fracture.

Keywords: tensile strength; flexural strength; friction stir welding; microstructure; dissimilar aluminium alloys

---

1. Introduction

Friction stir welding (FSW) is classified as one of the welding techniques that joins materials through heat generated during friction occurring between the tool shoulder and the workpieces [1]. The invention of this welding technique was based on the materials which were classified as un-weldable materials through conventional methods. Those classified materials included certain classes of aluminium alloys. Aluminium alloys are mostly used in many industries like aviation, shipbuilding, and automotive because of their light weight. The focus towards FSW has expanded such that it includes testing the capability of the technique in welding other materials that are outside the aluminium class. This includes the welding of copper and its alloys, titanium and magnesium and its alloys [2]. When the aluminium alloys are welded with the conventional technique, they are likely to have weld splitting on the joint line. The fusion welding of aluminium alloys is more difficult than the welding of steel due to their low melting point, softness, and so forth. The aluminium alloys sometimes bend and shrink when welded using conventional welding and those effects are caused by residual thermal stress [3].

In some materials, FSW possesses good metallurgical properties when compared to fusion welding, and this is caused by the microstructural modification that occurs during welding. Mishra and Ma [4] reported that to have a great weld and to avoid defects on the weld, it is important to take welding parameters, material flow, and heat generation into consideration during the welding process.

There are mainly four crucial steps that are involved during the performance of FSW, i.e., plunging, dwelling, welding, and pulling. The rotating tool is inserted into the butt joint slowly until the shoulder touches the surface of the workpieces (plunging). The plunged tool remains in one location for a certain period with the purpose of building up input heat (dwelling). The plunged tool moves along the plates being joined at a specified speed (welding). The rotating tool gets removed vertically from the welded plates soon after reaching the ends of the plates (pulling), and this normally leaves a hole that indicates the end of the weld [5].

There are several studies focusing on FSW of aluminium alloys [6]. Typical examples include the production of Al/NiTi composites by FSW assisted by electrical current, analysis of welding properties in FSW of AA6351 plates added with silicon carbide particles, and dynamics of rotational flow in FSW of aluminium alloys [7–9]. It has been reported that for the production of the good weld, the tool geometry and welding parameters play a very important role. This includes the rotation speed, traverse speed, tool tilt angle, and plunged depth. Plunge depth has been found to be a critical parameter in the heat generation and for proper consolidation of material without defects. The plunge depth was also identified as one of the parameters that plays an important role towards the microstructural arrangement of the joint [10].

Welding dissimilar materials is quite challenging when compared with similar welding materials due to the difference in mechanical properties and chemical composition of the base materials. To acquire better weld mechanical properties, the harder material must be placed on the retreating and softer material placed in the advancing side [11,12]. The tool geometry plays a very important role in welding dissimilar materials. Welding dissimilar alloys requires the use of different tool profiles such as threaded, squared, and triangular profiles to transfer the material from top of the joint to bottom and vice versa by stirring movement [13]. Kundu and Singh [14] reported that tool pin profile geometry plays an important role in weld quality, while the surface quality of the weld joint depends upon the tool tilt angle. The increase in tool tilt angle affects the flow and fill up of material during welding.

In most cases, the welding of dissimilar materials involves the welding of aluminium alloys, which are not far from each other in terms of mechanical properties, e.g., 5xxx will be welded together with 6xxx, 6xxx welded with 7xxx, etc. Recently, there are attempts that have been made in trying to weld the aluminium alloys that are mechanically far apart from each other, i.e., FSW of AA2024 to AA6061 [15]. This investigation used a single- and a dual-pin tool. The defect-free joint was obtained in all selected parameters except the welding speed beyond 90 mm/min. The onion rings were visible on the nugget region for joints produced using the dual-pin tool but absent on the single-pin tool. The highest ultimate tensile strength (UTS) was obtained with the dual-pin tool at a welding speed of 150 mm/min, whereas the single-pin produced the UTS at a welding speed of 90 mm/min. However, the UTS provided by single-pin was always less than the one produced by the dual-pin tool.

There are various aspects that have been studied through the use of dissimilar materials. This involves the analysis of the strain hardening behaviour on the friction stir welded dissimilar alloys, which are mechanically far apart from each other, i.e., 2024-T351 and 5083-H112, 2024-T351 and 7075-T651 [16]. This analysis was performed on two types of joints, i.e., 2024-T351 and 5083-H112, with 2024-T351 on the advancing side and 5083-H112 on the retreating side. The second joint was 2024-T351 and 7075-T651, with 2024-T351 on the retreating side and 7075-T651 on the advancing side. It was discovered that the strain-hardening rate of the AA7075/AA2024 joint was higher than that of the parent material, while the strain-hardening rate of the AA2024/AA5083 joint lay between those of the parent material. It was also found that the tensile properties of both joints were lower than those of the parent material. Xia-Wei et al. [17] did the microstructural analysis correlatively with mechanical properties of the FSW joint using dissimilar alloys. The lamellar structure in the bottom of the nugget zone was found to be more homogeneous and finer than other regions. The hardness on the copper side of the nugget was higher than that on the aluminium side. The UTS of the joint was found to be relatively lower than that of the base metal. The tensile morphology revealed ductile-brittle fracture mode.

Kumbhar and Bhanumurthy [18] did a comparative study on friction stir welding of similar to dissimilar aluminium alloys, i.e., AA5052 to AA6061, and AA6061 to AA6061. The similar and dissimilar joints were produced at various combinations of tool rotation speeds and tool traverse speeds. The microstructural analysis revealed that there was no rigorous mixing in the nugget region for both materials. The tensile properties of dissimilar materials (AA5052-AA6061) were much better compared to the properties of similar materials (AA6061-AA6061). Welding dissimilar aluminium alloys that are mechanically far apart has gained much attention and interest from researchers [19]. This includes the analysis of friction stir welding of dissimilar AA2017A-T451 and AA7075-T651 plates at a different tool rotation speed. The results revealed that the best tensile properties were achieved when AA2017A-T451 was on the retreating side. It was also established that the material that is located on the retreating side dominates the weld centre, and this is consistent with the results reported by other researchers [11,12,16]. Ranjith and Kumar [20] analysed the impact of joining two dissimilar aluminium alloys AA2014 T651 and AA6063 T651 by friction stir welding. They discovered that the tensile strength was better when the tool was offset towards AA2014 (advancing side). When it was offset towards AA6063 (retreating side), it resulted in insufficient heat generation on the advancing side, which then resulted in an incomplete fusion of AA2014. Sarsılmaz and Caydas [21] conducted a study on statistical analysis on mechanical properties of friction-stir-welded AA1050-H14/AA5083-H321 couples. The study investigated the effect of friction stir welding parameters focusing on rotational speed, traverse speed, and stirrer geometry. In their investigation, they discovered that traverse speed has a significant effect on UTS and nugget hardness. Their analysis also included the optimized welding parameters to be used in welding the said aluminium alloys.

The analysis of the effect of material positioning during FSW has gone outside the aluminium family. This includes the study which analyzed the effect of location variation in FSW of steel with different carbon content. It was discovered that the placement of the stronger material on the advancing side reduced the weld nugget size and increased the amount of martensite formation. The location of the strongest material on the advancing side led to higher temperature and stress due to the highest temperature on the advancing side [22]. It is evident from the literature that the FSW that involved materials that are mechanically apart involved mainly 2xxx as the weaker material. There are very few studies which utilized AA1xxx [21].

This paper reports on the mechanical properties of the weld produced by the friction stir welding technique using AA1050-H14 and AA5083-H111. The AA1050-H14 is mostly used in the chemical industry, automotive industry, reflectors, heat exchangers, and food industry [23]. The AA5083-H111 is widely used in the prevention of corrosion, hence used in shipbuilding. This alloy is also used in the automotive industry as well [24]. It was then crucial to analyze different aspects related to the joint formed from these two distinct materials. This was done so as to prepare the future application of these two materials in producing products and components. It then became crucial to analyze comparatively the mechanical behaviour of the joint in different locations. This type of analysis will give information regarding the best location for sampling the welds produced from the materials with unique properties and composition.

2. Experimental Procedure

The 6mm thick AA1050-H14 and AA5083-H111 plates were cut into 70 mm × 530 mm long strips using a guillotine cutting machine (see Figure 1).

Figure 1. Aluminium alloy plates prepared for welding.

The two dissimilar plates were fixed on the backplate of the semi-automated milling machine (FSW machine), as shown in Figure 2. The AA5083-H111 was kept on the retreating side throughout the experiments, while AA1050-H14 was on the advancing side. This kind of material location was followed based on the recommendation from the literature [11,12,16,19,25,26]. The plates were fixed on the machine's backplate by eight clamps. The function of clamps was to make sure the plates did not move apart when the rotating pin was in motion.

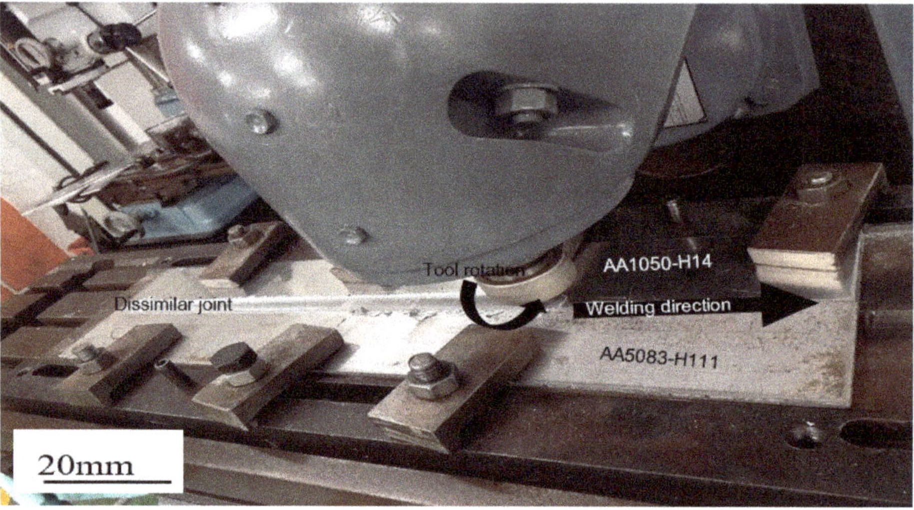

Figure 2. Friction stir welding machine.

The plates were then friction stir welded using a high carbon steel (H13) tool shown in Figure 3. The tool was machined using a lathe machine and was heat-treated to about 50HRC. The profile of the pin was triangularly threaded with a 20 mm shoulder diameter and 6 mm pin diameter. The triangular threaded pin had a 1mm pitch and the height of 5.8 mm. The FSW parameters used in this study are presented in Table 1, and these parameters were chosen based on various speed test combinations performed prior to the main welding. However, those trial results are not included in this work since they are not part of the paper's focus.

Table 1. FSW parameters.

Rotational Speed (rpm)	Traverse Speed (mm/min)	Tilt Angle (°)
1000	30	2

Figure 3. Friction stir welding tool.

Figure 4 shows the friction stir welding of 6 mm thick AA1050 and AA5083 plates. Stage 1 shows two plates ready for welding. Stage 2 shows the rotating tool plunged into the two pieces being welded, i.e., AA1050-H14 and AA5083-H111. Stage 3 shows the welded part of the plates. Stage 4 shows the end of FSW and the unplunging of the tool. The finished product is clearly shown in Figure 5. The nominal chemical composition for materials used in this study is presented in Table 2. The chemical composition was measured using the BELEC COMPACT PORT HLC spectrometer (Belec Spektrometrie Opto-Elektronik GmbH, Georgsmarrienhutte, Germany) according to ASTM E716-16 standard. These chemical compositions were in the range of those reported in the literature [16,21,27].

Table 2. Chemical composition of AA1050-H14 and AA5083-H111 (wt.%) [16,21,27].

Material	Si	Fe	Cu	Mn	Mg	Cr	Zn	Ti	Al
AA1050-H14	0.10	0.29	0.01	-	0.02	-	0.01	0.02	Balance
AA5083-H111	0.14	0.20	0.01	0.65	4.62	0.10	0.01	0.01	Balance

Figure 4. Friction stir welding process.

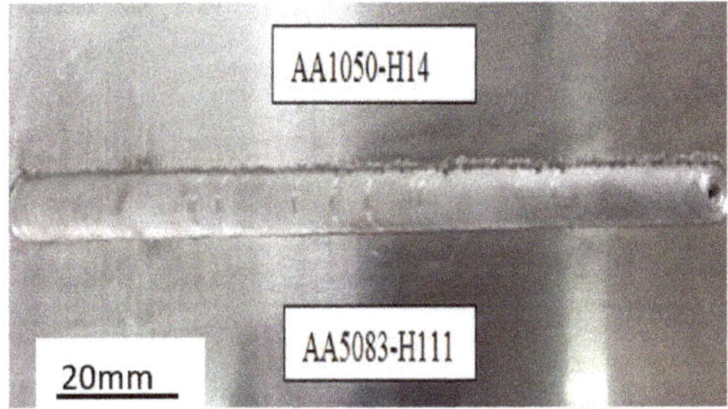

Figure 5. FSW joint.

3. Welded Joint Analysis

This section describes the mechanical and microstructural analysis of the welded joint. This includes the use of the tensile test machine, microstructure analysis, bending test, and scanning electron microscope (SEM). It should be noted that the friction stir welded joint had start, middle, and end points. The specimens were all labeled A, B, and C. Label A indicated the specimens cut at the beginning of the joint, specimens cut in the middle were labeled B, while C symbolized the specimens cut at the end of the joint. This format was followed throughout the performance of the tests.

3.1. Tensile Test

The machine that was used for the tensile test was the Hounsfield tensometer (universal testing machine). The tensile test was performed using the ASTM E8 standard. The tensile test specimens were cut perpendicular to the welding direction (see Figure 6). The tensile test specimens were designed according to the ASTM E8M-04 standard [28] for accurate dimensions. The specimens were cut using CNC wire cutters and the coordinates for cutting specimens were manually generated using 2-D drawings in Solidworks, shown in Figure 7. This method of cutting was selected because it does not induce heat during cutting. There were three tensile specimens extracted from different locations of the plate (see Figure 6). The extensometer was used to capture the data relevant to the joint.

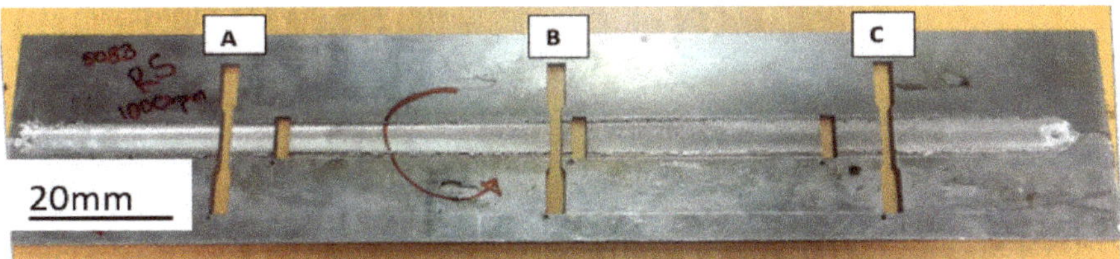

Figure 6. FSW plate showing specimen positioning.

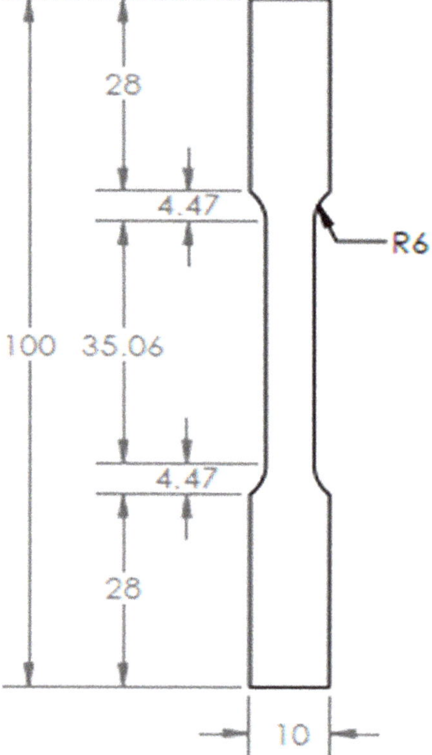

Figure 7. Tensile test specimen.

The specimens were tensile tested until they fractured.

3.2. Microstructure Analysis

The microstructure analysis was performed using the Nikon Eclipse L150 microscope (Nikon, Tokyo, Japan). The microstructure was observed under a polarized slider with an Axiocam 105 colour camera for acquiring the pictures. The specimens were cut into 26 × 8 × 6 mm using the CNC wire cutter and prepared for analysis using Keller's reagent etchant.

3.3. Bending Test

Three-point bending tests were conducted using the same Hounsfield tensometer previously used for the tensile tests. The performance of the bending test was based on the ASTM E290-97a standard. There were six rectangular-shaped specimens prepared for bending tests. The bending tests were performed on the face and the root of the joint. The three specimens presented in Figure 8 show the face of the joint (the surface that was in contact with the tool shoulder) and another three presented in Figure 9 show the root of the joint (the surface that was in contact with the backplate). For comparative purposes, the bending test was also performed on parent materials.

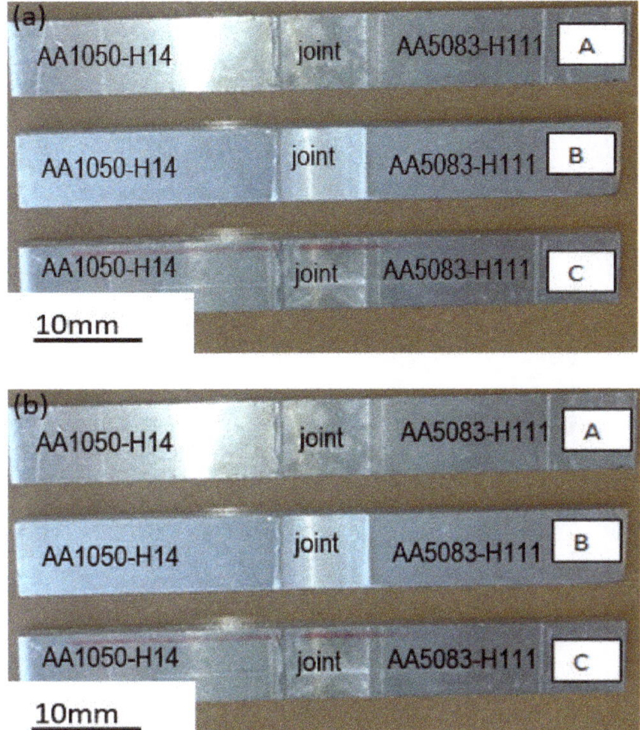

Figure 8. Bending specimens, (**a**) face; (**b**) root.

3.4. Scanning Electron Microscope (SEM)

The SEM used was the Zeiss Auriga (Carl Zeiss Microscopy GmbH Co., Germany). Prior to the analysis, the samples were coated with a layer of carbon to ensure sufficient conductivity during analysis. The results obtained are presented and explained in the next section.

4. Results and Discussion

The detailed discussions of the results obtained from the various tests performed are presented in this section. The results obtained include the tensile test, bending test, microstructure and SEM analysis.

4.1. Microstructure

The FSW joint was characterized by general unique zones as shown in Figure 9. Those zones included the stir zone (SZ) around the weld centre line, the thermo-mechanically affected zone (TMAZ) on both sides of the weld nugget/stir zone, the heat-affected zone (HAZ), which surrounded the TMAZ, and the non-affected base metal (BM) [12–18]. The regions originated from the material flow behaviour caused by the tool rotation. The macroscopic and microscopic patterns of the three specimens were similar, hence one pattern is presented to avoid duplication. Figure 9a shows the macroscopic view of the joint with its thermal zones. There are no defects visible on the joint. This suggests that the welding was performed successfully [21,29]. Figure 9b,c shows the micrographs of base metals AA1050-H14 and AA5083-H111, respectively. Figure 9d shows the microscopic view of the AA1050-H14 pulled towards the stir zone, and this morphology is called hook defect [30]. Figure 9e shows rotational traces of the materials at the stir zone. Figure 9f shows the magnified morphology of the stir zone. The magnified view shows the refined grain structures at the stir zone. The average grain size for the base metal AA1050-H14 was about 29 µm while the average grain size of AA5083-H111 base metal was about 7.3 µm. The average grain size for the stir zone was about 9.35 µm. The average grain size for the stir zone is close to the average grain size for AA5083-H111 base metal.

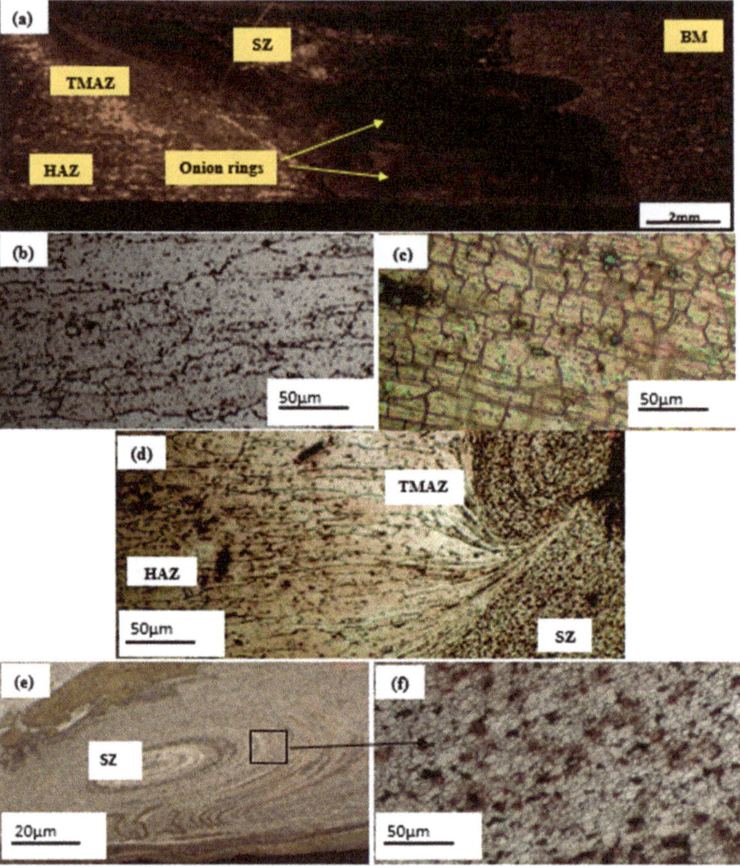

Figure 9. (**a**) Macrostructure of the welded joint; (**b**) microstructure of AA1050-H14; (**c**) microstructure of AA5083-H111; (**d**) thermo-mechanical affected zone; (**e**) stir zone; (**f**) grains of the stir zone.

4.2. Tensile Test

Figure 10 shows the fractured specimens post tensile tests. It was observed that the fracture occurred on the advancing side (AA1050) of the specimen. This behaviour suggests that the weld joint was mechanically stronger than the AA1050 alloy [11,12,28]. This also suggests that the welded joint was dominated by AA5083-H111, hence it was stronger [16,29–31].

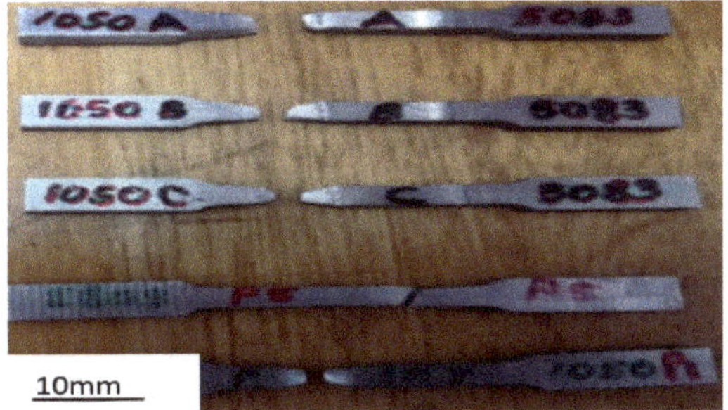

Figure 10. Fractured specimens.

Table 3 shows the results of the ultimate tensile stress (UTS) and percentage elongation. The AA1050 and AA5083 show the UTS of 104.89 MPa and 326.75 MPa, respectively, while the FSW specimens A, B, and C show 50.67 MPa, 66.47 MPa, and 63.19 MPa, respectively. It should be noted that all the specimens were fractured by the HAZ region of the AA1050-H14 side. The HAZ region is known to have a negative impact towards the UTS of a material due to the coarsened grains that are normally associated with this region [19,20,30–33]. This is suggested to be the cause of the lower UTS for the three specimens compared to the parent materials. This assumption is in line with the grain variations observed during microstructural analysis. The percentage elongation was found to be 19%, 25%, and 26% for specimen A, B, and C, respectively. These indicate that the FSW specimens B and C were more ductile compared to the specimen A and AA1050-H14 parent material, but less ductile compared to parent material AA5083.

Table 3. Tensile test results.

Specimen	Ultimate Tensile Strength (UTS) (MPa)	Percentage Elongation
A	50.67	19%
B	66.47	25%
C	63.19	26%
1050 Parent	104.89	23.5%
5083 Parent	326.75	33.5%

Figure 11 shows the tensile test-strain curve of AA1050-H14, AA5083, and FSW specimens A, B, and C. The UTS of the parent material AA5083 was larger than that of the parent material AA1050-H14 and of the FSW specimens. Specimen A, which was the first specimen from the welded plate, was the weakest while specimens B and C were close to each other. This behaviour was assumed to be caused by insufficient heat input at the beginning of the weld. The temperature stabilized from the middle to the end of the plate, hence improved UTS.

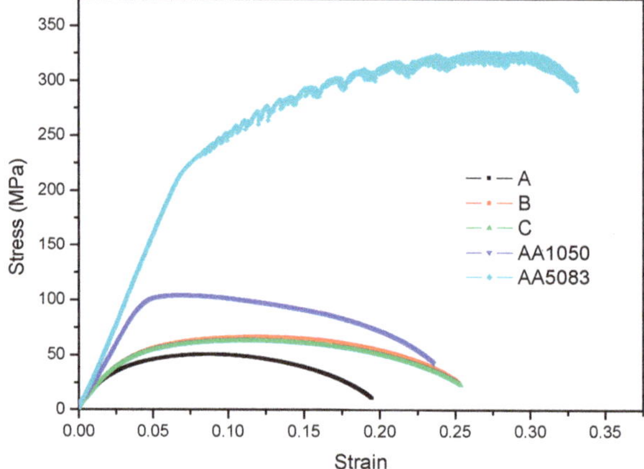

Figure 11. Stress-strain curve.

4.3. Bending Test

Figures 12 and 13 are the post-bending test specimens for FSW and parent materials. The reddish-brown line appearing on specimens in Figure 12a,b indicates the center of the joint. The face and the root bending occurred towards the advancing side of the joint. This behaviour suggests that the joint was mechanically stronger compared to AA1050-H14, hence bending occurred on the advancing side. This behaviour is in agreement with the behaviour observed in the tensile analysis. The post-bending results showed the tested specimens without failure. This means that the welded materials bonded well during welding.

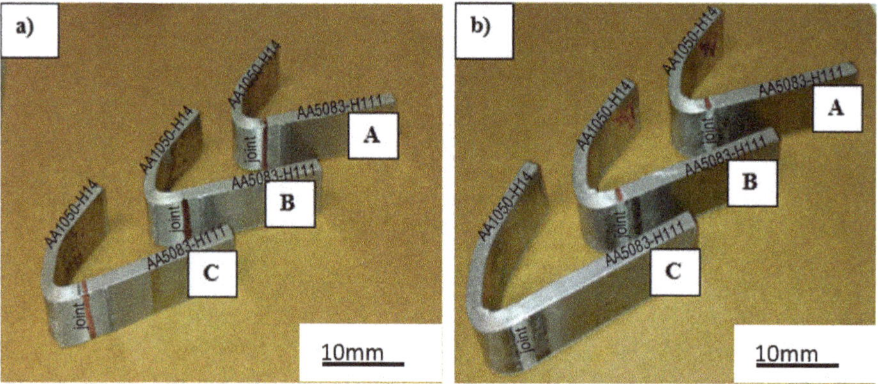

Figure 12. (**a**) Tested bending specimen (face), (**b**) Tested bending specimen (root).

Figure 13. Tested bending specimen (parent materials).

Figures 14 and 15 show bending stress and strain curves of the joint, together with parent materials. As it has been mentioned before that all the welded specimens bent on AA1050-H14, the flexural stresses for both face and root were within the range of that of AA1050-H14 (see Table 4). The average stress for face and root was 218.94 MPa and 259 MPa, respectively. These values suggest that the root side of the weld was stronger than that of the face. This could be caused by the fact that the lower side of the weld was exposed to the restricted downward movement due to bed backing plate. The flexural stress values were comparatively higher than the tensile values and this was due to higher temperatures involved during FSW.

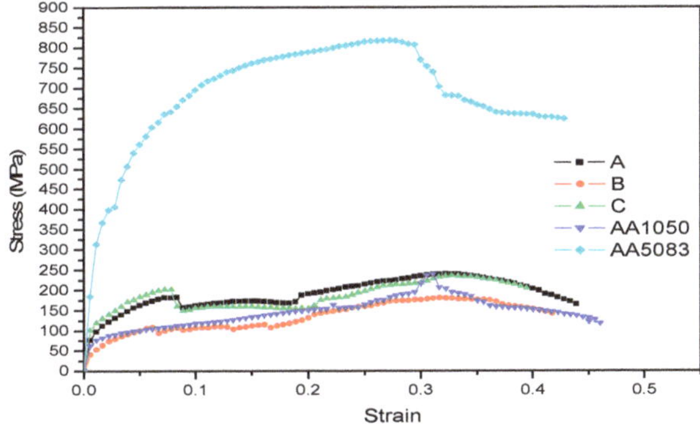

Figure 14. Bending stress—strain curves (face).

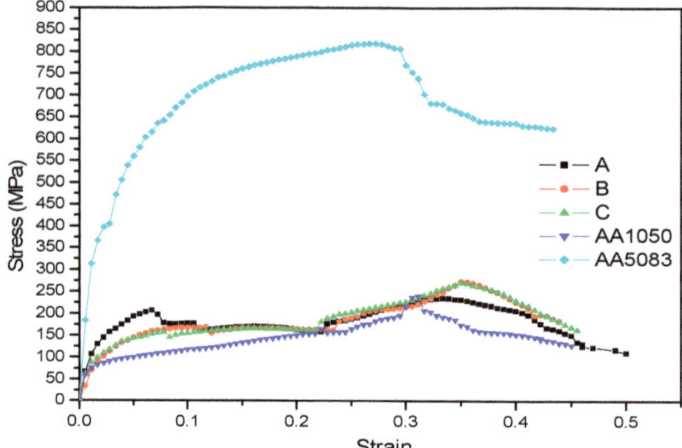

Figure 15. Bending stress—strain curves (root).

Table 4. Bending test results.

Specimen	Flexural Strength (MPa)	Flexural Strain
Face	-	-
A	240.56	0.43
B	180.38	0.41
C	235.88	0.39
Average stress	218.94	-
Root	-	-
A	234.75	0.5
B	272.25	0.44
C	270	0.45
Average stress	259	-
Parent Materials	-	-
AA1050-H14	240.19	0.46
AA5083	818.44	0.44

4.4. Scanning Electron Microscopy (SEM)

It should be noted that the fractured surface for base metals was studied comparatively with specimen C. The elimination of other specimens was due to the fact that there was no distinction on the surface morphology for the other specimens. Figures 16 and 17 show the surface morphologies of the base metals and weldment (joint). All the specimens showed a cup-like dimpled fracture, which is a characterization of ductile failure mode [30,32,33]. The similarity in surface fracture suggests that the ductility of the materials involved in the joint formation was preserved post welding even though there was some percentage of elongation variation.

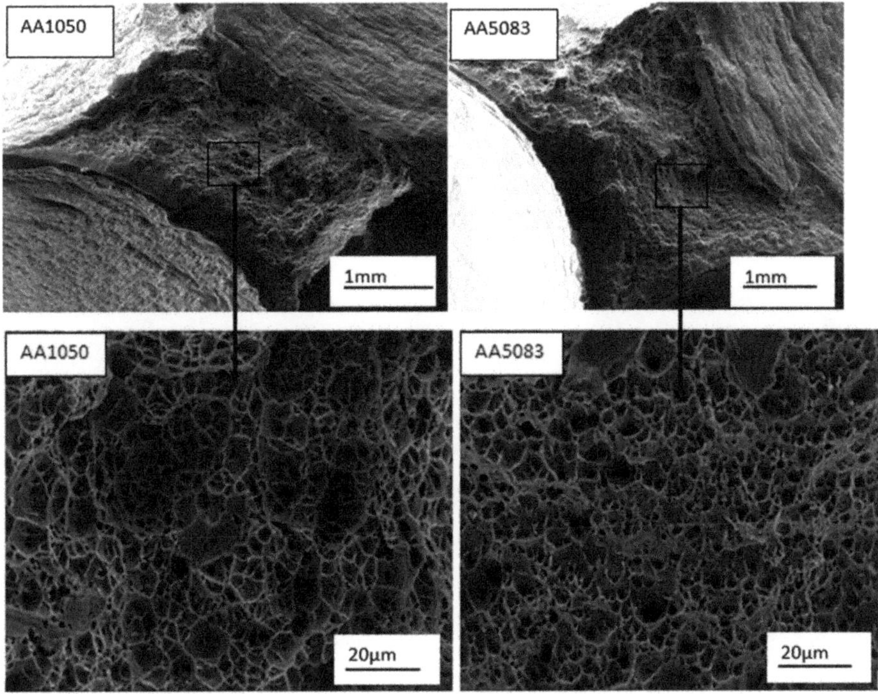

Figure 16. Micrograph of parent material.

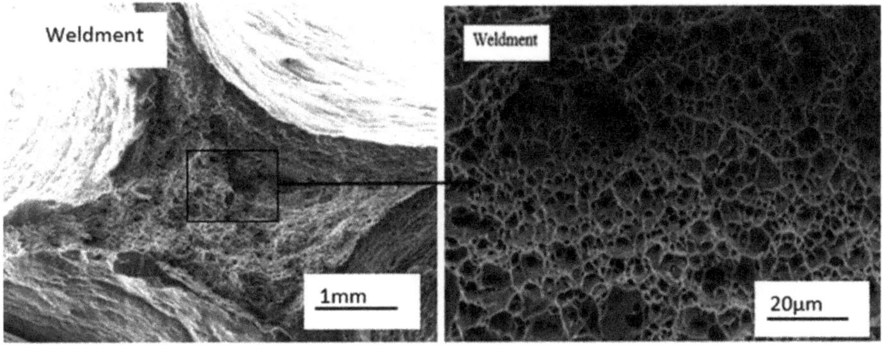

Figure 17. Micrograph of the welded specimen.

5. Conclusions

The absence of voids or defects on the macrostructure of the joint suggests the success of welding the two dissimilar aluminium alloys (AA1050-H14 and AA5083-H111). The UTS of the base metals was found to be higher than the UTS of the tested specimens. In addition to this, the specimen extracted at the beginning of the weld exhibited the lowest UTS compared to the specimens extracted from other locations of the weld. The ductility of the specimens fluctuated between that of the parent materials. The fracture location was found to be consistent with the one reported in the literature [11,12,16]. This kind of failure is due to the coarsened grain found in the location of failure (AA1050-H14 HAZ region). The characteristic hook defect on the side of the weaker material was also observed during the microstructural analysis [29–31,33]. The morphology of the fracture surface indicated ductile failure mode, which was characterized by cup-like dimples for all the specimens.

Author Contributions: Authors have contributed equally to this work. All authors have read and agreed to the published version of the manuscript.

Funding: This research received no external funding.

Acknowledgments: Authors would like to thank EM Masekwana for his technical assistance during welding.

Conflicts of Interest: The authors declare no conflict of interest.

References

1. Mishra, R.S.; Mahoney, W. Friction stir welding and processing. *ASM Int.* **2007**, *1*, 368.
2. Thomas, W.; Nicholas, E. Friction stir welding for the transportation industries. *Mat. Des.* **1997**, *18*, 269–273. [CrossRef]
3. Genevois, C.; Deschamps, A.; Denquin, A.; Boisneau-Cottignies, B. Quantitative investigation of precipitation and mechanical behaviour for AA2024 friction stir welds. *Act. Mat.* **2005**, *53*, 2447–2458. [CrossRef]
4. Mishra, R.S.; Ma, Z.Y. Friction stir welding and processing. *Mat. Sc. Eng.* **2005**, *50*, 1–78. [CrossRef]
5. Farias, A.; Batalha, G.F.; Prados, E.F.; Magnabosco, R. Tool wear evaluations in friction stir processing of commercial titanium Ti-6Al-4V. *Int. J. Sc.Techn. Fric. Lub. Wr.* **2013**, *302*, 1327–1333. [CrossRef]
6. Tiwari, S.K.; Shukla, D.K.; Chandra, R. Friction Stir Welding of Aluminium Alloys: A review. *Int. J. Mech. Aer. Ind. Mechat. Manuf. Eng.* **2013**, *7*, 1326–1331.
7. Liu, X.C.; Sun, Y.F.; Morisada, Y.; Fujii, H. Dynamics of rotational flow in friction stir welding of aluminium alloys. *J. Mat. Proc. Tech.* **2018**, *252*, 643–651. [CrossRef]
8. Oliveira, J.P.; Duarte, J.F.; Inácio, P.; Schell, N.; Miranda, R.M.; Santos, T.G. Production of Al/NiTi composites by friction stir welding assisted by electrical current. *Mat. Des.* **2017**, *113*, 311–318. [CrossRef]
9. Nallusamy, S. Analysis of Welding Properties in FSW Aluminium 6351 Alloy Plates Added with Silicon Carbide Particles. *Int. J. Eng. Res. Afr.* **2016**, *21*, 110–117. [CrossRef]
10. Soundarajan, V.; Yarrapareddy, E. Investigation of the friction stir lap welding of aluminium alloys AA 5182 and AA 6022. *J. Mat. Eng. Perf.* **2007**, *16*, 477–484. [CrossRef]
11. Sadeesh, P.; Rajkumar, P.; Avinash, N.; Arivazhagan, K.; Narayanan, S. Studies on friction stir welding of AA 2024 and AA 6061 dissimilar metals. *Proc. Eng.* **2014**, *75*, 145–149. [CrossRef]
12. Amancio-Filho, S.T.; Sheikhi, S.; dos Santos, J.F.; Bolfarini, C. Preliminary study on the microstructure and mechanical properties of dissimilar friction stir welds in aircraft aluminium alloys 2024-T351 and 6056-T4. *J. Mat. Proc. Techn.* **2008**, *206*, 132–142. [CrossRef]
13. Biswas, P.; Kumar, D.A. Friction stir welding of aluminium alloy with varying tool geometry and process parameters. *J. Eng. Man.* **2011**, *226*, 641. [CrossRef]
14. Kundu, J.; Singh, H. Friction stir welding of dissimilar Al alloys: Effect of process parameters on mechanical properties Friction stir welding of dissimilar Al alloys. *Eng. Sol. Mech.* **2016**, *4*, 125–132. [CrossRef]
15. Hou, W.; Shen, Y.; Huang, G.; Yan, Y.; Guo, C.; Li, J. Dissimilar friction stir welding of aluminium alloys adopting a novel dual-pin tool: Microstructure evolution and mechanical properties. *J. Man. Proc.* **2018**, *36*, 613–620. [CrossRef]
16. Niu, P.L.; Li, W.Y.; Chen, D.L. Strain hardening behaviour and mechanisms of friction stir welded dissimilar joints of aluminium alloys. *Mat. Let.* **2018**, *231*, 68–71. [CrossRef]
17. Xia-wei, L.I.; Da-tong, Z.; Cheng, Q.I.U.; Wen, Z. Microstructure and mechanical properties of dissimilar pure copper / 1350 aluminium alloy butt joints by friction stir welding. *Trans. Nonf. Met. Soc. China* **2012**, *22*, 1298–1306.
18. Kumbhar, N.T.; Bhanumurthy, K. Friction Stir Welding of Al 5052 with Al 6061 Alloys. *J. Met.* **2012**, *2012*, 1–7. [CrossRef]
19. Kopyściański, M.; Węglowska, A.; Pietras, A.; Hamilton, C.; Dymek, S. Friction Stir Welding of Dissimilar Aluminium Alloys. *K. Eng. Mat.* **2016**, *682*, 31–37. [CrossRef]
20. Ranjith, R.; Senthil Kumar, B. Joining of dissimilar aluminium alloys AA2014 T651 and AA6063 T651 by friction stir welding process. *W. Trans. App. Theo. Mech.* **2014**, *9*, 179–186.

21. Sarsılmaz, F.; Çaydaş, U. Statistical analysis on mechanical properties of friction-stir-welded AA 1050/AA 5083 couples. *Int. J. Adv. Man. Techn.* **2009**, *43*, 248–255. [CrossRef]
22. Choi, D.H.; Lee, C.Y.; Ahn, B.W.; Yeon, Y.M.; Park, S.H.C.; Sato, Y.S.; Kokawa, H.; Jung, S.B. Effect of fixed location variation in friction stir welding of steels with different carbon contents. *Sc. Techn. Weld. Join.* **2010**, *15*, 299–304. [CrossRef]
23. Mouhri, S.E.; Essoussi, H.; Ettaqi, S.; Benayoun, S. Relationship between microstructure, residual stress, thermal aspect in friction stir welding of AA1050. *Proc. Manuf.* **2019**, *32*, 889–894. [CrossRef]
24. Tsangaraki-Kaplanoglou, I.; Theohari, S.; Dimogerontakis, T.; Wang, Y.M.; Kuo, H.; Kia, S. Effect of alloy types on the anodizing process of aluminium. *Surf. Coat. Tech.* **2006**, *200*, 2634–2641. [CrossRef]
25. Sameer, M.D.; Birru, A.K. Mechanical and metallurgical properties of friction stir welded dissimilar joints of AZ91 magnesium alloy and AA 6082-T6 aluminium alloy. *J. Magn. All.* **2019**, *7*, 264–271.
26. Azeez, S.; Mashinini, M.; Akinlabi, E. Sustainability of friction stir welded AA6082 plates through post-weld solution heat treatment. *Proc. Manuf.* **2019**, *33*, 27–34. [CrossRef]
27. Fattah-alhosseini, A.; Naseri, M.; Gholami, D.; Imantalab, O.; Attarzadeh, F.R.; Keshavarz, M.K. Microstructure and corrosion characterization of the nugget region in dissimilar friction-stir-welded AA5083 and AA1050. *J. Mater. Sci.* **2019**, *54*, 777–790. [CrossRef]
28. Annual Book of ASTM Standards. *Standard Test Methods for Tension Testing of Metallic Materials*; ASTM International: West Conshohocken, PA, USA, 2014.
29. Koilraj, M.; Sundareswaran, V.; Vijayan, S.; Koteswara Rao, S.R. Friction stir welding of dissimilar aluminium alloys AA2219 to AA5083-Optimisation of process parameters using Taguchi technique. *Mat. Des.* **2012**, *42*, 1–7. [CrossRef]
30. Patel, V.; Li, W.; Wang, G.; Wang, F.; Vairis, A.; Niu, P. Friction Stir Welding of Dissimilar Aluminum Alloy Combinations: State-of-the-Art. *Metals* **2019**, *9*, 270. [CrossRef]
31. Jesusa, J.S.; Gruppelaara, M.; Costaa, J.M.; Loureiroa, A.; Ferreira, J.A.M. Effect of geometrical parameters on Friction Stir Welding of AA 5083-H111 T-joint. *Proc. Str. Int.* **2016**, *1*, 242–248. [CrossRef]
32. Sato, Y.; Urata, M.; Kokawa, H.; Ikeda, K. Hall–Petch relationship in friction stir welds of equal channel angular-pressed aluminium alloys. *Mat. Sc. Eng.* **2003**, *354*, 298–305. [CrossRef]
33. Li, P.; Chen, S.; Dong, H.; Ji, H.; Li, Y.; Guo, X.; Yang, G.; Zhang, X.; Han, X. Interfacial microstructure and mechanical properties of dissimilar aluminium/steel joint fabricated via refilled friction stir spot welding. *J. Man. Proc.* **2020**, *49*, 385–396. [CrossRef]

Article

Study and Characterization of EN AW 6181/6082-T6 and EN AC 42100-T6 Aluminum Alloy Welding of Structural Applications: Metal Inert Gas (MIG), Cold Metal Transfer (CMT), and Fiber Laser-MIG Hybrid Comparison

Giovanna Cornacchia * and Silvia Cecchel

DIMI, Department of Industrial and Mechanical Engineering, University of Brescia, via Branze 38, 25123 Brescia, Italy; s.cecchel@unibs.it

* Correspondence: giovanna.cornacchia@unibs.it; Tel.: +39-030-371-5827; Fax: +39-030-370-2448

Received: 18 February 2020; Accepted: 26 March 2020; Published: 27 March 2020

Abstract: The present research investigates the effects of different welding techniques, namely traditional metal inert gas (MIG), cold metal transfer (CMT), and fiber laser-MIG hybrid, on the microstructural and mechanical properties of joints between extruded EN AW 6181/6082-T6 and cast EN AC 42100-T6 aluminum alloys. These types of weld are very interesting for junctions of Al-alloys parts in the transportation field to promote the lightweight of a large scale chassis. The weld joints were characterized through various metallurgical methods including optical microscopy and hardness measurements to assess their microstructure and to individuate the nature of the intermetallics, their morphology, and distribution. The results allowed for the evaluation of the discrepancies between the welding technologies (MIG, CMT, fiber laser) on different aluminum alloys that represent an exhaustive range of possible joints of a frame. For this reason, both simple bar samples and real junctions of a prototype frame of a sports car were studied and, compared where possible. The study demonstrated the higher quality of innovative CMT and fiber laser-MIG hybrid welding than traditional MIG and the comparison between casting and extrusion techniques provide some inputs for future developments in the automotive field.

Keywords: aluminum alloy; fiber laser-MIG hybrid; CMT; MIG; hybrid joints; microstructure; hardness

1. Introduction

1.1. Background

Aluminum alloys have recently increased their employment in different engineering fields, especially for transport, due to their excellent properties including good corrosion resistance, high strength, good formability, and low density [1–4]. The current main requirements of the automotive field such as the necessity of reducing emissions, the improvement of vehicle performance, and the preservation of safety targets [5–7] imply a further need to improve and study lightweight structures fabricated from aluminum alloys [8–10]. The manufacturing of complex shapes, where welded joints are usually required, is even more challenging.

1.2. Aluminum Alloys Weldability

The welding of aluminum alloys is considered a slightly difficult process due to its high thermal and electrical conductivity, high thermal expansion coefficient, refractory aluminum oxide (Al_2O_3) formation tendency, and low stiffness. These characteristics, in general, make these alloys sensitive to defect formation that may lead to the loss of chemical, metallurgical, and mechanical properties. Typical welding defects in aluminum alloys are gas porosity, oxide inclusions and/or oxide filming, solidification (hot) cracking or hot tearing, reduced strength in the weld and heat affected zone (HAZ), lack of fusion, and reduced corrosion and electrical resistance. These defects determine reduced strength and corrosion in the fusion zone (FZ) and HAZ, with a general decrease in mechanical properties. These defects are generally reduced, providing efficient protection from the contamination of atmospheric gases to the weld pool or/and decreasing the influence of the weld thermal load by using welding processes with higher energy density [11,12].

These problems have to be faced, mainly due to the increasing employment of Al in complex vehicle parts, which implies a higher number of potential applications of Al hybrid structures. For example, during the last few years, both cast and wrought Al parts have been introduced in automotive complex shapes, which entails the need to join them into the final structure. Indeed, the dissimilar aluminum alloy joint can combine good strength and corrosion resistance, which is typical of this material, with exceptional castability used where complex sub-sections are needed, and excellent mechanical properties, achieved in other specific areas made of extruded parts.

1.3. Aluminum Alloys for Automotive Field

Against this background, the Al–Si–Mg alloy class is one of the most widely used for the production of aluminum casting components. In particular, the cast Al–7Si–Mg (EN AC 42100, also defined as A 356) alloy is widely used in automotive applications thanks to its high specific strength. Its microstructure consists of primary α (Al) grains and eutectic (Al–Si) structures. T6 heat treatment is normally used to obtain the desired mechanical properties. The solution treatment dissolves the β phase (Mg_2Si particles) in the Al matrix, homogenizes the alloying elements in the casting, and modifies the morphology of the eutectic structures [13–16]. Wrought aluminum is a widely used alloy in the automotive field, especially the heat-treatable 6xxx series, which is characterized by high strength and good corrosion resistance [17]. Two of the most commonly used are the wrought EN AW-6181-T6 and EN AW-6082-T6 aluminum alloys. These are age-hardenable alloys, thus their mechanical properties are mainly controlled by the hardening precipitates contained in the material. When the material is subjected to a solution heat treatment followed by a quenching and a tempering treatment, their mechanical properties reach their highest level. According to the literature [18–20], the T6 temper of the 6xxx alloys involves very thin precipitates, namely β''needle shaped precipitates, with a nanometric size and is partially coherent with the matrix. One of the most interesting characteristics of these alloys is the good weldability that, along with other properties, makes them very attractive in transport for complex structures assembled by welding [21,22]. Several works have studied the welding of aluminum and other alloys such as magnesium, steel, or titanium [23–26]. Only a limited number of scientific papers [27–32] have investigated the welding of dissimilar aluminum alloys together. These papers mainly deal with friction stir welding (FSW) [33,34], which, despite its potential, still has a high cost that needs to be improved in order to be used in industrial high volume applications. Wang et al. [35] studied the tensile properties and microstructure of a joined wrought EN AW-6181 aluminum alloy and vacuum high pressure die cast A356 aluminum alloy by using the metal inert gas (MIG) technique. The results showed that the low strengths of the A356-T6 alloy should be attributed to the absence of Mg-based intermetallic phase, coarse grain, and porosity, but the effect of the microstructure of the two base metals (BM) on the mechanical properties was not reported. An interesting study by Nie et al. [36] examined the microstructure, distribution of alloying elements, and mechanical properties of the wrought aluminum alloy 6061-T6 and cast aluminum alloy A356-T6 joined using a pulse MIG welding process. Additionally, the influence of welding speed on the microstructure and mechanical properties of the joints was investigated. They observed brittle Fe-rich phases in the partially melted zone and minimum hardness in the A356 aluminum alloy side. Some authors have recently tried to apply hybrid laser-arc welding to Fe–Al dissimilar joints [37], but in that case, the process was instable because of the significant difference in the thermal- and fluid-dynamic properties of the two metals. On the other hand, for this configuration, full penetration and low defectiveness were obtained by laser offset welding. Wang et al. applied laser welding with different beam oscillating modes on 5A06 aluminum alloy sheets and found that welding defects such as welding porosity could be improved by laser beam oscillation [38].

1.4. Welding Techniques Examined

In recent times, alternative welding techniques have been successfully applied to aluminum alloys.

MIG welding is an electric arc welding process that uses a continuously fed wire into the weld pool. It can be used to join long stretches of metal without stopping. Among the main advantages of this type of weld are that good quality welds can be produced much faster and there is flexibility for a wide variety of alloys. In addition, thanks to the gas protection, there is very little loss of alloying elements. Unfortunately, MIG welding cannot be used in vertical or overhead welding positions because of the high heat input, the fluidity of the weld puddle, and the complexity of the equipment [11,39].

CMT is a form of modified MIG welding based on the short-circuiting transfer process that guarantees interesting achievements such as process stability, reproducibility, and cost-effectiveness. This process differs from MIG welding only in the type of mechanical droplet cutting method that provides controlled material deposition and low thermal input by incorporating an innovative wire feed system coupled with a high-speed digital control. The two main advantages of the CMT process are the low heat input and the occurrence of short circuits in a stable controlled manner [40–43].

Laser welding is a very interesting joint process due to the high welding speed, smaller heat-affected zone (HAZ), and low deformation. Unfortunately, the high cooling rates can lead to the formation of hardening structures that increases hardness, decreases plasticity of the weld joint and HAZ, and increases the level of residual stresses. Fiber laser welding is one of many laser processes where the laser light is generated in a remote source and guided to the work piece by a flexible delivery optic fiber. The main benefits of this type of weld are the good beam quality, high precision control, lower heat input, lower electrical energy consumption, low cost of maintenance, and compact size. Welding with the laser technique requires alignment, fixation, and welding process control. This critical procedure can be solved by using the twin spot laser technique with filler wire or hybrid arc-laser welding, but will decrease the welding speed. One of the ways to solve this problem is the use of laser beam wobbling mode [11,12,44–50].

1.5. Aim of Work

The aim of the present study was to extend this last study and to investigate the feasibility, microstructural, and mechanical properties of EN AW-6181 or 6082-T6/EN and AC 42100-T6 aluminum alloy joints by considering other welding techniques. In particular, the present article analyzes and compares the metal inert gas technique (MIG), with cold metal transfer (CMT) and fiber laser-MIG hybrid welding through the evaluation of the final microstructure, the analyses of grain size, second-phase fraction, and dissolution of the precipitate in HAZ and FZ. Optical microscopy (OM) as well as the mechanical properties of welds (micro and macro hardness) were used. Junctions of both samples and a prototype frame of a sports car were examined and compared.

2. Materials and Methods

2.1. Base Metal and Filler Wire Materials

In order to be as reliable as possible, the analyses were carried out on samples machined from a prototype of an actual sports car frame composed of different aluminum alloys welded together. It is worthwhile noting that the opportunity to study actual automotive parts is extremely relevant to evaluate the real quality of the welds, especially considering that the automotive sector is highly demanding in terms of both market rules and safety regulations. Another relevant topic is the presence into the frame of joining between both wrought and cast alloys such as EN-AW 6181-T6, EN-AW

6082-T6, and EN AC 42100-T6. At the same time, in order to complete the research activity, samples were appositely made to reduce the data variation that could arise from a real frame investigation. In particular, extruded and casting bars were used to realize the same different combinations of the frame (EN AW-6181-T6 and EN AC 42100-T6).

During the welding, we used an additional UNI—ER 4043 (AlSi5) filler wire with a diameter of 1.2 mm to minimize either the porosity, the notching, or the cracking susceptibility of the joint. The chemical compositions of the base and filler materials selected in this study are reported in Table 1.

Table 1. Chemical composition of aluminum alloys.

Alloy	Si	Fe	Cu	Mn	Mg	Cr	Zn	Ti	Total Others El.	Al
EN AW-6181	0.8–1.20	0.45	0.10	0.15	0.60–1.0	0.10	0.20	0.10	0.15	Bal.
EN AW-6082	0.7–1.3	<0.50	<0.10	0.4–1.0	0.60–1.2	<0.25	<0.20	0.10	<0.05	Bal.
EN AC-42100	6.50–7.50	<0.60	<0.25	<0.35	0.20–0.45	-	<0.35	<0.25	0.15	Bal.
UNI – ER 4043	4.5–4.6	0.8	0.3	0.005	0.05	-	0.1	-	-	Bal.

2.2. Design of the Joint

The frame was entirely joined by a manual welding MIG. Different samples were obtained from cross sections of the components and can be summarized as follows:

- Two extruded (EN-AW 6082-T6)-extruded (EN-AW 6082-T6) welds (Figure 1a); and
- Two extruded (EN-AW 6082-T6)-cast (EN AC 42100-T6) welds (Figure 1b).

Additional samples made from the bars joining were realized in order to investigate a wider mixture of couplings, as follows:

- Cast bar with cast bar (indicated as C-C);
- Extruded bar with cast bar (indicated as E-C); and
- Extruded bar with extruded bar (indicated as E-E).

In addition, the comparison on a simpler geometry is useful to avoid potential influence on the results related to the different and complex shapes of the welding in the various sections of the frame. Each above-mentioned combination was welded with three different techniques: MIG, CMT, and the fiber laser-MIG hybrid. Thus, nine different kinds of samples were analyzed: three materials matching and three welding processes. The bars used were 3 mm thick with dimensions of 100 × 25 mm, as shown in Figure 2. Full penetration joints with zero gap were achieved and the welding direction was parallel to the sample's axis. Before welding, the oxide films were removed by using emery cloth and acetone. The welding machines used were the TPS 320i, Fronius RCU 5000i(Fronius Italia S.r.l., Verona, Italy), and TS 4.20 2D (TTM LASER SPA, Brescia, Italy) for the MIG, CMT, and fiber laser-MIG hybrid joints, respectively. All joints were performed in a protective argon atmosphere with a robotic setup and without preheating. A welding current of about 130–140 A and a welding voltage of approximately 18–20 V with a wire feeding rate of 6–8 m/min were used for MIG and CMT. The gas flow rate was 14.5, 15, and 20 l/min for MIG, CMT, and fiber laser-MIG hybrid joints, respectively. A wobble circular pattern with the speed of 8–10 mm/s and laser power of 3–4 kW was used for the fiber laser-MIG hybrid welding to achieve a wider weld beam and reduce the porosities. The wobbling frequency was about 100 Hz with amplitude of 40° and the torch angle from the surface was around 85°.

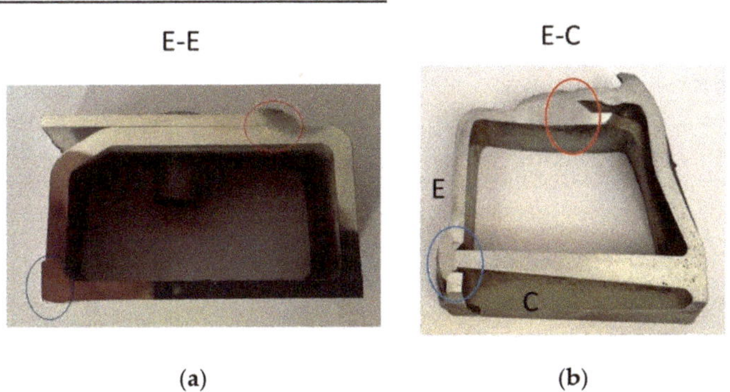

(a) (b)

Figure 1. Samples obtained from the sub-frames: (**a**) extruded (6082-T6)-extruded (6082-T6) welds, (**b**) extruded (6082-T6)-cast (EN AC 42100-T6) welds.

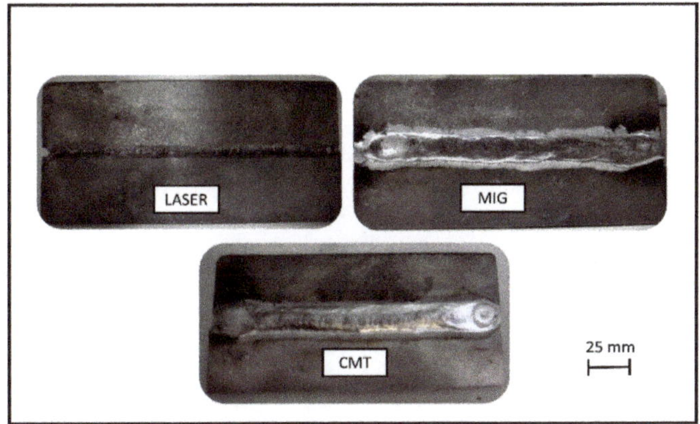

Figure 2. Example of the samples used for the different weld techniques.

2.3. Microstructure Characterization

In order to analyze the defects and microstructures of the joints, the different welded samples were examined using a Leica DMI 5000M (Leica Microsystem, Milan, Italy) optical microscope and LEO EVO-40 XVP Scanning Electron Microscope (SEM) equipped with Energy Dispersive Spectrometers (EDS) (LEO EVO 40, Carl Zeiss AG, Milan, Italy). The welds were observed using the Leica Application Suite (LAS 4.0, Leica Microsystem, Milan, Italy), which is an image processing software tool for image analysis that integrates a Leica automated microscope and digital camera. For metallographic observations, transversal sections of the samples were prepared with standard metallographic techniques (ground with SiC papers and polished with 1 μm diamond paste). The samples with the extruded part were etched with Keller's reagent in order to better investigate the defects and microstructures.

2.4. Hardness Test

Macro-hardness Rockwell scale F (HRF) was performed along a planar section of the welded joint while micro hardness Vickers (HV) was performed on the cross section of the welds. Hardness profiles were obtained by measuring the hardness at regular distances starting from the center of the weld and moving forward the base metal. For HRF measurements, a hardness tester Rockwell Rupac 500Mra

(Rupac srl Milano, Italy) with a 1.58 mm steel ball indenter diameter, load of 588 N (60 Kgf), and dwell time of 15 s was used, following the ASTM E 18-03 procedures [51]. At least three measurements were made and the average value was then considered. Vickers microhardness tests were carried out under 2.94 N (0.3 Kgf) load applied for 15 s by means of a Mitutoyo HM-200 hardness testing machine (Mitutoyo Italiana srl, Lainate, Italy, according to ASTM E92-16 [52] and ASTM E140-02 [53]. The microhardness profile was obtained in the center of the cross section after preliminary analyses confirming that the effect of the position was negligible for all three couplings of materials. The BM hardness was assumed to be 66 HRF, 65–70 HV for the 6xxx-T6 extrusion and 77 HRF, 85–90 HV for the EN AC 42100-T6 casting [54].

3. Results and Discussion

3.1. The Weld Geometry and Weld Defects

3.1.1. Frame Samples

Defects typical of aluminum weldings were observed including porosities and some incomplete penetration, as summarized in Figures 3 and 4. Figure 4 highlights some hot crackings (white circle) that occurred during MIG welding in the partially molten zone (PMZ). Indeed, the UNI ER-4043 filler wire increases the risk of liquation cracking because it decreases the local heat input and reduces the softening in the PMZ. Welds are usually inspected using liquid penetrant testing, which is a very valuable tool during new construction and in-service inspections. Figure 4 shows a liquid penetrant exam conducted during the quality tests of the present case study that revealed the presence of some incomplete penetrations of the joint. The geometries of the weld seam examined are essentially a "v" or "u" groove with a very different width, depending on the frame position.

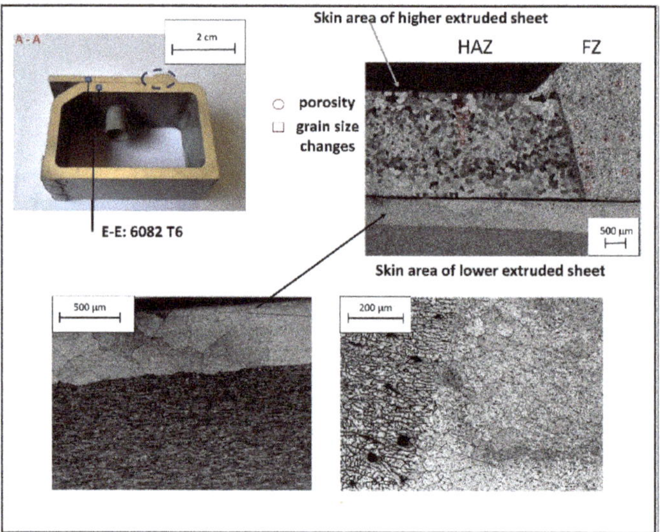

Figure 3. Defects and microstructures of metal inert gas (MIG) frame joints. E-E combination.

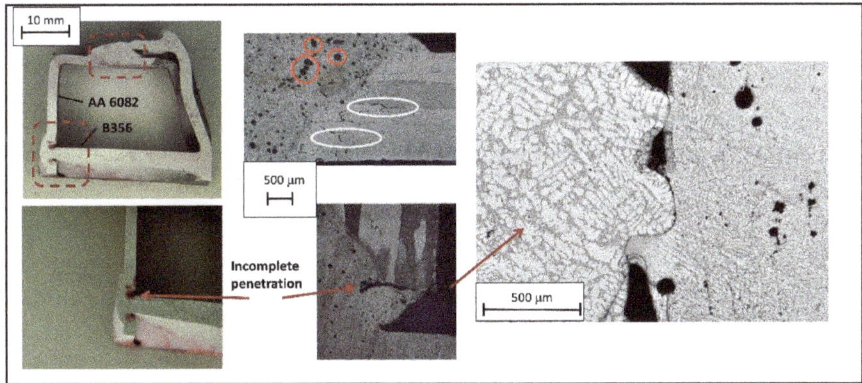

Figure 4. Defects and microstructures of MIG frame joints. E-C combination.

3.1.2. Bar Samples

Figure 5 shows an overview of the microstructure for each weld seam analyzed. These images were obtained from a collage of numerous local 50× microstructures of the weld cross section in order to guarantee high quality of the analysis. In Figure 5, different geometries of the weld seams, depending on the welding technology, can be noted: v and/or u groove and width from 6 to 8 mm; u shaped butt joint and width from 9 to 10 mm; and v groove and width from 2 to 4 mm for MIG, CMT and fiber laser welding, respectively.

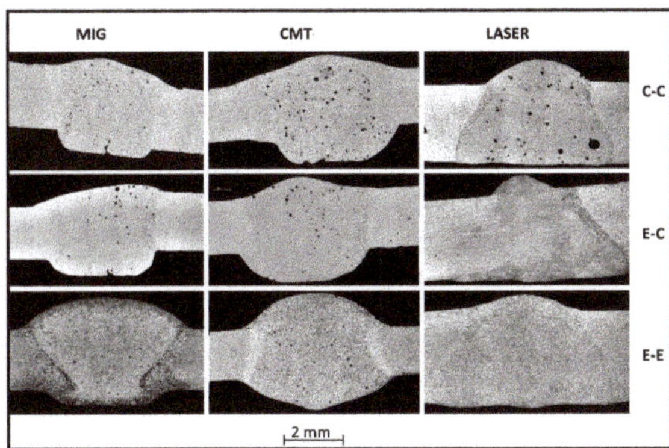

Figure 5. Weld geometry and defects of MIG, CMT (cold metal transfer), and fiber laser-MIG hybrid joints.

From a quality point of view, in the C-C combination, the CMT samples had the highest and most homogeneous diffusion of porosity in the fusion zone (FZ); MIG welds had a slightly smaller porosity in the FZ and the fiber laser-MIG hybrid welded samples only had some porosities. E-E and E-C were almost porosity free, thanks to the use of a wobbling head.

The average porosity diameter ranged from 21 µm to 145 µm and 18 µm to 120 µm for CMT and MIG, respectively. The fiber laser-MIG hybrid only had one significant porosity with a diameter of 270 µm, while a few others had dimensions similar to that in the CMT case.

It is remarkable to note that the overall highest level of porosity was observed in the joints with cast bars, especially in C-C. Indeed, castings have a high amount of Si, which reduces the thermal conductivity and increases the local heat near the welding seam. In addition, castings have more porosities than extrusions due to the lower solubility of hydrogen in solid aluminum than in liquid aluminum, which results in a diffusion of the entrapped hydrogen from the casting to the FZ during welding. This defect could be reduced by using vacuum casting in welding structures. It should also be noted that another defect was present, the incomplete penetration of the joint, mainly in the C-C combination of MIG welding. For the sake of clarity, incomplete joint penetration is defined as a condition of a weld where the filler metal does not extend through the entire joint thickness.

3.2. Material Characterization Results

3.2.1. Frame Samples

The 6082–T6 and 6181–T6 alloys, which are quite similar in composition and consequently also in mechanical behavior, have similar metallographic structures. In Figure 6, it is possible to observe the microstructure of the extruded 6181–T6 where the deformation direction and the consequent anisotropy are clearly visible. The grains are rather small thanks to the manganese fining properties.

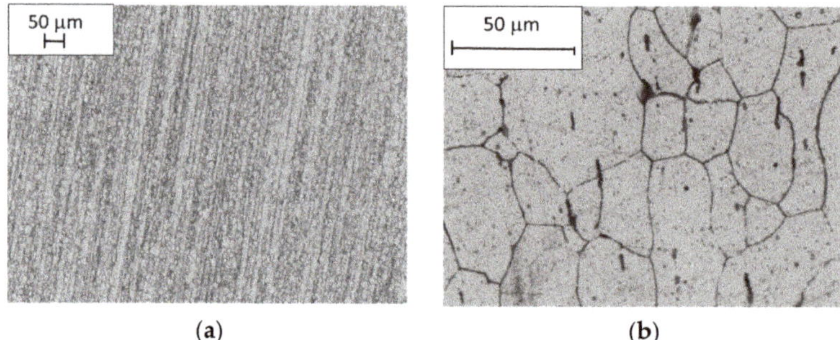

Figure 6. (**a**) Microstructure of the 6082-T6 alloy, BM. Keller etch, 500×; (**b**) Microstructure of 6181-T6 alloy, BM. Keller etch, 100×.

The EN AC 42100-T6 alloy microstructure is a typical example of an Al–Si alloy from a foundry. Cast alloys usually contain more alloying elements, therefore cooling can be quite long and a coarse-grained dendritic structure can be achieved. Like all foundry materials, it suffers from the presence of micro- and macro-segregation. Figure 7a shows the dendritic microstructure of the EN AC 42100-T6 BM, the large white dendrites (α-Al), and dark Al–Si eutectic in the space between the grains. Additionally, in this figure, it is possible to note some large shrinkage porosity, a frequent defect in the observed cast samples. In Figure 7b, the clear difference between the HAZ and FZ of the cast component side is highlighted. The PMZ was about 200–300 μm long. It should be also noted that during solidification, the material tends to also maintain a crystallographic continuity in the interface zone; in fact, the dendrites in the PMZ are developed starting from those already present in the cast. The microstructures at the interface between E-E and E-C are visible in Figures 3 and 4, respectively. Figure 3 reports the particular structure observed at the interface between the extruded material and the FZ. Starting from the extruded grains, the following changes occurred in the metallurgical structure: first planar, then cellular, and finally dendritic. These microstructural changes are a consequence of the growth rate effect at the solidification front. Moreover, the Keller etch highlighted two remarkably different structures on the extruded parts. In the upper one, the structure had coarse equiassic grains close to the outer side of the extrusion and was gradually thinner toward the center; the lower part had a completely different microstructure, probably due to the different reduction ratios and extrusion parameters. It is important to note that in both laminas, there was a skin zone on the outer faces.

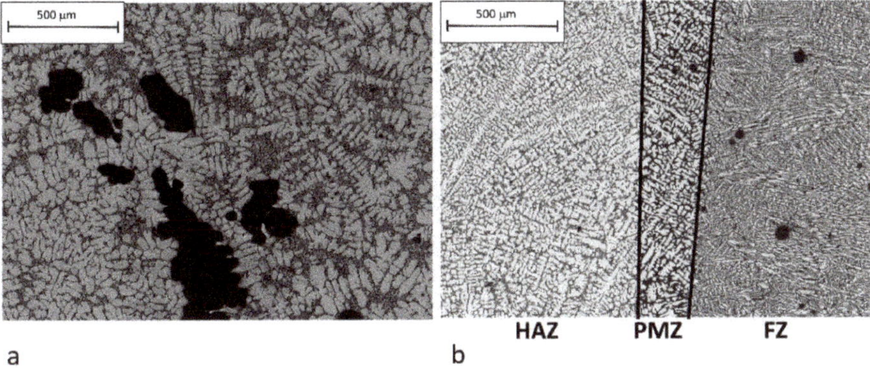

Figure 7. (**a**) Microstructure of the EN AC 42100-T6 alloy, 50×; (**b**) Microstructure present at the fusion zone (FZ)/cast component interface, 50×.

3.2.2. Bar Samples

The microstructural features of the interface between HAZ, PMZ, and FZ in all combinations for the three weld techniques are reported in Figures 8–10. In particular, the FZ of the MIG, CMT, and fiber laser joints was constituted by a dendritic structure of aluminum solid solution (α-Al) and Al–Si eutectic for all of the cases examined, with the dendrite arm spacing slightly wider in the MIG and CMT joints with respect to that of the fiber laser. Looking at the HAZ, it is possible to observe that the extruded portion of the bar contained few elongated grains. Regarding the cast bar, in the HAZ, the white dendritic Al primary phase that resulted extended from the PMZ zone and was longer than that in in the base metal; a grey spheroidized Al–Si eutectic was also present at the grain boundary.

For the C-C combination (Figure 8) the PMZ for the MIG and CMT techniques had a regular evolution and covered about 200–300 µm and 500 µm of the interface area, respectively. It should also be noted that the PMZ for the laser weld had an irregular profile and this area was smaller but had longer dendrites.

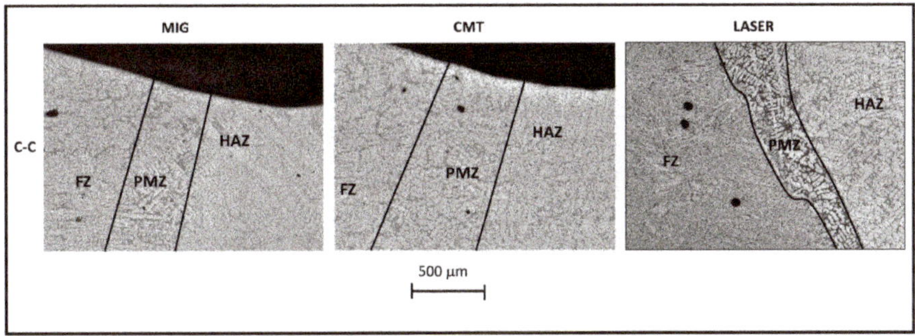

Figure 8. Macrostructures of the interface between the HAZ (heat-affected zone), PMZ (partially molten zone), and FZ (fusion zone) in the Cast-Cast combination (C-C), Keller etch, 50×.

For the E-E combination, it is possible to observe the PMZ, which was characterized by the presence of intermetallics visible even at low magnification (Figure 9). In this type of joint, the PMZ was slightly smaller than that in the C-C case, especially for the laser technique.

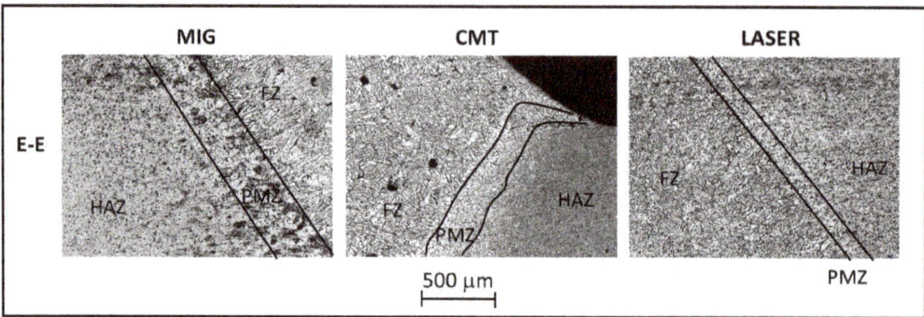

Figure 9. Macrostructures of the interface between the HAZ, PMZ, and FZ in the Extruded-Extruded (E-E) combination, Keller etch, 50×.

Figure 10 reports the E-C coupling; the extrusion–weld and the casting–weld interfaces are reported in the left and right side of the picture, respectively. The FZ of the joint consisted of a fine-grained dendrite structure formed by α-Al and Al–Si eutectic. Regarding the FZ/cast bar interface, the PMZ was due to the re-melting of the eutectic compound [55]. In this case, the grain size in PMZ was larger than that in the FZ, but the microstructure had a dendritic aspect with dimensions increasing from FZ to HAZ and a crystallographic continuity through the section due to the partial remelting. Looking at the extrusion–FZ interface, the change of structure was clearly less evident. The FZ microstructure in this part was similar to the cast alloy side. In the PMZ, the low melting point segregation phase was etched severely, was not uniform, and agglomerated near the FZ. In addition, the extrusion grains were more elongated in this configuration when compared to E-E, most likely because the casting side of the coupling implies an increase in local heat, due to the reduced thermal conductivity.

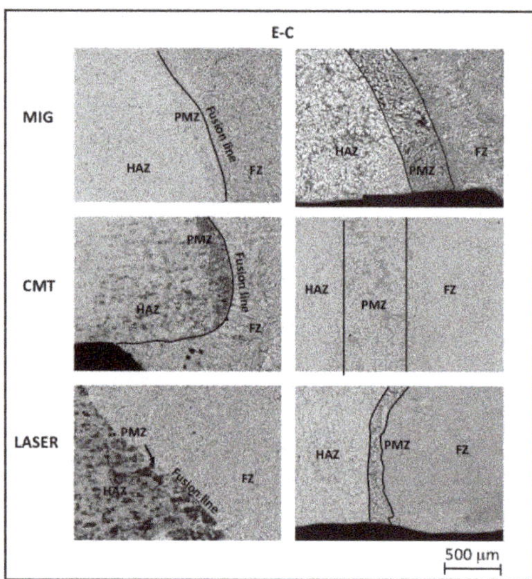

Figure 10. Macrostructures of the interface between the HAZ, PMZ, and FZ in the Extruded-Cast combination (E-C), Keller etch, 50×.

3.3. Hardness Distribution

From the hardness evolution depending on the distance from the welding axis, it is possible to acquire information on the temperature profiles reached in the HAZ. The Al alloys, as objects of study, can be approximated to a pseudo binary Al-Mg2Si system. For this class of alloys, the hardening metastable phase is subjected to a precipitation process that is generally reported in the following steps: β″ (hardening metastable phase), β′ (intermediate metastable phase), and β (equilibrium phase Mg_2Si). On the basis of previous work [56,57], the maximum hardness value was reached with the metastable phase β″ having a needle-like structure. The hardness contribution of phase β′, which had a rod morphology, was moderate, while that of the β equilibrium phase was almost negligible.

The allotropic transformations of Mg_2Si correlate to their transformation temperatures, identifiable in about 240 °C as the upper limit for β″ and approximatively 380° C for β′ (see Figure 11 scheme). Considering that the local temperature peaks reached during the welding process decrease as distance from the fusion line increases, the following HAZ subzones can be identified:

- Re-solubilization area (PMZ): In the PMZ, the high temperature (T > 450 °C) induces a complete dissolution of hardening precipitates. During cooling, according to times and temperature reached, the re-precipitation of hardening compounds could be possible.
- Intermediate area: The temperature exceeds 380 °C and the transformation of the equilibrium phase β takes place. Indeed, in this area, located at about 10 mm from the FZ, it is possible to observe the minimum hardness value.
- Over-aging area: 380 < T (°C) < 240. In these zones, the increase in hardness could be justified with the transformation of β″ in β′.
- Slightly altered zone: Temperatures do not exceed 240 °C, thus any marked over-aging phenomena of β″ can be avoided. In this area, the hardness value tends to be the base material, even if some hardness fluctuations are still present.

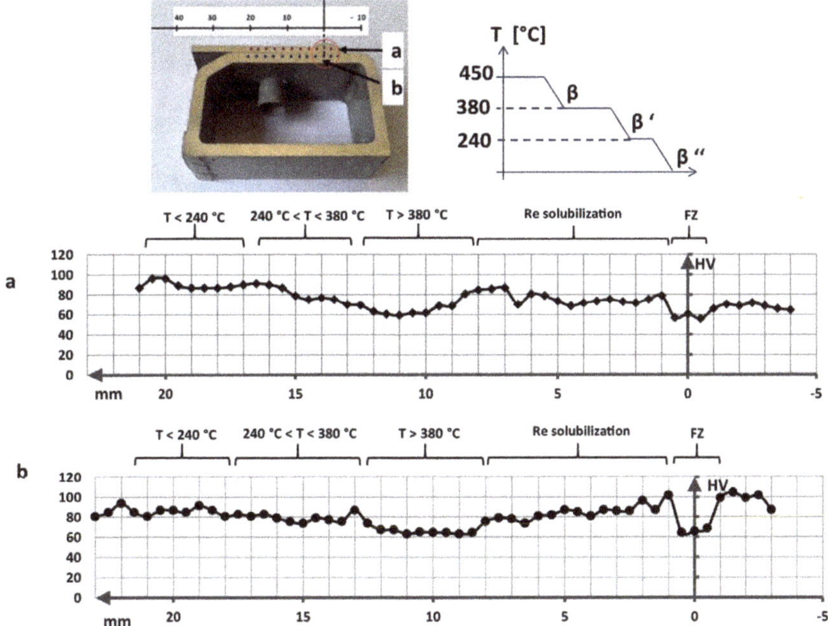

Figure 11. Evolution of the micro-hardness HV (hardness Vickers) in the E-E sub-frame combination as a function of the position along a transverse section of the weld (MIG joint).

3.3.1. Frame Samples

The micro-hardness results obtained for the E-E and E-C samples are summarized in Figures 11 and 12, respectively. Regarding the E-E sample, it can be observed that the FZ maintained a fairly constant hardness, around 60 HV, while there was first a hardness increase (about 90 HV) in the PMZ, followed by a decrease to 60 HV in the HAZ. Finally, the BM hardness was restored at about 20–25 mm from the fusion line for both the extrusion and casting interfaces. For the E-C sample, the HV profile of the extruded component showed a qualitative trend similar to the previous E-E analysis. In this case, however, the local minimum was approximately 80 HV; this suggests that during the welding process, the critical temperature of 380 °C was not reached, or not maintained long enough to complete the transformation of Mg_2Si in the equilibrium phase β. The HAZ was only about 12 mm from the fusion line, confirming the possibility that in this case, the welding process was faster than in the previous one. For the cast component, the micro-hardness profile was similar to that of the extrusion, but showed that the plastic deformation aluminum alloy was more sensitive to the thermal cycle than the cast one.

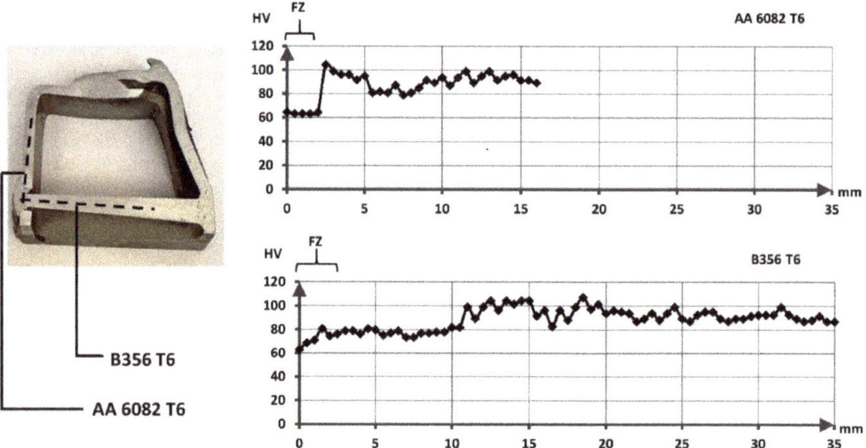

Figure 12. Evolution of the micro-hardness HV in the E-C sub-frame combination as a function of the position along a transverse section of the weld (MIG joint).

3.3.2. Bar Samples

In Figures 13–15, the macro and micro hardness results for the bar welded specimens are reported. Concerning the analysis of all samples, the HV profile in the MIG welded samples showed the strongest reduction of mechanical properties; the fiber laser-MIG hybrid welded samples showed the best behavior, while the CMT welded samples had intermediate characteristics. The use of UNI ER 4043 filler material, softer than EN AW 6082, certainly contributed to the mechanical properties, especially in the FZ. Regarding the E-E samples, the microhardness profiles confirmed the results and the interpretations obtained in the E-E frame case. In addition, a different behavior could be observed looking the macro-hardness HRF profiles. In this case, MIG and CMT had a very similar trend, with about 62 HRF in the FZ and little loss of hardness up to the HAZ, which was wider in the MIG technique case. For the laser technique, the weld area was the softest, there was a slight decrease at 5 mm, and the hardness finally tended to increase. These differences revealed in the FZ and PMZ are probably due to the diverse heat input from the MIG, CMT, and laser welding process on the external surface of the samples. Furthermore, the specimens were thinner compared to the frame sections. These considerations clarify why the heat exchange was greater in the bar surfaces, both with respect to the cross section of the same samples and to the joints of the frame. For the C-C joints, the micro-hardness profiles showed a clear definition of the BM, HAZ, and FZ. The trend was very similar for all techniques, but the hardness in the FZ center was very different, probably due to the presence of typical defects of the cast alloy used. Similar considerations can be advanced for hardness variations also observed in other areas of the samples. In the PMZ zone, the HV reached a hardness value of about 78–80 for MIG and CMT and 86 HV for the laser technique. In the HAZ, there was a slight drop of HV and afterward, the area far from the melt zone around 20 mm tended toward the stabilization in the HV value (50–55 HV for MIG and CMT, 80 HV for laser). The evolution of the macro-hardness HRF followed the trend already achieved in the case of the E-E combination. Noteworthy is the trend difference between CMT and MIG. Although the evolution was similar, the CMT welding showed a softer passage between the various zones. Once again, this fact is due to the different heat input during welding. It should be noted that MIG welding had drastic effects on the hardness, which reached values typical of an annealing. CMT welding had similar, but less pronounced effects. This could be caused by the high thermal input on thin samples, which caused an effect equivalent to an annealing treatment. This is probably also related to the high percentage of silicon (~7%) in this cast alloy, which entails a low material conductivity and a consequent difficult dissipation of the welding heat. On the other hand, during the laser technique, the heat input was highly localized in the FZ, thus this behavior was not observed.

Indeed, for the E-C coupling, the HV values confirmed the results of the tests conducted on the other two combinations and the E-C frame case studied. In particular, it was confirmed that the extruded component (Si~1%) better conducted the heat produced during welding, resulting in a reduced change in hardness [58–60]. Looking at the hardness profiles, considerations similar to E-E and E-C can be advanced for extrusion-welding and casting-welding interfaces, respectively. With regard to the HRF hardness test, the typical evolution was obtained, with an initial HRF peak in the PMZ, a subsequent decrease, and a gradual restoration of the properties as distance increased. It is worthwhile also noting that the HRF profile demonstrated the difference in heat exchange between the cast and extruded component during welding.

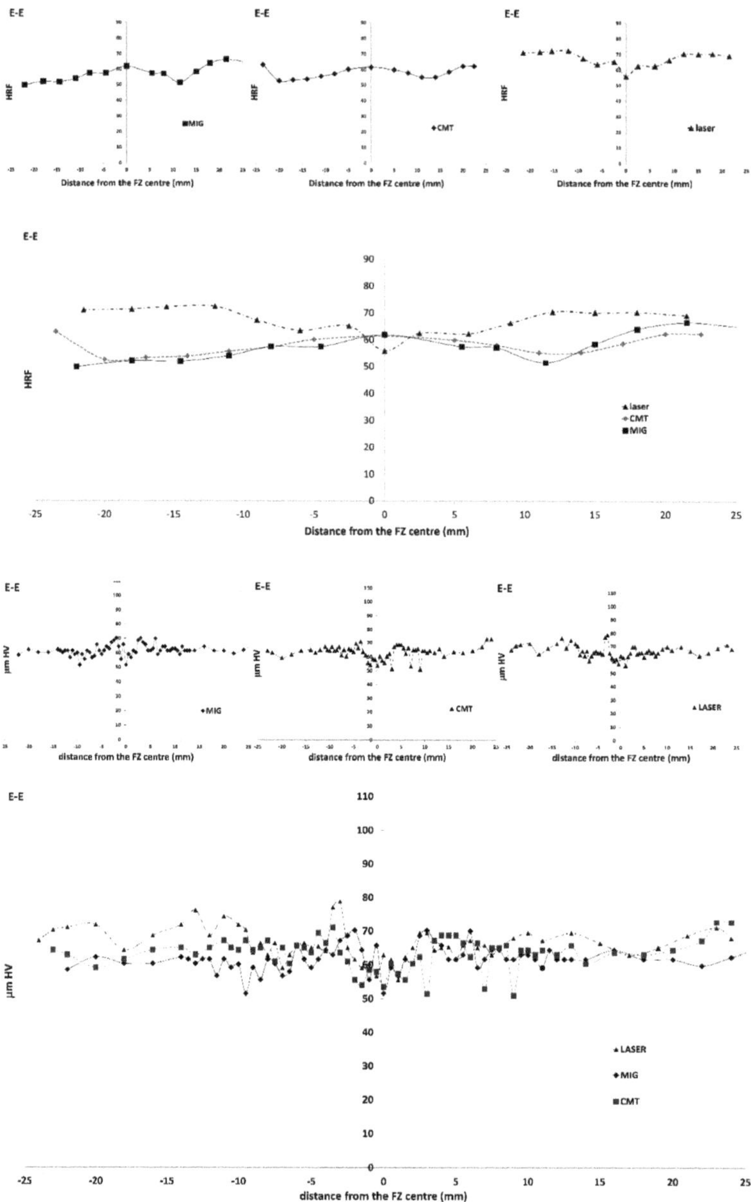

Figure 13. Evolution of the macro-hardness HRF and micro-hardness HV in the E-E combination of the weld for laser, CMT, and MIG.

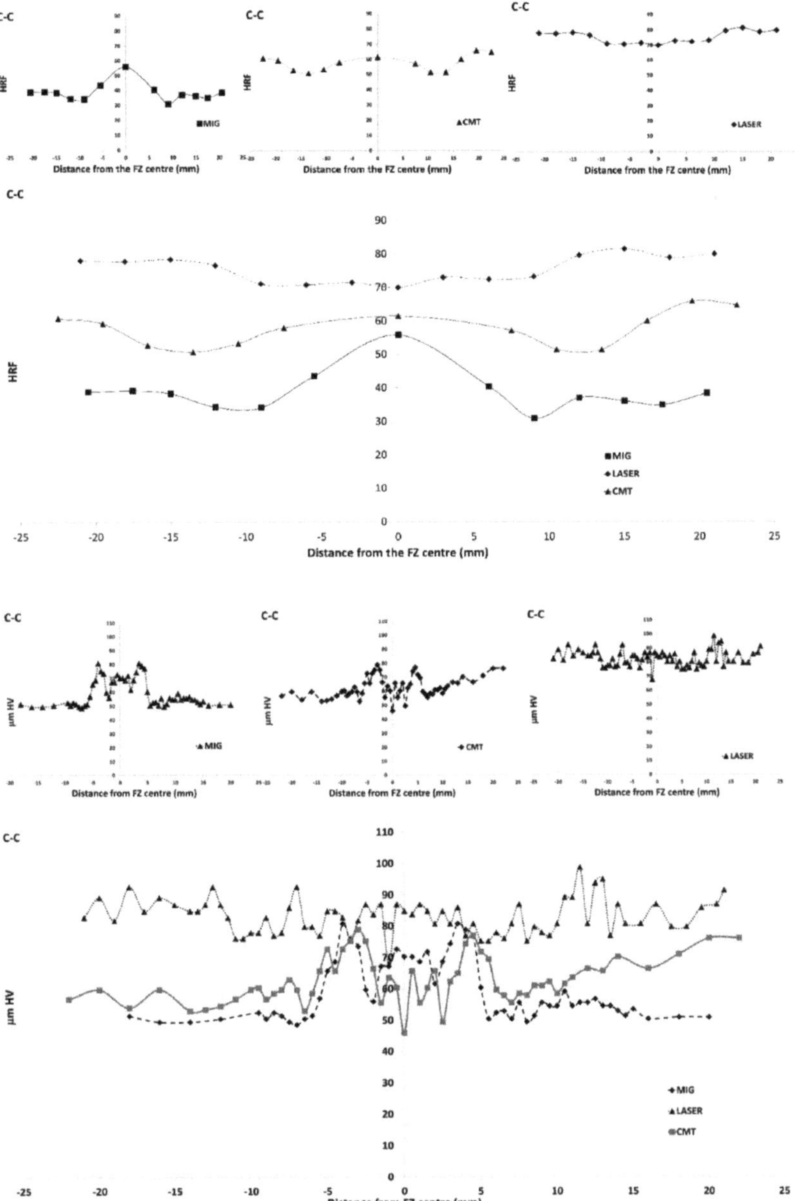

Figure 14. Evolution of the macro-hardness HRF and micro-hardness HV in the C-C combination of the weld for laser, CMT, and MIG.

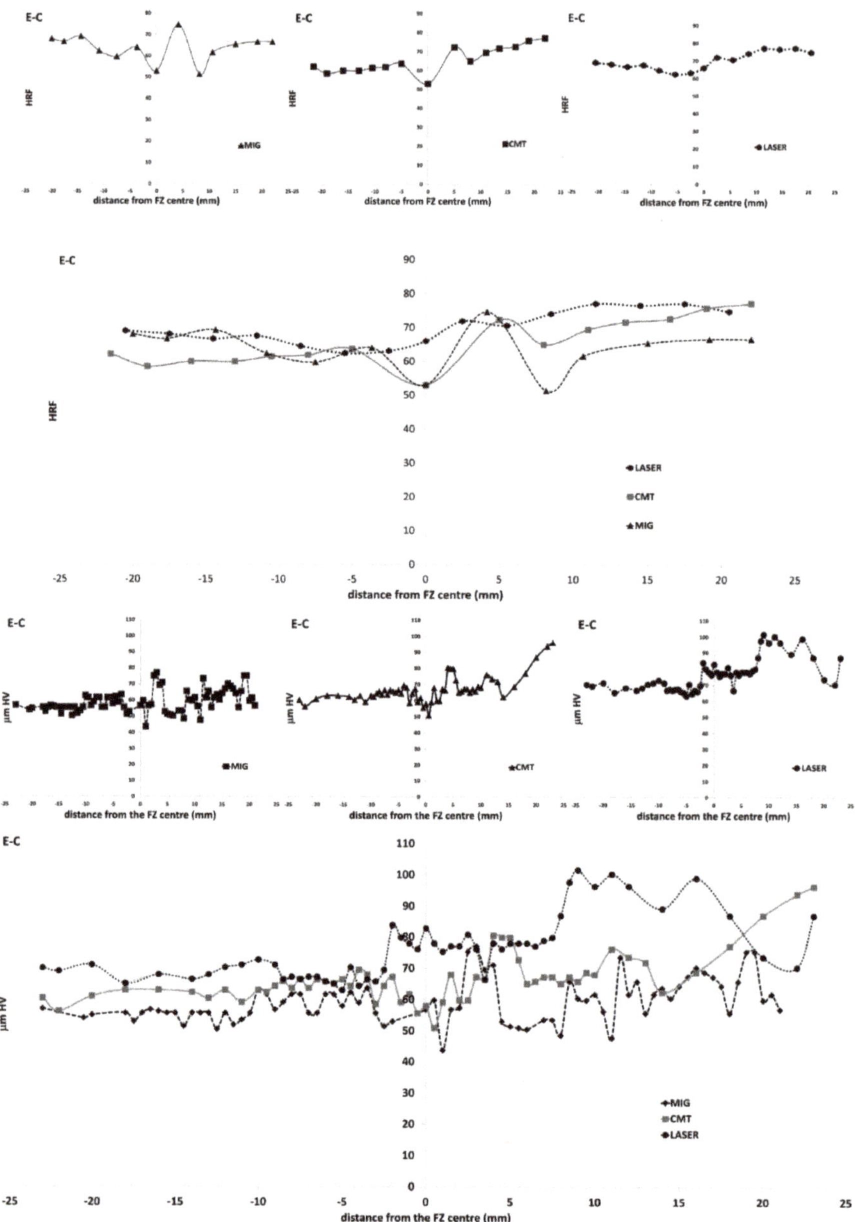

Figure 15. Evolution of the macro-hardness HRF and micro-hardness HV in the E-C combination of the weld for laser, CMT, and MIG.

3.4. SEM Analysis

The PMZ and HAZ are crucial areas in weld alloys that are hardened by artificial aging. The transformation phase along the welding profiles was explained in the previous section. In addition, in the literature [55,56], it is documented that the microstructure of the welded Al–Si alloys is characterized by the presence of elongated particles (dimensions of about 1–10 μm) and extremely fine particles (size approximately of 0.1–0.5 μm) dispersed in a matrix with uniformly distributed micropores. This is remarkable, since it has been reported that Si-rich precipitates are found due to the excess silicon in the alloys. The larger particles were identified as the (Fe, Mn)$_3$SiAl$_{12}$ compound, while the finer ones were Mg_2Si, the hardening phase obtained during thermal treatment (T6).

The structures of both EN AC 42100-T6 and 6000 series joints were revealed by the SEM/EDS investigations, as reported in Figure 16. The fusion zone had a chemical composition that was affected by the AA 4043 filler metal in addition to the welded materials. The chemical analysis of the white particles confirmed that these could be the compound (Fe, Mn)$_3$SiAl$_{12}$, with an average size about 3–5 μm. These particles were observed in both the extrusion and casting components, where the concentration was higher. The analysis of the PMZ highlights that the fiber laser-MIG hybrid samples had the highest presence of these particles with a finer morphology, while for the MIG and CMT techniques, the amount and dimensions were very similar.

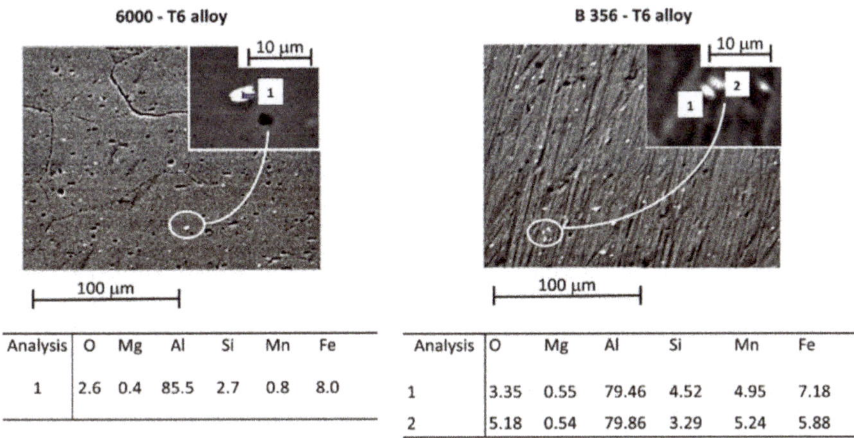

Analysis	O	Mg	Al	Si	Mn	Fe
1	2.6	0.4	85.5	2.7	0.8	8.0

Analysis	O	Mg	Al	Si	Mn	Fe
1	3.35	0.55	79.46	4.52	4.95	7.18
2	5.18	0.54	79.86	3.29	5.24	5.88

Figure 16. Some examples of the precipitate types in the PMZ: chemical composition in wt%.

4. Conclusions

The present paper studied different welding techniques applied to hybrid aluminum alloys (EN AW-6181 T6 and EN AC-42100 T6) joined each other in all possible combinations. First, a real case study was analyzed, which consisted of a sports-car frame where the cast EN AC-42100 T6 and EN AW-6082 T6 aluminum alloys were welded with the traditional MIG technique. The MIG, CMT, and fiber laser-MIG hybrid joints of EN AW-6181 T6 and EN AC-42100 T6 bars were studied to extend the study to other techniques that are difficult to find in real applications and compare the results in a more reliable way in simpler and more reproducible geometries. The microstructure and hardness properties were evaluated and the main conclusions can be summarized as follows:

- From a quality point of view, the typical defects of welded aluminum alloys such as porosity and the incomplete penetration were observed mainly in the frame welds. Then, cracks were noticed only in the frame joints, most likely due to a greater difficulty in heat dissipation for higher thicknesses and more complex geometries. In addition, typical casting defects, mainly shrinkage cavities, were found.
- Regarding the macrostructures, a greater quality was observed for the fiber laser joint. Then, the FZ, PMZ, and HAZ microstructures were observed. All joints had similar FZ microstructures, with the weld seam characterized by fine dendrites. The chemical composition of the fusion zone was affected by the addition of UNI ER 4043 filler. The PMZ was even observed at low magnification

and its width was greater in the C-C case. In general, the innovative techniques (CMT and fiber laser hybrid) allow a better microstructure to be obtained than the traditional technique (MIG).

- Hardness tests (HRF and micro-HV) demonstrated that the mechanical characteristics of the PMZ and HAZ were better for the innovative techniques due to reduced changes in their microstructure when compared to the MIG technique. In particular, the fiber laser-MIG hybrid technique showed the best behavior, and the CMT welded samples had intermediate characteristics. Considering the relationship between the local temperature peaks reached during the welding process and the hardness value obtained, four HAZ subzones were identified:
    - Re-solubilization area (PMZ, T > 450 °C) with a complete dissolution of hardening precipitates and subsequent re-precipitation related to cooling parameters;
    - Intermediate area (about 10 mm from FZ, T > 380 °C) where the β transformation takes place;
    - Over-aging area: (380 < T (°C) < 240) with the β″ to β′ transformation; and
    - Slightly altered zone (T < 240 °C) where any marked over-aging phenomena of β″ are avoided.
- The best behavior was observed, as expected, for the E-E samples, while the worst results were observed for the C-C combination, especially in the MIG welding, where the hardness values were typical of an annealing. This could be caused by both the high thermal input on thin samples and the high percentage of silicon (~7%) in this cast alloy (low material conductivity and difficult welding heat dissipation). Finally, the E-C coupling showed intermediate properties.

It can be concluded that the present study demonstrated the higher quality of innovative CMT and fiber laser-MIG hybrid welding over the traditional MIG, either in terms of metallurgical or mechanical properties. The properties achieved during these experiments could be useful information during the design of real automotive applications due to the increasing demand of highly loaded and lightened structures. Thus, there will be an increasing need for thinner and hybrid combinations of Al alloys that can be addressed by looking at the present database of CMT or laser-MIG hybrid welding results. In addition, the comparison between casting and extrusion techniques provided some input for future development in this field, which, for example, would require better thermal management of the welding input in order to increase the casting welding quality.

Author Contributions: Conceptualization, G.C., S.C.; Investigation, G.C., S.C.; Methodology, G.C., S.C.; Supervision G.C., S.C.; Writing—original draft, G.C.; Writing—review & editing S.C. All authors have read and agreed to the published version of the manuscript.

Acknowledgments: The authors are grateful to Ing. Enrico Lena, Ing Stefano Manesta, and Ing. Andrea Panvini for their collaboration in carrying out the experiments.

Conflicts of Interest: The authors declare no conflicts of interest.

References

1. Miller, W.S.; Zhuang, L.; Bottema, J.; Wittebrood, A.J.; De Smet, P.; Haszler, A.; Vieregge, A. Recent development in aluminium alloys for the automotive industry. *Mater. Sci. Eng. A* **2000**, *280*, 37–49. [CrossRef]
2. Cecchel, S.; Cornacchia, G.; Gelfi, M.A. A study of a non-conventional evaluation of results from salt spray test of aluminum High Pressure Die Casting alloys for automotive components. *Mater. Corros.* **2019**, *70*, 70–78. [CrossRef]
3. Heinz, A.; Haszler, A.; Keidel, C.; Moldenhauer, S.; Benedictus, R.; Miller, W.S. Recent development in aluminium alloys for aerospace applications. *Mater. Sci. Eng. A* **2000**, *280*, 102–107. [CrossRef]
4. ASM Specialty Handbook. *Aluminum and Aluminum Alloys*; Davis, J.R., Ed.; ASM International: Geauga, OH, USA, 1993; ISBN 978-0-87170-496-2.
5. Chindamo, D.; Lenzo, B.; Gadola, M. On the vehicle sideslip angle estimation: A literature review of methods, models and innovations. *Appl. Sci.* **2018**, *8*, 355. [CrossRef]
6. Hirsch, J. Aluminium in Innovative Light-Weight Car Design. *Mater. Trans.* **2011**, *52*, 818–824. [CrossRef]

7. European Commission. Proposal for a Regulation of the European Parliament and of the Council Amending Regulation (EC) No 443/2009 to Define the Modalities for Reaching the 2020 Target to Reduce CO2 Emissions from New Passenger Cars and Proposal for a Regulation of the European Parliament and of the Council Amending Regulation (EU) No 510/2011 to Define the Modalities for Reaching the 2020 Target to Reduce CO2 Emissions from new Light Commercial Vehicles, (European Commission, 2012). Available online: http://eur-lex.europa.eu/resource.html?uri=cellar:70f46993-3c49-4b61ba277319c424cbd.0001.02/DOC_1&format=PDF (accessed on 11 December 2019).
8. Helms, H.; Lambrecht, U. The potential contribution of light-weighting to reduce transport energy consumption. *Int. J. Life Cycle Assess* **2007**, *12*, 58–64.
9. Cecchel, S.; Ferrario, D.; Panvini, A.; Cornacchia, G. Lightweight of a cross beam for commercial vehicles: Development, testing and validation. *Mater. Des.* **2018**, *149*, 122–134. [CrossRef]
10. Kim, H.C.; Wallington, T.J. Life-Cycle Energy and Greenhouse Gas Emission Bene fits of Lightweighting in Automobiles: Review and Harmonization. *Environ. Sci. Technol.* **2013**, *47*, 6089–6097. [CrossRef]
11. ASM Metals Handbook. *Welding, Brazing and Soldering*, 10th ed.; American Society for Metals: Geauga, OH, USA, 1993; Volume 6.
12. Cornacchia, G.; Cecchel, S.; Panvini, A. A comparative study of mechanical properties of metal inert gas (MIG)-cold metal transfer (CMT) and fiber laser-MIG hybrid welds for 6005A T6 extruded sheet. *Int. J. Adv. Manuf. Technol.* **2017**, *94*, 2017–2030. [CrossRef]
13. Apelian, D.; Shivkumar, S.; Sigworth, G. Fundamental Aspects of heat treatment of cast Al−Si−Mg alloys. *AFS Trans.* **1989**, *97*, 727–742.
14. Faccoli, M.; Dioni, D.; Cecchel, S.; Cornacchia, G.; Panvini, A. An experimental study to optimize the heat treatment of gravity cast Sr-modified B356 aluminum alloy. *Trans. Nonferrous Met. Soc. China* **2017**, *27*, 1698–1706. [CrossRef]
15. Lados, D.A.; Apelian, D.; Wan, L. Solution treatment effects on microstructure and mechanical properties of Al−(1 to 13 pct)Si−Mg cast alloys. *Metall. Mater. Trans. B* **2011**, *42*, 171–180. [CrossRef]
16. Dioni, D.; Cecchel, S.; Cornacchia, G.; Faccoli, M.; Panvini, A. Effects of artificial aging conditions on mechanical properties of gravity cast B356 aluminum alloy. *Trans. Nonferrous Met. Soc. China* **2015**, *25*, 1035–1042. [CrossRef]
17. Morita, A. Aluminium alloys for automobile applications. In Proceedings of the ICAA-6, Toyohashi, Japan, 5–10 July 1998; Volume 1, pp. 25–32.
18. Andersen, S.J.; Zandbergen, H.W.; Jansen, J.; Traeholt, C.; Tundal, U.; Reiso, O. "The crystal structure of the β" phase in al-mg-si alloys. *Acta Mater.* **1998**, *46*, 3283–3298. [CrossRef]
19. Ikeno, K.M.S.; Sato, T. Hrtem Study of Nano-Precipitation Phases in 6000 Series Aluminum Alloys. *Sci. Technol. Educ. Microsc. An Overv.* **2003**, 152–162.
20. Maisonnette, D.; Suery, M.; Nelias, D.; Chaudet, P.; Epicier, T. Effects of heat treatments on the microstructure and mechanical properties of a 6061 aluminium alloy. *Mater. Sci. Eng. A Struct. Mater.* **2011**, *528*, 2718–2724. [CrossRef]
21. International Aluminium Institute. Primary Aluminium Production. 22/09/08. 2008. Available online: http://www.world-aluminium.org (accessed on 10 December 2019).
22. Voisin, P. Métallurgie extractive de l'aluminium. Techniques de l'ingénieur. *Matér. Mét.* **1992**, *2340*, 2340.
23. Liu, L.; Ren, D.; Liu, F. A Review of Dissimilar Welding Techniques for Magnesium Alloys to Aluminum Alloys. *Materials* **2014**, *7*, 3735–3757. [CrossRef]
24. Cao, X.; Jahazi, M.; Immarigeon, J.P.; Wallace, W. A review of laser welding techniques for magnesium alloys. *J. Mater. Process. Technol.* **2006**, *171*, 188–204. [CrossRef]
25. Abioye, T.E.; Olugbade, T.O.; Ogedengbe, T.I. Welding of Dissimilar Metals Using Gas Metal Arc and Laser Welding Techniques: A Review. *J. Emerg. Trends Eng. Appl. Sci.* **2017**, *8*, 225–228.
26. Fang, Y.; Jiang, X.; Mo, D.; Zhu, D.; Luo, Z. A review on dissimilar metals' welding methods and mechanisms with interlayer. *Int. J. Adv. Manuf. Technol.* **2019**, *102*, 2845–2863. [CrossRef]
27. Song, Y.; Yang, X.; Cui, L.; Hou, X.; Shen, Z.; Xu, Y. Defect features and mechanical properties of friction stir lap welded dissimilar AA2024–AA7075 aluminum alloy sheets. *Mater. Des.* **2014**, *55*, 9–18. [CrossRef]
28. Lean, P.P.; Gil, L.; Urena, A. Dissimilar welds between unreinforced AA6082 and AA6092/SiC/25p composite by pulsed-MIG arc welding using unreinforced filler alloys (Al–5Mg and Al–5Si). *J. Mater. Process. Technol.* **2003**, *143–144*, 846–850. [CrossRef]

29. Palanivel, R.; Mathews, P.K.; Dinaharan, I.; Murugan, N. Mechanical and metallurgical properties of dissimilar friction stir welded AA5083-H111 and AA6351-T6 aluminum alloys. *Trans. Nonferrous Met. Soc. China* **2014**, *24*, 58–65. [CrossRef]
30. Gungor, B.; Kaluc, E.; Taban, E.; Aydin, S.I.K. Mechanical and microstructural properties of robotic Cold Metal Transfer (CMT) welded 5083-H111 and 6082-T651 aluminum alloys. *Mater. Des.* **2014**, *54*, 207–211. [CrossRef]
31. Jonckheere, C.; de Meester, B.; Denquin, A.; Simar, A. Torque, temperature and hardening precipitation evolution in dissimilar friction stir welds between 6061-T6 and 2014-T6 aluminum alloys. *J. Mater. Process. Technol.* **2013**, *213*, 826–837. [CrossRef]
32. Ilangovan, M.; Boopathy, S.R.; Balasubramanian, V. Microstructure and tensile properties of friction stir welded dissimilar AA6061–AA5086 aluminium alloy joints. *Trans. Nonferrous Met. Soc. China* **2015**, *25*, 1080–1090. [CrossRef]
33. Uematsu, Y.; Tozaki, Y.; Tokaji, K.; Nakamura, M. Fatigue behavior of dissimilar friction stir welds between cast and wrought aluminum alloys. *Strength Mater.* **2007**, *40*, 138–141. [CrossRef]
34. Ghosh, M.; Husain, M.M.; Kumar, K.; Kailas, S.V. Friction stir-welded dissimilar aluminum alloys: Microstructure, mechanical properties, and physical state. *J. Mater. Eng. Perform.* **2013**, *22*, 3890–3901. [CrossRef]
35. Wang, M.; Zou, Y.D.; Hu, H.; Meng, G.; Cheng, P.; Chu, Y.L. Tensile properties and microstructure of joined vacuum die cast aluminum alloy A356 (T6) and wrought alloy 6061. *Adv. Mater. Res.* **2014**, *939*, 90–97. [CrossRef]
36. Nie, F.; Dong, H.; Chen, S.; Li, P.; Wang, L.; Zhao, Z.; Li, X.; Zhang, H. Microstructure and Mechanical Properties of Pulse MIG Welded 6061/A356 Aluminum Alloy Dissimilar Butt Joints. *J. Mater. Sci. Technol.* **2018**, *34*, 551–560. [CrossRef]
37. Casalino, G.; Leo, P.; Mortello, M.; Perulli, P.; Varone, A. Effects of Laser Offset and HybridWelding on Microstructure and IMC in Fe–Al DissimilarWelding. *Metals* **2017**, *7*, 282. [CrossRef]
38. Wang, Z.; Oliveira, J.P.; Zeng, Z.; Bu, X.; Peng, B.; Shao, X. Laser beam oscillating welding of 5A06 aluminum alloys: Microstructure, porosity and mechanical properties. *Opt. Laser Technol.* **2019**, *11*, 58–65. [CrossRef]
39. Manti, R.; Dwivedi, D.K.; Agarwal, A. Microstructure and hardness of AlMg-Si weldments produced by pulse GTA welding. *Int. J. Adv. Manuf. Technol.* **2008**, *36*, 263–269. [CrossRef]
40. Feng, J.; Zhang, H.; He, P. The CMT short-circuiting metal transfer process and its use in thin aluminium sheets welding. *Mater. Des.* **2009**, *30*, 1850–1852. [CrossRef]
41. Zapico, E.P.; Lutey, A.H.; Ascari, A.; Pérez, C.R.G.; Liverani, E.; Fortunato, A. An improved model for cold metal transfer welding of aluminium alloys. *J. Therm. Anal. Calorim.* **2018**, *131*, 3003–3009. [CrossRef]
42. Wang, J.; Feng, J.C.; Wan, Y.X. Microstructure of Al–Mg dissimilar weld made by cold metal transfer MIG welding. *J. Mater. Sci. Technol.* **2008**, *24*, 827–831. [CrossRef]
43. Pickin, C.G.; Young, K. Evaluation of cold metal transfer (CMT) process for welding aluminium alloy. *J. Sci. Technol. Weld. Join.* **2006**, *11*, 583–585. [CrossRef]
44. Katayamaa, S.; Kawahitoa, Y.; Mizutania, M. Elucidation of laser welding phenomena and factors affecting weld penetration and welding defects. *Phys. Procedia* **2010**, *5*, 9–17. [CrossRef]
45. Katayama, S.; Nagayama, H.; Kawahito, Y. Fiber laser welding of aluminium alloy. *J. Weld. Int.* **2009**, *23*, 744–752. [CrossRef]
46. Kuryntsev, S.V.; Gilmutdinov, A.K. The effect of laser beam wobbling mode in welding process for structural steels. *Int. J. Adv. Manuf. Technol.* **2015**, *81*, 1683–1691. [CrossRef]
47. Aalderink, B.J.; Pathiraj, B. Seam gap bridging of laser based processes for the welding of aluminium sheets for industrial applications. *Int. J. Adv. Manuf. Technol.* **2010**, *48*, 143–154. [CrossRef]
48. Hayashi, T.; Matsubayashi, K.; Katayama, S.; Abe, N.; Matsunawa, A.; Ohmori, A. Reduction mechanism of porosity in tandem twin-spot laser welding of stainless steel. *Weld. Int.* **2003**, *17*, 12–19. [CrossRef]
49. Yangchun, Y.; Wang, C.; Xiyuan, H.; Wang, J.; Shengfu, Y. Porosity in fiber laser formation of 5A06 aluminum alloy. *J. Mech. Sci. Technol.* **2010**, *24*, 1077–1082.
50. Chowdhury, S.H.; Chen, D.L.; Bhole, S.D.; Powidajko, E.; Weckman, D.C.; Zhou, Y. Fiber laser welded AZ31 magnesium alloy: The effect of welding speed on microstructure and mechanical properties. *Metall. Mater. Trans. A* **2012**, *43*, 2133–2147. [CrossRef]

51. ASTM International. *ASTM E18–03 (2003) Standard Test Methods for Rockwell Hardness and Rockwell Superficial Hardness of Metallic Materials*; ASTM International: West Conshohocken, PA, USA, 2003.
52. ASTM International. *ASTM E92–160 (2016) Standard Test Methods for Vickers Hardness and Knoop Hardness of Metallic Materials*; ASTM International: West Conshohocken, PA, USA, 2016.
53. ASTM International. *ASTM E140–02 (2002) Standard Hardness Conversion Tables for Metals Relationship among Brinell Hardness, Vickers Hardness, Rockwell Hardness, Superficial Hardness, Knop Hardness, and Scleroscope Hardness*; ASTM International: West Conshohocken, PA, USA, 2002.
54. EN AW-6082, Metra. Available online: http://www.metra.it/aluminium/tabellaLeghe/tabellaCatalogo6082.pdf (accessed on 8 October 2019).
55. Hwang, L.; Gung, C.; Shih, T. A study on the qualities of GTA-welded squeeze-cast A356 alloy. *J. Mater. Process. Technol.* **2001**, *116*, 101–113. [CrossRef]
56. Missori, S.; Montanari, R.; Sili, A. Caratterizzazione meccanica mediante prove fimec di giunti saldati in lega di Al 6082. *La Metall. Ital.* **2001**, *3*, 35–39.
57. Myhr, O.R.; Grong, O.; Fjaer, H.G.; Marioara, C.D. Modelling of the microstructure and strength evolution in Al–Mg–Si alloys during multistage thermal processing. *Acta Mat.* **2004**, *52*, 4997–5008. [CrossRef]
58. Tang, N.-K.; Chen, J.K.; Hung, H.-Y. Effect of silicon on thermal conductivity of Al-Si alloys. In Proceedings of the Materials Science and Technology Conference and Exhibition 2013, (MS&T'13), Montreal, PQ, Canada, 27–31 October 2013.
59. Zhilin, A.S.; Jianguo, L.; Yalunina, V.R.; Varlamenko, D.S.; Bykov, V.A.; Derevjankin, E.V. Influence of Silicon on Thermal Conductivity at Room Temperature of Al–Si–Fe Alloys. *KnE Eng.* **2018**, *3*, 294–297. [CrossRef]
60. Stadler, F.; Antrekowitsch, H.; Werner, F.; Kaufmann, H.; Riccardo, P.E.; Peter, U. The effect of main alloying elements on the physical properties of Al-Si foundry alloys. *Mater. Sci. Eng. A* **2013**, *560*, 481–491. [CrossRef]

Article

Laser Welding of ASTM A553-1 (9% Nickel Steel) (PART I: Penetration Shape by Bead on Plate)

Jaewoong Kim [1], Jisun Kim [1,*], Sungwook Kang [2] and Kwangsan Chun [3]

[1] Smart Manufacturing Process R&D Group, Korea Institute of Industrial Technology, Gwangju 61012, Korea; kjw0607@kitech.re.kr

[2] Transport Machine Components R&D Group, Korea Institute of Industrial Technology, Jinju 52845, Korea; swkang@kitech.re.kr

[3] Welding Engineering R&D Department, Industrial Application R&D Institute, Daewoo Shipbuilding & Marine Engineering Co., LTD., Geoje, Gyeongnam 53302, Korea; kschun@dsme.co.kr

* Correspondence: kimjisun@kitech.re.kr; Tel.: +82-62-600-6302

Received: 16 March 2020; Accepted: 3 April 2020; Published: 5 April 2020

Abstract: The International Maritime Organization (IMO) is tightening regulations, in order to reduce greenhouse gas emissions from ship operations. As a result, the number of vessels using Liquefied Natural Gas (LNG) as fuel has increased rapidly. At this time, ASTM A553-1 (9% nickel steel) is being used as a tank material for storing LNG as fuel, because it has higher strength than other cryogenic materials. Currently, shipyards are manufacturing LNG fuel tanks by using the Flux Cored Arc Welding (FCAW) method, using 9% nickel steel material. However, fabrication through FCAW welding has two drawbacks. The first is to use a welding electrode that is 20 times higher in cost than the base metal, and the second is that the total production cost increases because the thickness of the tank increases due to the strength drop near the Heat Affected Zone (HAZ) after welding. Laser welding, which does not require additional welding rods and has no strength reduction in the HAZ, can overcome the drawbacks of FCAW welding and ensure price competitiveness. In this study, it is confirmed the characteristics of the penetration shape of Bead on Plate (BOP) after various laser welding conditions as a basic study to apply laser welding to A553-1 welding. For this, penetration characteristics of A553-1, according to laser welding speed and power, which is a main factor of laser welding, are confirmed.

Keywords: laser welding; ASTM A553-1 (9% nickel steel); penetration shape; Bead on Plate (BOP)

---

1. Introduction

With the global environmental regulations to prevent climate change, regulations on emissions of ships are intensifying. In the shipbuilding industry, the use of Liquefied Natural Gas (LNG) for marine fuel is continuously studied as an alternative to existing fossil fuels [1–5]. By using LNG as fuel, CO_2 emission is reduced by about 20%, nitrogen oxides (NO_x) by 80%, sulfur oxides (SO_x) by 90% and particulate matter (PM) by 99%, in comparison to using Heavy Fuel Oil (HFO), which is the existing marine fuel. It enables Tier III compliance, which is the latest regulation from International Maritime Organization (IMO). In other words, in the case of ships using LNG as fuel, it is possible to operate eco-friendly vessels without the need for additional equipment, to reduce harmful exhaust gas when operating.

A total of four materials (9% Nickel, STS 304L, Al 5083-0 and Invar) are available for the low-temperature/cryogenic energy storage/transport vessels approved by International Code of the Construction and Equipment of Ships Carrying Liquefied Gases in Bulk (IGC Code) [6]. Moreover, mechanical properties are shown in Table 1.

Table 1. IGC Code list of cryogenic fuel tank material.

Material	Chemical Composition	Yield Strength (MPa)	Ultimate Tensile Strength (MPa)
9% Nickel	Fe-9Ni	>585	690–825
STS304L	Fe-18.5Cr-9.25Ni	>205	>585
Al5083-0	Al-4.5Mg	124–200	276–352
Invar	Fe-36Ni	230–350	400–500

A553-1 is used as a material for cryogenic tanks, such as LNG, and has recently been used as an LNG fuel tank material because of its relatively higher yield strength/tensile strength than other materials. There are several factors that determine the thickness of LNG fuel tanks, among which the minimum yield/tensile strength is an important factor. A553-1 has a weak point in that the thickness of the tank becomes thicker due to weakened strength at the weld after welding [7]. When laser welding is applied, the thickness of the tank can be reduced by increasing the strength of this part. The base material used in this study is A553-1, and the chemical composition is shown in Table 2.

Table 2. The chemical composition of ASTM A553 Type 1 (9% nickel steel).

Component	Percentage (wt.%)
Carbon, C	0.13 max
Manganese, Mn	0.90 max
Phosphorous, P	0.015 max
Silicon, Si	0.15–0.40
Sulfur, S	0.015 max
Nickel, Ni	8.5–9.5

Laser is the abbreviation of "Light Amplification by Stimulated Emission of Radiation" and refers to the welding method applying the output of the laser beam. Laser welding has been studied for application to various materials due to its low welding deformation, easy automation and its deep, narrow heat-affected part [8–22]. The purpose of this study was to analyze the penetration characteristics of A553-1, according to the welding speed and power among the laser-welding parameters, and to improve the price competitiveness, by securing the weldability with good mechanical strength in the future.

2. Experiments of ASTM 553-1 (9% Nickel Steel) Bead on Plate (BOP) by Laser Welding

2.1. Laser-Welding Equipment, Parameters of Experiment and Base Material

For the experiment, 5 kW fiber laser welder were used, and Figure 1 shows the laser-welding oscillator, incidental material, optical system and jig used in the experiment.

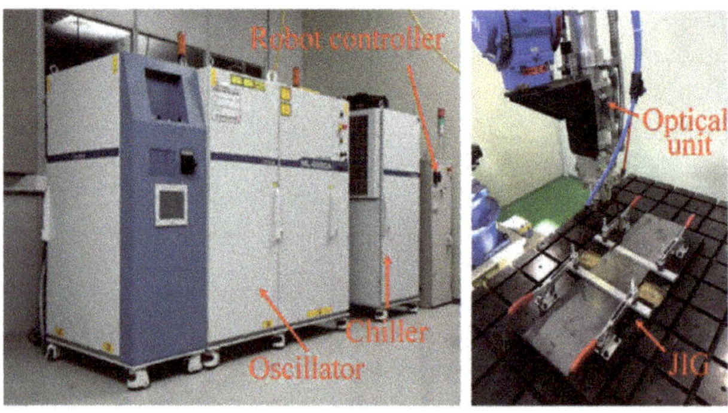

Figure 1. Fiber laser welding equipment.

The optical system used in this experiment has a spot diameter of 400 μm. The focus focal length was 148.8 mm, and the depth of focus was 6 mm.

Functionally controllable parameters in laser welding are approximately four parameters (welding speed, laser power, focus position and polarization), and most are optimized for the four conditions and used according to the situation. In addition, the alignment condition (thickness, gap and mismatch) of the welding material is also one of influences on the welding quality. The effect of each welding variable on welding quality is as follows:

(1) Laser power: The laser power is the most important factor in defining the penetration depth limit, and it is impossible to obtain a weld with a certain depth or more, even if the other conditions are optimal.
(2) Welding speed: In addition to the laser output, the welding speed has a close relationship with the welding heat input received by the welding position.
(3) Focus position (defocus): The focus position in the thickness direction is the point of impurity which places the energy focusing point at a specific position in the thickness direction of the material to be welded.
(4) Polarization (work angle and tilling angle): Optimization should be performed by adjusting the working angle and tilting angle according to the shape of the welding seam. In addition, welding must be performed at a constant angle, to protect the equipment according to the reflected light or absorption rate of the material.
(5) Alignment (thickness, gap and mismatch): Depending on the thickness of the welding material, the range of selection of the welding power is determined, since the welding power is related to the penetration depth. In laser welding, gaps play an important role in weld quality (bead shape, so it is important to select welding conditions according to gaps.

In this study, laser power and laser welding speed, which greatly affect the penetration, are controlled, and the penetration shape is observed after BOP.

Experiments are conducted to analyze the effect of welding conditions on the formation of the molten part and the shape of the bead through the BOP test. The penetration depth that can be seen through this experiment provides data that can correspond to the thickness of the base metal. The data on the width of the bead indicate the welding conditions that can be applied to open-gap welds. In order to achieve the above goal, experiments are performed with 2 parameters (laser power and welding speed) and collect data of bead geometry, penetration, micro image and effect of parameters. Each experimental BOP condition for A553-1 using the laser welding process is shown in Table 3.

Table 3. Experimental conditions of laser Bead on Plate (BOP).

Welding Parameters	Experimental Conditions
Laser power (kW)	3.0–5.0 (5 cases)
Welding speed (meter per minute, m/min)	0.3–3.0 (11 cases)
Defocus (mm)	0
Shielding gas	N_2, 15 L/min
Tilting angle, Working angle	0°

In order to analyze the laser-welding characteristics of A553-1, reference data for general laser welding are obtained through the welding experiments of the A36 (low carbon steel) which chemical composition is shown in Table 4 with some of same welding conditions. In this study, the laser weldability of carbon steel is not discussed in detail, but is used only as data to compare welding characteristics of simple A553-1.

Table 4. The chemical composition of A36 (low carbon steel).

Component	Percentage (wt.%)
Carbon, C	0.29
Iron, Fe	98.0
Manganese, Mn	0.80–1.20
Phosphorous, P	0.045
Silicon, Si	0.15–0.40
Sulfur, S	0.050
Copper, Cu	>0.20

2.2. Measurement of Bead Geometry

Macro cross-sectional inspection is a method to smoothly polish the surface of a welded part and perform a chemical solution treatment, to examine the structure, pore, penetration, HAZ and the like. Bead geometry measurements are made in all experiments and basically performed in BOP test, butt and fillet welding.

In this study, in order to analyze the effect of welding variables on the shape of the molten part shape, 9 measuring positions were selected, as shown in Figure 2.

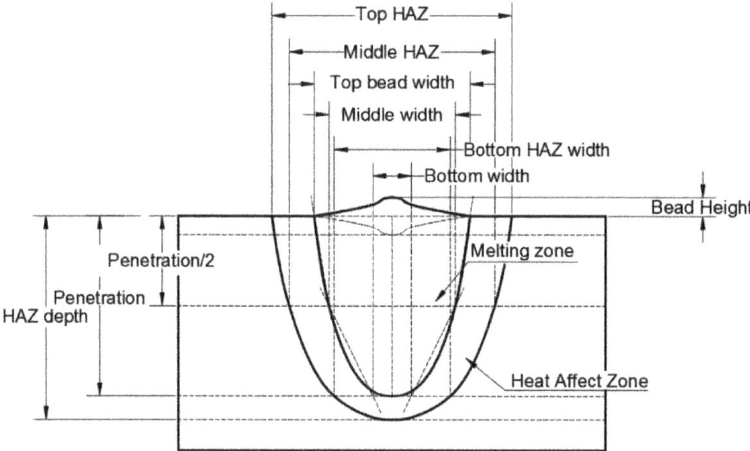

Figure 2. Definition of measurement site on bead cross-section.

A description of the cross-sectional measurement positions of the melted zone and the heat-affected zone is as follows:

(1) Top bead width: melted surface width of base material that can be observed with naked eyes.
(2) Top HAZ: the length of the HAZ of the base material surface (Top) observed through the micro-section.
(3) Penetration: the vertical depth of melting zone from surface of base metal.
(4) HAZ Depth: the vertical depth of HAZ from surface of base metal.
(5) Bead Height: the vertical length of the portion protruding above the surface of the base material after the melted portion is formed.
(6) Middle Width: the width of the melting zone at the midpoint of the penetration depth.
(7) Middle HAZ: the width of the HAZ at the midpoint of the penetration depth.
(8) Bottom Width: the width of the melting zone at the endpoint of the penetration depth.
(9) Bottom HAZ: the width of the HAZ at the endpoint of the penetration depth.

In this study, the size of welding specimen is 150 mm × 300 mm × 15mm. The laser power is stabilized at the center of the specimen; the sections from this location make it possible to show the representative section, as shown Figure 3. For measuring back-bead geometry, middle parts of horizontal axis of the experiment specimen were cut in 10 mm × 25 mm size, by a wire-cutting machine, and polished.

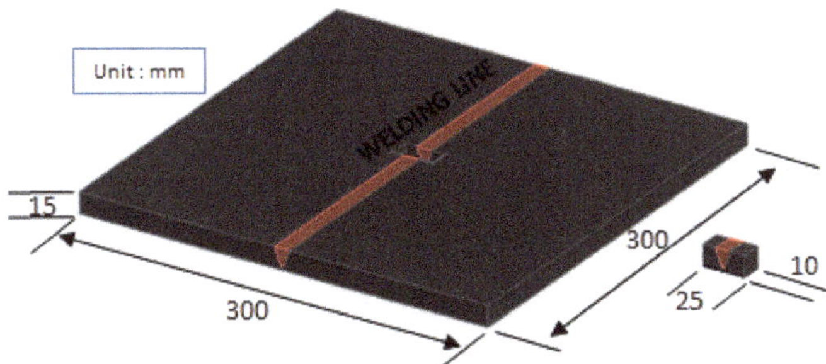

Figure 3. Welding specimen and coupon for cross-section observation.

To make the experiment specimen's bead geometry clearly visible, Nital (10% HNO_3 and Ethanol) solution were applied for the etching of the cross-section of specimens. Moreover, an optical microscope system was used for accurate measurement of bead geometry and actually measured cross-sectional bead geometries. Figure 4 shows the digital electronic microscope with 2 Mega pixels and photographed at ×60 magnification, associated with bead geometry measurement. The shape of the bead cross-section, the penetration depth, bead width and height are measured by matching the 0.1 mm mesh of the optical camera with the image.

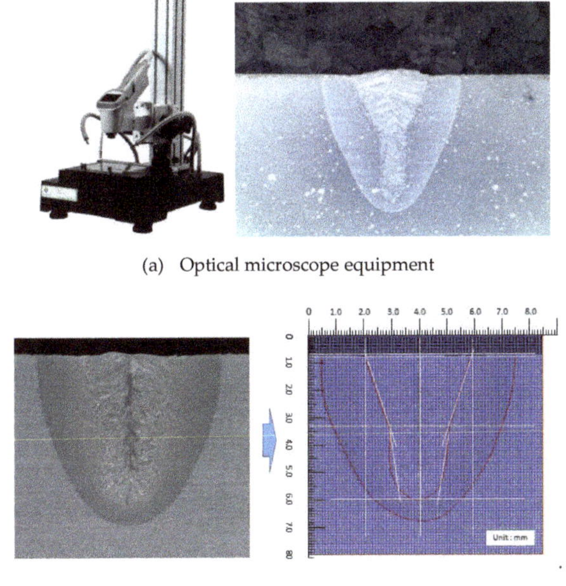

Figure 4. Equipment and method for measuring for bead geometry; (**a**) Optical microscope equipment, (**b**) Bead geometry measurement.

3. Results and Discussions

3.1. Measurement of Bead Penetration Shape

In this study, laser-welding power was experimented with on five cases (3, 3.5, 4, 4.5 and 5 kW), and laser-welding speed was experimented with 11 cases (0.3, 0.5, 0.8, 1.0, 1.2, 1.5, 1.8, 2.0, 2.2, 2.5 and 3.0 m/min) for A553-1. The total number of experiments was 55.

Figure 5 shows the results when the laser power was 3 kW and the welding speed was 0.5, 1.0, 1.5, 2.0 and 3.0 m/min, respectively.

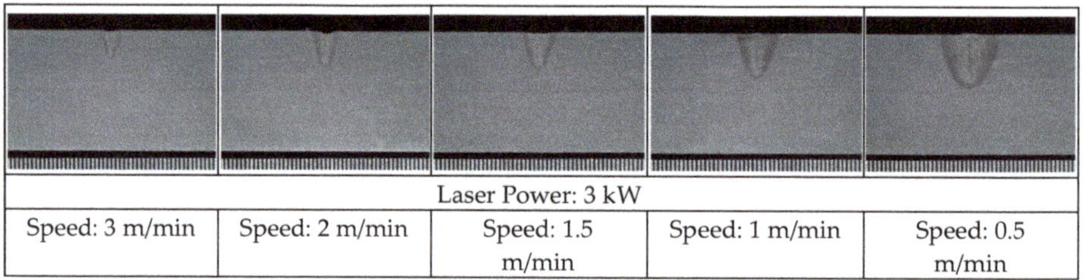

Figure 5. Bead geometry on BOP welding test (laser power: 3 kW).

Figure 6 shows the results when the laser power was 4 kW and the welding speed was 0.5, 1.0, 1.5, 2.0 and 3.0 m/min, respectively.

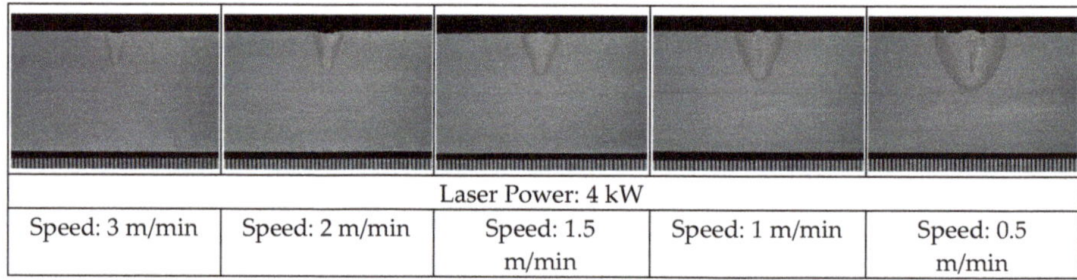

Figure 6. Bead geometry on BOP welding test (laser power: 4 kW).

Figure 7 shows the results when the laser power was 4 kW and the welding speed was 0.5, 1.0, 1.5, 2.0 and 3.0 m/min, respectively.

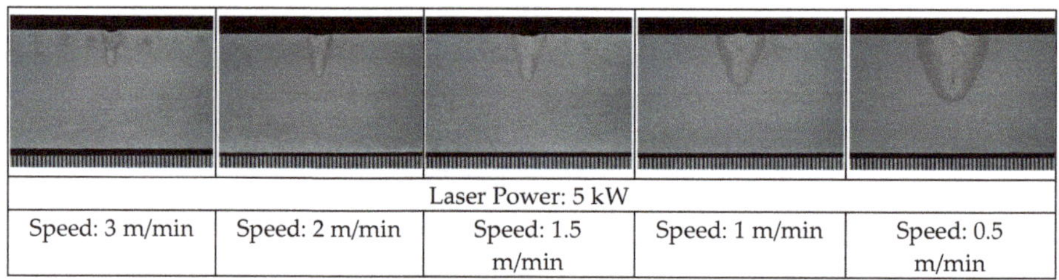

Figure 7. Bead geometry on BOP welding test (laser power: 5 kW).

In the case of the A553-1, which was the focus of this study, micro-cracks were observed in the center line of the welded section with the welding conditions of 5 kW power and 0.5 m/min speed. The internal porosity was confirmed, as shown on the left side of Figure 8. It is a pore that often occurs because of excessive heat, incomplete penetration, gravity-laser angle, etc. [23]. On the contrary, as shown on the right side of Figure 8, unsafe infusion occurred with the welding speed of 2.5 m/min, and the fine particles melted in an unstable molten state spatter.

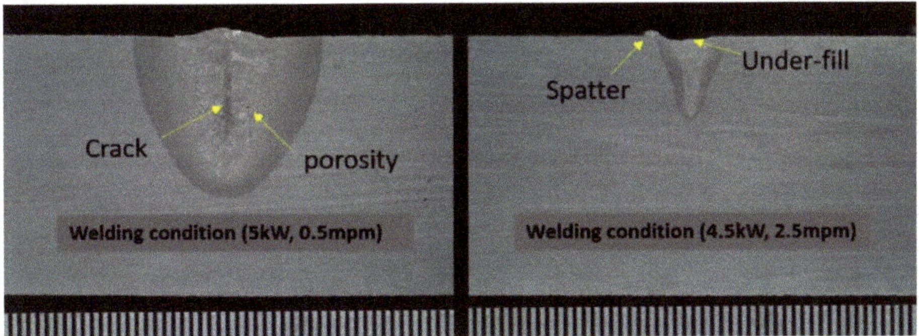

Figure 8. Crack, porosity, spatter and under-fill of A553-1 after laser welding.

3.2. *Effect of Laser Power*

In laser welding, laser power is one of the important control variables. It has been reported that the shape of the fused portion is controlled according to the laser power. In this study, its influence on three bead shape was analyzed. The difference in the shape of the beads, according to the laser power, is shown from Figures 9–11.

In the case of bead height, shown in Figure 9, the increase in bead height of A553-1 was not uniform as the laser power increased. Generally, as the welding power increases, the surface beads are under-filled, due to an increase in the amount of spatter generated. However, this result does not necessarily indicate under-fill as the laser power increases. Moreover, when compared with A36, it is confirmed that the height of bead is similar to that of laser power, despite the difference of physical properties of the two metals. There is variation in some sections (3.5 kW), but it can be concluded that the overall trend is almost consistent. It can be predicted that the laser power of both metals has no influence on the height of the weld bead.

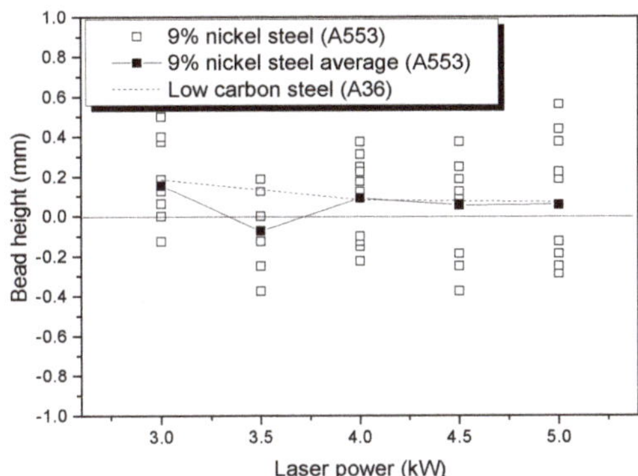

Figure 9. Height fluctuation of surface beads, according to laser power.

For A553-1 (the width of the surface bead shown in Figure 10), the width of the bead tends to widen as the laser power increases in the range of 3.5 to 4.5 kW. However, it does not continuously increase in proportion to the laser. In comparison with A36, on average, the bead width of A36 is narrower than the bead width of A553-1. This can be inferred from the fact that the direction of diffusion of the welding heat source is rapidly diffused in the thickness direction of the base material, so that the formation of the welded keyhole is very fast. The difference in the thermal conductivity coefficient of the two materials can be evidence of the above reasoning. The difference in thermal conductivity coefficient between the two materials is more than twice that of A36. Thermal conductivity coefficient of A553-1 and A36 are 17.43 and 52.7 (W/(m^2K)), respectively. When a heat source is supplied to material of A36 with a relatively high thermal conductivity characteristic, heat is transferred in all directions of the base material. It accelerates the continuous penetration of the heat source in the thickness direction of the base material. However, 9% nickel steel (A553) has low thermal conductivity, and as the energy source is concentrated on the surface, the width of the surface beads increases. In conclusion, it is evident that laser power is not an important parameter for the width of the bead.

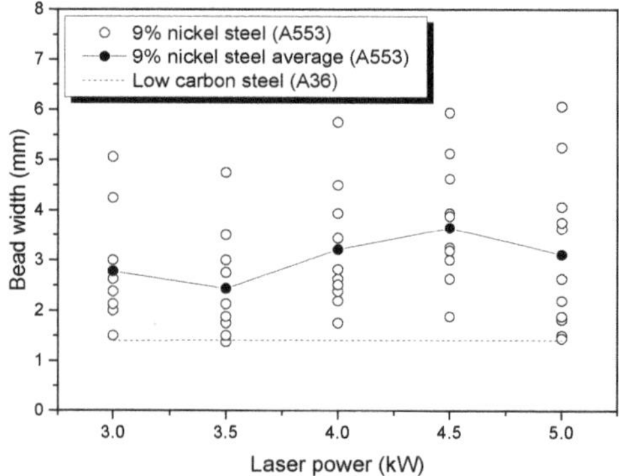

Figure 10. Width fluctuation of surface beads, according to laser power.

The most important penetration depth shown in Figure 11 is directly related to welding efficiency. If it is possible to obtain deep penetration at low power, it is very advantageous in terms of productivity. It is also important to determine what conditions need to be adjusted to ensure sufficient penetration depth. As shown in Figure 11, the relationship between penetration depth and laser power shows that the penetration depth linearly increases with increasing laser power. By comparing the penetration-depth changes of the two materials, it can be confirmed that two materials penetrations are formed at almost the same level. The penetration depth of the two materials is highly dependent on the laser power, despite the difference in properties of the two materials.

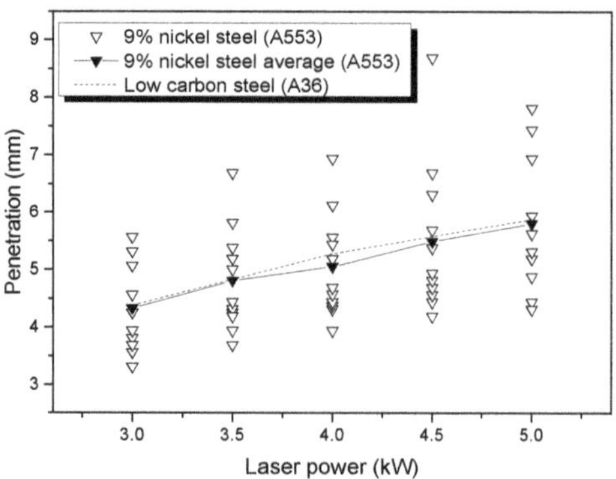

Figure 11. Penetration fluctuation of surface beads, according to laser power.

Figure 12 shows that penetration depth increases of A553-1 with increasing output at the same welding speed (1.8 m/min) and same defocus (0, surface). This result shows that laser-power control is required to control the penetration depth efficiently. Further, since the influence of the laser power on the width and height of the surface beads is small, it can be a means for independently controlling the penetration depth.

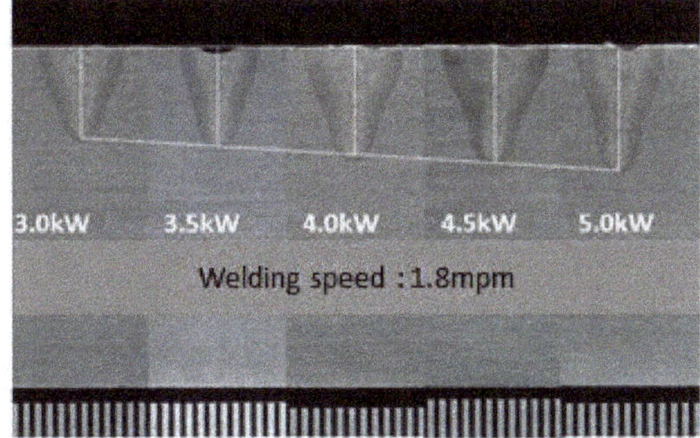

Figure 12. Variation of penetration depth, according to laser-power change by macro cross-sectional shape analysis.

3.3. *Effect of Laser Speed*

It is found that the laser power at A553-1 and A36 welds is related only to the penetration depth and does not affect the melt width and height.

However, as can be seen from Figures 13–15, as the welding speed increases, the bead height, width and penetration depth are reduced. In particular, the change in welding speed shows only the characteristics of A553-1.

Figure 13 clearly shows the difference in bead height between A36 and A553-1. As the welding speed increases, A553-1 continues to decrease in bead height, resulting in under-fill. In the case of A553-1, surface beads without under-fill are formed at a speed of 1.2 m/min or higher. On the other hand, at the high speed of 1.5 to 3.0 m/min, it is confirmed that the under-fill appeared as an average number of times. However, in the case of A36, under-fill does not occur, but the bead of uniform height is formed even when the welding speed is changed. A36 also under-fill under certain conditions, but on average, it proves the above. The occurrence of under-fill is described later. In this way, A553-1 can confirm the formation of bead height, depending on the welding speed.

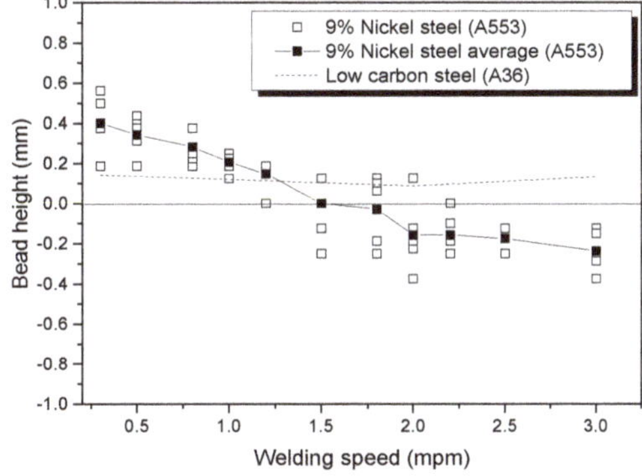

Figure 13. Height fluctuation of surface beads, according to welding speed.

In the case of the width of the weld bead shown in Figure 14, A553-1 showed a tendency to decrease as the height of the bead changes, with the change of the welding speed, but it is confirmed that the A36 is kept constant despite the change of the welding speed. This is the laser-welding characteristic ofA553-1, and it is a result of proving that the welding speed is one of the important parameters for bead width control of A553-1.

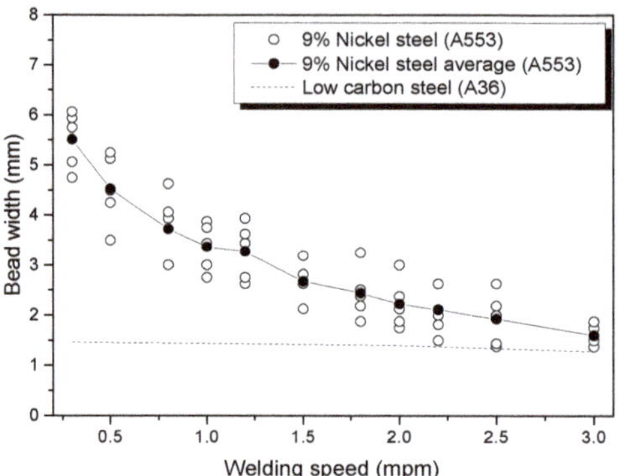

Figure 14. Width fluctuation of surface beads, according to welding speed.

For the depth of penetration shown in Figure 15, both A553-1and A36 have the same results. As the welding speed increases, it is similar that the penetration depth decreases in both materials, but on average, it is confirmed that the penetration depth of A36 is formed deeper under the same welding conditions. These results show that the welding speed is controlled to control the penetration depth in general A36 welding. However, in the case of A553-1, it is possible to control both the welding depth and the width and height of the surface bead. This is evidence that welding speed has a large impact on heat input in A553-1 laser welding. This is because the size of the melted portion decreases proportionally as the amount of heat input decreases.

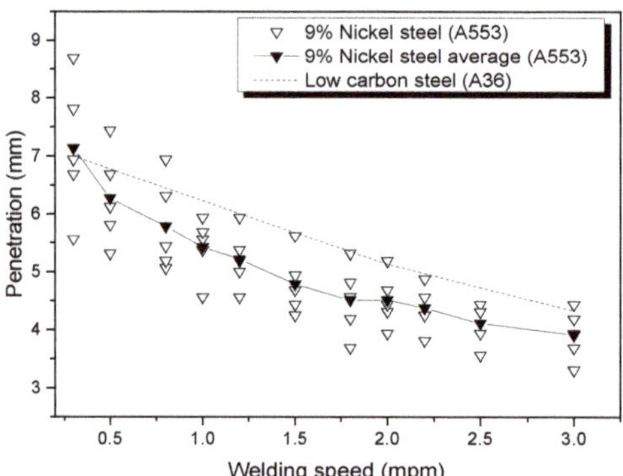

Figure 15. Penetration fluctuation of surface beads, according to welding speed.

The under-fill is reported to be caused by the synergy of the physical phenomenon of the molten part [24]. This is due to the complexities of the melt, such as volumetric shrinkage, surface tension, gravity, vapor pressure and phase transformation [25]. In this study, the cause of under-fill is explained by the phenomenon of the flow path of the exposure plasma vapor generated in the molten part. Although the steam generated by vaporization or organic plasma must be discharged to the outside through the occurrence of the keyhole of the welded portion, the discharge channel is clogged by the fast welding speed. Consequently, the molten pool explosion (spatter) is generated, and the surface depression cannot be filled again. To prevent weld under-fill, welding must ultimately be performed at a speed equal to or less than the constant speed. Figure 16 illustrates the under-fill phenomenon occurring in the high-speed section, when the welding power is constant at 3.5 kW and only the speed changes. Figure 17 is a graph that identifies the boundary of the under-fill zone, depending on the welding speed. It can be observed from this graph that the under-fill phenomenon occurs at a constant speed in all laser-power sections.

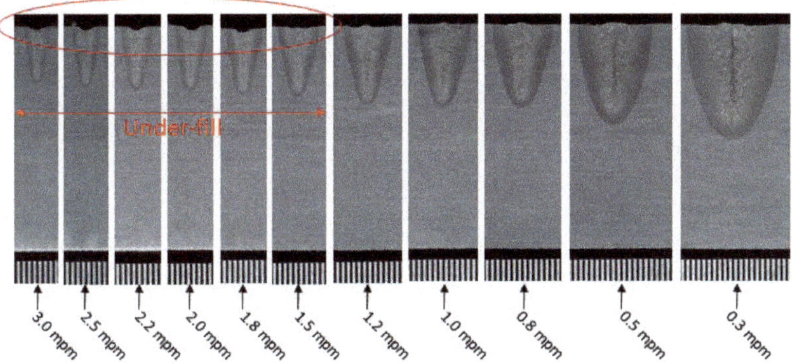

Figure 16. Under-fill phenomenon of A553-1 weld in high-speed section.

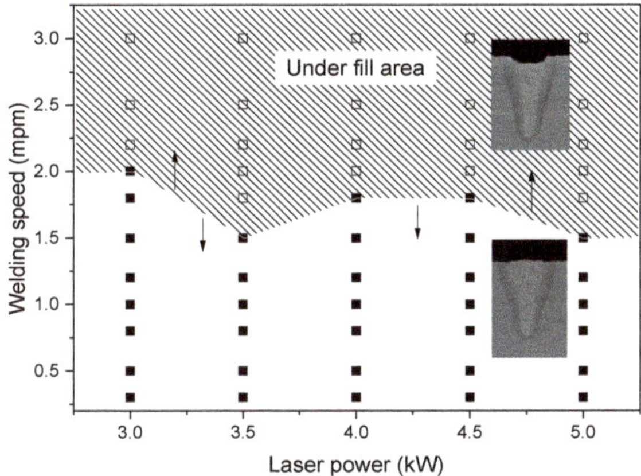

Figure 17. Under-fill section, according to welding speed.

4. Conclusions

In this study, the penetration shape of 9% nickel steel was confirmed by welding power and welding speed, which are the main laser-welding factors. The results of this study can be summarized as follows:

(1) The increase in bead height and bead width of 9% nickel steel was not uniform as the laser power increased. However, the penetration depth of the two materials was highly dependent on the laser power, despite the difference in properties of the two materials. In the case of A553-1, when the welding power increased from 3 to 5 kW, the average penetration depth increased from 4.3 to 5.7 mm, according to Figure 11.

(2) In the case of laser welding of A553-1, as the welding speed increased, the bead height, width and penetration depth were reduced. To prevent weld under-fill, welding must ultimately be performed at a speed equal to or less than the constant speed. It was confirmed that the occurrence of under-fill was dependent on the welding speed, and the result are observed in a specific range (over 1.5 m/min), according to Figure 17.
(3) Based on the contents of laser-welding penetration of the A553-1 material obtained in this study, optimized conditions of laser welding will be studied. Moreover, we will verify weld zone by the mechanical test of yield, tensile and impact, under cryogenic conditions, in PART II of this study.

Author Contributions: For research articles with several authors, a short paragraph specifying their individual contributions must be provided. The following statements should be used "Conceptualization, J.K. (Jaewoong Kim) and J.K. (Jisun Kim); methodology, J.K. (Jaewoong Kim); software, J.K. (Jisun Kim); validation, J.K. (Jaewoong Kim); formal analysis, S.K.; resources, K.C.; data curation, S.K.; writing—original draft preparation, J.K. (Jaewoong Kim); writing—review and editing, J.K. (Jaewoong Kim); visualization, S.K. and K.C.; project administration, J.K. (Jaewoong Kim); All authors have read and agreed to the published version of the manuscript.

Funding: This research has been conducted with the support of the Ministry of Trade, Industry and Energy, Republic of Korea as "Development of small and medium LNG fuel storage module for coastal ship" in the material parts technology development project.

Conflicts of Interest: The authors declare no conflict of interest.

References

1. Schinas, O.; Butler, M. Feasibility and commercial considerations of LNG-fueled ships. *Ocean Eng.* **2016**, *122*, 84–96. [CrossRef]
2. Yoo, B.Y. Economic assessment of liquefied natural gas (LNG) as a marine fuel for CO2 carriers compared to marine gas oil (MGO). *Energy* **2017**, *121*, 772–780. [CrossRef]
3. Thomson, H.; Corbett, J.J.; Winebrake, J.J. Natural gas as a marine fuel. *Energy Policy* **2015**, *87*, 153–167. [CrossRef]
4. Kim, B.E.; Park, J.Y.; Lee, J.S.; Kim, M.H. Study on the Initial Design of an LNG Fuel Tank using 9 wt.% Nickel Steel for Ships and Performance Evaluation of the Welded Joint. *J. Weld. Join.* **2019**, *37*, 555–563. [CrossRef]
5. Na, K.B.; Lee, C.I.; Park, J.H.; Cho, S.M. A Comparison of Hot Cracking in GTAW and FCAW by Applying Alloy 625 Filler Materials of 9% Ni Steel. *J. Weld. Join.* **2019**, *37*, 357–362. [CrossRef]
6. Lee, J.S.; You, W.H.; Yoo, C.H.; Kim, K.S.; Kim, Y.I. An experimental study on fatigue performance of cryogenic metallic materials for IMO type B tank. *Int. J. Nav. Arch. Ocean* **2013**, *5*, 580–597. [CrossRef]
7. Kim, J.H.; Shim, K.T.; Kim, Y.K.; Ahn, B.W. Fatigue Crack Growth Characteristics of 9% Ni Steel Welded Joint for LNG Storage Tank at Low Temperature. *J. Weld. Join.* **2010**, *28*, 45–50. [CrossRef]
8. Zhu, Y.; Li, C.; Zhang, L. Effects of Cryo-Treatment on Corrosion Behavior and Mechanical Properties of Laser-Welded Commercial Pure Titanium. *Mater. Trans.* **2014**, *55*, 511–516. [CrossRef]
9. Cui, H.; Lu, F.; Tang, X.; Yao, S. Reinforcement Behavior in Laser Welding of A356/TiB2p MMCs. *Mater. Trans.* **2012**, *53*, 1644–1647. [CrossRef]
10. Tsumura, T.; Murakami, T.; Nakajima, H.; Nakata, K. Numerical Simulation of Laser Fusion Zone Profile of Lotus-Type Porous Metals. *Mater. Trans.* **2006**, *47*, 2248–2253. [CrossRef]
11. Kim, J.W.; Jang, B.S.; Kim, Y.T.; Chun, K.S. A study on an efficient prediction of welding deformation for T-joint laser welding of sandwich panel PART I: Proposal of a heat source model. *Int. J. Nav. Arch. Ocean* **2013**, *5*, 348–363. [CrossRef]
12. Han, C.M.; Choi, H.W. FEM Analysis of the High Pressure Applied to an Impeller Blade during Laser Welding. *J. Weld. Join.* **2019**, *37*, 299–305. [CrossRef]
13. Lee, M.Y.; Kim, J.J. A Study for the Characteristics of Laser Welding on Over-lap Joint of Thin Magnesium Alloy Sheet. *J. Weld. Join.* **2019**, *37*, 293–298. [CrossRef]
14. Kang, M.J.; Choi, W.S.; Kang, S.H. Ultrasonic and Laser Welding Technologies on Al/Cu Dissimilar Materials for the Lithium-Ion Battery Cell or Module Manufacturing. *J. Weld. Join.* **2019**, *37*, 52–59. [CrossRef]
15. Park, J.U.; Lee, J.B.; An, G.B.; Kim, S.M.; Seo, H.W. Characteristic of Welding Rotational Deformation in Laser Welding of Thin Steel Sheet. *J. Weld. Join.* **2018**, *36*, 1–7. [CrossRef]

16. Jin, B.J.; Park, M.H.; Yun, T.J.; Kim, I.S.; Park, K.Y.; Kim, Y.; Yang, H.J. Optimization of Disk Laser Welding Parameters in Pure Ti Using Taguchi Method. *J. Weld. Join.* **2018**, *36*, 34–40. [CrossRef]
17. Kim, Y. Laser Welding Characteristic of Dissimilar Metal for Aluminum to Steel. *J. Weld. Join.* **2017**, *35*, 16–22. [CrossRef]
18. Zhao, H.; Qi, H. Numerical Simulation of Transport Phenomena for Laser Full Penetration Welding. *J. Weld. Join.* **2017**, *35*, 13–22. [CrossRef]
19. Kang, M.J.; Kim, C.H.; Kim, Y.M. Evaluation of the Laser Weldability of Inconel 713C alloy. *J. Weld. Join.* **2017**, *35*, 68–73. [CrossRef]
20. Zeng, Z.; Yang, M.; Oliveira, P.J.; Song, D.; Peng, B. Laser welding of NiTi shape memory alloy wires and tubes for multi-functional design applications. *Smart Mater. Struct.* **2016**, *25*, 085001. [CrossRef]
21. Zeng, Z.; Oliveira, P.J.; Yang, M.; Song, D.; Peng, B. Functional fatigue behavior of NiTi-Cu dissimilar laser welds. *Mater. Des.* **2017**, *114*, 282–287. [CrossRef]
22. Oliveira, P.J.; Zeng, Z.; Berveiller, S.; Bouscaud, D.; Braz Fernandes, F.M.; Miranda, R.M.; Zhou, N. Laser welding of Cu-Al-Be shape memory alloys: Microstructure and mechanical properties. *Mater. Des.* **2018**, *148*, 145–152. [CrossRef]
23. Chen, C.; Shen, Y.; Gao, M.; Zeng, X. Influence of welding angle on the weld morphology and porosity in laser-arc hybrid welding of AA2219 aluminum alloy. *Weld. World* **2020**, *64*, 37–45. [CrossRef]
24. Squillace, A.; Prisco, U.; Ciliberto, S.; Astarita, A. Effect of welding parameters on morphology and mechanical properties of Ti–6Al–4V laser beam welded butt joints. *J. Mater. Process. Tech.* **2012**, *212*, 427–436. [CrossRef]
25. Pakmanesh, M.R.; Shamanian, M. Optimization of pulsed laser welding process parameters in order to attain minimum underfill and undercut defects in thin 316L stainless steel foils. *Opt. Laser. Technol.* **2018**, *99*, 30–38. [CrossRef]

Article

Effects of Arc Length Adjustment on Weld Bead Formation and Droplet Transfer in Pulsed GMAW Based on Datum Current Time

Peizhuo Zhai [1], Songbai Xue [1,*], Jianhao Wang [1], Weizhong Chen [2], Tao Chen [1] and Shilei Ji [2]

[1] College of Materials Science and Technology, Nanjing University of Aeronautics and Astronautics, Nanjing 210016, China; zhaipz@nuaa.edu.cn (P.Z.); wangjh@nuaa.edu.cn (J.W.); taocmsc@nuaa.edu.cn (T.C.)

[2] Kunshan Huaheng Welding Co., Ltd., Kunshan 215347, China; wz.chen@huahengweld.com (W.C.); sl.ji@huahengweld.com (S.J.)

* Correspondence: xuesb@nuaa.edu.cn

Received: 25 April 2020; Accepted: 18 May 2020; Published: 20 May 2020

Abstract: The characteristics of weld bead formation and droplet transfer in pulsed gas metal arc weld (GMAW) with different arc lengths were studied by changing the base current time in this work. The results showed that it was easier to cause short circuits and spatters with a short arc. However, the deviation between the deepest point of penetration and the center of bead will be aggravated with the increase of arc length. In addition, more than 90% "one drop per pulse" (ODPP) transfer mode can be obtained when the pulse parameters were selected properly. However, the short arc trended to rise the proportion of "multiple drops per pulse" (MDPP), and the long arc trended towards increasing the proportion of "one drop per multiple pulses" (ODMP). Additionally, with the growth of the arc in the projected transfer zone, the penetration tended to become shallower because of the increase of arc heat dissipation, the fall of arc energy density, and droplet impact force. Overall, the strategy of choosing suitable arc length of pulsed GMAW was summarized: in order to obtain high-quality bead formation and weld joints, a shorter arc in the projected transfer zone was recommended.

Keywords: pulsed GMAW; droplet transfer; weld bead formation; droplet impact force

---

1. Introduction

Arc welding technology is one of the most widely used welding methods in industrial production at present, having the advantages of energy concentration, high efficiency, easy realization of automation and wide range of material applicability [1,2]. It has already been successfully applied in the joining of steels [3], aluminum alloys [4], nickel alloys [5], copper alloys [6], high entropy alloys [7], and so on. Additionally, industrial automation is an important direction for the future development of the manufacturing industry [8,9]. To meet the demands of productivity, pulsed gas metal arc weld (GMAW) is a good choice in industrial automation and robot welding for the advantages of controllable heat input, all position welding and no spatter [10–12]. In addition, the droplet transfer process, a research hotspot around the world, plays a crucial role in the welding quality. In pulsed GMAW, "one droplet per pulse" (ODPP) has been recognized as the most ideal transfer mode by many researchers [13–15].

The pulse peak current (I_p) and pulse peak current time (t_p) are generally regarded as the main parameters that affect ODPP mode. Moreover, the power law relationship ($I_p^n t_p$ = constant) is applied in lots of studies to determinate the peak current and time [16–18]. Additionally, Wu [15] put forward a six-parameter pulse waveform and obtained a mode of ODPP with lower heat input, proving the important influence of the droplet-detachment current and time on the droplet transfer. In addition, the base current is usually very small, and only plays the role of a stabilizing arc. As a result, a projected or spray type metal transfer at low average current can be obtained in pulsed GMAW.

Additionally, under the condition of short arc and low current, short circuiting transfer is also applied using constant voltage direct-current (DC) welding power source. However, the spatters are inevitable in short circuiting transfer, because of the electric explosion of the liquid bridge before the arcing period [19,20]. Compared with short circuiting transfer, the arc length could be easily adjusted, and no spatter welding seam could be obtained using pulsed current, leading to the advantages that the step of removing spatter on the base metal can be avoided and the manufacturing efficiency can be increased.

However, the base current time (or datum current time) has not been regarded as an influential parameter for ODPP in previous studies, but rather a parameter for adjusting the arc length. As a matter of fact, the arc length directly affects the shape of the arc, as well as the heat transfer and heat dissipation mode of the droplets [21–23]. For example, a certain arc space is required for a droplet from growth to detaching the wire. If the arc length is too short to provide sufficient space, the droplet would contact the molten pool but still on the wire, contributing to a short circuit [24]. However, most of the current research focuses on how to stabilize the arc length, while the study of how to choose the suitable arc length and the effect of different arc length on droplet transfer is ignored.

Therefore, in this work, different arc lengths were obtained by only changing the base time and fixing the other parameters. Then, the weld bead formation and the process of droplet transfer under different arc lengths were investigated. Eventually, the strategy of choosing suitable arc length of pulsed GMAW was found. The research results are helpful for optimizing the welding parameters, improving the quality of weld formation, and promoting the further application of pulsed GMAW in industrial automation production and robot welding field.

2. Welding and High-Speed Camera System

In this work, the electrode wire was connected to the positive pole of the welding power, while the workpiece was connected to the negative pole. For the convenience of the camera to shoot the arc and droplets, the welding torch was placed stationary over the workpiece, which can be moved by the motion platform at a constant speed.

Additionally, because the temperatures and rates of metal transfer are extremely high, the non-contact high-speed camera method has been widely used in droplet transfer. Figure 1 displays the schematic diagram of the welding and high-speed camera system, which can acquire synchronous voltage, current and image signal. The arc voltage was measured between contact tube and workpiece. The high-speed camera system was used at 10,000 fps and the sampling rate of the data acquisition card was 30 kHz. To lower the interference of bright arc, a laser light, whose wavelength is 850 nm, was used as a backlight. Other wavelengths of arc were weakened by a narrow-band filter centered at 850 nm between the arc and the camera. The laser light, arc, narrow-band filter and camera were all placed on the same horizontal line. As a result, the brightness of the arc was filtered to an appropriate range and the droplets can be clearly seen on the computer screen.

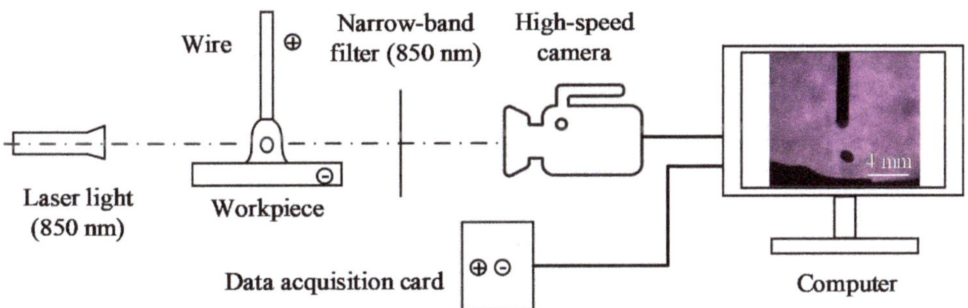

Figure 1. Schematic diagram of the welding and high-speed camera system.

3. Experimental Procedure

The bead formation and droplet transfer are more affected by waveform parameters [25,26]. Q235, a typical mild steel, was selected by researchers to study the formation, arc behavior and droplet transfer, like Wu [15], Ghosh [27,28], and so on. In our work, 4-mm-thick mild steel (Q235) sheets were used as workpieces, whose dimensions were 150 mm × 50 mm. Thus, mild steel wire (ER50-6) with a diameter of 1.2 mm was used as the electrode wire and the CTWD (contact tube-to-work distance) was selected as 20 mm. As shielding gas, 82% Ar + 18% CO_2 mixture gas with a flow rate of 15 L/min was selected. The speed of the wire feed and the speed of welding were 3 m/min and 3 mm/s, respectively. With the purpose of stable ODPP transfer, pulse current waveform showed in Figure 2 was selected according to Ref. [15]. In Figure 2, t_0, t_1 and t_2 are the start point of the pulse peak current period, the droplet-detachment current period and the base current, respectively. Over 90% droplets are ODPP transfer at middle arc length (about 5 mm) in terms of corresponding high-speed photographs. Typical high-speed photographs of ODPP transfer mode are displayed in Figure 3.

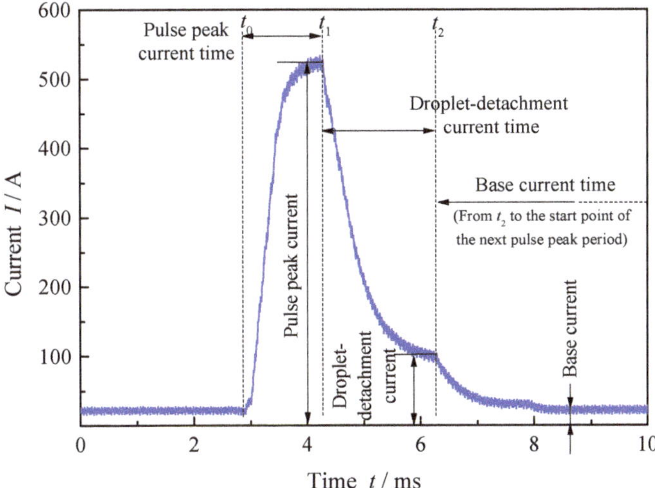

Figure 2. Main parameters of pulse current waveform.

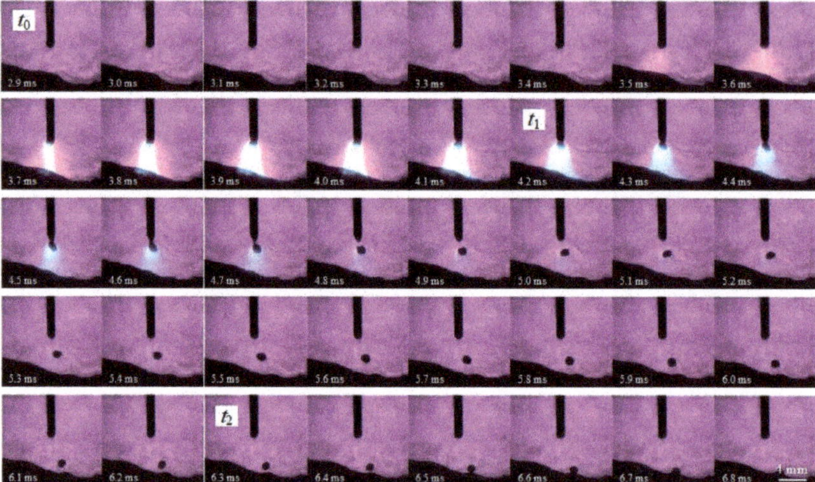

Figure 3. Typical high-speed photographs of "one droplet per pulse" (ODPP) transfer mode.

In the high-speed camera system, in order to observe the droplet process more clearly, a narrow-band filter was used to weaken the brightness of the arc. As a result, when the current was below 300 A, the arc was almost filtered out, and was difficult to observe (see Figure 3). However, the morphology of the arc can be seen clearly with the high current and arc brightness, such as in photographs near t_1. Additionally, the anode spot of arc is always over the droplet when the current is above a critical current level, so the arc length is not affected by the droplet near t_1. Therefore, for the purpose of more accurate measurement, the arc length near t_1 was selected to be measured in this pulse period. Then, the average arc length of a random 30 pulses is calculated as the arc length with these welding parameters.

Based on the pulse current waveform in Figure 2, the arc length was adjusted from less than 2 mm to more than 10 mm by only changing the base current time, without altering other parameters. Then, welding tests were performed on the piecework with different arc length. The parameters table of the pulse waveform and the experiment are presented in Table 1.

Table 1. The parameters table of pulse waveform and experiment.

Test	Base Current Time t_b/ms	Pulse Cycle Time T/ms	Pulse Frequency f/Hz	Average Current I_a/A	Average Arc Length L_a/mm	Standard Deviation of Arc Length σ_a/mm
1	12.3	15.6	64	79	1.5	0.37
2	11.5	14.8	68	82	2.3	0.49
3	11.1	14.4	69	84	3.1	0.22
4	10.3	13.6	74	88	5.0	0.29
5	9.1	12.4	81	94	7.9	0.21
6	7.9	11.2	89	103	10.4	0.62

4. Effects of Arc Length on Welding Formation

During the welding process in Tests 1 and 2, spatters and popping noises occurred from time to time. However, if the arc length were a little longer than that in Test 2, such as in Test 3, those spatters and popping noises would disappear. By comparing the high-speed photographs, it was found that the arc lengths were longer in Tests 3–6, and droplets were all in the projected transfer mode presented in Figure 3, meaning that droplets went into the molten pool after detaching from the wire. However, the arc lengths were shorter in Tests 1 and 2. Not all droplets were in the projected transfer mode, and some short circuits were found. Therefore, in terms of the modes of droplet transfer, the tests in this experiment can be divided into "the partial projected transfer zone" (Tests 1 and 2) and "the projected transfer zone" (Tests 3–6).

4.1. The Partial Projected Transfer Zone

The welding formation of Tests 1 and 2 in the partial projected transfer zone are displayed in Figure 4. It can be seen that the surface formation of the weld bead in Test 2 was better than that in Test 1. The edge of bead was crooked, and there are lots of big spatters around the bead in Test 1. In contrast, the edge of the bead was relatively straight, and there were few big spatters in Test 2. The centerline of penetration was useful to measure, but a larger difference can be noted looking at the area of the fusion zone. Although the highlighted profile looks similar in terms of cross section formation, the weld penetration of Test 2 was only 0.1 mm deeper than that of Test 1. Moreover, from the perspective of the brown welding fumes covering the surface of the workpiece, the fumes were obviously fewer and mainly distributed within 1 cm around the weld bead in Test 2, in which the bright steel plate far away from the weld bead was clearly visible. However, the surface of the workpiece was almost covered by fumes, and the regions of bright steel plate were nearly invisible in Test 1.

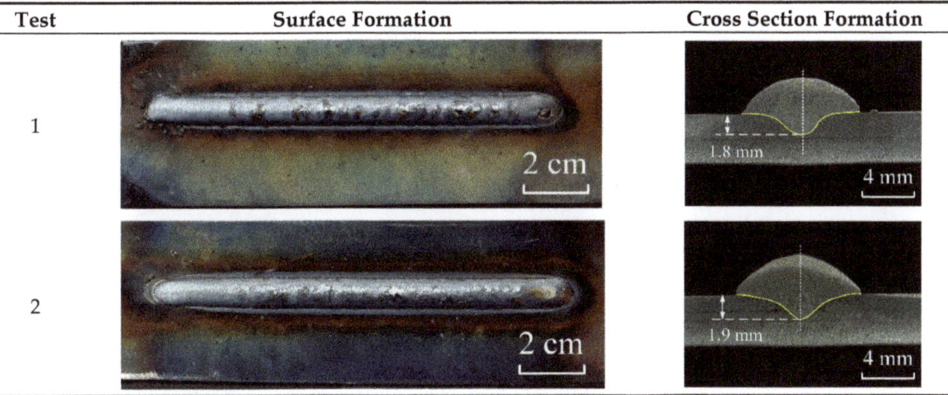

Figure 4. The formation of the weld bead in the partial projected transfer zone.

According to the high-speed photographs of Tests 1 and 2, except for projected transfer, two types of short circuit [24] were observed: normal short circuit (see Figure 5) and instantaneous short circuit (see Figure 6). It is necessary to point out that the short circuit phenomenon in pulsed GMAW is different from that in short circuiting transfer using a constant direct-current (DC) power source. According to Ref. [29], these short circuits in pulsed GMAW, as in Figures 5b and 6b, are also called meso-spray transfer. During meso-spray transfer, the position of necking is between the undetached droplet and solid wire, followed by short circuits. However, during short circuiting transfer using constant DC power source, the position of necking is at the liquid bridge, which is formed after short circuits.

During the normal short circuit (Figure 5), the neck between the undetached droplet and the wire gradually shrank, and the undetached droplet contacted the molten pool at the time P_1, with a dramatic fall of voltage. Then, from the next three frames of P_1 (3.1–3.3 ms), it can be seen that the bright arc was extinguished because of the short circuit and the minimum position of short circuiting liquid bridge was broken by explosion. A few spatters flied out of the explosive position, with an unstable arc and an abnormal voltage peak.

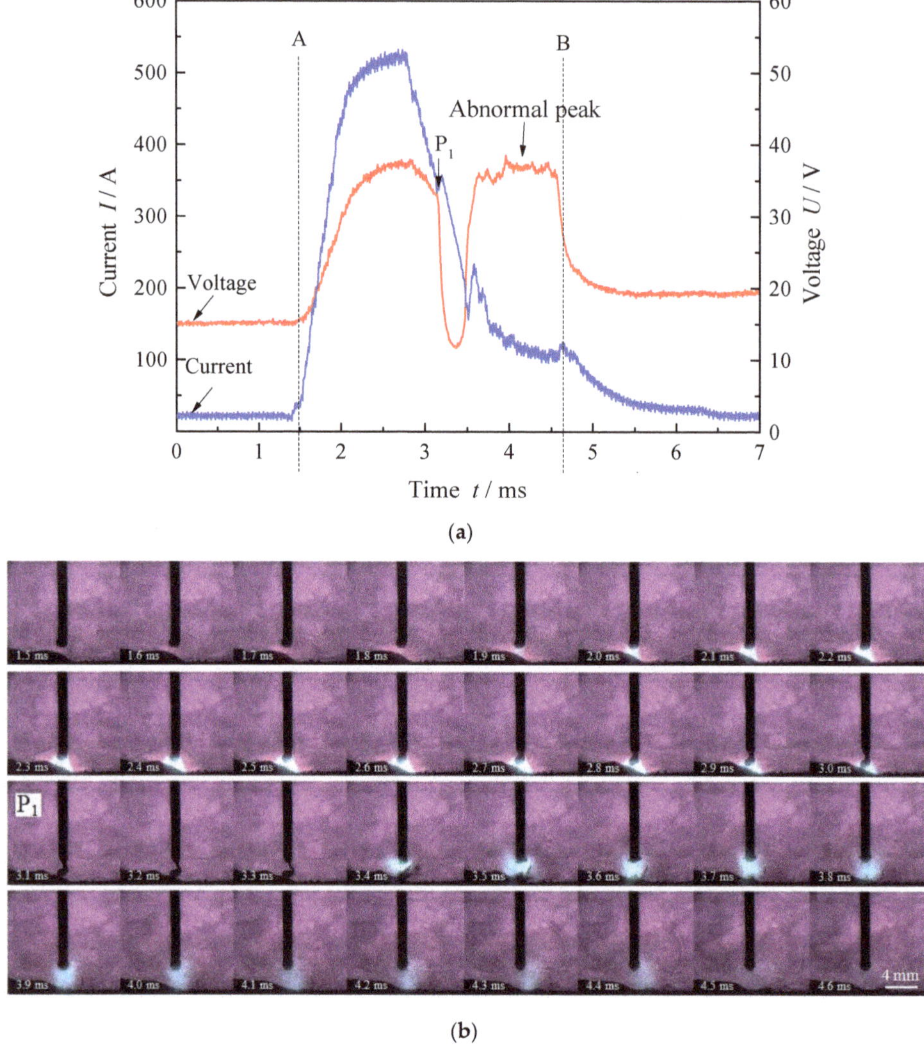

(**a**)

(**b**)

Figure 5. The normal short circuit: (**a**) The welding current and voltage waveform; (**b**) the high-speed photographs of interval AB.

When the arc was reignited, an unstable arc and an abnormal voltage peak were found. This is because at the time of arc reignition after short circuiting, cathode spots were reformed in the center of the weld pool surface. During the process of reforming the arc and cathode spots, the concentration of the cathode spots and the increase of the cathode surface work function led to a rise in the potential gradient across the cathode fall space and the adjoining contraction space. Consequently, the arc voltage became abnormally high despite the short arc length, which can be regarded as a signal of arc reignition after short circuiting [24,30].

A similar phenomenon is also shown in Figure 6. Even though the duration of the short circuit was too short to be captured by the 10,000 fps high-speed camera, the arc behavior and the abnormal voltage peak can indicate an instantaneous short circuit at the time Q_1 in Figure 6. During the process of instantaneous short circuit, the undetached droplet contacted the molten pool and was bounced off instantaneously at the time Q_1. The high current near the pulse peak current leaded to huge impact force of explosion, which contributed to the split of droplet and various big spatters.

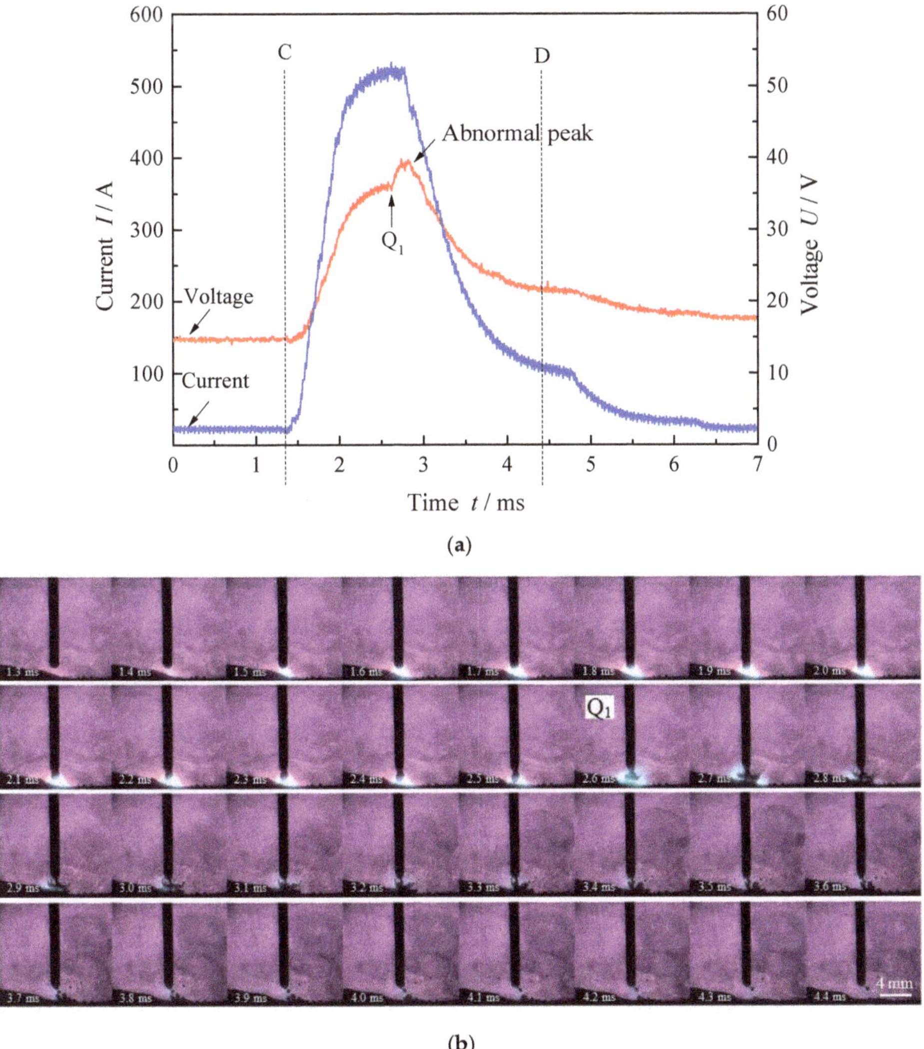

Figure 6. The instantaneous short circuit: (**a**) The welding current and voltage waveform; (**b**) the high-speed photographs of interval CD.

To research the cumulative effects of micro droplet transfer on macroscopic weld bead formation, the percentages of various types of droplet transfer were calculated based on each 200 pulses of Tests 1 and 2 (shown in Figure 7). Even through the average arc length in Test 1 (1.5 mm) was only 0.8 mm shorter than that in Test 2 (2.3 mm), their percentages of various types of droplet transfer were totally different. From Test 1 to Test 2, the percentage of projected transfer rise sharply from less than 10% to almost 90%. However, the proportion of the normal short circuits and the instantaneous short circuits fall from 61% and 30% to 9.5% and 2.5%, respectively. Moreover, based on the analysis of Figures 5 and 6, above, a few small spatters resulted from the normal short circuits, while a large number of spatters (especially the big spatters) resulted from the instantaneous short circuits. The longer arc allowed the droplets more time to gain enough energy to fully liquefy, so that they could impact the molten pool without short circuiting and causing explosions. Therefore, spatters declined obviously when the arc length grew a little from Test 1 to Test 2.

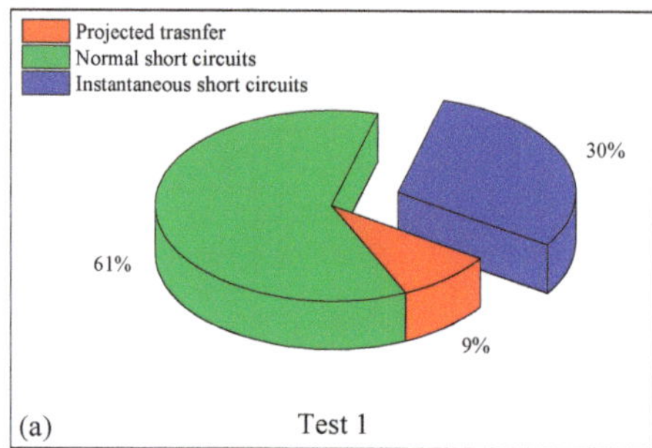

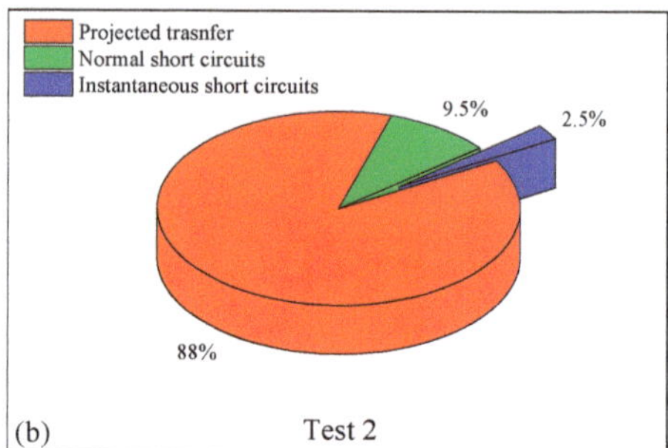

Figure 7. The percentages of various types of droplet transfer in the partial projected transfer zone: (**a**) Test 1; (**b**) Test 2.

4.2. The Projected Transfer Zone

If the arc length were long enough, the droplets would have enough space to drop into the molten pool after detaching from the wire. As a result, short circuits would disappear, and all the droplets would come into the projected transfer zone. The welding formations of Tests 3–6 in the projected transfer zone are displayed in Figure 8. As the arc length increased, the surface formation gradually deteriorated. The specific performances are as follows: the surfaces of Tests 3 and 4 were relatively straight and beautiful. The bead of Test 5 was generally smooth, but the edges of the bead were slightly crooked, with about a 1 mm offset. The surface formation of Test 6 is pretty poor. It can be clearly seen in Test 6 that the bead was uneven, and the edges were severely twisted, being visibly asymmetrical. Dozens of droplets dropped directly outside the molten pool and became large spatters near the bead. In addition, from the perspective of the brown welding fumes covered on the surface of the workpiece, the fumes were almost parallel to the bead on both sides. As the arc grew, there were more and wider fumes.

The relationship of arc length and parameters of cross section formation in the projected transfer zone is shown in Figure 9. With the increase in arc length, the penetration depth and weld reinforcement declined marginally, whereas the weld width rose a little. The greater spread of the plasma arc can cause the energy to be dispersed over a larger area. Figure 10 is the gray-scale image (256 levels) of long and short arc morphology captured at the same peak current time. The average gray levels of arc near molten pool (box region) in Figure 10a,b were 254.82 and 223.25, respectively. As we know, the higher the gray level is, the higher the arc brightness is, and the higher the current density is. Therefore, the current density of shorter arc near molten pool in Figure 10a was higher than that of the longer arc in Figure 10b. It can be also seen in Figure 10 that the arc width will increase with the increase of arc length, and the arc will be more dispersed, contributing to the expansion of heat source radius and heat dissipation of arc. As a result, the penetration depth declined and weld width rise. Additionally, because the speed of wire feed and welding are constant, weld reinforcement declined with the wider weld width.

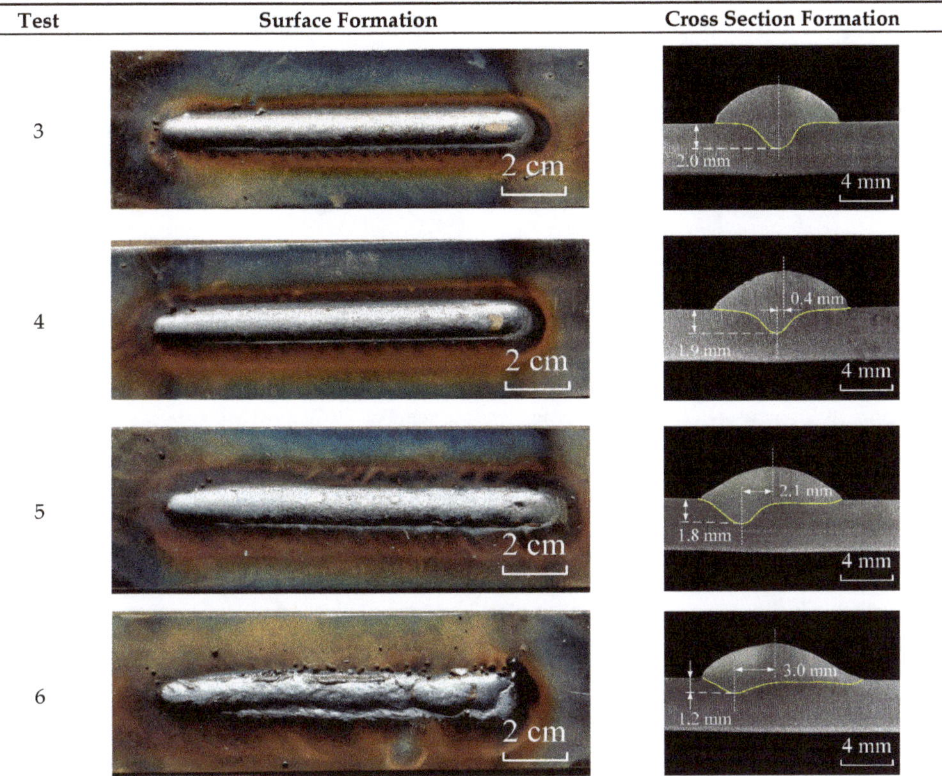

Figure 8. The formation of weld bead in the projected transfer zone.

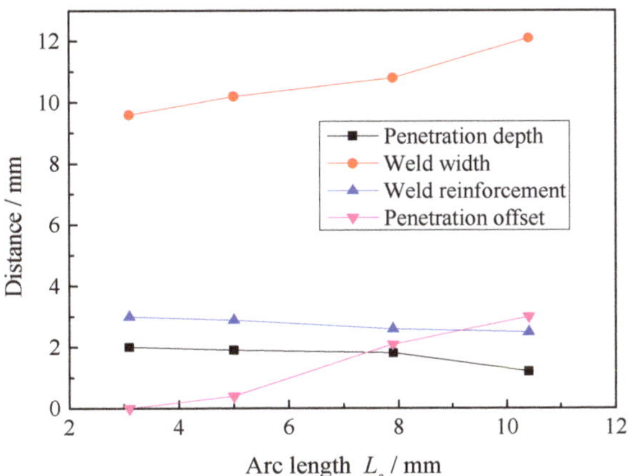

Figure 9. The relationship of arc length and parameters of cross section formation in the projected transfer zone.

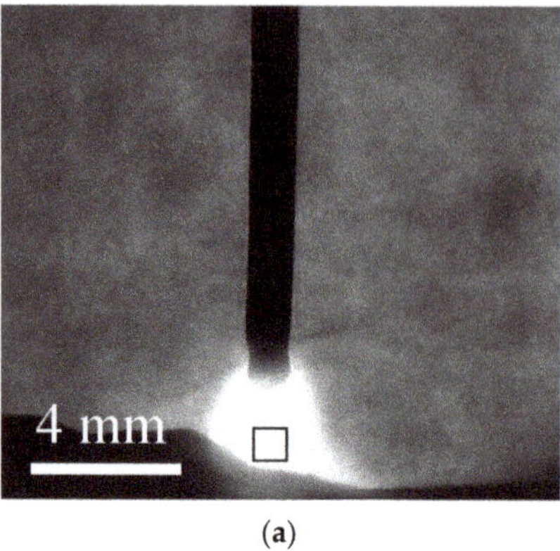

(**a**)

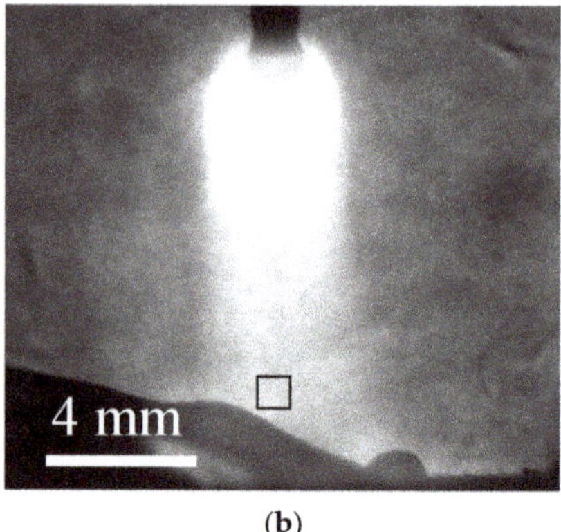

(**b**)

Figure 10. The gray-scale images of short and long arc morphology captured at the same peak current time: (**a**) Test 3; (**b**) Test 6.

Moreover, in Figure 8, the offset between the deepest point of penetration and the center of weld bead (penetration offset for short) significantly increased with the growth of arc length. The wire was not completely straight when fed out from the contact tube, because it was bent into a disc shape before welding for convenience of transportation and storage. Additionally, during real welding operation, wire inevitably inclines to a certain extent, which is not completely perpendicular to the plane of the workpiece. The end of the wire might be perpendicular to the workpiece, but the inclination of wire would increase with the growth of arc length. In addition, the droplets tend to move along the axis of welding wire in the projected transfer [23]. Therefore, the direction of droplet movement cannot be completely perpendicular to the plane of the workpiece. The rise of arc length increased the distance between wire tip and molten pool. Under the same transverse velocity, droplets could deviate from the center of the weld bead more greatly with the longer arc. Since the heat and mass carried by the droplets deviated from the center were transferred to the molten pool on one side, and the increasing magnetic fields generated by the plasma column likely caused rotation, contributing to an asymmetric driving force within the molten pool, the symmetry of penetration was broken, eventually leading to penetration offset. A similar phenomenon was also observed in Ref. [31]. A more serious deviation of the droplets and a greater penetration offset would be obtained with a longer arc length. Overall, a phenomenon was observed whereby the penetration offset in pulsed GMAW became more sensitive with the increase of arc length.

5. Effects of Arc Length on Droplet Transfer

There are three types of projected transfer in pulse GMAW: "multiple drops per pulse" (MDPP), "one drop per pulse" (ODPP), and "one drop per multiple pulses" (ODMP). ODPP was recognized as the most ideal transfer mode, and was mainly affected by the pulse peak current (I_p) and time (t_p), but not base time (t_b), by many researchers [13–15]. In addition, t_b was believed to be a parameter for adjusting the arc length. However, in this work, when I_p and t_p were fixed, it was found that the arc length had crucial effects on droplet transfer as well as the droplet impact force by only changing t_b.

5.1. The Effects of Arc Length on the Types of Projected Transfer

By observing the high-speed photographs of different tests in Figure 11, even if the parameters of the current waveform remained unchanged, the droplets did not belong 100% to ODPP. The mixture of MDPP and ODPP was displayed in Test 3 and Test 4, but no ODMP was found. In Test 5 and Test 6, the mixture of ODPP and ODMP replaced the mixture of MDPP and ODPP, and MDPP was no longer to be seen. Additionally, most of the droplets were in the form of one big droplet and one small droplet in the MDPP, as shown in Figure 11a,b. The diameter of the big droplet in the MDPP was similar to the droplet in the ODPP, but those droplets were smaller than that in the ODMP.

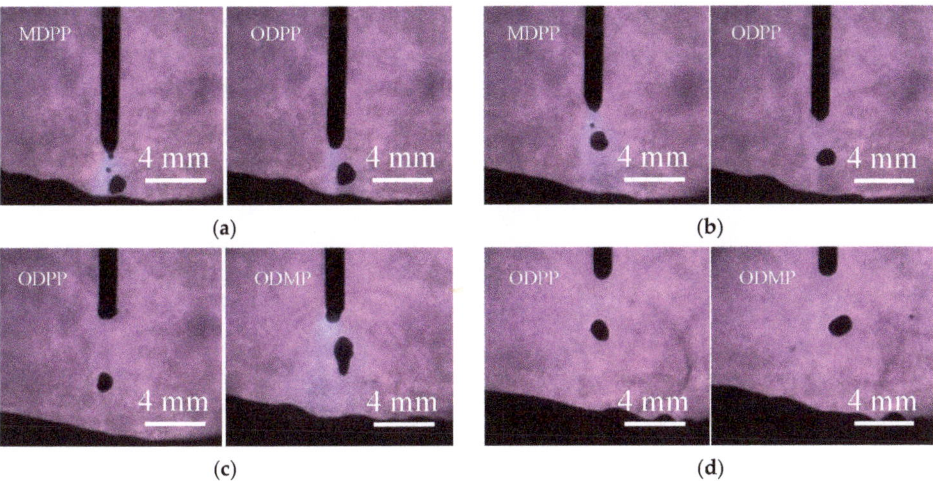

Figure 11. The typical droplet morphology of different types of projected transfer: (**a**) Test 3; (**b**) Test 4; (**c**) Test 5; (**d**) Test 6.

The three types of projected transfer were calculated based on 200 pulses in each test, and the statistical data is shown in Figure 12. The *Y*-axis φ in Figure 12 is the percentage of the pulses' number, rather than the percentage of the droplets' number. It can be found that ODPP was the main mode for each test. The proportion of ODPP of Test 4 was the highest in these tests, and was over 95%. However, ODPP occupied the least in Test 6 (less than 70%). The arc length grew gradually from Test 3 to Test 6. In terms of whole tendency, the short arc trended to rise the proportion of MDPP, but the long arc trended to rise the proportion of ODMP. According to the Static Force Balance Theory [32], the detaching force is the sum of the electromagnetic force (F_{em}), the gravitational force (F_g), and the plasma drag force (F_d), while the retaining force (F_γ) is the surface tension. As described in Section 4.2, the arc would be more dispersed with the growth of length, contributing to the expansion of heat source radius and heat dissipation. Thus, when the peak energy of a single pulse current was fixed, the smaller detaching force was obtained by a single droplet with the long arc. Consequently, more pulses were needed to increase the detaching force and help the droplet to detach from wire. Because the droplet size was bigger and the gravitational force grew, which contributes to ODMP process and vice versa. However, due to the suitable arc length in Test 4, the appropriate detaching arc force was matched with the ODPP condition and the highest proportion of ODPP was observed.

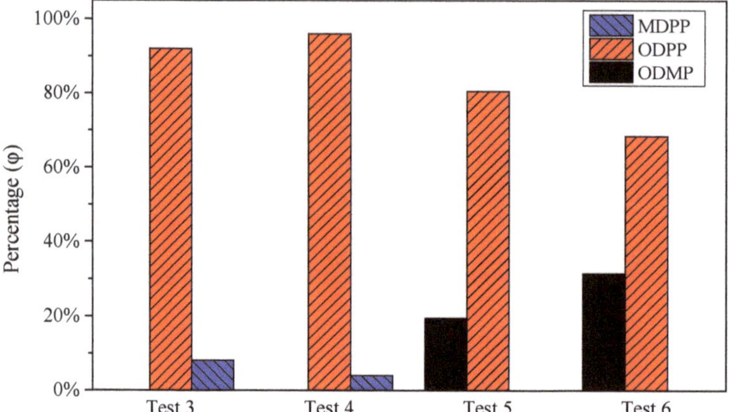

Figure 12. The statistical data of the three types of projected transfer.

5.2. The Effects of Arc Length on the Droplet Impact Force

The droplet impact force is the cumulative effect of droplets. If the types of droplet transfer were changed, the droplet impact force would be different as well [33]. According to Ref. [34], the droplet was considered approximately to be a sphere and the droplet impact force (P_d) was regarded as a force per unit area or pressure, which was relative to the droplet mass (m_d), velocity (v_d), frequency (f_d) and diameter (d_d), as displayed in Equation (1).

$$P_d = \frac{4m_d v_d f_d}{\pi d_d^2}. \quad (1)$$

According to the mass formula of the sphere, the m_d can be calculated by Equation (2).

$$m_d = \frac{\rho \pi d_d^3}{6}. \quad (2)$$

Therefore, Equation (1) can be further written as Equation (3) (density $\rho = 7.8$ g/cm^3).

$$P_d = \frac{2\rho v_d f_d d_d}{3}. \quad (3)$$

However, Equation (3) is only applicable to the single-mode droplet transfer. For the mixture of different types of projected transfer, the droplet impact force should be the sum of each droplet impact force in MDPP, ODPP and ODMP (Equation (4)).

$$P_d = P_{d,MDPP} + P_{d,ODPP} + P_{d,ODMP}. \quad (4)$$

The droplet frequency of MDPP, ODPP and ODMP are noted as f_{MDPP}, f_{ODPP} and f_{ODMP}, while the percentage of the pulses' number are noted as φ_{MDPP}, φ_{ODPP} and φ_{ODMP} respectively. Therefore, the droplet frequency of ODPP equals the percentage of the pulses' number (φ) times the frequency of pulse current wave (f) (Equation (5)).

$$f_{ODPP} = \varphi_{ODPP} f. \quad (5)$$

In addition, the frequency of the big droplet and the small droplet in MDPP (see Figure 11) are equal to the frequency of current pulses (Equation (6)). Multiple sized droplets can be found in the MDPP process, but the biggest droplet was much larger than the others (see Figure 11). Thus, the droplet impact force was approximately regarded as that of the biggest droplets in MDPP process.

$$f_{MDPP} = \varphi_{MDPP} f. \quad (6)$$

In the ODMP process, a droplet may detach from the wire after severe pulses. After observation and calculation from the high-speed photographs, the average pulses for every MDPP are 2.2 and 2.3 in Test 5 and Test 6, respectively. Thus, the droplet frequency in Test 5 and Test 6 can be calculated by Equations (7) and (8).

$$f_{ODMP} = \varphi_{ODMP} f / 2.2. \quad (7)$$

$$f_{ODMP} = \varphi_{ODMP} f / 2.3. \quad (8)$$

Moreover, the percentage of the pulses' number (φ) and the frequency of pulse current wave (f) can be read from Figure 12 and Table 1, respectively. The image-processing method in Ref. [35] was used to estimate the droplet velocity (v_d) and diameter (d_d). By using Equations (1)–(8) simultaneously, the average droplet impact force of each test was obtained in Figure 13. It can be seen that the droplet impact force of Test 3 was the highest, while the droplet impact force of Test 6 is the lowest. The former was about 50% larger than the latter, indicating that the arc length has an important influence on the droplet impact force. The droplet impact force declined with the growth of the arc length. In addition, the main part of the droplet impact force was contributed from ODPP.

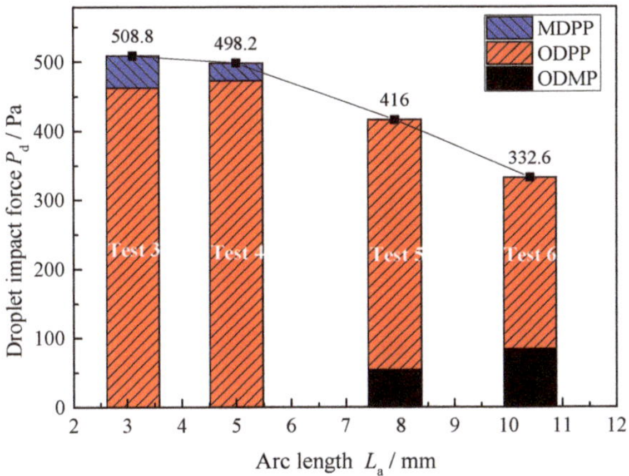

Figure 13. The average droplet impact force of projected transfer.

6. Discussion

The characteristics of weld bead formation and droplet transfer in pulsed GMAW with different arc lengths (about 1–11 mm) by changing the base current time were studied in this work. Because enough arc space was needed when a droplet detached from the wire and dropped into the molten pool, the mixture of projected transfer and short circuits would be obtained if the arc length were shorter than 3 mm. However, spatters would inevitably be produced from short circuits, especially instantaneous short circuits (Section 4.1). To reduce the spatters and improve the weld bead formation, the shortest arc is recommended to be limited to avoid the partial projected transfer zone.

Additionally, in the projected transfer zone, the quality of weld bead surface formation became worse gradually with the increase of the arc length. When the arc length was over 10 mm, dozens of droplets directly dropped outside the molten pool and became large spatters near the bead (Section 4.2). In addition, the penetration offset in pulsed GMAW will become more sensitive with the increase of arc length. The penetration offset was less than 0.5 mm when the arc length was no more than 5 mm. However, when the arc length exceeded 10 mm, the penetration offset can reach 3 mm, which is not conducive to the positioning and alignment of the bead.

On one hand, Figure 10 in Section 4.2 indicates that the increase of arc length contributes to the fall of energy density and increase of heat dissipation of arc. On the other hand, Figure 13 in Section 5.2 implies that the droplet impact force declined with the growth of the arc length. According to Ref. [34], large droplet impact force can promote the increase of weld penetration. Therefore, because of the above three factors (the increase of arc heat dissipation, the fall of arc energy density and droplet impact force), the penetration tended to become shallower.

As a result, it is suggested that the arc length should be limited to a shorter range (less than 5 mm) in the projected transfer zone. So that the welding formation can be improved, the penetration offset can be reduced and the larger penetration can be obtained. Additionally, Table 1 shows that the average current of shorter arc was smaller, which can reduce heat input and save electric energy when compared with the longer arc.

In our preliminary experiment, the tensile properties of welded joints were also affected under difference arc lengths. According to AWS: B4 specifications [36], the tensile test specimens were prepared from the butt joints in the perpendicular direction to the welded seam. Figure 14 shows the average tensile strengths and elongations of the welded joints with different arc lengths in the projected transfer zone. It can be seen that the tensile strengths were approximately constant. However, with the growth of arc length, the elongations fell gradually. This implied that the short arc is also beneficial for obtaining weld joints with high elongations.

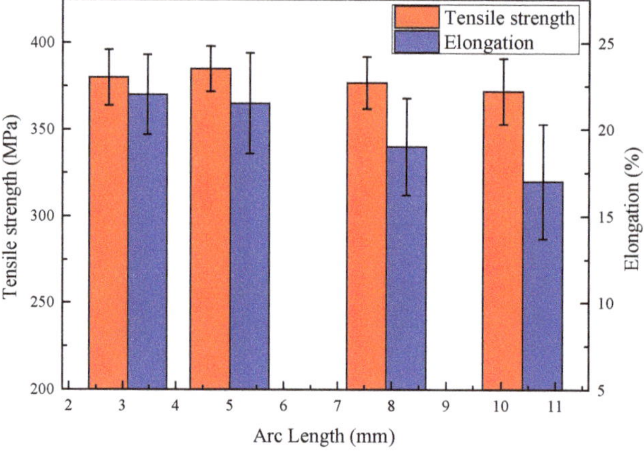

Figure 14. The tensile strengths and elongations of the welded joints with different arc lengths in the projected transfer zone.

In addition to the above discussion, the power law relationship ($I_p^n t_p$ = constant) has been used in many studies to determine the peak current and time of ODPP transfer mode [16–18], ignoring the influence of the base current time. However, in this work, it was found that the types of droplet transfer in pulsed GMAW changed a lot by only adjusting the base current time, even though the peak current and time were constant. More than 90% (but not 100%) ODPP transfer mode can be obtained when the pulse parameters were selected properly. The arc length can be adjusted by the base current time. The short arc trended to rise the proportion of MDPP, and the long arc trended to rise the proportion of ODMP.

7. Conclusions

(1) The arc length plays a crucial role in the quality of weld bead formation. In this work, the partial projected transfer zone can be obtained in pulsed GMAW when the arc length was shorter than 3 mm, resulting in the increasement of instantaneous short circuits and inevitable spatters. However, when arc length became longer than 5 mm in the projected transfer zone, the asymmetry and offset of penetration can be enlarged a lot.

(2) For the purpose of obtaining high-quality weld bead formation and weld joints, a shorter arc in the projected transfer zone was recommended. However, the partial projected transfer zone should be avoided when the arc length was short, in order to reduce unnecessary spatters. For example, for the ER50-6 wire with diameter of 1.2 mm used in this work, the appropriate arc length was about 3–5 mm. The base current time should be adjusted according to the appropriate arc length range.

(3) Peak current and peak current time are not were parameters that determine ODPP transfer mode in pulsed GMAW, which was also affected by the arc length. More than 90% (but not 100%) ODPP transfer mode can be obtained when the pulse parameters were selected properly. The short arc trended to rise the proportion of MDPP, and the long arc trended to rise the proportion of ODMP.

(4) The droplet impact force with different arc lengths in the projected transfer zone was calculated. The droplet impact force declined with the growth of the arc length. The increase of arc heat dissipation, the fall of arc energy density, and droplet impact force contributed to the result in that the penetration tended to become shallower in the projected transfer zone.

Author Contributions: Methodology, P.Z.; software, P.Z. and J.W.; investigation, P.Z., W.C., T.C., and S.J.; writing—original draft preparation, P.Z.; writing—review and editing, S.X. and P.Z.; supervision, S.X.; project administration, S.X.; funding acquisition, S.X. All authors have read and agreed to the published version of the manuscript.

Funding: This work was funded by the National Natural Science Foundation of China, grant No. 51675269 and the Priority Academic Program Development of Jiangsu Higher Education Institutions (PAPD).

Conflicts of Interest: The authors declare no conflict of interest.

References

1. Pépe, N.; Egerland, S.; Colegrove, P.A.; Yapp, D.; Leonhartsberger, A.; Scotti, A. Measuring the Process Efficiency of Controlled Gas Metal Arc Welding Processes. *Sci. Technol. Weld. Join.* **2011**, *16*, 412–417. [CrossRef]
2. Rout, A.; Deepak, B.B.V.L.; Biswal, B.B. Advances in Weld Seam Tracking Techniques for Robotic Welding: A Weview. *Robot. Comput. Integr. Manuf.* **2019**, *56*, 12–37. [CrossRef]
3. Praveen, P.; Yarlagadda, P.K.D.V.; Kang, M.J. Advancements in Pulse Gas Metal Arc Welding. *J. Mater. Process. Technol.* **2005**, *164–165*, 1113–1119. [CrossRef]
4. Huang, L.; Hua, X.; Wu, D.; Jiang, Z.; Li, F.; Wang, H.; Shi, S. Microstructural Characterization of 5083 Aluminum Alloy Thick Plates Welded with GMAW and Twin Wire GMAW Processes. *Int. J. Adv. Manuf. Technol.* **2017**, *93*, 1809–1817. [CrossRef]
5. Oliveira, J.P.; Barbosa, D.; Fernandes, F.M.B.; Miranda, R.M. Tungsten Inert Gas (TIG) Welding of Ni-rich NiTi Plates: Functional Behavior. *Smart Mater. Struct.* **2016**, *25*, 3. [CrossRef]
6. Oliveira, J.P.; Crispim, B.; Zeng, Z.; Omori, T.; Braz Fernandes, F.M.; Miranda, R.M. Microstructure and Mechanical Properties of Gas Tungsten Arc Welded Cu-Al-Mn Shape Memory Alloy Rods. *J. Mater. Process. Technol.* **2019**, *271*, 93–100. [CrossRef]
7. Oliveira, J.P.; Curado, T.M.; Zeng, Z.; Lopes, J.G.; Rossinyol, E.; Park, J.M.; Schell, N.; Braz Fernandes, F.M.; Kim, H.S. Gas Tungsten Arc Welding of as-rolled CrMnFeCoNi High Entropy Alloy. *Mater. Des.* **2020**, *189*, 108505. [CrossRef]
8. Pan, M.; Linner, T.; Pan, W.; Cheng, H.; Bock, T. A Framework of Indicators for Assessing Construction Automation and Robotics in the Sustainability Context. *J. Clean. Prod.* **2018**, *182*, 82–95. [CrossRef]

9. David, S.A.; Chen, J.; Gibson, B.T.; Feng, Z. Intelligent Weld Manufacturing: Role of Integrated Computational Welding Engineering. In *Transactions on Intelligent Welding Manufacturing*; Chen, S., Zhang, Y., Feng, Z., Eds.; Springer Singapore: Singapore, 2018; Volume 1, pp. 3–30.
10. Xu, Y.; Fang, G.; Lv, N.; Chen, S.; Zou, J.J. Computer Vision Technology for Seam Tracking in Robotic GTAW and GMAW. *Robot. Comput. Integr. Manuf.* **2015**, *32*, 25–36. [CrossRef]
11. Xu, Y.; Lv, N.; Fang, G.; Du, S.; Zhao, W.; Ye, Z.; Chen, S. Welding Seam Tracking in Robotic Gas Metal Arc Welding. *J. Mater. Process. Technol.* **2017**, *248*, 18–30. [CrossRef]
12. Kozakov, R.; Gött, G.; Schöpp, H.; Uhrlandt, D.; Schnick, M.; Häßler, M.; Füssel, U.; Rose, S. Spatial Structure of the Arc in a Pulsed GMAW Process. *J. Phys. D Appl. Phys.* **2013**, *46*, 224001. [CrossRef]
13. Pal, K.; Pal, S.K. Effect of Pulse Parameters on Weld Quality in Pulsed Gas Metal Arc Welding: A Review. *J. Mater. Eng. Perform.* **2011**, *20*, 918–931. [CrossRef]
14. Ghosh, P.K.; Goyal, V.K.; Dhiman, H.K.; Kumar, M. Thermal and Metal Transfer Behaviours in Pulsed Current Gas Metal Arc Weld Deposition of Al–Mg Alloy. *Sci. Technol. Weld. Join.* **2006**, *11*, 232–242. [CrossRef]
15. Wu, C.S.; Chen, M.A.; Lu, Y.F. Effect of Current Waveforms on Metal Transfer in Pulsed Gas Metal Arc Welding. *Meas. Sci. Technol.* **2005**, *16*, 2459–2465. [CrossRef]
16. Amin, M. Pulse Current Parameters for Arc Stability and Controlled Metal Transfer in Arc Welding. *Metal Constr.* **1983**, *15*, 272–278.
17. Rajasekaran, S. Weld Bead Characteristics in Pulsed GMA Welding of AI-Mg Alloys. *Weld. J.* **1999**, *78*, 397s–407s.
18. Rajasekaran, S.; Kulkarni, S.D.; Mallya, U.D.; Chaturvedi, R.C. Droplet Detachment and Plate Fusion Characteristics in Pulsed Current Gas Metal Arc Welding. *Weld. J.* **1998**, *77*, 254s–269s.
19. Pinchuk, I. Stabilization of Transfer and Methods of Reducing the Spattering of Metal in CO_2 Welding with a Short Arc. *Weld. Res. Abroad* **1982**, 33–35.
20. Kang, M.J.; Kim, Y.; Ahn, S.; Rhee, S. Spatter Rate Estimation in the Short Circuit Transfer Region of GMAW. *Weld. J.* **2003**, *82*, 238s–247s.
21. Harwig, D.D.; Dierksheide, J.E.; Yapp, D.; Blackman, S. Arc Behavior and Melting Rate in the VP-GMAW Process. *Weld. J.* **2006**, *85*, 52s–62s.
22. Hertel, M.; Rose, S.; Füssel, U. Numerical Simulation of Arc and Droplet Transfer in Pulsed GMAW of Mild Steel in Argon. *Weld. World* **2016**, *60*, 1055–1061. [CrossRef]
23. Chen, C.; Fan, C.; Cai, X.; Lin, S.; Yang, C. Analysis of Droplet Transfer, Weld Formation and Microstructure in Al-Cu Alloy Bead Welding Joint with Pulsed Ultrasonic-GMAW Method. *J. Mater. Process. Technol.* **2019**, *271*, 144–151. [CrossRef]
24. Tong, H.; Ueyama, T.; Tanaka, M.; Ushio, M. Observations of the Phenomenon of Abnormal Arc Voltage Occurring in Pulsed Metal Inert Gas Welding of Aluminum Alloy. *Sci. Technol. Weld. Join.* **2005**, *10*, 695–700. [CrossRef]
25. Joseph, A.; Farson, D.; Harwig, D.; Richardson, R. Influence of GMAW-P Current Waveforms on Heat Input and Weld Bead Shape. *Sci. Technol. Weld. Join.* **2013**, *10*, 311–318. [CrossRef]
26. Palani, P.K.; Murugan, N. Selection of Parameters of Pulsed Current Gas Metal Arc Welding. *J. Mater. Process. Technol.* **2006**, *172*, 1–10. [CrossRef]
27. Ghosh, P.K.; Dorn, L.; Devakumaran, K.; Hofmann, F. Pulsed Current Gas Metal Arc Welding under Different Shielding and Pulse Parameters; Part 1: Arc Characteristics. *ISIJ Int.* **2009**, *49*, 251–260. [CrossRef]
28. Ghosh, P.K.; Dorn, L.; Devakumaran, K.; Hofmann, F. Pulsed Current Gas Metal Arc Welding under Different Shielding and Pulse Parameters; Part 2: Behaviour of Metal Transfer. *ISIJ Int.* **2009**, *49*, 261–269. [CrossRef]
29. Li, Z.; Wang, Y.; Yang, L.; Li, H. Meso Spray Transfer in GMAW of Aluminum and its Control. In Proceedings of the 2009 4th IEEE Conference on Industrial Electronics and Applications, Xi'an, China, 25–27 May 2009; pp. 3148–3151.
30. Choi, J.H.; Lee, J.Y.; Yoo, C.D. Simulation of Dynamic Behavior in a GMAW System. *Weld. J.* **2001**, *80*, 239s–246s.
31. Zhang, Z.; Xue, J.; Jin, L.; Wu, W. Effect of Droplet Impingement on the Weld Profile and Grain Morphology in the Welding of Aluminum Alloys. *Appl. Sci.* **2018**, *8*, 1203. [CrossRef]
32. Kim, Y.S.; Eagar, T.W. Analysis of Metal Transfer in Gas Metal Arc Welding. *Weld. J.* **1993**, *72*, 269s–278s.
33. Zhu, F.L.; Tsai, H.L.; Marin, S.P.; Wang, P.C. A Comprehensive Model on the Transport Phenomena during Gas Metal Arc Welding Process. *Prog. Comput. Fluid Dyn.* **2004**, *4*, 99–117. [CrossRef]

34. Wu, C.; Dorn, L. The Influence of Droplet Impact on Metal Inert Gas Weld Pool Geometry. *Acta Metall. Sin.* **1997**, *33*, 774–780.
35. Zhai, P.; Xue, S.; Chen, T.; Wang, J.; Tao, Y. An Image-Processing Method for Extracting Kinematic Characteristics of Droplets during Pulsed GMAW. *Appl. Sci.* **2019**, *9*, 5481. [CrossRef]
36. Mirzaei, M.; Arabi Jeshvaghani, R.; Yazdipour, A.; Zangeneh-Madar, K. Study of Welding Velocity and Pulse Frequency on Microstructure and Mechanical Properties of Pulsed Gas Metal Arc Welded High Strength low Slloy Steel. *Mater. Des.* **2013**, *51*, 709–713. [CrossRef]

Article

Investigation on the Dynamic Behavior of Weld Pool and Weld Microstructure during DP-GMAW for Austenitic Stainless Steel

Tao Chen [1], Songbai Xue [1,*], Peng Zhang [1], Bo Wang [2], Peizhuo Zhai [1] and Weimin Long [2]

[1] College of Materials Science and Technology, Nanjing University of Aeronautics and Astronautics, Nanjing 210016, China; taocmsc@nuaa.edu.cn (T.C.); mstzhangpeng@nuaa.edu.cn (P.Z.); zhaipz@nuaa.edu.cn (P.Z.)

[2] Institute of Advanced Brazing Materials and Technology, China Intelligent Equipment Innovation Institute (Ningbo) Co., Ltd., Ningbo 315700, China; wangbo4175@126.com (B.W.); brazelong@163.com (W.L.)

* Correspondence: xuesb@nuaa.edu.cn; Tel.: +86-8489-6070

Received: 15 May 2020; Accepted: 2 June 2020; Published: 5 June 2020

Abstract: The influence of heat and droplet transfer into weld pool dynamic behavior and weld metal microstructure in double-pulsed gas metal arc welding (DP-GMAW) was investigated by the self-designed high-speed welding photography system. The heat input, the arc pressure, the droplet momentum and impingement pressure were measured and calculated. It was found that the arc pressure is far less than the droplet impingement pressure. The heat input and droplet impingement pressure per unit time acting on weld pool were proportional to the current pulse frequency, which fluctuated with thermal pulse. The size and oscillation amplitude of the weld pool had noticeable periodic changes synchronized with the process of heat input and droplet impingement. Compared to the microstructure of pulsed gas metal arc welding (P-GMAW) weld metal, that of DP-GMAW weld metal was significantly refined. High oscillation amplitude assisted the enhancement of weld pool convection, which leads to more constitutional supercooling. The heat input and shear force during the peak of thermal pulse causing dendrite fragmentation which provided sufficient crystal nucleus for the growth of equiaxed grains and the possibility of grain refinement. The effects of current parameters on welding behavior and weld metal grain size are investigated for further understanding of DP-GMAW.

Keywords: double-pulsed gas metal arc welding (DP-GMAW); droplet impingement pressure; weld pool oscillation; grain refinement; constitutional supercooling

---

1. Introduction

As a widely used spatter-free welding technology, pulsed gas metal arc welding (P-GMAW) can achieve directional transition of spatter-free droplets with low heat input through current pulse [1]. However, current pulses with the constant frequency of P-GMAW could not effectively stir the weld pool with the heat-sensitive and high viscosity liquid metal such as stainless steel and aluminum alloys, which often results in the formation of structure defects such as coarse grains, pores and cracks [2]. To solve this problem, many arc-based welding techniques have been developed for advanced materials joining [3,4]. DP-GMAW was developed based on P-GMAW to assisting weld pool oscillation [5]. By periodically changing the output current, double-pulsed gas metal arc welding (DP-GMAW) leads to the periodic change of current pulse frequency. Through the whole process, not only the stable transfer mode of "one drop per pulse" can be obtained [6], but also the frequency oscillation and stirring effect of weld pool can be obviously improved [7]. Therefore, to some extent, DP-GMAW can refine the grains [2], reduce the cracking sensitivity [8,9] and porosity of welds [3,4] and improve the weld formation and joint performance.

Figure 1 shows the waveform of DP-GMAW, and the double pulse period consists of a peak thermal period and a base thermal period. The heat and mass transfer process of the heat pulse is determined by the waveform parameters of double-pulse current such as pulse frequency, current difference and duty ratio of two phases. To study the evolution rules of DP-GMAW weld formation and microstructure, various methods were carried out. Yao and Zhou et al. systematically investigated the influence of current waveform parameters of DP-GMAW on the weld waviness of austenitic stainless steel [10]. They discussed the regularity of DP-GMAW weld formation by using gray theory analysis [11] and further optimized the welding parameters of austenitic stainless steel [12]. Compared with P-GMAW, DP-GMAW has a wider adjustment range, broader root gap configuration and stronger solute agitation with the same heat input rate. Wang et al., suggested that increasing the frequency of thermal pulse of DP-GMAW could reduce dendrite size [13]. Wang studied the influence of the current amplitude of thermal pulse on the geometry, cooling rate, solidification parameters of the aluminum alloy weld pool and weld metal grain size from both experimental and numerical simulation aspects. It was proved that DP-GMAW could increase the cooling rate of the weld pool with the same heat input [14]. In the investigation on welding procedure of ferritic stainless steel and austenitic stainless steel, Shen [8] and Devakumaran [9] both confirmed that DP-GMAW could effectively inhibit the growth of HAZ (Heat Affected Zone) grains and promote the transformation of columnar grains to equiaxed grains in the weld zone. Anhua Liu et al. analyzed the dynamic process of weld pool shape of the aluminum alloy with the aid of high-speed camera [15]. The results showed that the size of weld pool changed synchronously with the frequency the thermal pulse, and the addition of thermal pulse obviously changed the behaviors of weld pool. As the frequency of thermal pulse increased, the grain size of weld metal decreased, and the eutectic Mg_2Si precipitates in the weld zone were evenly distributed.

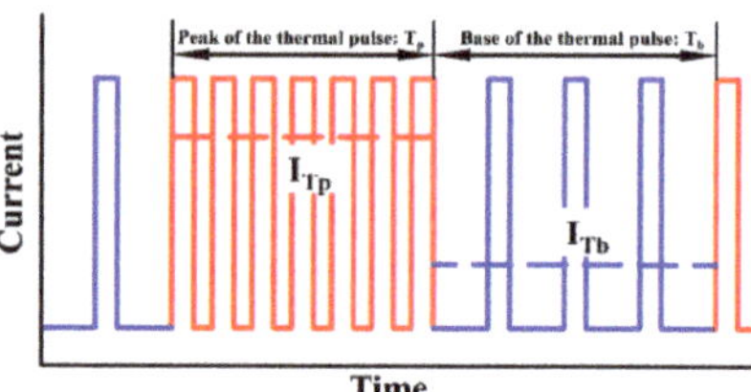

Figure 1. Schematic diagram of current waveform of double-pulsed gas metal arc welding (DP-GMAW).

It should be pointed out that the heat and mass transfer process during welding determines the dynamic behavior of the weld pool. Besides, the solidification behavior of weld pool is greatly controlled by the dynamic behavior of weld pool and welding heat input process. So far, studies on the effect of DP-GMAW on grain refinement of the weld mainly concentrated on the optimizing welding procedure. However, little research was conducted on the dynamic characteristics of weld pool in DP-GMAW, and the relationship between the dynamic behavior of weld pool and welding metal microstructure can hardly be established. Therefore, extensive research work needs to be carried out to analyze the influence of waveform parameters of DP-GMAW on the dynamic behavior of weld pool and its relationship with welding metal microstructure.

In this paper, with the help of a laboratory-made high-speed welding photography system, the influence of waveform parameters of DP-GMAW on weld pool oscillation behavior of austenitic stainless steel was studied. The purpose of this study is to explore the internal relationship between weld pool oscillation behavior and welding metal microstructure and explain the action mechanism of grain refinement.

2. Materials and Methods

2.1. Experiment System

The experimental system used in this study consisted of a welding system, high-speed photographic system and an image processing system, as shown in Figure 2. An 850 nm laser source and some 850 nm near-infrared filters equipped on camera were used in a high-speed photographic system to eliminate the light of strong arc. A high-speed camera was located in different positions to obtain weld pool images with various visual angles. The side-view of the weld pool during welding was recorded in position 1, as shown at the position of high-speed camera 1 in Figure 2. The top-view of the weld pool was recorded in position 2, as shown at the position of highspeed camera 2 in Figure 2. Otto Arc MIG-500DP (OTTO Arc, Shanghai, China) was selected as a welding power source. Welding position was PA (Flat position, as per ISO 6947). The frequency of an electric signal acquisition system was 5 × 105 Hz.

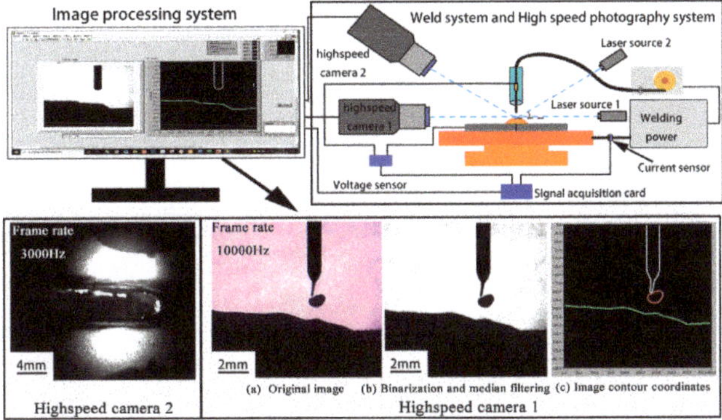

Figure 2. Experiment system and image processing steps: (**a**) original image; (**b**) binarization and median filtering; (**c**) image contour coordinates.

An image processing system based on LabView (LabVIEW 2017, National Instruments, Austin, TX, USA) was developed to capture the outline of droplet and pool surface from the side-view picture. The image processing flow is shown in Figure 2a–c.

2.2. Algorithm to Extract Characteristics of Pool Oscillation, Droplet Transfer and Arc Profile

The outlines of droplet and pool surface captured by the image processing system are shown in Figure 2c. The contour coordinates were deformed with the droplet transfer and the pool oscillation. The dynamic information of the droplet and weld pool can be obtained by tracing the contour coordinates as a function of time, as shown in Figure 3.

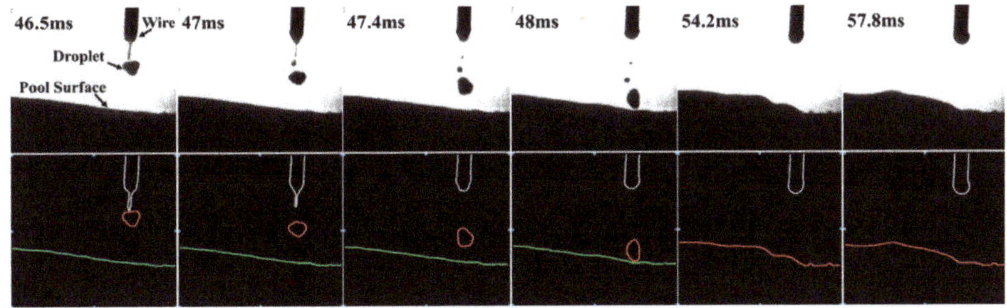

Figure 3. Contour extraction of droplets and weld pool.

A reference point (A) was defined on the weld pool surface to trace the pool surface. The x coordinate of the reference point is constant; the fluctuation of the y coordinate of the reference point is the direct information about the weld pool oscillation. To avoid the hindrance of droplet transition to the surface contour extraction of the weld pool, the reference point (A) was located on the weld pool surface 1.8 mm from the center of welding wire, as shown in Figure 4.

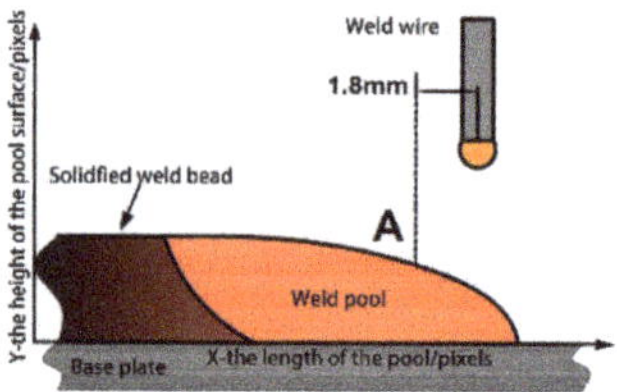

Figure 4. Position of the reference point: A.

The center of gravity of the droplet as approximated to the average value of its contour coordinate, as calculated by Equation (1) [1].

$$(x_{Gt}, y_{Gt}) = (\frac{\sum_1^n x_{nt}}{n}, \frac{\sum_1^n y_{nt}}{n}) \tag{1}$$

(x_{Gt}, y_{Gt}) is the coordinate of the center of droplet at t, (x_n, y_n) are the coordinates of the droplet contour, n is the number of contour pixels. The droplet diameter can be calculated by Equation (2) [1].

$$D_{droplet} = \sqrt{(4 \times S_{droplet})/\pi} \tag{2}$$

$D_{droplet}$ is the equivalent diameter of droplets, $S_{droplet}$ is the area of droplet profile calculated by the number of pixels surrounded by the droplet contour line. Droplet velocity can be calculated by measuring the center coordinate of droplet in continuous photographs, as shown in Equation (3).

$$V_{droplet} = \frac{\sqrt{(x_{Gt1} - x_{Gt2})^2 + (y_{Gt1} - y_{Gt2})^2}}{|t_2 - t_1|} \tag{3}$$

(x_{Gt1}, y_{Gt1}) and (x_{Gt2}, y_{Gt2}) are the coordinates of the center of droplet at t_1 and t_2.

Changing the exposure time and the number of filters of high-speed camera can obtain different shooting effects. Increasing the exposure time and the number of filters equipped on the camera can cause the background light stronger than the arc light, which can filter the arc light, as shown in Figure 5a. Reducing exposure time and the number of filters leads to a higher arc light intensity than the background light intensity, and a precise arc contour can be obtained, as shown in Figure 5b. The arc characteristics were defined by its root diameter (D_R) and projected diameter (D_P) during pulse on the period, as schematically shown in Figure 5b.

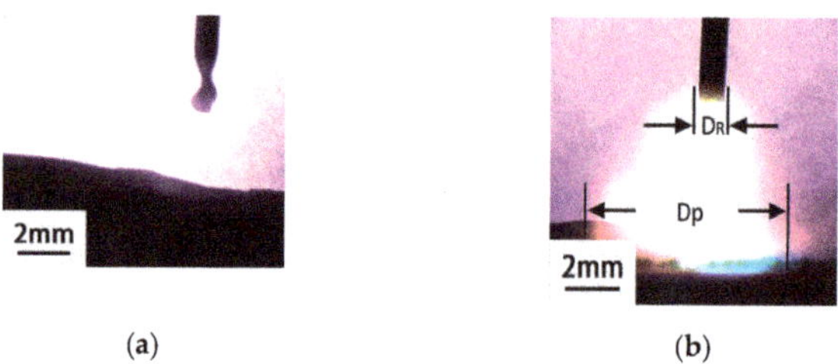

Figure 5. (**a**) Complete filtering out of the arc; (**b**) typical nature of arc profile.

2.3. Calculation of Cooling Rate, Growth Rate and Thermal Gradient

The cooling rate of weld can be evaluated by the two equations shown below [6].

For thick plate:

$$C_R = 2 \times \pi \times k \times (T_C - T_0)/H_{net} \tag{4}$$

For thin plate:

$$C_R = 2 \times \pi \times k \times \rho \times c \times t^2 \times (T_C - T_0)^3/{H_{net}}^2 \tag{5}$$

where C_R = cooling rate (K·s^{-1}), k = thermal conductivity = 15 W·m^{-1}·K^{-1}, ρ = density = 7850 Kg·m^{-3}, c = specific heat = 500 J·Kg^{-1}·K^{-1}, t = plate thickness(mm), T_C = peak temperature = 1534.15 K, T_0 = final temperature = 300.15 K and H_{net} = heat input rate(J·m^{-1}). The relative plate thickness factor was derived to select the proper cooling rate Equation [6]:

$$\tau = t \times \sqrt{\rho \times c \times (T_C - T_0)/H_{net}} \tag{6}$$

Equation (4) is applicable when $\tau \geq 0.75$, else Equation (5) is appropriate. In this study, 3 mm thin base plate was used, τ of all the studied conditions was lower than 0.75. Therefore, Equation (5) was selected to calculate the cooling rate. The dendrite growth rate in the solidification zone at the end of the weld is calculated as follows [6]:

$$R = v cos\theta \tag{7}$$

where R = growth rate (mm/s), v = welding speed (mm/s) = 3.5 mm/s, θ = the angle between the normal to solidification front and the welding direction. The calculation equation of the thermal gradient of weld pool (G, K × mm^{-1}) without considering the convection of the weld pool is as follows [6]:

$$G = C_R/R \tag{8}$$

2.4. Sample Fabrication

A commercial 304 stainless steel plate of 200 mm × 150 mm × 3 mm was used as a base plate to prepare the bead on plate welds, using 308L stainless steel wire of 1.2 mm (nominal diameter) as the electrode. The weld groove shape is "I" with no gap. The chemical composition of the base plate and filler wire is given in Table 1. Gas composed of 98% argon and 2% O_2 was used as shield gas (20 L/min). The contact tip to base plate distance was 15 mm. In this paper, the influences of current waveform parameters on the weld pool behavior and microstructure of DP-GMAW were studied, which are thermal pulse frequency (TPF), thermal pulse current change (ΔI) and duty ratio of thermal pulse peak phase (D_{Tp}), respectively. As a comparison, the weld pool behavior and weld metal microstructure of P-GMAW were also studied. Welding parameters are listed in Table 2.

Table 1. Material characteristics of the base plate and welding wire.

Materials	C	Si	Mn	Cr	Ni	S	P	N	Mo
304	≤0.08	≤1	≤2	18–20	8–10.5	≤0.03	≤0.03	≤0.1	-
316L	≤0.03	≤1	≤2	16–18	10–14	≤0.03	≤0.045	-	2–3

Table 2. Welding parameters.

No.	Process	I_{Tp} (A)	I_{Tb} (A)	TPF (Hz)	ΔI (A)	D_{Tp} (%)	V	Speed (mm/s)	Penetration
1	DP	130	90	0.5	40	50	22.5	20	Full
2	DP	130	90	1	40	50	22.5	20	Full
3	DP	130	90	2	40	50	22.5	20	Full
4	DP	130	90	3	40	50	22.5	20	Full
5	DP	130	90	2	40	20	22.5	20	Full
6	DP	130	90	2	40	35	22.5	20	Full
7	DP	130	90	2	40	70	22.5	20	Full
8	DP	130	105	2	25	50	22.5	20	Full
9	DP	130	75	2	55	50	22.5	20	Full
10	DP	130	60	2	70	50	22.5	20	Full
11	P	90		-	-	-	22.5	20	Full
12	P	110		-	-	-	22.5	20	Full
13	P	130		-	-	-	22.5	20	Full

Note: I_{Tp} = the average current at the peak of the thermal pulse; I_{Tb} = the average current at the base of the thermal pulse; TPF = thermal pulse frequency; $\Delta I = I_{Tp} - I_{Tb}$, D_{Tp} = duty ratio of thermal pulse peak phase; V = voltage; Speed = weld speed; DP = DP-GMAW, P = P-GMAW.

3. Results and Discussion

3.1. Effect of Arc and Droplet Transfer on Weld Pool

During the welding process, the arc pressure and the impingement of droplet agitate weld pool intensify the convection in the weld pool. Hence, the study on heat and mass transfer process definitely is the premise of that on the dynamic behavior of pool in DP-GMAW. In P-GMAW, the current pulse frequency is constant. In DP-GMAW, the frequency of the current pulse changes periodically. Typical electrical signal waveforms of P-GMAW and DP-GMAW.

A thermal pulse cycle of DP-GMAW consists of peak period, base period and transition period of thermal pulse, as shown in Figure 6. The peak and base period of thermal pulse differ in the frequency and the peak of current pulse, leading to variational heat input, arc pressure and droplet impingement force. Therefore, it is necessary to analyze the heat input and force acting on the weld pool during a single current pulse period, then analyze the variational process of the heat input and the force acting on the pool during the whole thermal period of DP-GMAW. The thermal nature of the DP-GMAW weld pool is largely controlled by the heat content of the droplet and the arc heating. Assuming that the current distribution is homogeneous on the arc projection plane on the pool surface, the arc heat during one current pulse (E_{arc}) is expressed as follows [16]:

$$E_{arc} = \int_0^{1/f} I(V_w - \varphi)dt \tag{9}$$

where V_w is the cathode voltage when the cathode material is stainless steel and φ is the electronic work function of stainless steel. Y Yokomizu [17] points out that for stainless steel, V_w is about 16.7 v, φ is 4.77 v. P_{arc} is the instantaneous power of heat input of arc to weld pool. The heat content of the droplet ($E_{droplet}$) can be estimated by the following Equation [14]:

$$E_{droplet} = \rho h \frac{4}{3}\pi \left(\frac{D_d}{2}\right)^3 \tag{10}$$

where ρ is the density of the liquid stainless steel, h is the enthalpy of the droplet and D_d is the diameter of the droplet. Considering overheating of the droplets, the assumed temperature before the droplets enter the molten pool is 2900 K, and the enthalpy of the droplets h [16] is:

$$h = \int_{300}^{2900} C_p dT \tag{11}$$

where C_p is the specific heat capacity of the droplet and T is the temperature. According to the research data of Davim [18], h = 1.578 × 106 J/kg. The total heat input (E_{total}) and the power of heat input (P_{total}) acting on weld pool in a single current pulse period are:

$$E_{total} = \rho h \frac{4}{3}\pi\left(\frac{D_d}{2}\right)^3 + \int_0^{1/f} I(V_w - \varphi)dt \tag{12}$$

$$P_{total} = E_{total} \times f \tag{13}$$

where f is the instantaneous frequency of the current pulse. The arc force (F_{arc}) during welding process is as follows [18]:

$$F_{arc} = \frac{\mu}{4\pi} I^2 \log\frac{D_d}{D_R} \tag{14}$$

where μ is the space permeability of the arc area, μ = 4.073 × 10^{-4} N/A^2 [19]. The momentum of the droplet ($p_{droplet}$) and the droplet impingement pressure on weld pool (P_d) are shown as follows [16]:

$$p_{droplet} = \frac{4}{3}\pi\left(\frac{D_{droplet}}{2}\right)^3 \rho\, V_d \tag{15}$$

$$P_d = \frac{2f\rho D_d V_d}{3} \tag{16}$$

where V_d is the velocity of the droplet. Combined with the above formula, the arc behavior, droplet transition behavior, heat input and force acting on the weld pool during a current pulse period were analyzed, as shown in Figures 7 and 8. The arc size increases first and then decrease with the change of current during a current pulse, as shown in Figure 8a. At the beginning of the current pulse, the current and voltage rise rapidly to the peak, the size of the arc, the pressure and the heat power of the arc acting on weld pool rises synchronously. During peak time, current, arc size, arc pressure rise to the maximum in the pulse period. The arc transmits most of the heat to the weld pool during peak time, as shown in Figure 8b. With the decrease of current, the size of arc decreases gradually. The arc pressure and the heating power also decreases synchronously with current. While the size of the arc increased after the time of the droplet detached from the wire. It was caused by metal vapor concentration increasing in arc space when the droplets detach from the wire, and the phenomenon of arc jumping is also one of the factors, as shown in Figure 7(5,6). While this increase of arc size has no obvious effect on arc force and heating process at low current. The heat contained by the droplet and its impact force acting on weld pool can be calculated by Equations (5) and (11). Based on the above data, the total heat input, the arc pressure and the droplet impingement force acting on the weld pool in a single pulse period can be calculated, as shown in Figure 8b.

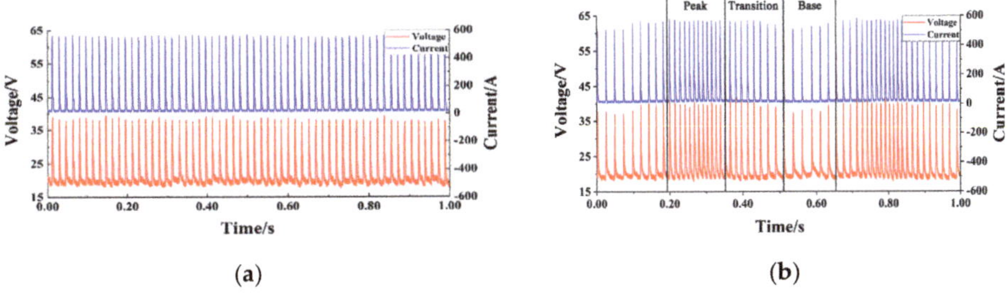

Figure 6. Welding electrical signal waveform of (**a**) P-GMAW(No.12); (**b**) DP-GMAW(No.3).

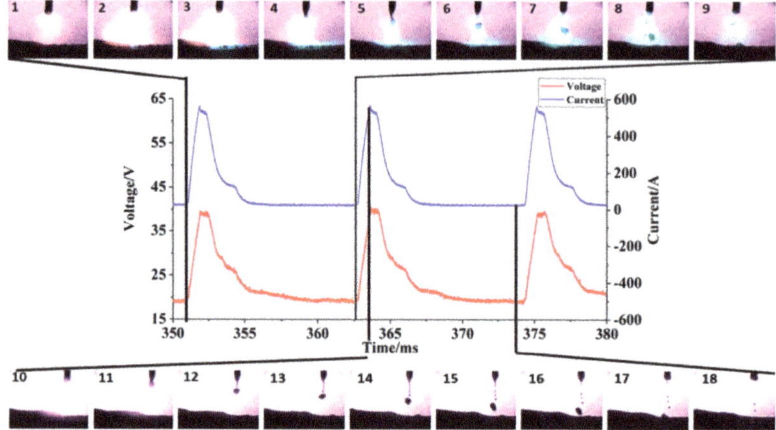

Figure 7. Combination of arc behavior, droplet transition behavior and electrical signals: 1–9 are the arc profile, 10–18 are images of the droplet transfer process (No.3).

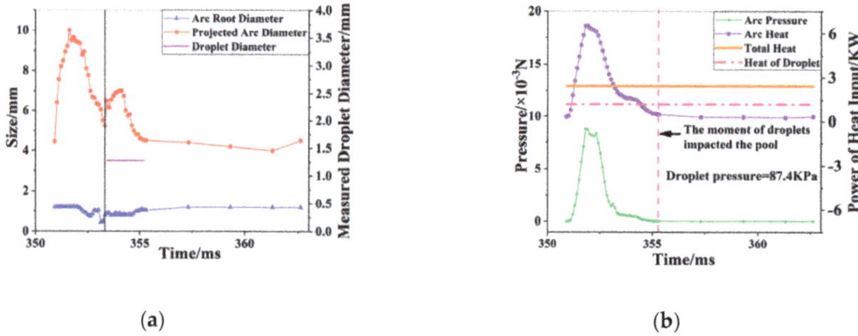

(**a**) (**b**)

Figure 8. (**a**) Arc size and droplet diameter; (**b**) heat input and pressure acting on weld pool.

DP-GMAW process periodically adjusts the output welding current, pulse waveform and pulse frequency vary with the current on the basis of "one droplet one pulse". Figure 9 shows the peak pulse current and pulse frequency with different output welding current. With the increase of the output current, the pulse frequency increases significantly, while the peak current of the pulse increases slightly. The main way to control output current for DP-GMAW process is to adjust the current pulse frequency. The purpose of adjusting the peak of the current pulse is to maintain the stability of the transition under different welding currents.

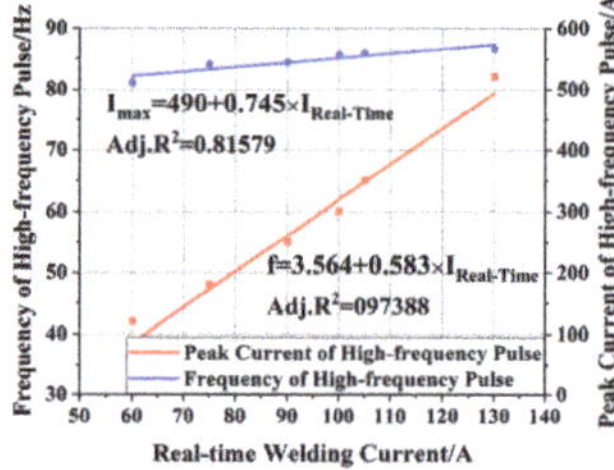

Figure 9. Peak current and frequency of pulses with different welding currents.

Figure 10 shows the heat content, the power of heat input, the pressure of arc, the droplet impingement force and momentum per unit current pulse with different output welding current. As shown in Figure 10a, the heat content of droplet was similar under different output welding current the arc heat had the same trend. Due to the uneven composition of the welding wire and the unstable wire feed speed during welding process, the size of the droplets cannot be consistent. However, the difference in droplet size with different average currents were small due to the consistency of the current pulse waveform. However, the heating power acting on weld pool increased obviously due to a clear growth of current pulse frequency. The momentum carried by the droplets were similar under different output welding currents, as shown in Figure 10b. Many studies can contribute to explaining it. The research of Emanuel [1]. Found that the droplet speed depends on the ratio between base to peak current of P-GMAW, which is different from the traditional GMAW. A slight change in the peak current could not significantly affect the droplet velocity. P.K. Ghosh's study [18] came to a similar conclusion that the diameter and speed of droplet in P-GMAW process predominantly depends upon I_p irrespective of mean current and arc voltage. The increase of droplet transition frequency leads to an obvious increase of equivalent droplet impact force, as shown in Figure 10b. It is worth noticing that the arc pressure acting on the weld pool was much less than the droplet impingement force. Other researchers have described similar phenomena that liquid waves in P-GMA welding are triggered primarily by the impact of droplet, not by arc pressure [20]. Therefore, this paper only considers the droplet impingement force on weld pool.

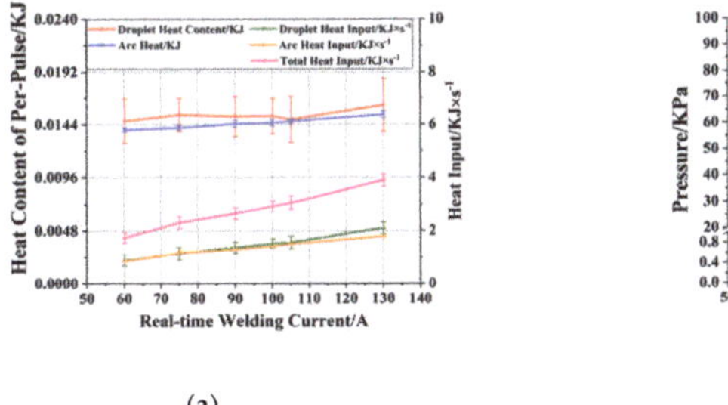

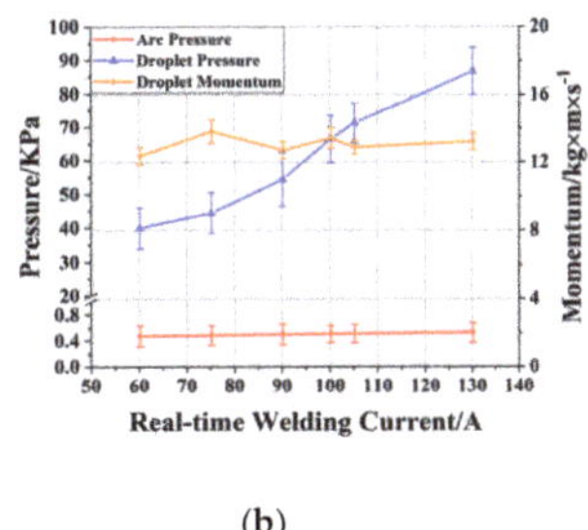

(a) (b)

Figure 10. Heat input (**a**), pressure and momentum (**b**) of arc and droplet during single current-pulse with different welding currents.

Figure 11 shows the heat input and the droplet impingement force acting on the weld pool during the thermal pulse period with different ΔI. As shown in Figure 11, the process of thermal and pressure acting on the weld pool demonstrated the characteristic periodically fluctuation same with the thermal frequency. The heat input rate of the different stage of the thermal pulse were the average heat input rate during T_p or T_b. This paper compares the heat input rate and the thermal gradient of weld pool (G) at different period under all parameters, as shown in Table 3. It should be noted that the thermal gradient of weld pool (G) along the welding direction calculated by Equation (7) is without considering the pool convection.

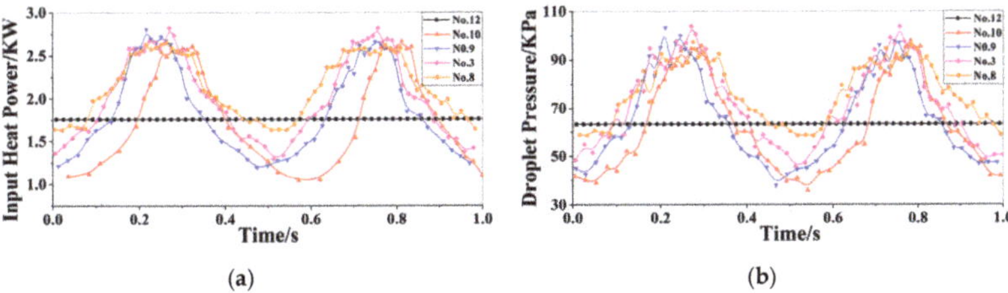

(a) (b)

Figure 11. Changes of heat input (**a**) and droplet impingement force (**b**) in thermal pulse of DP-GMAW.

Table 3. Heat input rate (KJ/m) and the thermal gradient of weld pool (K × mm^{-1}) with different parameters.

No.	1	2	3	4	5	6	7	8	9	10	11	12	13
T_p	68.2	72.5	66.2	68.7	71.5	66.7	71.8	82	60.6	52.1	-	-	-
T_b	137	134.9	136	133.8	135.6	133.8	135.7	133	133	133	-	-	-
Avg.	102.6	103.7	101	101.2	84.3	90.2	116.5	107	96.8	92.6	71	102.2	142
G_{Tp}	67.3	59.6	71.4	66.3	61.2	70.4	60.7	46.5	85.2	115.2	62	30.4	15.5
G_{Tb}	16.7	17.2	16.9	17.5	17.0	17.5	17.0	17.8	17.7	17.6			

3.2. The Behavior Characteristics of Weld Pool in Double-Pulsed GMAW

In the process of DP-GMAW, the thermal and pressure acting on the weld pool of T_P and T_B are significantly different, the dynamic behavior of weld pool were varied with thermal frequency (F). For a better understanding of the influence of DP-GMAW current waveform parameters on the dynamic behavior of austenite stainless steel weld pool, the profile and oscillation characteristics under different current waveform parameters were recorded and extracted. In order to simplify the analysis process, two typical current waveform parameters (No.3 and No.12) were selected to summarize the weld pool profile.

The heat and the mass transferred to the weld pool during the period of a single current pulse can hardly affect the shape of weld pool significantly. In P-GMAW, the heat input and the droplet impingement force acting on weld pool are constant during welding process, and the profile of the weld pool caused by them remains stable, as shown in Figure 12a. In DP-GMAW, the length and the width of weld pool in the base period of thermal phase were obviously smaller than that in the peak period, as shown in Figure 12b,c. The smaller heat input and less metal deposition lead to rapid shrinkage of pool size during base period. The pool trailing edge shrank and separated from the solidified bead boundary of the weld, as shown in Figure 12c. This is the main factor that forms bead surface ripple. The shrinkage and expansion of the weld pool outline is mainly affected by the fluctuation of heat transfer and mass transfer in T_p and T_b period of thermal pulse.

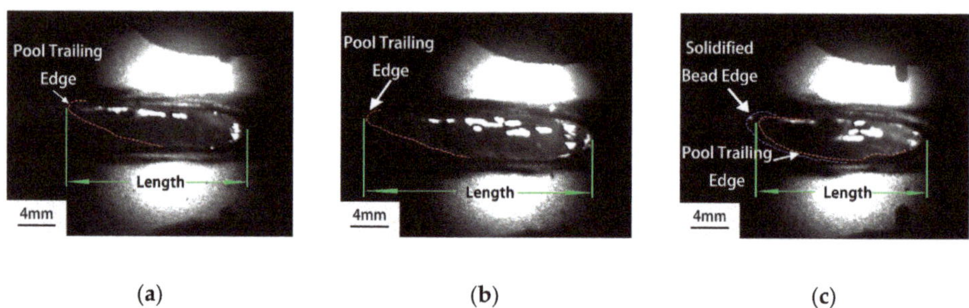

Figure 12. The variation of weld pool profile in thermal pulse: (**a**) P-GMAW (No.12); (**b**) peak period of thermal pulse of DP-GMAW (No.3); (**c**) Base period of thermal pulse of DP-GMAW (No.3).

The length variation of the weld pool under different parameters of the current waveform are shown in Figure 13. Figure 13a shows the weld pool length at T_p and T_b with different TPF. During the P-GMAW welding process, the weld pool can be regarded as a constant. The blue and red curves of Figure 13a are the curve of weld pool length at T_p and T_b, respectively, with the thermal frequency. The descending trend of blue curve and the ascending red curve indicates that pool length at T_p is gradually shortened while that at Tb is gradually increased with the increase of heat pulse frequency. It indicates that with the increase of thermal frequency, the period of T_p decreases, and the heat accumulation and droplet transition acting on the weld pool gradually decrease, resulting in the decrease of the weld pool length at T_p. While the differences of the heat and droplet transfer amount into the weld pool at T_p and T_b stage gradually decreases, the pool length difference between T_p and T_b gradually decreases but both close to the length of P-GMAW.

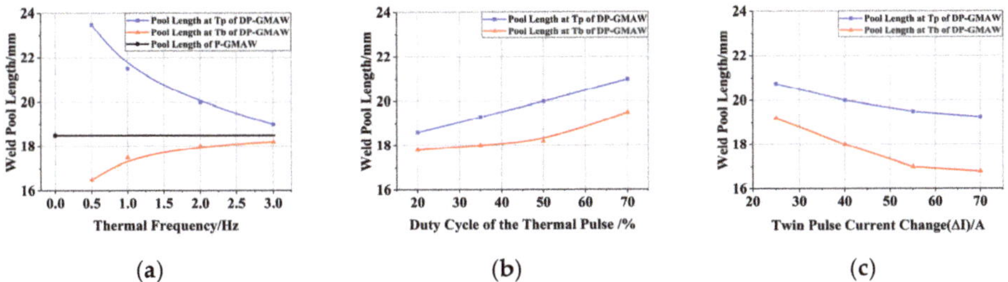

(a) (b) (c)

Figure 13. Variation of weld pool length with (**a**) heat thermal frequency; (**b**) duty cycle of the thermal pulse; (**c**) current amplitude of thermal pulse.

Figure 13b shows the pool length at T_p and T_b with different D_{Tp}. With larger D_{Tp}, the total heat input and droplet transition acting on the weld pool increase, the pool length at T_p and T_b both increasing. It has to be noticed that as D_{Tp} increases, the difference between the pool length at T_p and T_b always increases first and then decreases. Small D_{Tp} could cause longer T_b with fixed thermal pulse frequency, the heat accumulation in T_p was too small to significantly expand the size of weld pool. A larger D_{Tp} could cause smaller T_b, shorter low heat input time (T_b) could not cause a significant reduction in the size of the pool. Too large or too small D_{Tp} could result in small length difference.

Figure 13c shows the pool length at T_p and T_b with different ΔI. With larger ΔI, the total heat input and droplet transition acting on the weld pool decrease, the pool length at T_p and T_b both decreasing. Larger ΔI increases the differences of heat accumulation and amount of deposited metal between the T_p and T_b.

The oscillation process of weld pool was recorded by the method mentioned in Section 2.2. Two typical current waveform parameters (No. 3 and No. 11) were selected to summarize the oscillation process of the weld pool, as shown in Figure 14. The fluctuation in the amplitude of the pool oscillation during the P-GMAW welding process was constant (Figure 14a), while that during the DP-GMAW was obvious (Figure 14b). That is the oscillation amplitude of weld pool in T_p is much larger than that in T_b, since the droplet impingement force of P-GMAW and the size of weld pool were stable (Figures 11b and 13a), the amplitude of pool oscillation can be regarded as a constant.

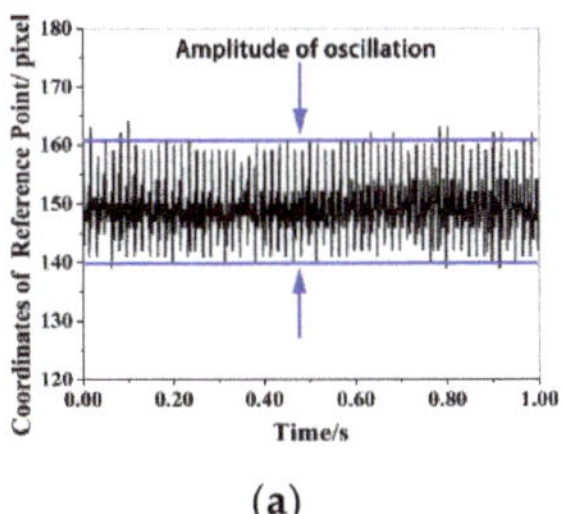

(a)

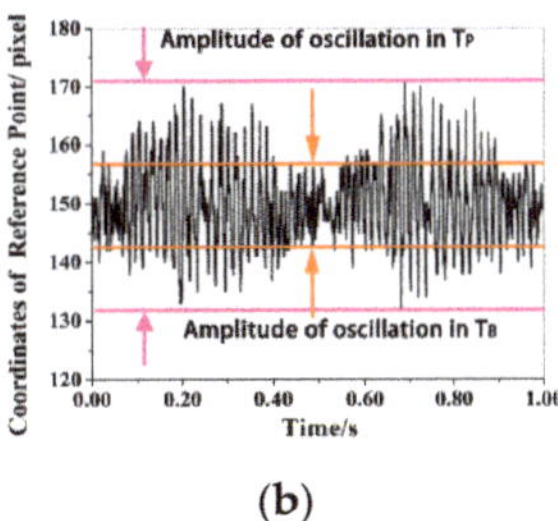

(b)

Figure 14. Height of reference point A as a function of time: (**a**) P-GMAW(No.12); (**b**) DP-GMAW(No.3).

In DP-GMAW, the fluctuation process of oscillation amplitude is similar to the process of thermal and pressure on the weld pool, as shown in Figure 11. The droplet impingement force on the weld pool increases significantly when the high-frequency current pulses increase the volume of weld pool during T_p period, which is the main reason for the larger oscillation amplitude of the weld pool. The size of the weld pool and the droplet impingement force simultaneously resulting in a decrease in the amplitude of oscillation during Tb period. The variation of oscillation amplitude of weld pool under different parameters of the current waveform are shown in Figure 15.

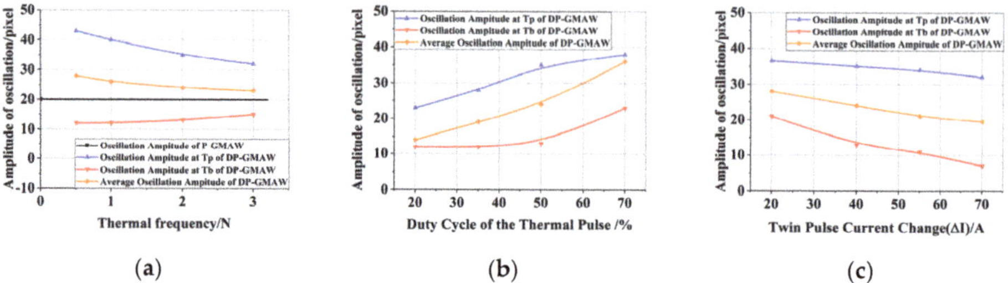

Figure 15. Variation of weld pool oscillation amplitude with (**a**) heat thermal frequency; (**b**) duty cycle of the thermal pulse; (**c**) current amplitude of thermal pulse.

The curves of the weld pool amplitude in T_p and T_b under different thermal pulse frequencies had a similar trend to the weld pool length, as shown in Figures 13a and 15a. Increasing the frequency of thermal pulses led to a reduction in the difference between the size of the weld pool at T_p and T_b. As for the oscillation amplitude at T_p, the decrease in the size of the weld pool can decrease the oscillation amplitude under the same droplet impingement force. The increase in the weld pool size at T_b can increase the oscillation amplitude, as shown in the red line of Figure 15a. However, there is no denying the fact that the average oscillation amplitude of DP-GMAW was greater than that of P-GMAW.

Larger duty cycle of the thermal pulse (D_{Tp}) can trigger greater oscillation amplitude with larger pool size and longer high-frequency droplet impingement. too large or too small D_{Tp} could result in small amplitude difference of the oscillation amplitude within T_p and T_d, as shown in Figure 15b.

Larger ΔI will increase the difference between the pool size and the droplet impingement force in T_p and T_b, leading to a greater oscillation amplitude difference. However, the decrease of total heat input decreased the average amplitude of weld pool, as shown in Figure 15c.

So, as to what is known, DP-GMAW weld pool behavior is more complicated compared with P-GMAW. During switching from T_p to T_b, the pool size experiences "expanding–shrinking" variation, the change of the oscillation amplitude of weld pool is synchronized with the pool size.

3.3. *Effect of Process Parameters on Microstructures*

The microstructure of fusion weld metal is controlled by the solidification behavior of weld pool [21]. Metallographic observation of all weld cross-sections with different welding parameters was conducted, and no defects such as incomplete fusion and porosity can be found. As mentioned above, the thermal pulse of DP-GMAW caused significant fluctuations in weld pool size and oscillation amplitude during welding process. It resulted in significant differences in welding microstructure between P-GMAW and DP-GMAW. The weld cross-sections, typical microstructures and heat affected zone (HAZ) of different weld processes (No.3 and No.12) are shown in Figures 16 and 17.

The microstructure of the P-GMAW weld was composed of coarse austenite (γ) columnar structures and small part of equiaxed austenite grain. The equiaxed crystal regions are distributed in the gaps between the ends of the columnar crystal regions, there is no obvious boundary between the equiaxed crystal regions and the columnar crystal regions, as shown in Figure 16b. The ferrite (δ) morphology was skeleton-shaped, distributed in the columnar austenite grain gap.

Compared to P-GMAW, the microstructure of DP-GMAW weld exhibited an obvious refinement of microstructure along with large distribution of equiaxed crystal regions. A clear boundary emerged between the equiaxed crystal area and columnar crystal area, as shown in Figure 17b. Although the structure of austenite columnar crystals was significantly refined, it can be found that the ferrite size increased significantly by comparing the ferrite morphology in austenite gap between P-GMAW and DP-GMAW, as shown in Figures 16d and 17d.

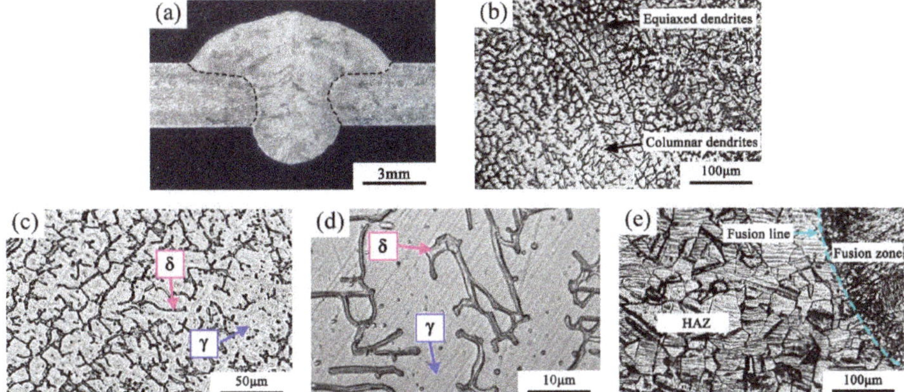

Figure 16. Optical micrographs of microstructure and heat affected zone (HAZ) of P-GMAW (No.12): (**a**) weld cross-section; (**b**) typical microstructure of P-GMAW weld, (**c**) coarse austenite columnar structures, (**d**) ferrite morphology and (**e**) HAZ.

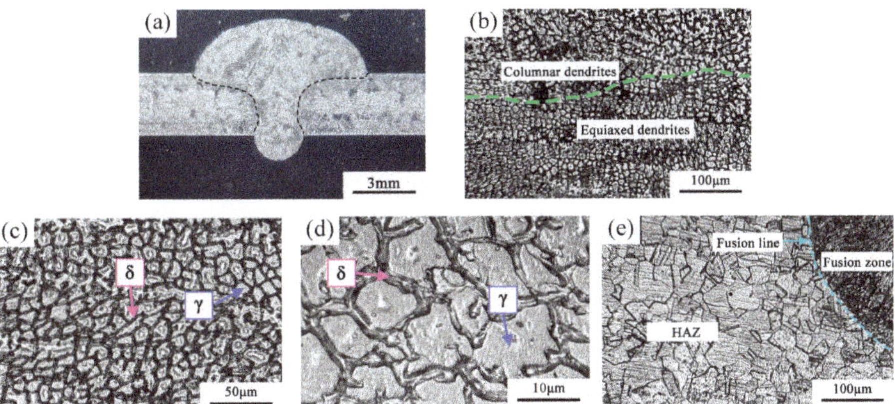

Figure 17. Optical micrographs of microstructure and HAZ of DP-GMAW (No.3): (**a**) weld cross-section; (**b**) typical microstructure of DP-GMAW weld, (**c**) austenite equiaxed dendrites structures, (**d**) ferrite morphology and (**e**) HAZ.

Figure 18 is a schematic sketch explaining how the thermal pulse of DP-GMAW helps grain refining. The dynamic behavior of the weld pool and the fluctuation of solidification parameters are the main factors that trigger the refinement of structure of DP-GMAW. The thermal pulse of DP-GMAW caused synchronized periodical fluctuations in heat input and pool amplitude, as shown in Figure 18a. Low heat input led to the large temperature gradient G of the weld pool during T_b. Ferrite was the primary grain in weld pool, ferrite columnar dendrites dominate with less constitutional supercooling (the area surrounded by T_L and T_{actual}), as shown in Figure 18b.

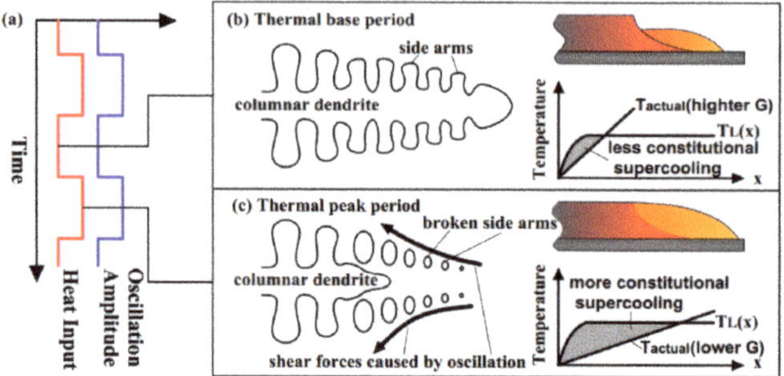

Figure 18. Thermal pulse helping grain refining: (**a**) process of the heat input and pool oscillation amplitude in DP-GMAW; (**b**) constitution supercooling in thermal base period; (**c**) constitution supercooling in thermal peak period.

The high heat input during T_p can reduce the temperature gradient G in weld pool. The oscillation amplitude of weld pool was greatly increased, resulting in the enhancement of convection, which is expected to reduce the temperature gradient G in weld pool, leading to more constitutional supercooling (the area surrounded by T_L and T_{actual}) in the weld pool, as shown in Figure 18c. The peak thermal period can cause reheating and melting of dendrite arms, thus hindering the further crystal growth and causing dendrite fragmentation [22]. At the same time, the shear stress produced by the enhanced convection in weld pool aggravates the dendrite fragmentation during T_p. The broken dendrite particles provide the necessary crystal nucleus for the liquid metal crystallization. Excessively constitutional supercooling was beneficial to the dendrite fragments survival and grow into equiaxed grains.

The equiaxed ferrite grains in the freshly solidified weld transforms into austenite by diffusion transformation in subsequent T_b of thermal pulse. While the fast weld cooling speed reduced transition time of "$\delta \rightarrow \gamma$", both diffusion of ferrite-forming elements and austenite-forming elements were suppressed. It is the main factor leading to increasing the size and the content of ferrite structure in DP-GMAW weld microstructure compared to P-GMAW, as shown in Figure 16c,d and Figure 17c,d.

The microstructure of HAZ of weld joints prepared by P-GMAW and DP-GMAW are shown in Figures 16e and 17e. There is no obvious difference in the grain size of HAZ between the weld joints prepared by P-GMAW and DP-GMAW respectively. Similar heat input rate between welding parameters of No.11 and No.3 may be the main reason. The average size of weld microstructure and HAZ with different welding parameters were measured using the intercept method (as per ASTM E112-10), the values are statistic presented in Figure 19a. When the base metal of HAZ is heated, the microstructure undergoes a process of recrystallization, new undistorted equiaxed grains appear in the microstructure and gradually replace distorted grains. After the recrystallization is completed, continue to heat up or prolong the elevated temperature holding time could make the grain continue to grow. The size of HAZ in weld joint mainly depends on the heat input rate, Large heat input rate leads to longer elevated temperature holding time of HAZ, which is conducive to the diffusion and recrystallization process of tissue, thus leading to serious growth of grains. The fitting curve between grain size of HAZ and heat input rate is shown in Figure 19b. It can be found that the heat input rate has a good linear relationship with the grain size of HAZ, while no obvious correlation could be found between the grain size of HAZ and the thermal pulse of DP-GMAW.

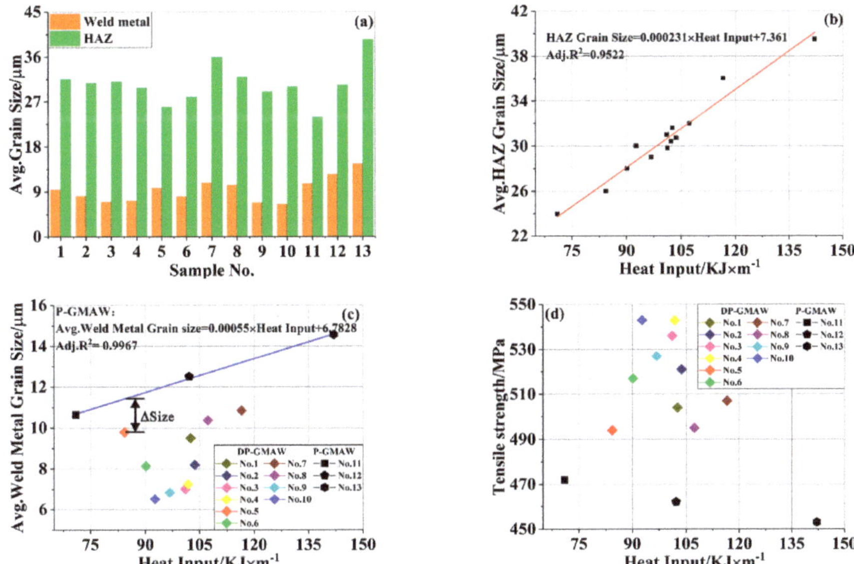

Figure 19. (**a**) Average grain size of weld metal and HAZ; (**b**) grain size of HAZ with different heat input; (**c**) grain size of weld metal with different heat input; (**d**) tensile strength with different heat input.

The increase of heat input rate could also coarsen the weld metal grain. The weld grain size has a good linear relationship with the heat input rate in P-GMAW [6,23], as shown in Figure 19c. In order to evaluate the beneficial effect of thermal pulse of DP-GMAW on grain refinement of weld metal microstructure with different welding parameters. "ΔSize" is used to characterize the degree of grain refinement, "ΔSize" is the difference between the size of DP-GMAW welding grain and P-GMAW welding grain under the same heat input. The grain size of P-GMAW welding was estimated by the linear fitting equation between the heat input and weld metal grain size of P-GAMW, as shown in Figure 19c. A larger "ΔSize" represents the more significant effect of thermal pulses on grain refinement. The transverse tensile test of the weld joints prepared by P-GMAW and DP-GMAW with different heat input is given in Figure 19d. It is observed that the tensile test of DP-GMA weld joint is higher than those of P-GMA weld joint due to the thermal pulse. As for the hardness of the weld joints, Ping Yao found that the variation characteristic of the hardness was approximately the same as that of grain size of weld joints of the 304 stainless steel prepared by P-GMAW and DP-GMAW [12].

Figure 20a shows "ΔSize" with different thermal pulse frequencies. In contrast, the average grain size of double pulse weld metal microstructure at thermal frequency = 2 was the smallest. Increasing the frequency of thermal pulses can effectively refine the grains of the weld metal with the given heat input rate. Figure 20b shows "ΔSize" with a different duty cycle of thermal pulse. A low duty cycle leads to the short duration of large constitutional supercooling of the weld pool, and the dendrite arms could not be melted off fully. Insufficient dendritic fragments can be the equiaxed nuclei in the weld pool. A large duty cycle leads high heat input which promote coarse-grain. Figure 20c shows "ΔSize" with different thermal pulse current change. Under the premise of maintaining the stability of the arc, the larger the current difference, the more effective the grain refinement of the thermal pulse. However, large current difference could result in poor weld formation. Therefore, increasing the current difference of thermal pulse cannot simultaneously obtain the fine grain and the good weld formation. It is necessary to analyze the actual situation in the application.

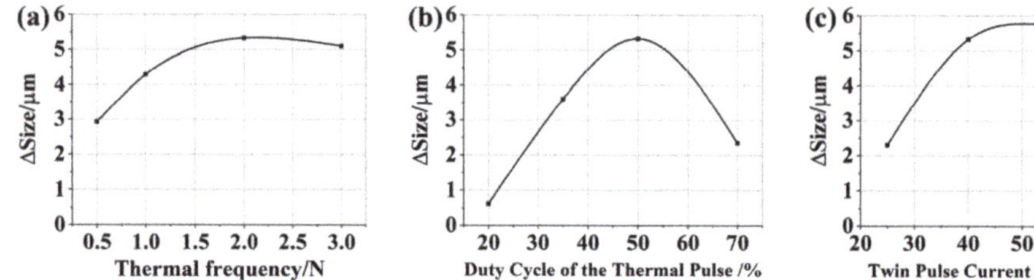

Figure 20. Variation of ΔSize with (**a**) heat thermal frequency; (**b**) duty cycle of the thermal pulse; (**c**) current amplitude of thermal pulse.

4. Conclusions

In this paper, a high-speed photography system and image processing technology were used to extract the characteristics of arc profile, the droplet transfer and the weld pool oscillation. The influence of DP-GMAW welding parameters on heat input, pressure acting on weld pool, weld pool size and oscillation amplitude have been calculated. Additionally, the internal relation between weld pool behavior and microstructure was analyzed. The conclusions are as follows:

(1) In contrast with P-GMAW, the length and the oscillation amplitude of the weld pool show periodic changes within one thermal pulse of DP-GMAW. The thermal pulse led to remelting and resolidification of the weld bead near the pool trailing edge which shrank and separated from the solidified bead boundary of the weld during switching from T_p to T_b.

(2) Welding pool oscillation caused by the thermal pulse enhances the weld pool convection, which can help dendrite fragmentation, thus providing sufficient crystal nucleus for liquid metal crystallizing. The convection can reduce the temperature gradient of the weld pool and increase constitutional supercooling of the weld pool to promote equiaxed grains surviving and growing.

(3) The size of HAZ in the weld joint mainly depends on the heat input rate. Thermal pulse of DP-GMAW has an insignificant effect on the grain size of HAZ.

Author Contributions: Methodology, T.C.; software, T.C.; investigation, T.C., P.Z. (Peng Zhang), writing—original draft preparation, T.C., P.Z. (Peng Zhang); writing—review and editing, S.X., T.C., B.W., P.Z. (Peizhuo Zhai); supervision, S.X.; project administration, S.X., W.L.; funding acquisition, S.X. All authors have read and agreed to the published version of the manuscript.

Funding: This work was funded by the National Natural Science Foundation of China, grant No.51675269 and the Priority Academic Program Development of Jiangsu Higher Education Institutions (PAPD).

Acknowledgments: The authors gratefully acknowledge the financial supports by the National Science Foundation of China under Grant numbers 51675269, as well as the Priority Academic Program Development of Jiangsu Higher Education Institutions (PAPD).

Conflicts of Interest: The authors declare no conflict of interest.

References

1. Dos Santos, E.B.F.; Pistor, R.; Gerlich, A.P. Pulse profile and metal transfer in pulsed gas metal arc welding: Droplet formation, detachment and velocity. *Sci. Technol. Weld. Join.* **2017**, *22*, 627–641. [CrossRef]
2. Liu, A.; Tang, X.; Lu, F. Study on welding process and prosperities of AA5754 Al-alloy welded by double pulsed gas metal arc welding. *Mater. Des.* **2013**, *50*, 149–155. [CrossRef]
3. Oliveira, J.P.; Curado, T.M.; Zeng, Z. Gas tungsten arc welding of as-rolled CrMnFeCoNi high entropy alloy. *Mater. Des.* **2020**, *189*, 108505. [CrossRef]
4. Oliveira, J.P.; Curado, T.M.; Zeng, Z. Microstructure and mechanical properties of gas tungsten arc welded Cu-Al-Mn shape memory alloy rods. *J. Mater. Process. Technol.* **2019**, *271*, 93–100. [CrossRef]
5. da Silva, C.L.M.; Scotti, A. The influence of double pulse on porosity formation in aluminum GMAW. *J. Mater. Process. Technol.* **2006**, *171*, 366–372. [CrossRef]

6. Sen, M.; Mukherjee, M.; Singh, S.K.; Pal, T.K. Effect of double-pulsed gas metal arc welding (DP-GMAW) process variables on microstructural constituents and hardness of low carbon steel weld deposits. *J. Manuf. Process.* **2018**, *31*, 424–439. [CrossRef]
7. Wang, L.L.; Wei, H.L.; Xue, J.X.; DebRoy, T. Special features of double pulsed gas metal arc welding. *J. Mater. Process. Technol.* **2017**. [CrossRef]
8. Zhang, H.; Hu, S.; Shen, J.; Ma, L.; Yin, F. Microstructures and mechanical properties of 30Cr-4Mo ferritic stainless steel joints produced by double-pulsed gas metal arc welding. *Int. J. Adv. Manuf. Technol.* **2015**, *80*, 1975–1983. [CrossRef]
9. Mathivanan, A.; Senthilkumar, A.; Devakumaran, K. Pulsed current and dual pulse gas metal arc welding of grade AISI: 310S austenitic stainless steel. *Def. Technol.* **2015**, *11*, 269–274. [CrossRef]
10. Yao, P.; Zhou, K.; Tang, H. Effects of Operational Parameters on the Characteristics of Ripples in Double-Pulsed GMAW Process. *Materials* **2019**, *12*, 2767. [CrossRef] [PubMed]
11. Yao, P.; Zhou, K.; Lin, H.; Xu, Z.; Yue, S. Exploration of Weld Bead Forming Rule during Double-Pulsed GMAW Process Based on Grey Relational Analysis. *Materials* **2019**, *12*, 3662. [CrossRef] [PubMed]
12. Yao, P.; Zhou, K.; Huang, S. Process and Parameter Optimization of the Double-Pulsed GMAW Process. *Metals* **2019**, *9*, 1009. [CrossRef]
13. Wang, L.L.; Wei, H.L.; Xue, J.X.; DebRoy, T. A pathway to microstructural refinement through double pulsed gas metal arc welding. *Scri. Mater.* **2017**, *134*, 61–65. [CrossRef]
14. Wang, L.; Jin, L.; Huang, W.; Xu, M.; Xu, J. Effect of thermal frequency on AA6061 aluminum alloy double pulsed gas metal arc welding. *Mater. Manuf. Process.* **2016**, *31*, 2152–2157. [CrossRef]
15. Liu, A.; Tang, X.; Lu, F. Weld pool profile characteristics of Al alloy in double-pulsed GMAW. *Int. J. Adv. Manuf. Technol.* **2013**, *68*, 2015–2023. [CrossRef]
16. Pang, J.; Hu, S.; Shen, J.-Q.; Wang, P. Arc characteristics and metal transfer behavior of CMT+P welding process. *J. Mater. Process. Technol.* **2016**, *238*, 212–217. [CrossRef]
17. Yokomizu, Y.; Matsumura, T.; Sun, W.Y.; Lowke, J.J. Electrode sheath voltages for helium arcs between non-thermionic electrodes of iron, copper and titanium. *J. Phys. D Appl. Phys.* **1998**, *31*, 880–883. [CrossRef]
18. Ghosh, P.K.; Dorn, L.; Kulkarni, S.; Hofmann, F. Arc characteristics and behavior of metal transfer in pulsed current GMA welding of stainless steel. *J. Mater. Process. Technol.* **2009**, *209*, 1262–1274. [CrossRef]
19. Murphy, A.B. The effects of metal vapour in arc welding. *J. Phys. D Appl. Phys.* **2010**, *43*, 434001. [CrossRef]
20. Yudodibroto, B.Y.B.; Hermans, M.J.M.; den Ouden, G.; Richardson, I.M. Observations on droplet and arc behavior during pulsed GMAW. *Weld. World* **2009**, *53*, R171–R180. [CrossRef]
21. Li, J.; Sun, Q.; Liu, Y.; Zhen, Z.; Sun, Q.; Feng, J. Melt flow and microstructural characteristics in beam oscillation superimposed laser welding of 304 stainless steel. *J. Manuf. Process.* **2020**, *50*, 629–637. [CrossRef]
22. Yuan, T.; Luo, Z.; Kou, S. Grain refining of magnesium welds by arc oscillation. *Acta Mater.* **2016**, *116*, 166–176. [CrossRef]
23. Verma, J.; Taiwade, R.V. Effect of welding processes and conditions on the microstructure, mechanical properties and corrosion resistance of duplex stainless steel weldments—A review. *J. Manuf. Process.* **2017**, *25*, 134–152. [CrossRef]

Article

Quality Assessment Method Based on a Spectrometer in Laser Beam Welding Process

Jiyoung Yu [1], Huijun Lee [1], Dong-Yoon Kim [2], Munjin Kang [2,*] and Insung Hwang [2,*]

1 Research institute, Monisys Co., Ltd., 775, Gyeongin-ro, Yeongdeungpo-Gu, Seoul 07299, Korea; susagye@naver.com (J.Y.); manian12@hanmail.net (H.L.)
2 Joining R & D Group, Korea Institute of Industrial Technology, 156 Gaetbeol-ro, Yeonsu-Gu, Incheon 21999, Korea; kimdy@kitech.re.kr
* Correspondence: moonjin@kitech.re.kr (M.K.); hisman@kitech.re.kr (I.H.); Tel.: +82-32-850-0215 (M.K.)

Received: 9 June 2020; Accepted: 23 June 2020; Published: 24 June 2020

Abstract: For the automation of a laser beam welding (LBW) process, the weld quality must be monitored without destructive testing, and the quality must be assessed. A deep neural network (DNN)-based quality assessment method in spectrometry-based LBW is presented in this study. A spectrometer with a response range of 225–975 nm is designed and fabricated to measure and analyze the light reflected from the welding area in the LBW process. The weld quality is classified through welding experiments, and the spectral data are thus analyzed using the spectrometer, according to the welding conditions and weld quality classes. The measured data are converted to RGB (red, green, blue) values to obtain standardized and simplified spectral data. The weld quality prediction model is designed based on DNN, and the DNN model is trained using the experimental data. It is seen that the developed model has a weld-quality prediction accuracy of approximately 90%.

Keywords: deep neural network; high strength steel; laser beam welding; penetration; quality assessment; spectrometer

1. Introduction

The advent of Industry 4.0 has brought significant enhancements to manufacturing processes, which are now based on smart and autonomous systems, and are incorporated with data and machine learning (ML) [1]. The enhancements to smart manufacturing processes, especially concerning welding technology, must be extended and established, in the context of modern manufacturing [2]. Among various welding technologies, laser beam welding (LBW) has a higher precision, productivity, flexibility, effectiveness, and numerous other advantages, which include a deeper penetration, higher welding speed, and lower distortion, compared to other welding technologies, owing to the properties of the power sources in LBW [3–6]. In addition, the LBW is a suitable joining technology for welding dissimilar metals and advanced materials such as NiTi alloy [7–9]. The LBW presents a substantial potential for manufacturing applications [10], and more study on the LBW process is required to increase the benefits of the LBW. In particular, there seems to be a need of research in the field of monitoring and controlling the welding process and weld quality, which is attributed to the complexity of LBW systems and the characteristics of the LBW process. Considering that the technologies in this field are essential to develop a smart manufacturing system for LBW, extensive research must be carried out to achieve this objective.

The monitoring of a process, and consequently the quality assessment, can be divided into three categories: pre-process, in-process, and post-process [11]. The pre-process monitoring concerns the weld seam tracking before welding or ahead of the laser heat source. The in-process monitoring focuses on the monitoring of the phenomena on the welded zone during welding, such as the keyhole's shape stability. The post-process monitoring, generally, is concerned with the weld defect detection, and the measurement of the form of the weld seam, after welding, or behind the weld pool. In the pre-process and post-process monitoring methods, ultrasonic and camera-based techniques have been dominantly used. In the case of in-process monitoring, a wider range of measuring methods, including optical (ultraviolet, visual, infrared) and acoustic detectors, X-ray radiography, and camera-based methods, have been used to adequately deal with welding phenomena generated by the high-energy-density laser.

According to Stavridis et al. [10], the in-process quality assessment, i.e., in-process monitoring of LBW, can be further sub-divided into four parts, based on the monitoring techniques used, such as image processing, acoustic emission, X-ray radiography, or optical signal techniques. Vision and thermal images are mainly used in image processing techniques to observe the weld pool [12], keyhole [13], plume, and spatters [14]. The acoustic emission techniques have been applied for the measurement of the melting, vaporization, plasma generation, keyhole formation [15], and propagation of cracks [16], and the X-ray techniques have been used to observe the welding phenomena and defects, including slag inclusions, blow holes, incomplete penetration, and undercuts [17]. In the case of optical signal techniques, they use vision systems (charge-coupled device and complementary metal–oxide–semiconductor cameras), photodiodes [18], and spectrometers (which are highly related with the topic of the study).

Sibillano et al. [19] studied the dynamics of the plasma plume, produced in LBW of 5083 aluminum alloy, with the help of correlation spectroscopy, and presented the results of the influence of welding speed on the loss of alloying elements. Rizzi et al. [20] investigated the spectroscopic signals, produced by the laser-induced plasma optical emission, together with energetic and metallographic analyses of CO_2 laser-welded stainless-steel lap joint, using the response surface methodology (RSM). This statistical approach allowed the study of the influence of the laser beam power and laser welding speed, on the plasma plume electron temperature, joint penetration depth, and melted area. Konuk et al. [21] used a spectrometer to collect the optical emissions of the welding area, and calculate the electron temperature, and the data measured and calculated were used to determine the weld quality and to control the laser power. Sebestova et al. [22] designed a sensor to monitor the pulsed Nd:YAG laser welding process, based on the measurement of the plasma electron temperature, and this sensor was used to detect the weld penetration depth. Zaeh and Huber [23] investigated the radiation emission of the laser-induced plume during LBW of aluminum and steel alloys, and they developed a system to control the chemical composition of the melt pool, during the active welding process. Chen et al. [24] proposed a spectroscopic method based on a support vector machine (SVM) and artificial neural network (ANN) for the detection and classification of fiber laser welding defects. The spectral data captured by the spectrometer was processed by selecting sensitive emission lines and extracting features of the evolution of the spectral data; the SVM and ANN models were designed based on the processed spectral data. It has been verified that these two models were quite effective in detecting and classifying the weld defects. Zhang et al. [25] designed a multiple-sensor system that includes an auxiliary illumination visual sensor system, an ultraviolet- and a visible-band visual sensor system, a spectrometer, and two photodiodes to capture the real-time welding signal and determine the laser welding quality. A deep learning framework based on the stacked sparse autoencoder (SSAE) was established to model the relationship between the multi-sensor features and their corresponding welding statuses; a genetic algorithm (GA) was used to optimize the parameters of the SSAE framework. Lee et al. [26] developed an in-situ monitoring system using a spectrometer for laser welding on galvanized steel. They applied Fisher's criterion to rank several features for extraction of the most valuable features, and then used the K-nearest neighbors and SVM algorithms to classify welding conditions related to the welding defects. Although some studies have been carried out to investigate weld quality assessment using a spectrometer in LBW process, till date, little attention has been paid to the quality assessment methodology, in LBW, based on the spectrometer and ML algorithms.

This study presents a deep neural network (DNN)-based quality assessment method in spectrometry-based LBW. We designed a spectrometer, that can measure and analyze the light reflected from the welding area in LBW process. The weld quality of LBW was classified through welding experiments, and the spectral data were analyzed using the spectrometer, according to the welding conditions and weld quality classes. Subsequently, the spectral data were converted to CIE 1931 RGB values, to standardize and simplify the spectral data. A weld quality prediction model was designed based on DNN, and the DNN model was trained using the experimental data. The weld-quality prediction accuracy of the model was calculated.

The remainder of the paper is organized as follows. Section 2 describes the experimental procedure and setup. The obtained results are discussed in Section 3, and Section 4 concludes the paper with a brief summary.

2. Experimental System and Procedure

2.1. Spectrometer Development

Figure 1 shows the schematic and prototype of the spectrometer designed in this study, which includes a collimator, an optical fiber, a connector, a reflective diffraction grating, a focusing mirror, and a complementary metal-oxide-semiconductor (CMOS) linear sensor. The optical fiber, through which the light enters, is connected to the collimator with an SMA (SubMiniature version A) connector. Since the round-to-linear fiber optic bundle (model No.: BFL200HS02, Thorlabs Inc., Newton, MA, USA) is used as the optical fiber, there is no need to add a separate slit as an entrance aperture. This indicates that the output terminal of the fiber optic bundle, i.e., the linear bundle, acts as the slit. The collimator (model No.: PAF2S-7A, Thorlabs Inc., Newton, MA, USA) is used to collimate the light, coming from the fiber optic bundle, and then send a collimated beam towards the grating. The reflective diffraction grating (model No.: GR13-0605, Thorlabs Inc., Newton, MA, USA) splits the photons coming from the collimator, depending on the wavelength, and then spreads the light across the focusing mirror (concave). The focusing mirror directs the light, at each wavelength, onto the CMOS linear image sensor. The line camera (model: USB line camera 8M, Coptonix GmbH, Berlin, German) which consists of a main circuit board, and the CMOS linear image sensor (model No.: S11637-1024Q, Hamamatsu, Hamamatsu City, Japan) is used as the detector. Each pixel of the CMOS linear image represents a portion of the spectrum that is translated into a measurable value, by a spectroscopy software. A dedicated software was also developed to operate the spectrometer, display the measured data, and store measurement results. The spectral response range of the developed spectrometer was 225–975 nm, and its sampling frequency was 5 kHz.

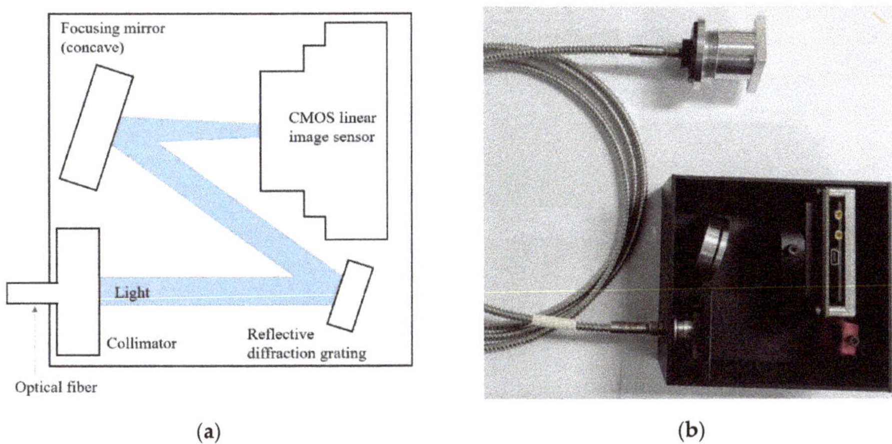

Figure 1. Configuration of the developed spectrometer: (**a**) schematic; (**b**) designed prototype.

2.2. Experimental Setup and Material

The experiment was conducted using a 4-kW disk laser welding system (model No.: TRUMPF HLD 4002, procured from Trumpf GmbH + Co. KG, Ditzingen, Germany). Figure 2 shows a diagram of the experimental setup, which was constructed by reference to the study of You and Katayama [27]. The scanner laser head was equipped with a six-axis robot arm, and was connected with two optical fibers. One of them was the round-to-linear fiber optic bundle connected to the spectrometer, and the other was the optical fiber for laser beam input transmission from the laser power source. In the former case, the scanner laser head and the fiber optic bundle were assembled with a connector, shown in Figure 3, which also has a function to condense the light entering the scanner laser head. The light emission of visible light and laser light from the welding area was transmitted to the spectrometer, through the scanner laser head (having a partially transmitting mirror) and optical fiber. The transmitted visible light emission and laser beam reflection were collimated by the collimator, and detected by the line camera. The image data obtained by the CMOS linear image sensor were sampled and digitalized at a frequency of 5 kHz. Subsequently, the processed data were collected, displayed, and stored by the computer software. Figure 4 shows a plot of the spectrum intensity variation, in LBW of lap joint configuration, with a welding speed of 1.5 m/min and a laser power of 3500 W. To test the weld quality and collect data for development of weld quality prediction models, the experiment was carried out using typical automotive high strength steel (uncoated 780 MPa-grade dual-phase (DP) steel) sheets of 1.2 mm thickness. The workpiece was cut to dimensions of 100 × 30 mm, and welding was conducted in the overlap joint configuration.

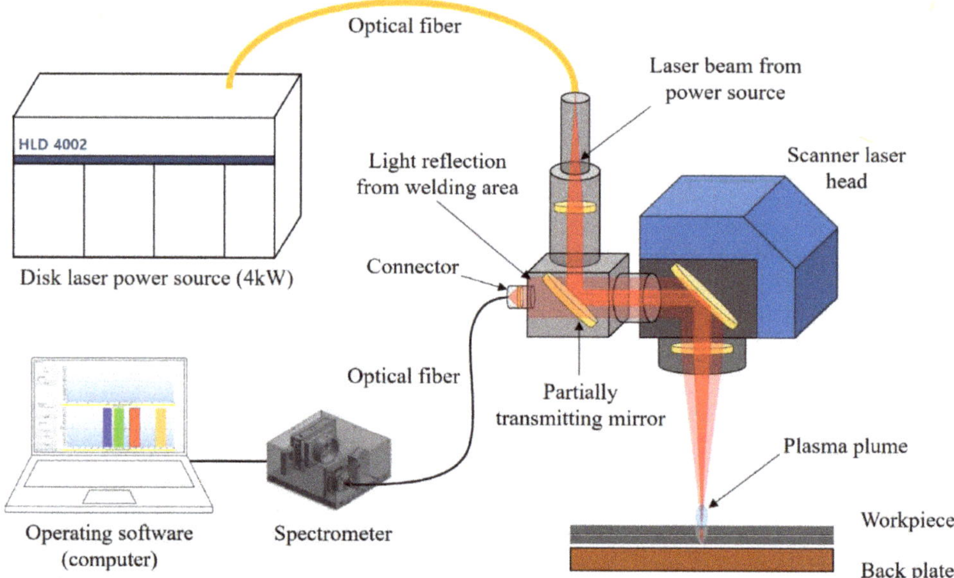

Figure 2. Schematic of experimental setup.

Figure 3. Experimental setup and installation of the spectrometer.

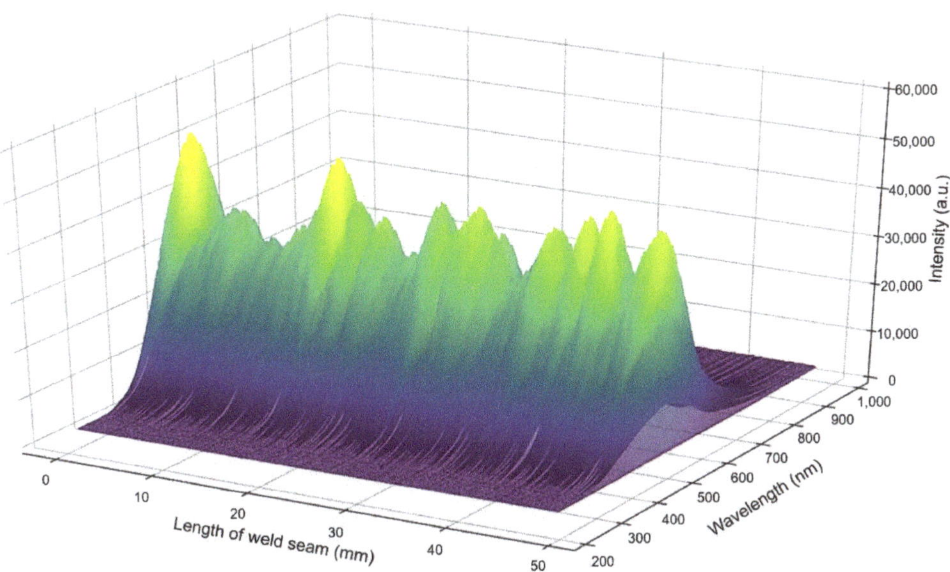

Figure 4. Spectrum intensity variation, in the LBW of a lap joint configuration, with a welding speed 1.5 m/min and a laser power of 3500 W (thickness of sheet is 1.2 mm).

2.3. Experimental Procedure

In this study, we tried to classified the weld quality into four types as shown in Figure 5:
(Class 1) Unwelded: No weld is formed between the workpieces
(Class 2) Incomplete penetration: Weld is formed between the workpieces but not full penetration
(Class 3) Full penetration: Weld fully penetrates all workpieces
(Class 4) Unwelded by a gap, between the workpieces in lap joint configuration

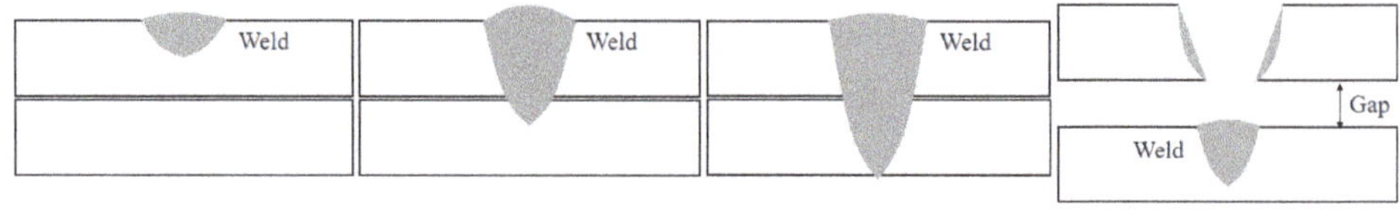

[Class 1: Unwelded] [Class 2: Incomplete penetration] [Class 3: Full penetration] [Class 4: Unwelded by a gap]

Figure 5. Schematics of four weld qualities defined in this study.

To evaluate the weld quality and as well as to obtain experimental data, which can be qualified according to the predefined types and are enough to develop a quality prediction model, the welding parameters were initialized with values presented in Table 1. For the laser power and the welding speed, welding was carried out twice (2 replicates) in all possible conditions (21 combinations), at a gap of 0 mm. At a gap of 0.8 mm, welding was carried out only at laser powers of 3000 and 4000 W for all welding speeds, which means that welding was performed under six conditions for two replicates. All welding runs were carried out with a length of 50 mm in the lap joint configuration. The diameter of a focused laser beam was 0.6 mm, and the wavelength of disk laser was 1030 nm. The defocused distance was fixed at 0 mm. Argon (Ar) inert gas was used as shielding gas, and the flow rate of the shield gas was set to 20 L/min. The metal back plate was set under the workpieces as shown in Figure 1.

Table 1. Experimental factors and their values.

Welding Parameter	Value
Laser power (W)	1000, 1500, 2000, 2500, 3000, 3500, 4000
Welding speed (m/min)	1.5, 2.0, 2.5
Gap (mm)	0, 0.8

3. Results and Discussion

3.1. Weld Quality Evaluation

In this study, we evaluated the weld quality, by examining the cross sections of the weld joints, shown in Table 2. Three pre-defined types of joints were observed: unwelded, incomplete penetration, and full penetration, according to different welding speeds and laser power. The "unwelded" type was mainly observed in case of a relatively low power range of 1000–1500 W, and the type "incomplete penetration" was observed in the range of 1500–2000 W. The type "full penetration" was observed in the range of 2500–4000 W. Because the back plate was used in this study (shown in Figure 2), it seems that burn-through was not created when a high laser power was used. That might have led to a larger range of the "full penetration" type. Figure 6 shows the top, bottom, and side views of the weldment, having a gap of 0.8 mm. Although a welding speed of 1.5 m/min and laser power of 3000 W produced a "full penetration" joint, in the case of zero gap (Table 2), a weld joint was not seen, in the case of 0.8 mm gap. This is because the molten metal was insufficient to fill the gap, which was larger in size. This results in the formation of the last type of joint, "unwelded," which is due to the unfilled gap between the workpieces in the lap joint configuration.

Table 2. Cross-sections of welded joint according to the laser power and welding speed.

Laser Power (W)	Welding Speed (m/min)		
	1.5	**2.0**	**2.5**
1000	Unwelded	Unwelded	Unwelded
1500	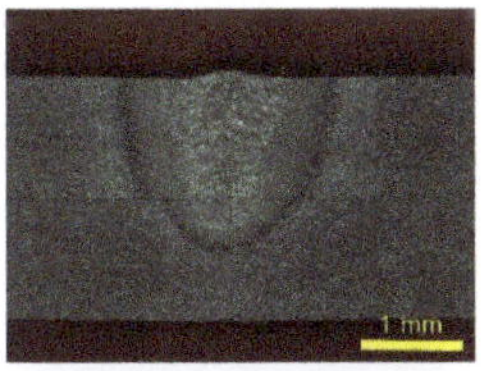 1 mm	Unwelded	Unwelded
2000	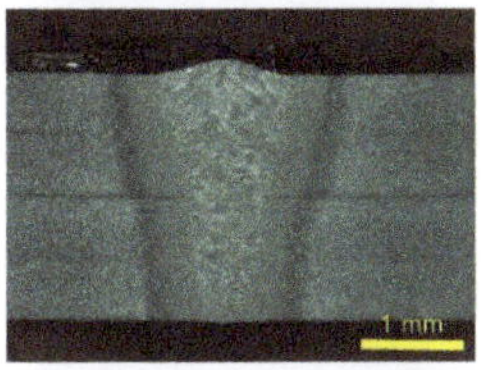 1 mm	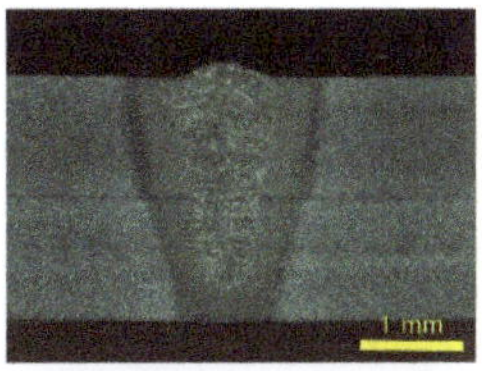 1 mm	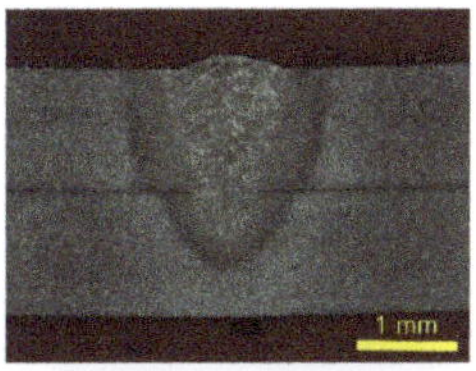 1 mm
2500	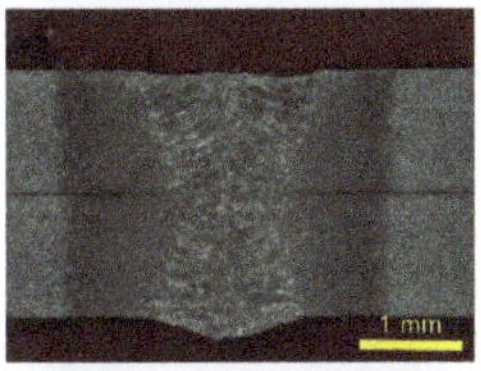 1 mm	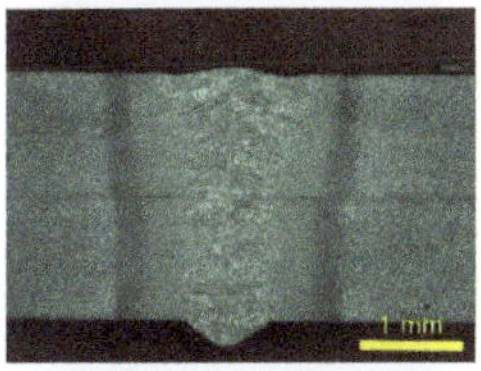 1 mm	 1 mm
3000	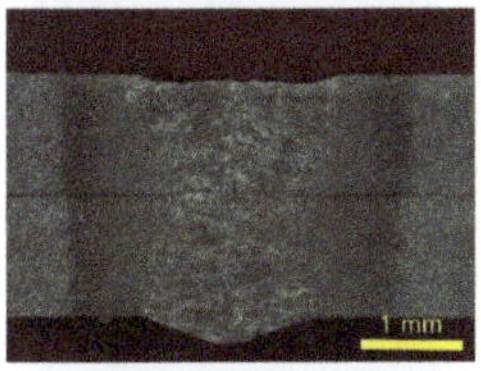 1 mm	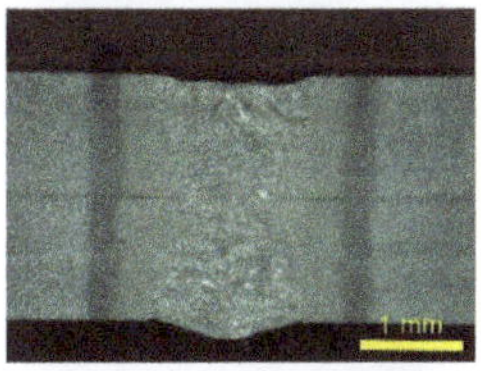 1 mm	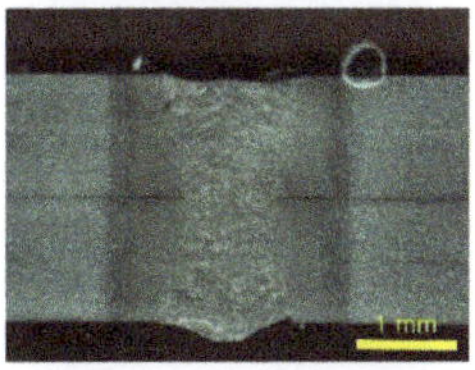 1 mm
3500	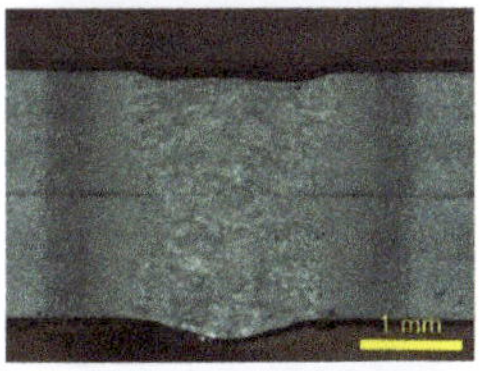 1 mm	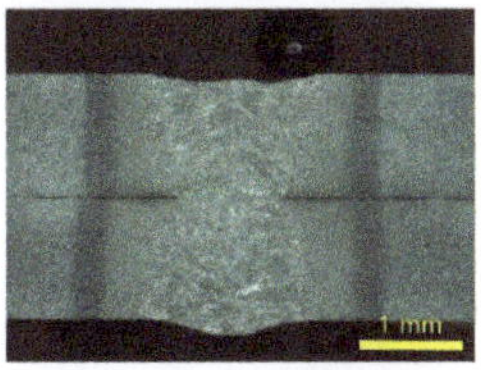 1 mm	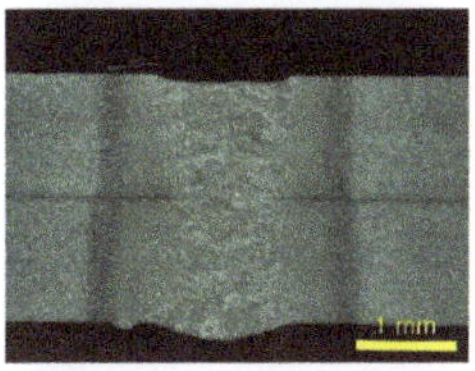 1 mm
4000	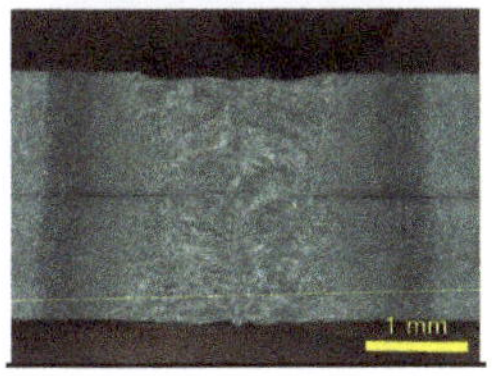 1 mm	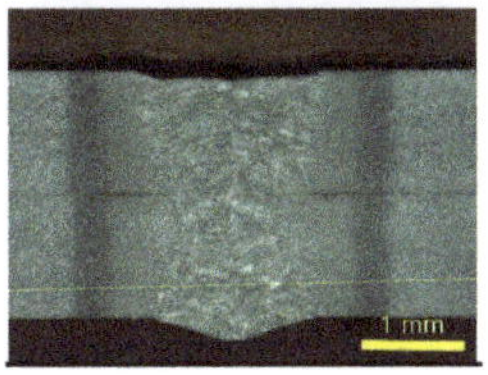 1 mm	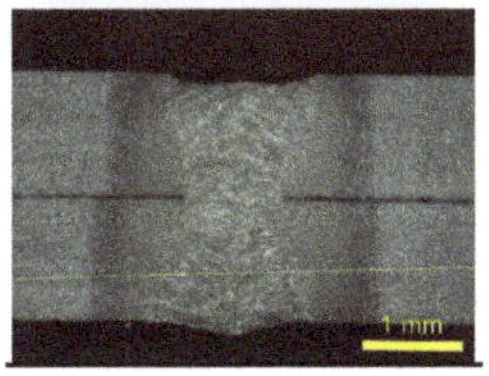 1 mm

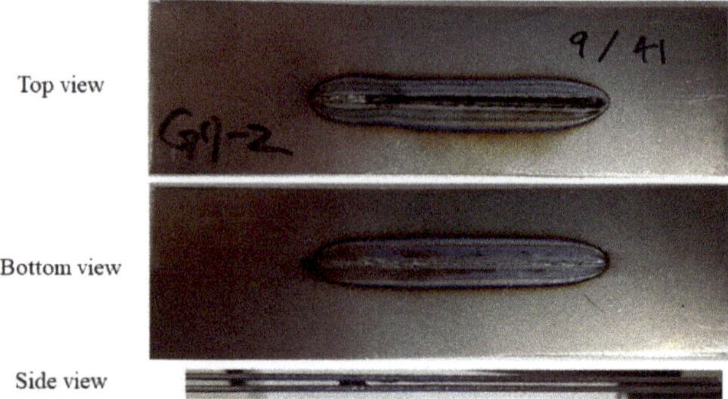

Figure 6. Workpiece appearance with a gap of 0.8 mm after welding with a welding speed 1.5 m/min and a laser power of 3000 W: top view, bottom view, side view.

3.2. Signal Analysis

Table 3 shows the color-coded images, which helps to visualize a 3D image of the spectrum intensity (Figure 4) in a 2D form of the spectrum during 0.2 s (0.4–0.6 s). In the color-coded image, the horizontal and vertical axes represent the time and wavelength, respectively, and the intensity is mapped in color. The used color scale was the same used by Viridis [28], and the minimum and maximum values of the color scale were set as 0 and 65535, respectively. It was seen that higher intensities were observed, mainly in the wavelength range of 350–650 nm. In general, when the laser power was kept constant, a higher welding speed increases the intensity. In terms of the size of weld bead, i.e., the penetration and width, this signifies that smaller the weld bead, more was the light, reflected into the spectrometer. Similarly, in case of constant welding speeds, the higher the welding power, the larger was the intensity. However, in this case, even if the size of the weld bead increased, the intensity increased, as the laser power increased. This means that the intensity is influenced more by the laser power, than the weld bead size. These trends were quite clearly observed, in the range of 1000–2500 W of the laser power. However, these trends were rarely observed in all values of laser power more than 3000 W, and rather, all color-coded images in these conditions were almost identical. Additionally, in these conditions, the size of the weld beads, at each welding speed, was almost identical, regardless of the laser power. The growth of the weld bead seems to saturate at a laser power above 3000 W. It is thought that two experimental results, observed when the laser power was more than 3000 W, are related to each other. Table 4 presents the color-coded images of spectrum at all conditions, at a gap of 0.8 mm, and the other configurations except the gap setting are same as those of Table 3. In this case, the pattern of the spectrum intensity was irregular, and the intensity level was smaller, compared to those of Table 3. This result indicates that the irregular geometry of the welding area, caused by the gap, caused a higher irregular reflection of the visible light and laser light. The 3D images (Figure 4) and color-coded images (Tables 3 and 4) may be difficult to be interpret, and their color and form can also change quite drastically, when compared to a referenced color scale and other configurations. Therefore, in this study, we used the color space created by the International Commission on Illumination in 1931 (CIE 1931 color space), to convert the measured spectrum data to a standard form. The measured spectrum data were converted into CIE 1931 XYZ values, and then these values were further converted into CIE 1931 RGB values by Colour, an open-source Python package, providing algorithms and datasets for color science [29]. The RGB format is effective for showing the features of data measured by the spectrometer and it consists of much smaller data converted using the spectrum data, which allows to estimate models and handle data, with less computation time. Additionally, the maximum value of wavelength for each sample was searched and added.

Table 3. Color-coded images of spectrum, according to the laser power and welding speed; in each image, the horizontal axis is time (s), the vertical axis is wavelength (nm), and intensity is mapped in color (color scale: Viridis, date from [28]).

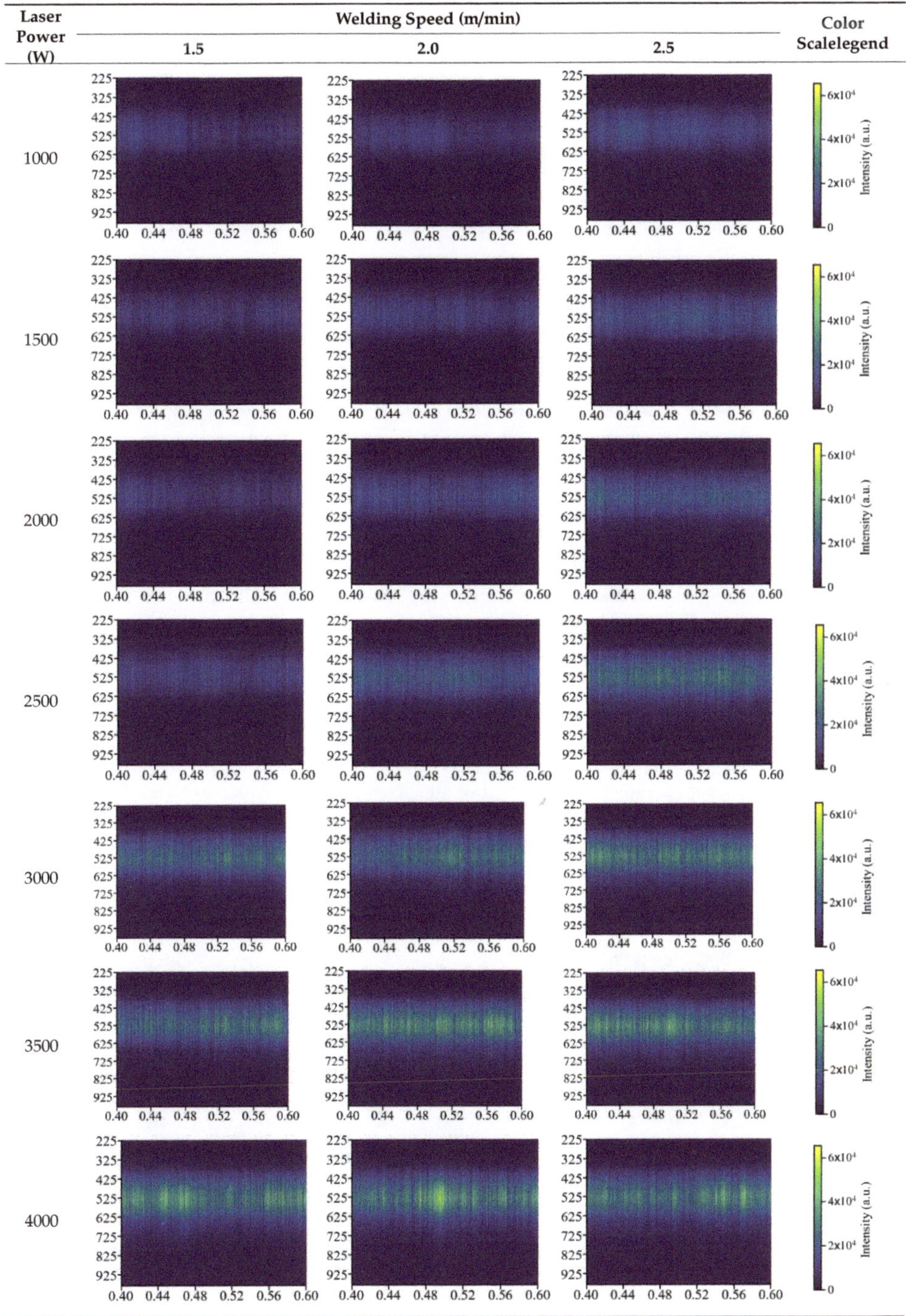

Table 4. Color-coded images of spectrum at a gap of 0.8 mm according to the laser power and welding speed: in each image, the horizontal axis is time (s), the vertical axis is wavelength (nm), and intensity is mapped in color (color scale: Viridis, data from [28]).

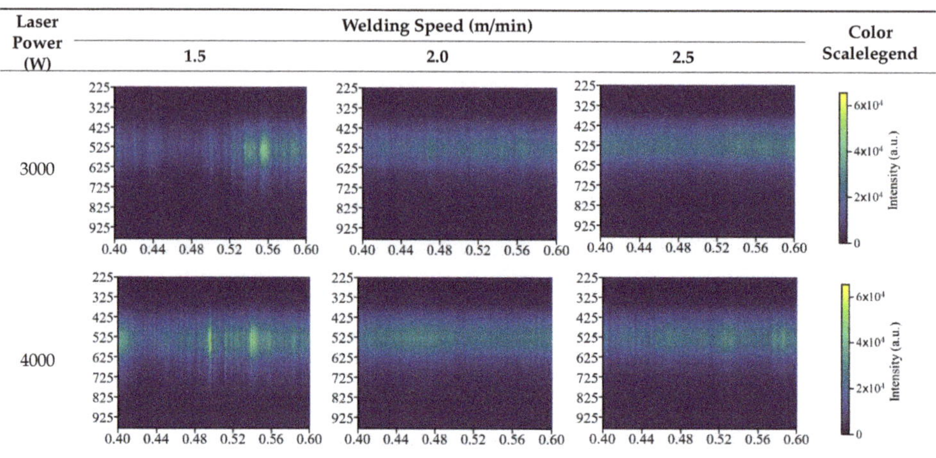

To investigate the effect of the weld penetration on the spectrum intensity, the maximum wavelength and RGB values, at a laser power of 2000 W and different welding speeds, are shown in Figure 7. It is seen that the welding heat input (expressed in kJ/mm) increases, as the welding speed decreases, when the laser power was kept constant. Likewise, the weld penetration increased with increasing heat input, as the welding speed decreased (Table 2). The RGB value, which is converted from the intensity data, decreased when the penetration was increased, which is similar to the results shown in Table 3. It might be inferred that the amount of light reflected from the welding area decreases, as the penetration depth increases, when the laser power is limited to 3000 W, as used in this study. The maximum values of the wavelength were found to be distributed in the range of 500–540 nm, and these values in the case of welding speed 1.5 m/min were slightly lower, when a different welding speed was used. Figure 8 shows the maximum wavelength and RGB values wavelength, at a welding speed of 2.5 m/min, according to different laser powers. The RGB values increased, as the laser power increased, even though the size of weld bead increased due to the high laser power. In addition, the RGB values were almost same, in the conditions of the laser power more than 3000 W. These results were found to be in good agreement, with those of Table 3. The maximum values of wavelength were distributed in the range of 530–540 nm. Figure 9 shows the RGB values and maximum wavelength at a gap of 0.8 mm, and a 3000 W laser power, according to the welding speed. In this case, the maximum wavelength and RGB values were almost same as those shown in Figure 8c, which had conditions of a 3000 W laser power and a 2.5 m/min welding speed. However, overall waveforms of the RGB curves were found to be uneven and irregularly-shaped, and the variations of the RGB and maximum wavelength values were relatively small, as compared to those of the cases when the laser power was above 3000 W. For this reason, the green curve is clearly separated from the red and blue curves, as seen in Figure 9.

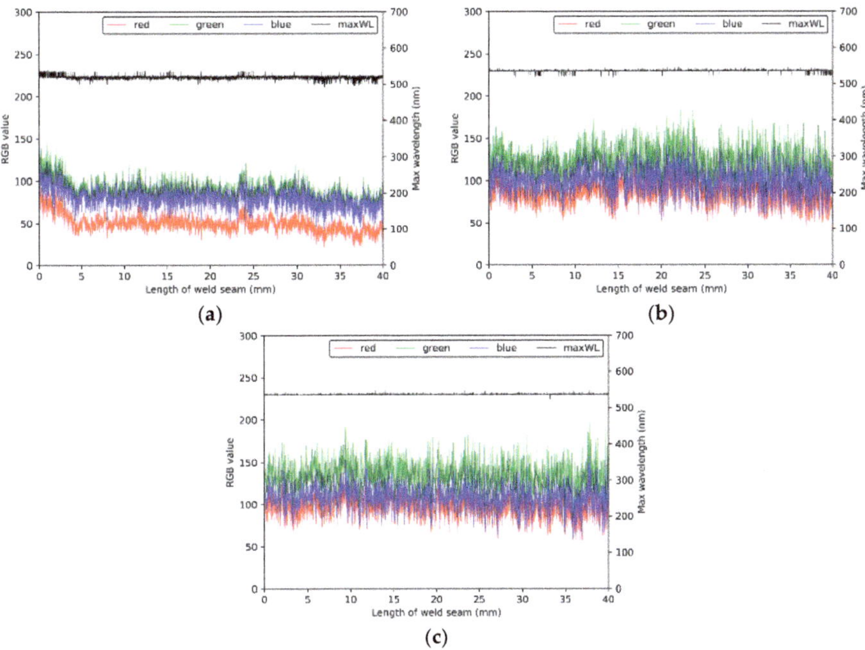

Figure 7. RGB values and maximum wavelength at laser power of 2000 W: welding speed of (**a**) 1.5 m/min; (**b**) 2.0 m/min; (**c**) 2.5 m/min.

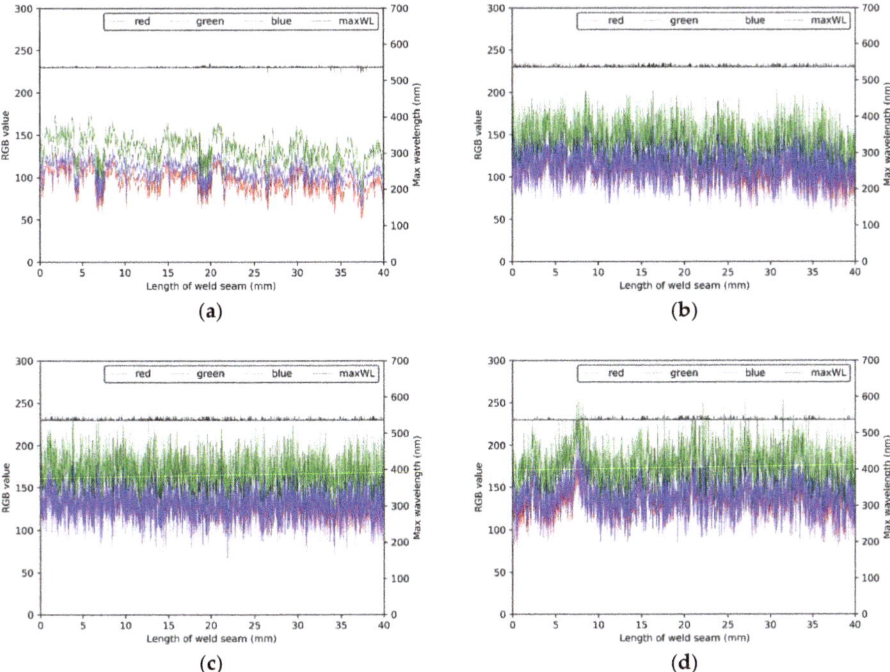

Figure 8. RGB values and maximum wavelength at welding speed of 2.5 m/min: laser power of (**a**) 1000 W; (**b**) 2000 W; (**c**) 3000 W; (**d**) 4000 W.

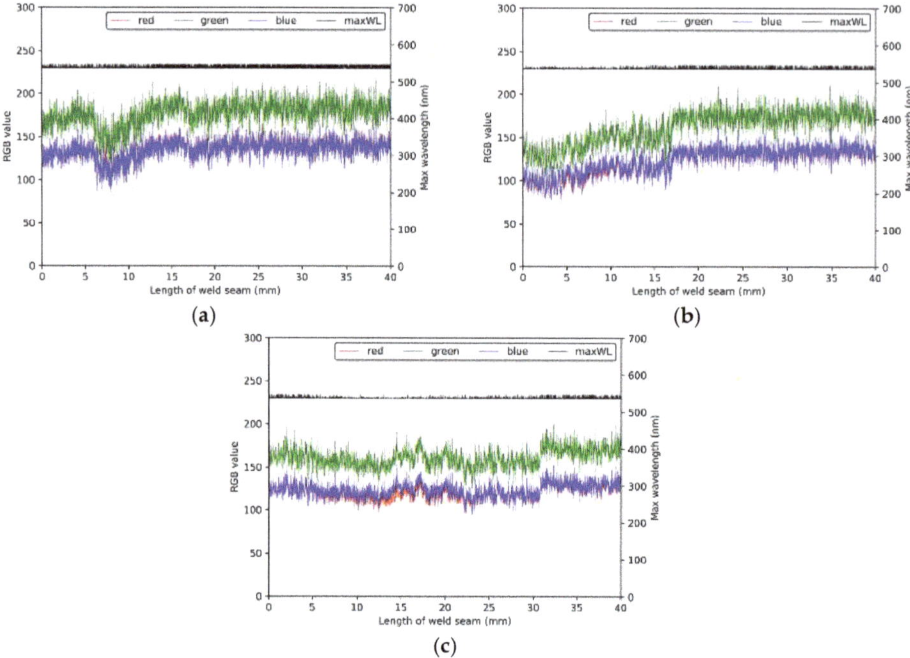

Figure 9. RGB values and maximum wavelength, at 0.8 mm gap and 3000 W laser power: welding speed of (**a**) 1.5 m/min; (**b**) 2.0 m/min; (**c**) 2.5 m/min.

3.3. Weld Quality Prediction Model

To predict the weld quality, a DNN model was used as the prediction model. As shown in Figure 10, we modeled a neural architecture, as a seven-hidden-layer DNN, and assigned 256 neurons to each of the hidden layer. The neurons of every preceding layer were fully connected, to those of the succeeding layer. The eight values, which were the average and standard deviation of the RGB values and the maximum wavelength, were used as the inputs to the network (see Nomenclature). Especially, considering the weld size and speed of welding, the model was designed to predict the weld quality, per 0.5 mm of the weld length. The outputs are the four joint types, described in Section 2.3: unwelded (Y_1), incomplete penetration (Y_2), full penetration (Y_3), and unwelded by a gap (Y_4). The other specifications of the DNN implementation model is summarized in Table 5. All the hidden layers utilized rectified linear unit (ReLU) functions, to calculate their respective intermediate outputs. The output layer had four nodes, and the output layer with the Softmax activation function computed and returned the type probability with respect to each of the output types. The prediction of the weld quality, by the model, was determined based on the type probability. A total of 5400 data sets (4320 training sets and 1080 test sets) were used to generate the DNN model. Each dataset contained the given eight input variables, and four output variables. The backpropagation algorithms were carried out with a batch size of 50, with 500 epochs. Figure 11 shows the training results, including the cost, training accuracy, and test accuracy. After training, the cost value was calculated as 0.2806, the training accuracy was 0.8993, and the test accuracy was 0.9083.

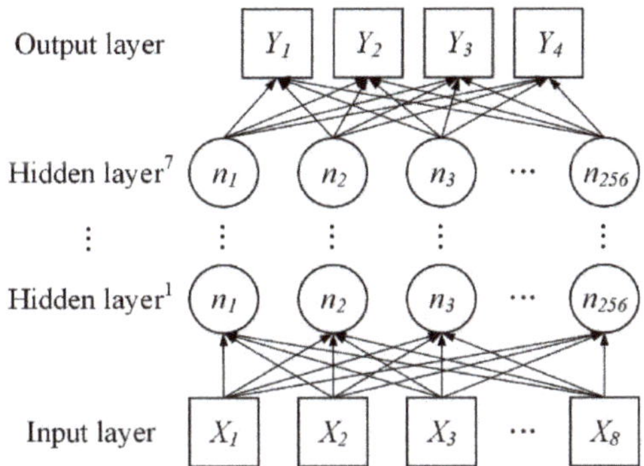

Figure 10. Structure of the deep neural network (DNN) model to classify the weld quality.

Table 5. Summary of DNN model implementation.

Item	Description		
Structure	**Input Layer**	**Hidden Layer**	**Output Layer**
Number of nodes	8	256	4
Learning rate		0.001	
Epoch		1000	
Batch size		100	
Activation function		ReLU	
Function of output layer		Softmax	
Cost function		Cross-entropy	
Optimizer		Adam optimizer	

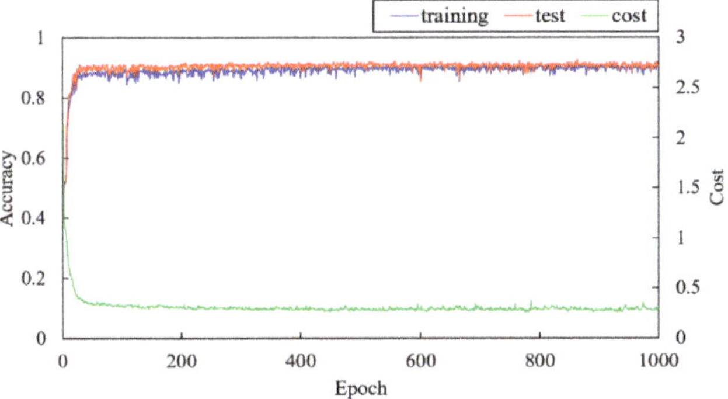

Figure 11. Training results: training accuracy, test accuracy, and cost.

To test the developed model, we produced 100 test datasets, for each class, by carrying out an additional welding experiment. Figure 12 shows the result of testing the developed DNN model using 400 test datasets, the confusion matrix and prediction result according to weld seam length. In Figure 12, the vertical and horizontal axes represent the predicted value of the DNN model and the data sequence, respectively. The four symbols, green, gray, red, and blue, represent the class to which each dataset belongs. It is seen that most of the errors occurred owing to the false prediction of Y_2 as Y_1. That is, 22 of the 100 datasets of class 2 (Y_2) made false predictions, and among them, the 21 datasets were predicted as class 1 (Y_1). Although Y_1 and Y_2 are divided into unwelded and incomplete penetration, respectively, in this study, Y_1 and Y_2 can be the same case, which is not full penetration. In particular, in the lap joint configuration of the two same metal sheets, if the penetration depth is less than 50% of the sum of the thickness of the two sheets, the weld quality is classified Y_1. If the penetration depth is more than 50% but less than the sum of thickness of the two sheets, the weld quality is classified Y_2. It is considered that this relationship between Y_1 and Y_2 causes this error. In addition, most errors occurred at the beginning of welding, which could be due to the relatively unstable spectral signal at the beginning of welding. In the case of Y_4, 9 of the 100 datasets made false predictions. The errors, which occur, by predicting Y_4 as the other classes, might be due to excessively irregular signal of Y_4. In the case of Y_3, there is no error in predicting the weld quality because its signal was stable and had a distinct distribution in the RGB space. In summary, the biggest prediction error was seen to have occurred in predicting Y_2. This error could be explained by the relatively small amount of available training data needed to increase the prediction accuracy of Y_2. That is, it is assumed that there were insufficient training data that otherwise could have made the quality prediction model more accurate in classifying Y_2 and Y_1. As presented in Tables 3 and 4, the range of the welding conditions in which the data for Y_2 can be obtained is smaller than that of other classes, and these welding conditions exist in the range of laser power of 1500–2000 W. If sufficient training data is obtained through more detailed experiments in the same range of laser power, the obtained data can reduce the prediction errors by more clearly learning a prediction model the boundary of Y_2 and Y_1.

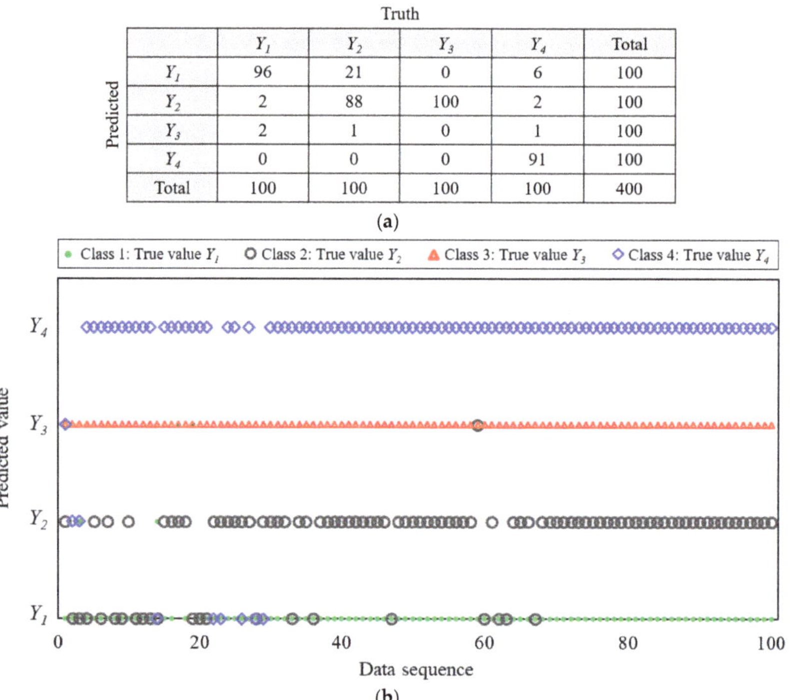

Truth

Predicted	Y_1	Y_2	Y_3	Y_4	Total
Y_1	96	21	0	6	100
Y_2	2	88	100	2	100
Y_3	2	1	0	1	100
Y_4	0	0	0	91	100
Total	100	100	100	100	400

(**a**)

(**b**)

Figure 12. Verification results: (**a**) confusion matrix; (**b**) prediction result according to data sequence.

4. Conclusions

In this study, we proposed the DNN-based quality assessment method based on a spectrometer in the LBW process, conducted the relevant experiments to analyze the features of the measured data, and verified the developed method. Notable developments and outcomes from this study are as follows.

- We designed and developed a spectrometer that can measure and analyze the light reflected from the welding area in an LBW process. The spectral response range of the developed spectrometer was 225–975 nm, and its sampling frequency was 5 kHz.

- The spectral data were converted to the CIE 1931 RGB color space to analyze the features of the spectral data and obtain the standardized and simplified spectral data.
- The prediction model that can classify the weld quality for LBW using data measured by the spectrometer was also developed. The weld quality prediction model was designed based on DNN, and the DNN model was trained using the converted RGB data and maximum frequency values. The developed model had a weld quality prediction accuracy of approximately 90%.

The results of this study substantially contribute to the state-of-the-art, regarding the automation of the welding process and monitoring technology, for LBW processes. Despite our study's contributions, some limitations are worth noting. Although our quality prediction model, based on the spectrometer, effectively estimates the weld quality of laser welding, the application of the model is limited to the scope of this study. Future work will focus on the generalization of the quality assessment method proposed in this study. For this purpose, we will collect sufficient data on various materials and welding conditions, and then further improve the data processing techniques and quality assessment algorithms using advanced ML algorithms. In parallel, we will also improve our spectrometer and its software developed in this study.

Author Contributions: The research presented here was carried out in collaboration between all authors. Conceptualization, M.K., I.H. and H.L.; data curation, D.-Y.K. and J.Y.; formal analysis, J.Y. and D.-Y.K.; investigation, J.Y. and D.-Y.K.; methodology, J.Y.; software, H.L. and J.Y.; validation, J.Y. and D.-Y.K.; writing—original draft, J.Y.; writing—review and editing, M.K. and J.Y.; funding acquisition, I.H.; supervision, I.H. and H.L.; project administration, M.K. All authors have read and agreed to the published version of the manuscript.

Funding: This work was supported by funding from the Korea Institute of Industrial Technology.

Acknowledgments: This study has been conducted with the support of the Korea Institute of Industrial Technology as "Development of intelligent root technology with add-on modules (kitech EO-20-0017)".

Conflicts of Interest: The authors declare no conflict of interest.

Nomenclature

ML	Machine learning
LBW	Laser beam welding
RSM	Response surface methodology
SVM	Support vector machine
ANN	Artificial neural network
GA	Genetic algorithm
SSAE	Stacked sparse autoencoder
DNN	Deep neural network
CMOS	Complementary metal-oxide-semiconductor
DP	DP Dual phase
Ar	Argon
CIE	International commission on illumination
ReLU	Rectified linear unit
X_1	Average value of red light during per 0.5 mm of weld length
X_2	Standard deviation of red light per 0.5 mm of weld length
X_3	Average value green light per 0.5 mm of weld length
X_4	Standard deviation of green light per 0.5 mm of weld length
X_5	Average value blue light per 0.5 mm of weld length
X_6	Standard deviation of blue light per 0.5 mm of weld length
X_7	Average value of the maximum wavelength per 0.5 mm of weld length
X_8	Standard deviation of blue light per 0.5 mm of weld length
Y_1	Unwelded (Class 1)
Y_2	Incomplete penetration (Class 2)
Y_3	Full penetration (Class 3)
Y_4	Unwelded by a gap (Class 4)

References

1. Tmforum. Available online: https://inform.tmforum.org/archive/2020/03/the-fourth-industrial-revolution-manufacturing-and-beyond/ (accessed on 4 June 2020).
2. Chryssolouris, G.; Papakostas, N.; Mavrikios, D. A perspective on manufacturing strategy: Produce more with less. *CIRP J Manuf. Sci. Technol.* **2008**, *1*, 45–52. [CrossRef]
3. Abdullah, H.A.; Siddiqui, R.A. Concurrent laser welding and annealing exploiting robotically manipulated optical fibers. *Opt. Laser Eng.* **2002**, *38*, 473–484. [CrossRef]
4. Tsoukantas, G.; Salonitis, K.; Stournaras, A.; Stavropoulos, P.; Chryssolouris, G. On optical design limitations of generalized two-mirror remote beam delivery laser systems: The case of remote welding. *Int. J. Adv. Manuf. Technol.* **2007**, *32*, 932–941. [CrossRef]
5. Mohammed, G.R.; Ishak, M.; Ahmad, S.N.A.S.; Abdulhadi, H.A. Fiber laser welding of dissimilar 2205/304 stainless steel plates. *Metals* **2017**, *7*, 546. [CrossRef]
6. Javaheri, E.; Lubritz, J.; Graf, B.; Rethmeier, M. Mechanical properties characterization of welded automotive steels. *Metals* **2020**, *10*, 1. [CrossRef]
7. Zeng, Z.; Oliveira, J.P.; Yang, M.; Song, D.; Peng, B. Functional fatigue behavior of NiTi-Cu dissimilar laser welds. *Mater. Des.* **2017**, *114*, 282–287. [CrossRef]
8. Oliveira, J.P.; Fernandes, F.M.B.; Schell, N.; Miranda, R.M. Shape memory effect of laser welded NiTi plates. *Funct. Mater. Lett.* **2015**, *8*, 1550069. [CrossRef]
9. Zeng, Z.; Yang, M.; Oliveira, J.P.; Song, D.; Peng, B. Laser welding of NiTi shape memory alloy wires and tubes for multi-functional design applications. *Smart Mater. Struct.* **2016**, *25*, 085001. [CrossRef]
10. Stavridis, J.; Papacharalampopoulos, A.; Stavropoulos, P. Quality assessment in laser welding: A critical review. *Int. J. Adv. Manuf. Technol.* **2018**, *94*, 1825–1847. [CrossRef]
11. Stournaras, A.; Stavropoulos, P.; Salonitis, K.; Chryssolouris, G. Laser process monitoring: A critical review. In Proceedings of the 6th International Conference on Manufacturing Research, Uxbridge, UK, 9–11 September 2008; pp. 425–435.
12. Huang, W.; Kovacevic, R. A laser-based vision system for weld quality inspection. *Sensors* **2011**, *11*, 506–521. [CrossRef]
13. Kim, C.-H.; Ahn, D.-C. Coaxial monitoring of keyhole during Yb:YAG laser welding. *Opt. Laser Technol.* **2012**, *44*, 1874–1880. [CrossRef]
14. Gao, X.-D.; Wen, Q.; Katayama, S. Analysis of high-power disk laser welding stability based on classification of plume and spatter characteristics. *Trans. Nonferrous. Metals Soc. China* **2013**, *23*, 3748–3757. [CrossRef]
15. Purtonen, T.; Kalliosaari, A.; Salminen, A. Monitoring and adaptive control of laser processes. *Phys. Procedia* **2014**, *56*, 1218–1231. [CrossRef]
16. Zeng, H.; Zhou, Z.; Chen, Y.; Luo, H.; Hu, L. Wavelet analysis of acoustic emission signals and quality control in laser welding. *J. Laser Appl.* **2001**, *13*, 167–173. [CrossRef]
17. Du, D.; Cai, G.-R.; Tian, Y.; Hou, R.-S.; Wang, L. Automatic inspection of weld defects with x-ray real-time imaging. *Lect. Notes Contrib. Inf.* **2007**, *362*, 359–366.
18. Park, Y.W.; Park, H.; Rhee, S.; Kang, M. Real time estimation of CO_2 laser weld quality for automotive industry. *Opt. Laser Technol.* **2002**, *34*, 135–142. [CrossRef]
19. Sibillano, T.; Ancona, A.; Berardi, V.; Schingaro, E.; Parente, P.; Lugarà, P.M. Correlation spectroscopy as a tool for detecting losses of ligand elements in laser welding of aluminium alloys. *Opt Laser Eng.* **2006**, *44*, 1324–1335. [CrossRef]
20. Rizzi, D.; Sibillano, T.; Pietro Calabrese, P.; Ancona, A.; Mario Lugarà, P. Spectroscopic, energetic and metallographic investigations of the laser lap welding of AISI 304 using the response surface methodology. *Opt Laser Eng.* **2011**, *49*, 892–898. [CrossRef]
21. Konuk, A.R.; Aarts, R.G.K.M.; Veld, A.J.H.; Sibillano, T.; Rizzi, D.; Ancona, A. Process control of stainless steel laser welding using an optical spectroscopic sensor. *Phys. Procedia* **2011**, *12*, 744–751. [CrossRef]
22. Sebestova, H.; Chmelickova, H.; Nozka, L.; Moudry, J. Non-destructive real time monitoring of the laser welding process. *J. Mater. Eng. Perform.* **2012**, *21*, 764–769. [CrossRef]
23. Zaeh, M.F.; Huber, S. Characteristic line emissions of the metal vapour during laser beam welding. *Prod. Eng.* **2011**, *5*, 667–678. [CrossRef]

24. Chen, Y.; Chen, B.; Yao, Y.; Tan, C.; Feng, J. A spectroscopic method based on support vector machine and artificial neural network for fiber laser welding defects detection and classification. *NDT E Int.* **2019**, *108*, 102176. [CrossRef]
25. Zhang, Y.; You, D.; Gao, X.; Wang, C.; Li, Y.; Gao, P.P. Real-time monitoring of high-power disk laser welding statuses based on deep learning framework. *J. Intell. Manuf.* **2020**, *31*, 799–814. [CrossRef]
26. Lee, S.H.; Mazumder, J.; Park, J.; Kim, S. Ranked Feature-Based Laser Material Processing Monitoring and Defect Diagnosis Using k-NN and SVM. *J. Manuf. Process.* **2020**, *55*, 307–316. [CrossRef]
27. You, D.; Gao, X.; Katayama, S. Multiple-optics sensing of high-brightness disk laser welding process. *NDT E Int.* **2013**, *60*, 32–39.
28. CRAN. Available online: https://cran.r-project.org/web/packages/viridis/vignettes/intro-to-viridis.html/ (accessed on 4 June 2020).
29. COLOUR. Available online: https://colour.readthedocs.io/en/develop/ (accessed on 4 June 2020).

Article

Study on the Relationship between Root Metal Flow Behavior and Root Flaw Formation of a 2024 Aluminum Alloy Joint in Friction Stir Welding by a Multiphysics Field Model

Jian Luo [1,2,3,*], Jiafa Wang [2], Hongxin Lin [2], Lei Yuan [4], Jianjun Gao [2] and Haibin Geng [2,3]

1 State Key Laboratory of Advanced Welding and Joining, Harbin Institute of Technology, Harbin 150001, China
2 College of Mechanical Engineering and Automation, Fuzhou University, Fuzhou 350116, China; wjf13174517152@163.com (J.W.); linhongxin_edu@163.com (H.L.); gjj410zd@fzu.edu.cn (J.G.); genghb@fzu.edu.cn (H.G.)
3 State Key Laboratory of Solidification Processing, Northwestern Polytechnical University, Xi'an 710072, China
4 Beijing Special Vehicle Research Institute, Beijing 100072, China; yuanlei110119@126.com
* Correspondence: luojian@fzu.edu.cn; Tel.: +86-591-22866262

Received: 29 May 2020; Accepted: 2 July 2020; Published: 8 July 2020

Abstract: In friction stir welding (FSW), many defects (such as kissing bond, incomplete penetration, and weak connection) easily occur at the root of the welded joint. Based on the Levy–Mises yield criterion of the Zener–Hollomon thermoplastic constitutive equation, a 3D thermal–mechanical coupled finite element model was established. The material flow behavior and the stress field at the root area of a 6 mm thick 2024-T3 aluminum alloy FSW joint were studied. The influence of pin length on the root flaw was investigated, and the formation mechanism of the "S line" defects and non-penetration defects were revealed. The research results showed that the "S line" defect forms near the bottom surface of the pin owing to the insufficiently mixed material from the advancing side (AS) and retreating side (RS) near the weld center. The non-penetration defect forms near the bottom surface of the workpiece owing to the insufficient driving force to make the material flow through the weld center. With the continual increase of pin length, the size of the "S line" defect and non-penetration defect reduces, and finally, the defect-free welded joint can be obtained with an optimized suitable length of the pin in this case.

Keywords: friction stir welding; welding seam root; metal flow behavior; root flaw; pin length; Incomplete penetration; weak connection

---

1. Introduction

Friction stir welding (FSW) is a novel solid phase bonding technique developed by The Welding Institute (TWI) in 1991 [1]. During welding, heat input generated by the friction heat between the tool and the workpiece, and the plastic deformation of the welded metal changes the welded metal into a thermoplastic state. The plastic metal generates plastic flow and forms a closed joint under the combined action of the pin and the shoulder [2]. FSW is of high quality with a low welding heat input, no filler metal, no smoke, and no weld splash during the welding process, and it is of a green, environmentally friendly nature. FSW technology is often used for welding magnesium alloy, aluminum alloy, and other light alloy materials [3–5]. More importantly, in recent years, FSW technology shows great advantage in joining dissimilar alloys such as Al–Mg–Si/Al–Zn–Mg alloy [6], Al/NiTi alloy [7], and Al/Steel [8]. However, FSW involves the changes of temperature, adhesive shear force, and metal flow behavior, which is a complex thermal–fluid coupled process. The root of the welded joint often forms a weak connection or kissing bond of "S" line and non-penetration defects due to insufficient fluidity of plastic materials and heat input.

The weak-connection defect at the root of the weld has a great influence on the mechanical properties of the FSW welded joint. Weak connection defects will reduce the plasticity of the joint. When there is a weak connection defect at the root of the weld, the fracture mode in the tensile test will be brittle. For well-formed welded joints, the fracture mode is plastic [9–11]. Jolu [12–14] investigated the effect of root flow on the tensile and fatigue behavior. They found that the root flaw acts as a crack initiation site during the tensile test, which reduces the yield strength of the welded joint by 40% and the ultimate tensile strength by 20%. In the fatigue life test, they found that the tilting angle of the weak connection can influence the fatigue life of the welded joint. When the weak-connection defect is strongly tilted with respect to the loading direction, the fatigue life of the joint is higher. Zhou [15] studied the effect of weak connection defects on the fatigue strength of the FSW joint. They found that when the strain rate is 0.1/s, the fatigue life of intact weld seams is 21 to 43 times longer than that of the weld seams with weak connection defects at the root, which is owing to the fact that fatigue cracks can generate from the weak connection line, resulting in a great reduction of the fatigue life of the weld. Kadlec [16] found that the size of the root flaw significantly affects the fatigue life of the welded joint. When the size of root flaw is 315 um, it does not affect the fatigue strength of the connection. However, the 670 um size weak connection defect reduces the fatigue life of the joint to 91% of the base metal.

Due to the influence of root defects on the mechanical properties of FSW joints, many scholars began to pay attention to the formation mechanism of root defects and the relationship between the root defects and process parameters. Sato [10,17] and Okamura [18] studied the weak connection of the FSW welded joint of aluminum alloy. They found that "S line" defects originated from oxide film on the butt surface of the workpiece. During the FSW process, broken oxide particles form a black flow trace originated from the retreating side and extending to the advancing side. "S" shape weak connection lines distribute continuously in the welding direction, forming a weak bonding surface. Chen [19] used the numerical simulation method to quantitatively research the bonding behavior of the FSW joint. The numerical simulation results show that the weak connection defect is caused by the insufficient fluidity of the plastic material and heat input at the root of the welding seam, which leads to insufficient bonding pressure at the butt surface of the weld. Luo [20] studied material flow in the FSW process based on a computational fluid dynamic model; they found that the ratio of rotation speed to welding speed has an effect on the material flow behavior around the stirring pin, which is related to the defect formation in the FSW welded joint. Moussawi [21] studied the friction stir welding of DH36 and EH46 steels at different welding speeds and rotation speeds. They found that the high-speed movement of the stirring pin will result in insufficient fluidity of the weld material and weak connection defect at the root of the welded joint. Hou [22] and Zhou [23] found that the root flaws occur under high welding speed and low rotation speed owing to the insufficient heat input and material flow.

To eliminate the root flaw of the FSW welded joints, some scholars adopted the improved FSW method such as electricity-assisted friction stir welding (EAFSW) [24–26], bobbin tool friction stir welding (BTFSW) [27–29], and ultrasonic-assisted friction stir welding (UAFSW) [30,31], although the complex design of the stirring tool or the additional equipment induced by these methods increases the cost of the FSW process. In traditional FSW, some scholars believe that the root defects are caused by the improper selection of the pin length. Since the length of the pin has an important influence on the welding heat input, it also determines the forging behavior and extrusion pressure of the tool on the plastic material in the welding process [32]. However, if the pin is too long, its length will exceed the thickness of the base metal, which makes the backing plate adhere to the base material. If the pin is too short, a non-penetration defect will appear at the bottom surface of the workpiece [33]. Mandache [34] investigated the effect of pin length on the formation of defects at the root of the welding seam. They found that with the increase of the pin length, the size of the non-penetration defect decreases gradually, but the size of the defect of the weak connection fluctuates in a wavy manner.

Weak connection and non-penetration defects at the root of the weld have seriously affected the quality of the weld, but the existing studies have not revealed the formation mechanism of defects at the root of the weld from the flow mechanism, and there is no in-depth study on the mechanism of eliminating defects at the root with the length of tool pin. In this paper, the metal flow behavior at the root of the weld and the formation mechanism of weak connection defects were studied by the numerical simulation method, the influence of different tool pin lengths on the plastic metal flow behavior at the root of the weld was investigated, and the optimized suitable length of the pin with a non-defects welded joint was identified.

2. Models

2.1. Geometry and Material Model

A 2024-T3 aluminum alloy plate with a length of 120 mm, a width of 35 mm, and a thickness of 6 mm was used for the geometry model. The shoulder diameter of the stirring tool is 16 mm with concentric ring grooves. The diameter of the top and root of the pin is 6 mm and 3 mm, respectively. The length of the pin is L, which changes from 5.8 to 6.0 mm. The conical angle θ changes from 14.5° to 14° with the pin length changing from 5.8 to 6.0 mm. Figure 1 shows the diagram of the workpiece and the stirring tool with the corresponding size.

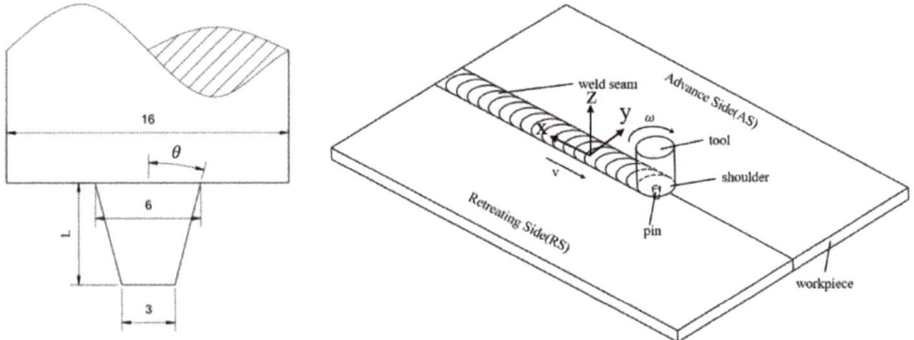

Figure 1. Schematic diagram of tool size and welding sample.

The chemical composition of 2024-T3 aluminum alloy is shown in Table 1. The thermal conductivity, specific heat capacity, thermal expansion coefficient, elastic modulus, Poisson's ratio, and other physical parameters of aluminum alloy 2024 are shown in Figure 2.

Table 1. Chemical composition of aluminum alloy 2024.

Chemical Component	Si	Fe	Cu	Mn	Mg	Ni	Zn	Ti	Al
Content/wt %	0.5	0.5	3.8	0.3	1.3	0.1	0.3	0.09	Bal

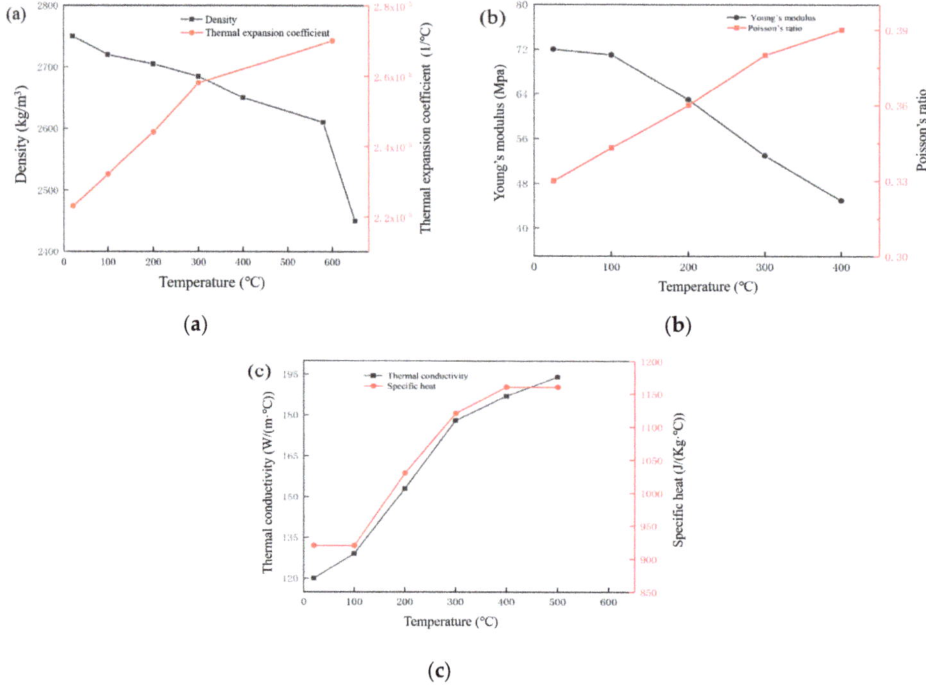

(a) (b)

(c)

Figure 2. Thermophysical properties of aluminum alloy 2024 with temperature change. (**a**) Density and thermal expansion coefficient; (**b**) Young's modulus and Poisson's ratio; (**c**) thermal conduction coefficient and specific heat capacity [35].

2.2. Heat Generation Model

In the process of FSW, welding heat input includes three parts: (1) friction between the shoulder and the workpiece; (2) friction between the pin and the workpiece; and (3) plastic deformation of the material. According to the minimum torque required by the tool to overcome the rotation of friction, the heat produced by friction on the shoulder, the heat produced by friction on the side of the pin, and the heat produced by friction on the bottom of the pin can be calculated as:

$$Q_{shoulder} = \frac{4}{3}\pi^2 \mu PN\left({R_s}^3 - {R_2}^3\right) \tag{1}$$

$$Q_{pin-side} = \frac{4\pi^2 N\mu P}{3sin\alpha}\left(R_2^3 - R_3^3\right) \tag{2}$$

$$Q_{pin-bottom} = \frac{4}{3}\pi^2 \mu PNR_3^3 \tag{3}$$

where $Q_{shoulder}$, $Q_{pin-side}$, $Q_{pin-bottom}$ represent heat generation at the shoulder, heat generation at the side of the pin, and heat generation at the bottom of the pin, respectively. μ is the friction coefficient, P is the welding pressure, N is the rotating speed, R_s is the shoulder radius, R_2 is the root radius of the pin, R_3 is the end radius of the pin, and α is the cone angle of the pin.

In the process of FSW, the material is subjected to severe plastic deformation. Most of the deformation work of the material is transferred to the surrounding area in the form of heat energy. This part of the deformation work accounts for about 90% of the deformation work, while the rest of the work is used in the plastic deformation of the material. The heat source density generated by plastic deformation is:

$$q_p = \overline{\sigma}\alpha_p\dot{\overline{\varepsilon}} \tag{4}$$

where q_p is the heat source density of the plastic deformation of the material, $\overline{\sigma}$ is the equivalent stress, α_p is the heat conversion efficiency, which is set as 0.9, and $\dot{\overline{\varepsilon}}$ is the equivalent strain rate.

Therefore, the heat produced by plastic deformation of metal materials is:

$$Q_v = \int_v \overline{\sigma}\alpha_p\dot{\overline{\varepsilon}}dV \tag{5}$$

where Q_v represents the plastic deformation heat production of the material with volume V.

2.3. Material Flow Model

A welding material is defined as a viscoplastic material. The mechanical response of viscoplastic materials can be defined by the constitutive relationship, hardening law, yield criterion, and flow rule. This paper has some reasonable assumptions without affecting the research results. First, the welding stage studied is a steady state—that is, the heat production and material flow reach a stable state. Second, the plastic softening material is a non-Newtonian, incompressible viscoplastic fluid. Third, elastic strains are negligible.

2.3.1. Flow Rule

The flow rule defines the change rate of various strain rate components during plastic deformation of the material, which is expressed in Equation (6):

$$\varepsilon_{ij}^p = \lambda\frac{\partial g}{\partial\sigma_{ij}} \tag{6}$$

where g is a scalar function of the invariants of the deviating stress and is called plastic potential, and λ is a positive proportionality constant. When g is equal to the yield function $f(\sigma_{ij})$, Equation (6) is the associated flow rule and can be rewritten as Equation (7) [24]:

$$\frac{\dot{\varepsilon}_x^p}{\sigma_x'} = \frac{\dot{\varepsilon}_y^p}{\sigma_y'} = \frac{\dot{\varepsilon}_z^p}{\sigma_z'} = \frac{\dot{\gamma}_{xy}^p}{2\tau_{xy}} = \frac{\dot{\gamma}_{yz}^p}{2\tau_{yz}} = \frac{\dot{\gamma}_{zx}^p}{2\tau_{zx}} \tag{7}$$

where $\dot{\varepsilon}_x$ can be expressed in Equation (8):

$$\dot{\varepsilon}_x = \{\sigma_x - \frac{1}{3}(\sigma_x + \sigma_y + \sigma_z)\}\dot{\lambda}. \tag{8}$$

The $\dot{\varepsilon}_y$ and $\dot{\varepsilon}_z$ have similar equations. Similarly, the shear strain rate can be expressed as:

$$\dot{\varepsilon}_{xy} = \frac{\dot{\gamma}_{xy}}{2} = \tau_{xy}\dot{\lambda}. \tag{9}$$

The $\dot{\varepsilon}_{yz}$ and $\dot{\varepsilon}_{zx}$ have similar equations. The scaling factor in Equations (8) and (9) can be derived from the work-hardening criterion as follows:

$$\dot{\lambda} = \frac{3\dot{\overline{\varepsilon}}}{2\overline{\sigma}} \tag{10}$$

where effective strain rate epsilon $\dot{\bar{\varepsilon}}$ is:

$$\dot{\bar{\varepsilon}} = \sqrt{\frac{2}{3}\{\dot{\varepsilon}_{ij}\dot{\varepsilon}_{ij}\}}. \tag{11}$$

2.3.2. Constitutive Law

The constitutive law of the material describes the relationship between the flow stress of the material and the plastic deformation, the strain rate, and the temperature, which can be summarized by the functional equation:

$$\sigma = f(\dot{\varepsilon}, T) \tag{12}$$

where σ is the rheological stress in the plastic deformation process of material; $\dot{\varepsilon}$ is the strain rate; and T is the deformation temperature. Rheological stress can be calculated by the Zener–Hollomon formula. The specific expression is:

$$\sigma(T, \dot{\varepsilon}) = \sigma_p \sin h^{-1}\left[\left(\frac{Z}{A}\right)^{\frac{1}{n}}\right] \tag{13}$$

where, A, σ_p, and n are the material parameters, and Z is the Zener–Hollomon parameter. The specific expression is:

$$Z = \dot{\varepsilon} exp\left(\frac{Q}{RT}\right) \tag{14}$$

where Q is the activation energy independent of temperature and R is the gas constant. The specific parameters are shown in Table 2.

Table 2. 2024 aluminum alloy material parameters [36].

Parameter	Value
A	2.29×10^{11} s^{-1}
n	5.46
Q	178.0 kJ/mol
σ_p	47.7 MPa
R	8.314 J/mol·K

2.4. Boundary Conditions

2.4.1. Thermal Boundary Condition

The initial temperature of the workpiece, the tool, and the environment is defined as 25 °C. During the welding process, the heat transfer between the workpiece, tool, and the environment includes convection and radiation heat transfer, as shown in Equation (15). This heat transfer mode is defined on all surfaces except the root and butt surfaces of the workpiece. The heat transfer between the workpiece and the tool is defined as convective heat transfer, as shown in Equation (16). The heat transfer between the workpiece and the bottom plate is also defined as convection, as shown in Equation (17).

$$Q_{wa} = \sigma_b \varepsilon_b \left(T_w^4 - T_a^4\right) + h_a(T_w - T_a) \tag{15}$$

$$Q_{wt} = K_{wt}\frac{\partial T}{\partial z} = h_{wt}(T_w - T_t) \tag{16}$$

$$Q_{wb} = K_{wb}\frac{\partial T}{\partial z} = h_{wb}(T_w - T_b) \tag{17}$$

2.4.2. Mechanical Boundary Condition

All degrees of freedom of the workpiece are constrained to prevent rigid displacement and rotation of the workpiece during welding. Therefore, the velocity in the z direction of the bottom surface of the workpiece is defined to be 0. The velocity in x and y directions on the side of the workpiece is 0. The tool is defined to rotate around the z-axis in the positive direction—that is, counter-clockwise. When the welding is at the plunging stage, the speed of the tool is along the z-axis. The speed of the tool is in the x-direction during welding.

2.5. FEM Computational Mesh

In the process of FSW, large plastic deformation occurs in the welding area, and the welding temperature field and stress and strain field change violently. In order to truly simulate the stirring action of the workpiece material in the welding process and visualize the plastic flow behavior of the material, the grid size of the welding seam area must be small enough. However, in the whole welding process, the welding parts are larger, and if the whole workpiece meshes with the same grid size, the calculation will be very large, which will seriously reduce the simulation efficiency. Therefore, in this model, the workpiece grid is divided into two types of grid: free quadrilateral grids used for welding plates, and free triangular grids used for the area around the stirring tool. In order to further reduce the calculation amount while ensuring the calculation accuracy, only the grid around the stirring tool is refined. In local grid refinement, through the absolute size control grid refinement of the grid size, the minimum grid size is set to 0.004 mm, and the largest size is set to 1.4 mm, so the accuracy of the grid is sufficient to reflect the plastic flow behavior of materials, the temperature field, and stress–strain field of the weld area. The grid size of the remaining area of the workpiece is set to 2.4 mm to improve computing efficiency. In this paper, the effect of the back plate on the workpiece is considered through mechanical and thermal boundary conditions, so the backing plate is not meshed to improve computing efficiency. The mesh partitioning of the model is shown in Figure 3.

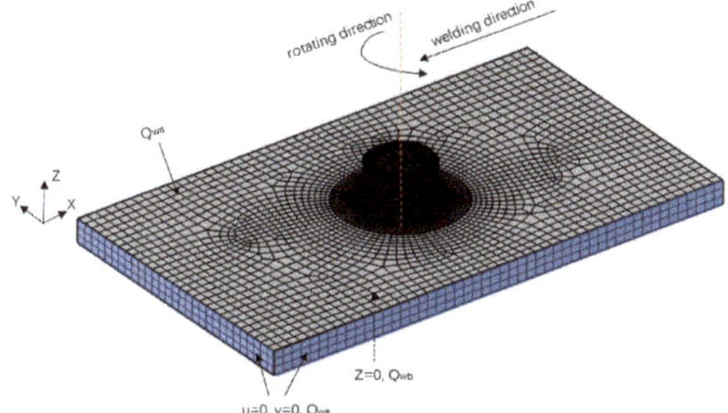

Figure 3. Mesh of the simulation domain.

2.6. Model Validation

To verify the established FSW model, a validation simulation was conducted according to the previous experiment [33]. In the validation simulation, the size of workpiece, technological parameters, and material are the same as those in the referenced experiment. The accuracy of the established FSW model was studied by comparing with the actual heat cycle curve of the measurement points located 8 mm from the weld center under different welding speeds. Figure 4 shows the heat cycle curve of experiment and simulation. By comparing the results between simulation and experiment, it can be seen that the heat cycle curves obtained by this model are basically consistent with those obtained by the actual welding test. Therefore, this model is accurate and feasible.

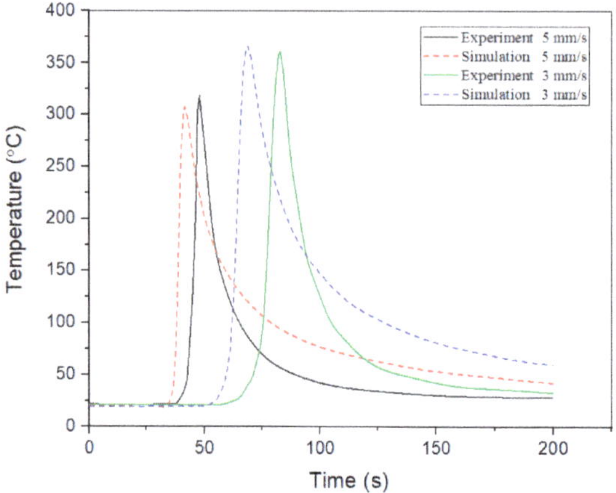

Figure 4. Comparison results of thermal cycling curves at different welding speeds obtained by simulation and experiment.

3. Results and Discussions

3.1. Analysis of Fluidity at the Bottom of the Pin

3.1.1. Velocity Distribution

Figure 5 shows the "S line" and non-penetration defect formed at the root of the weld due to insufficient plastic material flow. The material at the root of the weld can be divided into two different zones according to the flow behavior. As can be seen from Figure 5, at Zone I, the materials flow parallel to the bottom of the pin, and the materials on the advancing side (AS) and retreating side (RS) flow in the opposite direction and mix on the RS side. However, materials on both sides fail to be fully mixed, resulting in the appearance of a weak connection "S line" at the root of the weld. As for Zone II, near the weld center, all materials flow in the same direction (RS→AS), while almost all materials stop at the weld center, leading to the "vertical line" shape defect, i.e., non-penetration defect. In fact, the area under the pin bottom is called SWZ (the Swirl Zone) in Zeng's work [37], which refers to Zone I + Zone II in this paper. In their experimental research of the root flaws in a 6 mm 2014Al-T6 FSW joint under a pin length of 5.73 mm, the "S line" defects appear at the weld root and extend to the stirring zone under a wide range of process parameters. The shape of the "S line" defects obtained in this paper in Figure 5 is similar with that observed in Figure 3 of Zeng's work. A similar root flaw was also observed in a 6 mm DH36 steel FSW joint under a pin length of 5.7 mm in Al-Moussawi's work [21].

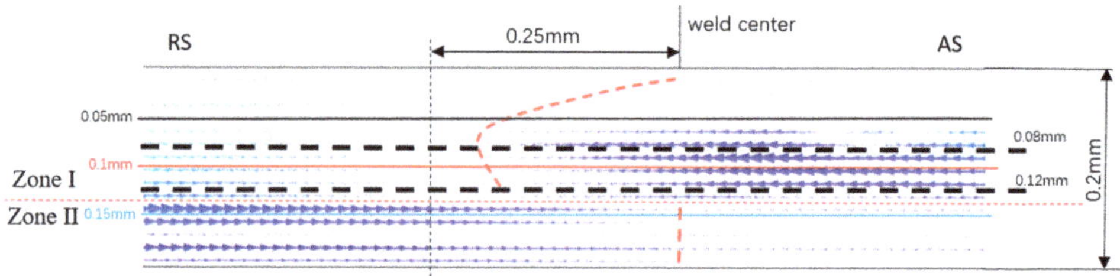

Figure 5. Simulation result of the plastic flow at the root of the weld.

To explore the fluidity of plastic metal at the root of the FSW joint, the velocity distribution of the plastic material in Figure 5 was measured under different depths (0.05, 0.1, and 0.15 mm under the pin). Figure 6a–c shows the measurement results for the velocity distribution in the x, y, and z-directions, respectively. The horizontal axis represents the distance to the center of the weld, in which the positive axis is AS, the negative axis is RS, and the vertical axis represents the partial velocity.

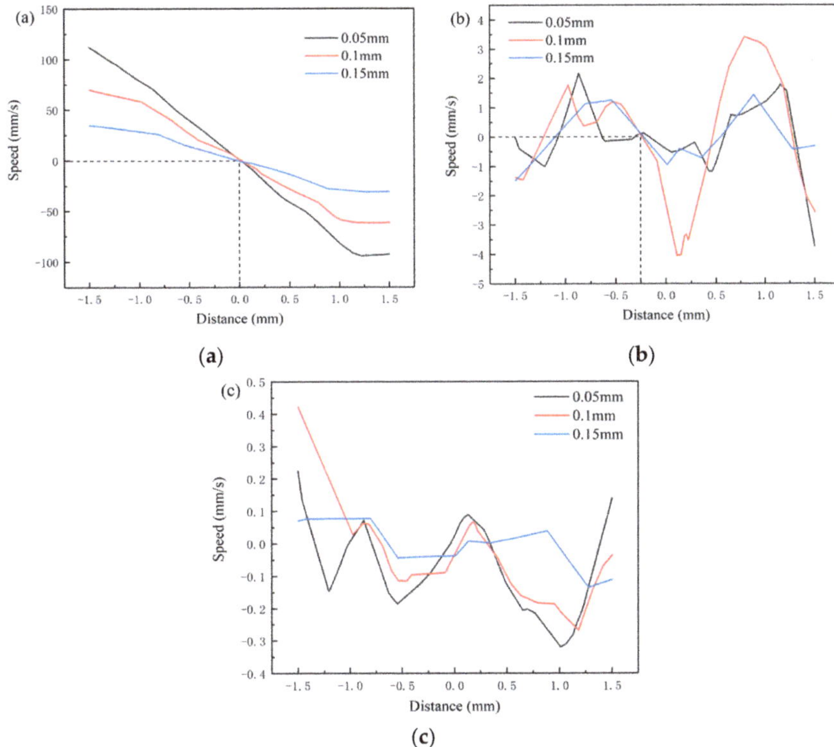

Figure 6. Speed curves of different distances from the bottom of the pin (1000 rpm, 120 mm/min, 5.8 mm pin length). (**a**) Velocity in the x-direction; (**b**) velocity in the y-direction; (**c**) velocity in the z-direction.

In Figure 6a, it can be seen from the figure that the velocity of the plastic metal in the x-direction takes the center of the weld (y = 0) as the boundary line, advancing side (AS) plastic metal flows in the welding direction (v < 0), and retreating side (RS) plastic metal flows in the opposite direction of the welding (v > 0). The closer it is to the end face of the pin (0.05 mm), the better the fluidity of the plastic metal in the x-direction and the faster the flow speed.

As can be seen in Figure 6b, near 0.25 mm from the weld center on the RS side, the transverse flow velocity at different depths is close to 0, indicating that the plastic flow in this area is poor. As a result, weak connection defects can easily form near the center line of the weld due to insufficient material fluidity. However, at the middle position between the end of the pin and the bottom of the workpiece (0.1 mm), the fluidity is relatively good. AS and RS plastic metal move in the opposite direction toward the center of the weld line and meet at RS (x = −0.25).

In addition, the maximum velocity of the plastic metal at the root of the weld is only 0.45 mm/s in the depth direction (z direction), as shown in Figure 6c. Therefore, the plastic metal at the root of the weld basically has almost no flow in the depth direction. It can also be found that the velocity distribution in Zone II is always the smallest in three directions compared with zone I, indicating the poorer fluidity of material in Zone II.

3.1.2. Driving Force Distribution

At the bottom of the pin, the pressure between the plastic metal is an important driving force for the flow of the plastic metal. Furthermore, the shear action of the tool against the plastic metal is also one of the driving forces for the flow of the plastic material. Therefore, the combined effect of pressure and shear stress on the plastic material directly affects the fluidity of the material. In order to investigate the distribution of the pressure and shear stress, a series of measurement points were set at the bottom of the pin, as shown in Figure 7. In Figure 7, the distance from the end face of the pin is 0.02 mm, 0.05 mm, 0.08 mm, 0.1 mm, 0.12 mm, 0.15 mm, and 0.18 mm, respectively.

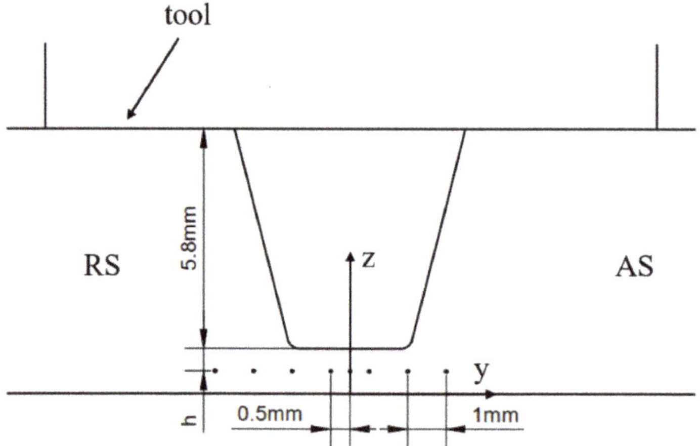

Figure 7. Distribution of test points at the root of the weld.

Figure 8a shows the distribution of pressure at the root of the weld. In Figure 8, as the depth increases, the pressure reduces. Furthermore, there is a low-pressure zone near the center line on the RS side, and the lowest pressure is reached at y = −0.5. The uneven distribution of pressure near the center line causes the plastic metal to flow from both sides of AS and RS to the center of the weld, which is consistent with the velocity distribution in Figure 5. However, at the low-pressure zone, the pressure is not enough to fully mix the material from both sides of AS and RS, leading to the weak connection.

Figure 8b shows the distribution of shear stress at the root of the weld. In Figure 8b, near the weld center line (y = 0), it can be found that the shear stress is close to 0 when the distance from the bottom surface of the pin is too short (0.02–0.05 mm) or too long (0.15–0.18 mm). However, when the distance is moderate (0.08–0.12 mm), the shear stress reaches a relatively high level. Therefore, plastic metal near the bottom surface of the pin and the bottom surface of the workpiece has poor fluidity, while the plastic metal between these has strong fluidity. According to Figure 5, it can be found that the location of the left-leaning line of the "S line" is at the depth of 0.08–0.12 mm, and this is consistent with the depth range of the high shear stress, which indicates that the formation of the left-leaning line is related to the change of the shear stress. In Figure 8c, it can also be found that the difference of shear stress between y = −0.5 and y = 0 is smaller when the depth is in the range of 0.08–0.12 mm. Furthermore, combined with Figure 8a,b and Figure 5, it can be found that at the Zone II (h > 0.15 mm), the pressure and shear stress are relatively lower than that at Zone I, which is consistent with the velocity distribution at Zone II, as shown in Figure 5.

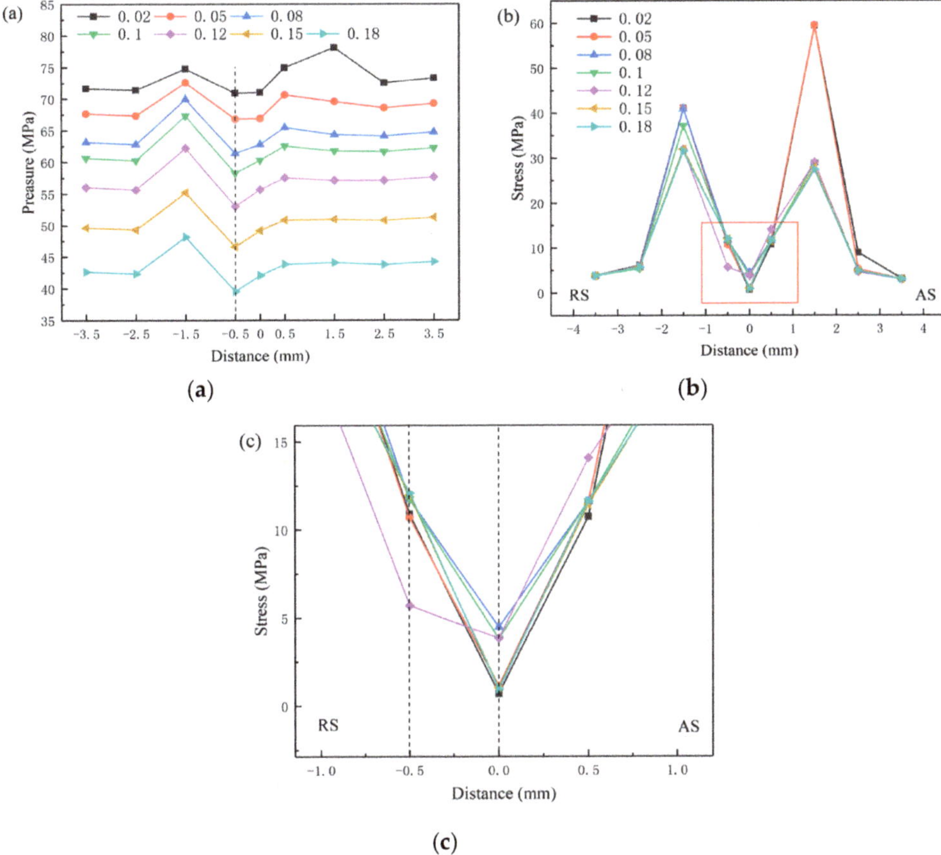

Figure 8. Distribution of the driving force at a different distances from the end face of the pin (1000 rpm, 120 mm/min). (**a**) Pressure distribution; (**b**) shear stress distribution; (**c**) partial enlargement of (b).

3.2. *Effect of Pin Length on the Formation of Root Flaw*

3.2.1. Effect of Pin Length on the Material Flow

In order to solve the defects of a weak connection and non-penetration at the root of FSW joints of aluminum alloy, the relationship between welding allowance and root flaws was investigated. Figure 9 shows the velocity distribution under a pin length of 5.8, 5.85, 5.9, 5.95, and 6.0 mm, respectively. Figure 10 shows the velocity distribution in the y-direction at the half depth between the bottom surface of the pin and the bottom surface of the workpiece.

When the length of the pin is 5.8 mm or 5.85 mm, the "S line" defect appears at the root of the FSW weld seam, as shown in Figure 9a,b. As the pin length increases to 5.9 mm, the "S line" shape defect changes to the "right-tilting line" shape defect (Figure 9c), and finally the "S line" defect is eliminated when the pin length is 5.95 mm (Figure 9d). In Figure 10, as the pin length increases from 5.8 to 5.95 mm, the velocity difference between RS and AS near the weld center (Figure 10 mark I, II) becomes larger, which indicates a larger driving force, leading to the material being more fully mixed near the weld center.

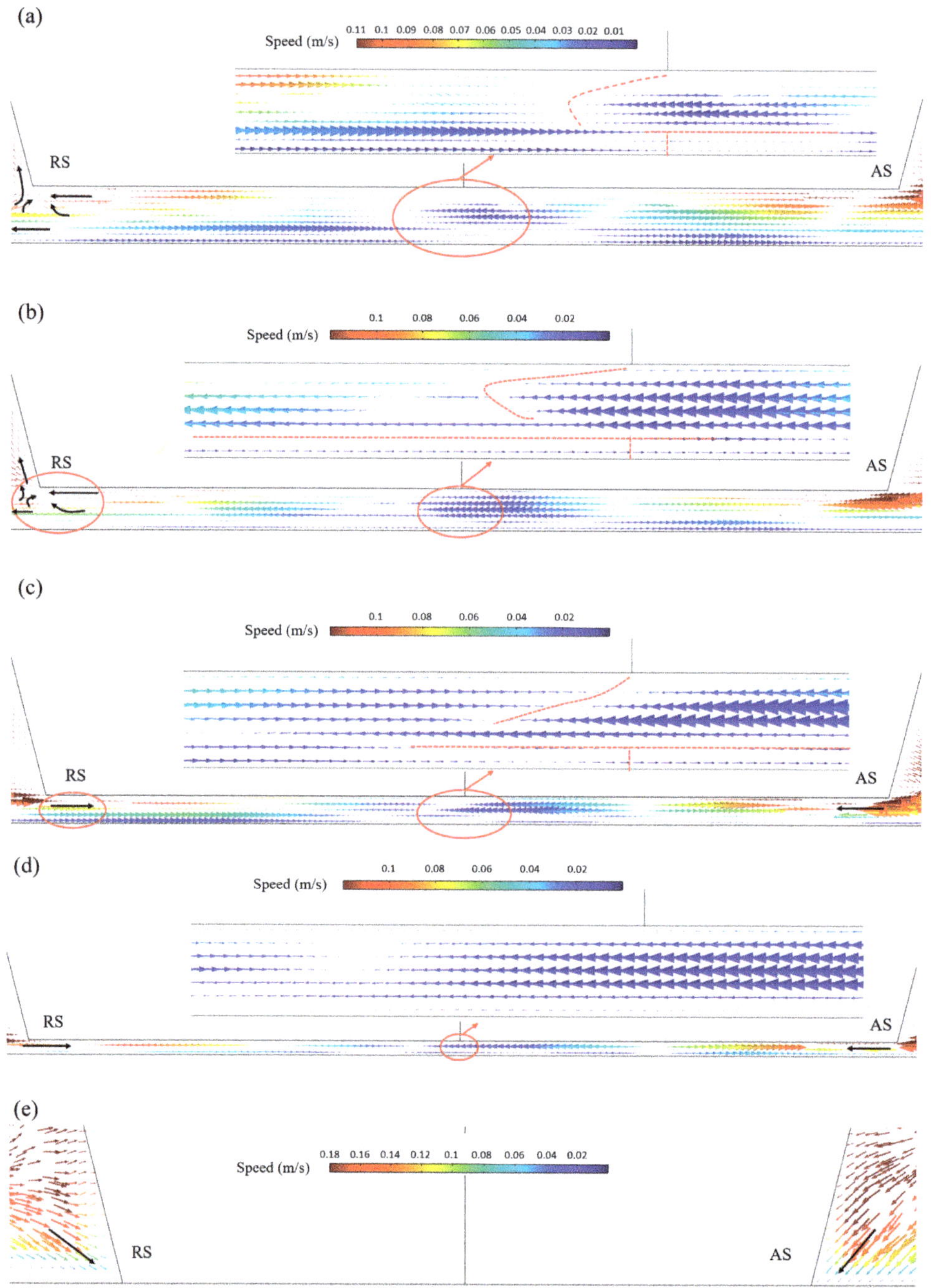

Figure 9. Effect of pin length on plastic metal flow at the root of the weld (1000 rpm, 120 mm/min). (**a**) 5.8 mm; (**b**)5.85 mm; (**c**) 5.9 mm; (**d**) 5.95 mm; (**e**) 6.0 mm.

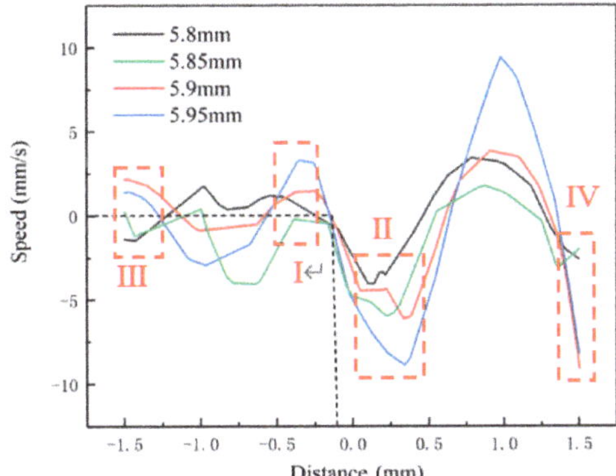

Figure 10. Velocity distribution in the y-direction in Figure 5 (1000 rpm, 120 mm/min, 5.8 mm pin length).

It should be noted that the material flow formation near the weld center changes when the pin length increases. When the pin length increases from 5.8 to 5.85 mm, near the weld center, the material flow direction changes from RS→AS to AS→RS under the "S line", which causes more material transfer from AS to RS, and the "S line" further extends to the RS side. When the pin length increases to 5.9 mm, the "S line" changes to the "right-tilting line", because the material on the right side of "left-tilting line" all flows to the RS side. Finally, the material on the right side of the "right-tilting line" all flows to the RS side, and the defect-free welded joint is obtained when the pin length is 5.95 mm.

On the other hand, the material flow form on the margin of the pin also significantly changes when the pin length increases. As the pin length increases from 5.85 to 5.9 mm, the material flow on the left margin of the pin changes from vortex flow (marked red circle in Figure 9b) into the transverse flow (marked red circle in Figure 9c), and the flow direction changes from left to right, which can also be observed in Figure 10 mark III. It can also be found that the material flow form on the right margin of the pin remains in transverse flow when the pin length increases, while the velocity increases significantly as shown on the mark IV in Figure 10.

When the pin length is equal to the thickness of the workpiece, there is no welding allowance, and the welding plate does not exist in weak connections and non-penetration defects; as can be seen from Figure 9e, the plastic materials on both sides flow to the root of the weld. In this case, the pin directly contacts with the backing plate, which is easy to weld the workpiece together with the working table, and the bottom of the welding plate is poorly formed. Therefore, zero welding allowance is not recommended in practical production.

When the pin length is 5.9 mm, there is still a weak connection and non-penetration defects at the root of the weld. Therefore, the plastic metal flow behavior near the root of the weld center was investigated in detail when the pin length was 5.91–5.95 mm, as shown in Figure 11. When the pin length increases from 5.91 to 5.93 mm, the "S line" defects (red dashed line in the Figure 11) and non-penetration defects still exist. It should be noted that with the increase of the pin length, the width of the "S line" and the angle between the tangent line of the "S line" and the bottom surface of pin gradually reduces, as shown in Figure 11a–c, which may be attributed to the increased shear stress (see the Section 3.2.3) and speed difference between the RS and AS sides near the weld center (Figure 12 mark I, II). When the pin length increases to 5.94 mm (Figure 11d), the plastic metal flows through the center line under the end surface of the pin, which eliminates the "S line" defect. However, the fluidity near the bottom surface of the weld is still insufficient, so that the non-penetration defect still exists. The non-penetration defect is finally eliminated when the pin length is increased to 5.95 mm (Figure 11e).

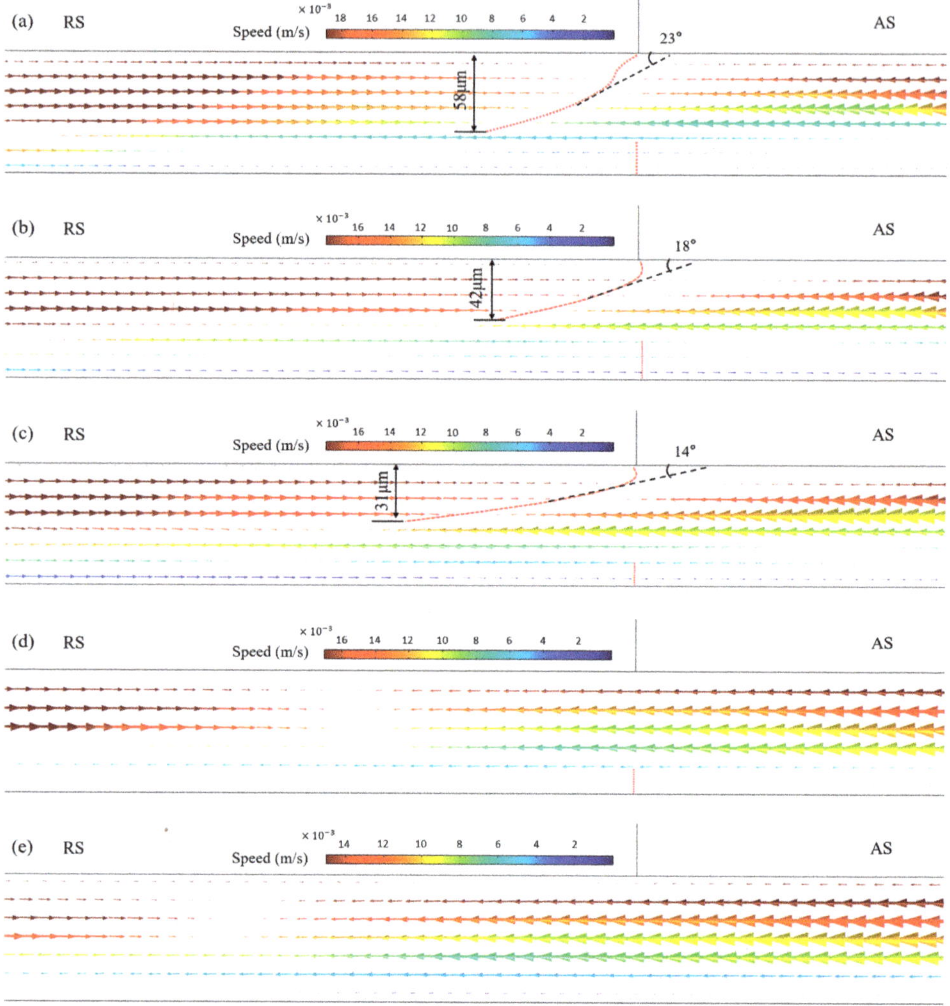

Figure 11. Influence of root welding allowance on plastic metal flow behavior at the root of the welding seam (1000 rpm, 120 mm/min). (**a**) 0.09 mm; (**b**) 0.08 mm; (**c**) 0.07 mm; (**d**) 0.06 mm; (**e**) 0.05 mm.

3.2.2. Effect of Pin Length on the Heat Input

Figure 13 shows the temperature of the top and bottom surface of the weld under different pin lengths. It is obvious from the figure that the welding heat input is basically unchanged when the pin length changes in a small range (0.2 mm). Therefore, the temperature of the plastic metal at the root of the weld is basically unchanged when welding with different lengths of the pin.

We continually increased the length of the pin by a step of 0.01 mm and studied the plastic metal flow at the bottom of the weld with the length of the pin at 5.96 mm, 5.97 mm, 5.98 mm, and 5.99 mm. Figure 14 shows the temperature field distribution of the weld cross-section when the pin length exceeds 5.95 mm. As can be seen from the Figure 14, when the pin length is 5.95 mm, the highest temperature of the weld (423 °C) appears on the upper surface of the weld, as shown in Figure 14a. However, when the pin length is greater than 5.95 mm, the maximum temperature of the weld appears at the center of the bottom of the weld. When the pin length increases from 5.96 mm to 5.99 mm, the maximum temperature increases from 439 to 481 °C, as shown in Figure 14b–e [38]. As aluminum alloy is a low-melting alloy, if the weld temperature is too high, the recrystallization of the weld will be coarsened. According to the Hall–Petch principle, the coarsening of the weld will reduce the joint strength.

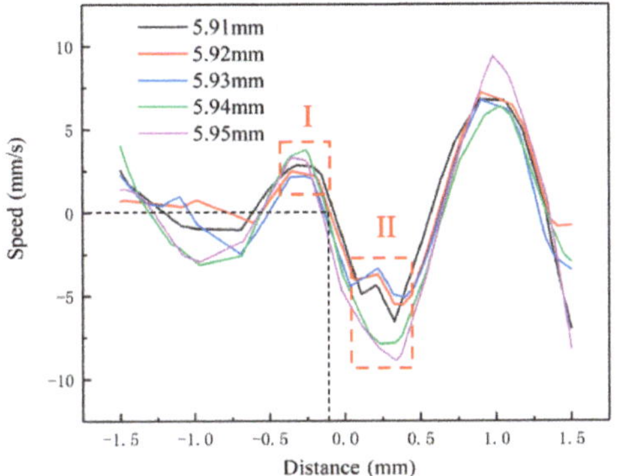

Figure 12. Flow velocity curves of plastic metal at the root of joint with different pin lengths (1000 rpm, 120 mm/min). (**a**) 0.05 mm per step from 5.8 mm to 5.95 mm; (**b**) 0.01 mm per step from 5.91 mm to 5.95 mm.

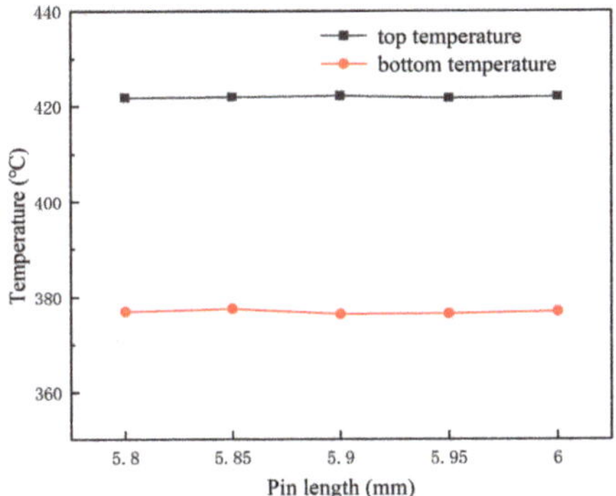

Figure 13. Influence of pin length on temperature field of the top and bottom surface of the weld (1000 rpm, 120 mm/min).

Although the temperature at the bottom of the weld increases with the increase of the length of the pin, when the length of the pin exceeds 5.95 mm, the flow velocity of the plastic metal at the bottom of the weld does not increase accordingly, as shown in Figure 15 (within the two black dashed lines). At the center of the weld, the flow rate of the plastic metal is basically the same, about 5 mm/s. Therefore, when the length of the pin exceeds 5.95 mm, further increasing the length of the pin will not improve the flow capacity of the plastic metal at the bottom of the joint, but it will sharply increase the temperature at the bottom of the weld, leading to the coarsening of the recrystallization tissue and the reduction of the joint strength. On the other hand, if the pin is too long, it will pass through the whole thickness of the workpieces and directly contact with the hard backing plate or the FSW machine substrate. In the friction stir welding process, the two will produce intense friction and wear, resulting in the rapid damage of the pin and the FSW machine substrate. At the same time, the pin itself also changes significantly in length, shape, and structure due to rapid friction and wear, and the stable FSW welding process becomes difficult to control, which leads to the deterioration of welding forming and affects the quality of welded joints greatly. Therefore, for 6 mm thick 2024 aluminum alloy, a 5.95 mm pin is most suitable for welding, which can eliminate the "S line" and non-penetration defects at the bottom of the welding seam.

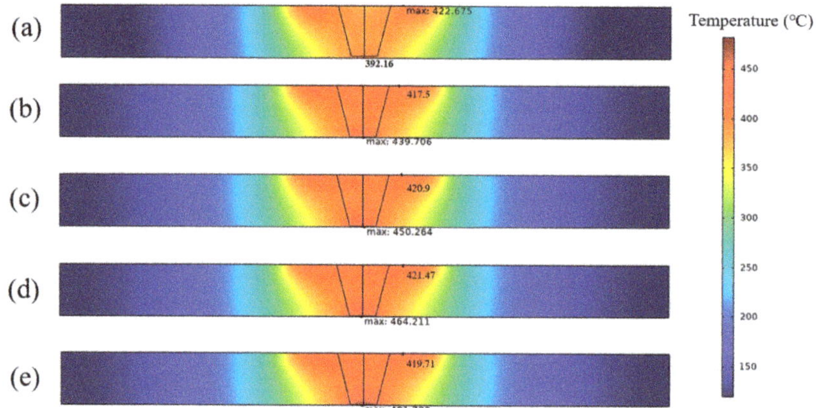

Figure 14. Temperature field of weld cross-section at different lengths of pin (1000 rpm, 120 mm/min, 6 mm thick workpiece) (**a**) 5.95 mm; (**b**) 5.96 mm; (**c**) 5.97 mm; (**d**) 5.98 mm; (**e**) 5.99 mm.

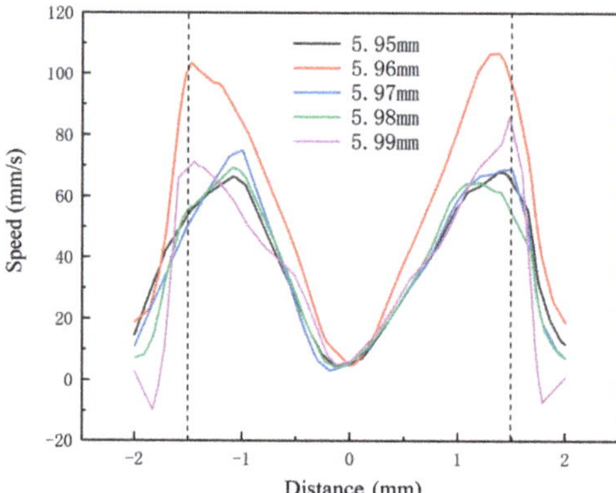

Figure 15. Distribution curve of flow velocity of plastic metal at the bottom of weld seams with different lengths of pin (1000 rpm, 120 mm/min).

In fact, studying the different pin lengths is not only necessary for the engineering design and use of the stirring pin; more importantly, the pin length itself is varied in the actual FSW process owing to the wear of the stirring pin. Hence, the material flow behavior under the pin bottom is also influenced by the change of pin length. Therefore, studying the different pin lengths is to consider the inevitable wear of the stirring pin in the actual FSW process and its effect on the formation of the FSW joint, which is of great significance. The issues related to the wear of the stirring pin will be discussed in our research work and papers in the future.

3.2.3. Effect of Pin Length on Stress Field

Although slight adjustment of the length of the pin can increase the temperature of the weld root, it has little effect on the improvement of material fluidity near the weld center, as outlined in Section 3.2.2. However, the shear action of stirring on plasticized metal will change as the length of the pin increases, as shown in Figure 16. At the root of the weld, the shear stress near the center of the weld is the lowest, which indicates that the plastic metal flow in this area is the weakest. When the length of the pin is increased from 5.91 mm to 5.95 mm with 0.01 mm per step, the shear effect of stirring on the plastic metal at the root of the weld would be gradually strengthened. Therefore, the fluidity of the plastic metal at the root of the weld also increases, which results in the gradual improvement of the forming quality at the root of the weld. When the length of the pin was 5.95 mm, a weld without root defects could be obtained.

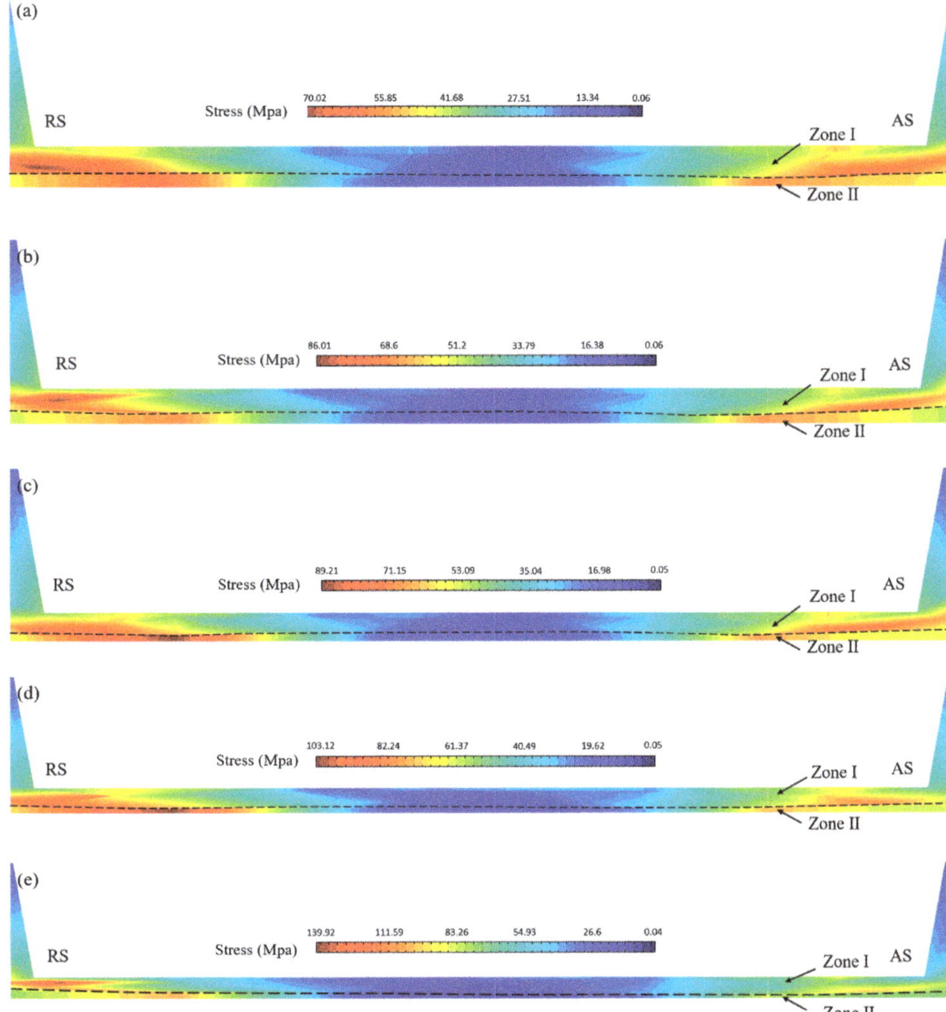

Figure 16. Distribution of shear stress at the root of the joint at different pin lengths (1000 rpm, 120 mm/min). (**a**) 5.91 mm; (**b**) 5.92 mm; (**c**) 5.93 mm; (**d**) 5.94 mm; (**e**) 5.95 mm.

It should be noted that there is a sharp transition of the shear stress near the bottom surface of the workpiece, as shown by the black dash line in Figure 16. The sharp transition of the shear stress illustrates that the material flow behavior in zones I and II (marked in Figure 16a) is different. In fact, this can be explained by the difference of the constraint condition and the driving force of the material in zones I and II. In zone I, the material is constrained by the bottom surface of the pin and the material in zone II. The material flow in zone I is driven by the rotation of the bottom surface of the pin, and the pressure and shear stress are all higher in zone I according to Figure 8a,b. Meanwhile, in zone II, the material is constrained by the bottom surface of the workpiece and the material in zone I. The material flow in zone II is driven by the material flow in zone I, and the pressure and shear stress are lower than that in zone I according to Figure 8a,b, resulting in a slower velocity in zone II, which is consistent with the velocity distribution in Figure 6b. Therefore, non-penetration defects easily generate near the weld center in zone II due to the poorest fluidity and no material mixing in this area. With the increase of the pin length, the thickness of the zone II reduces, and it is close to 0 near the weld center when the pin length increases to 5.95 mm, so that the non-penetration defect is eliminated.

Figure 17 shows the distribution curve of shear stress at the half depth between the bottom surface of the pin and the bottom surface of the workpiece under different pin lengths. In Figure 17a, when the pin length increases from 5.8 to 5.95 mm with 0.05 mm per step (Figure 17a), the shear stress on the RS side near the weld center ($-0.5 < y < 0$) significantly increases, which makes the material mixed more fully in this area where the "S line" defect appears. A similar trend can be observed as the pin length increases from 5.9 to 5.95 mm with 0.01 mm per step, as shown in Figure 17b. As the length of the pin changes, it should be some relationship between the increasing shear stress (Figure 17b) and the morphotype of the "S line" shown in Figure 11a–c. That is to say, in Figure 17b, as the shear stress gradually increases with the increase of pin length, the width and the tilting angle of "S line" gradually reduces according to Figure 11a–c, which also indicates that the shear stress plays a dominant role in eliminating the "S line" defect.

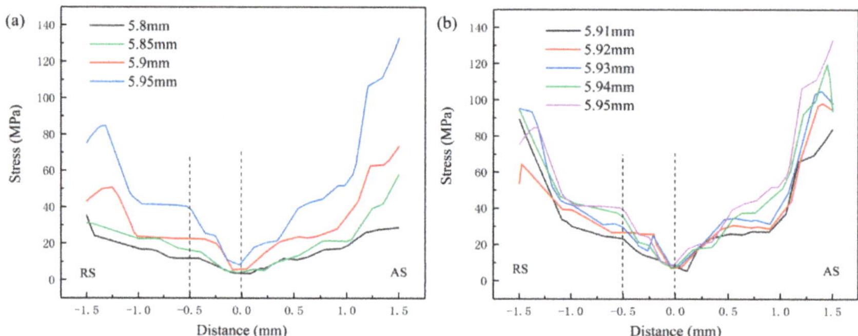

Figure 17. Distribution curves of shear stress at the root of joint with different pin lengths (1000 rpm, 120 mm/min). (**a**) 0.05 mm per step from 5.8 to 5.95 mm; (**b**) 0.01 mm per step from 5.91 to 5.95 mm.

External cooling was applied to the bottom of the workpiece to reduce the bottom temperature of the weld and explore the plastic metal flow behavior at the bottom of the weld at different temperatures. In order to change the temperature of the plastic metal at the bottom of the weld, five groups of simulations were carried out to cool the weld floor. The temperature change curve of the bottom of the weld is given in Figure 18. At the welding parameters of 1000 rpm and 120 mm/min, the temperature at the bottom of the weld gradually decreased from 377 to 302 °C, and the plastic metal flow behavior at the bottom of the weld was observed. The flow behavior of plastic metal at the bottom of the weld is shown in Figure 19. When the temperature at the bottom of the weld changes, the fluidity of the plastic metal in the weld basically does not change. Near the center of the pin bottom, plastic metal has little flow. With the increase of the distance to the end face of the pin bottom, the flow of plastic metal gradually increases. RS and AS plastic metal flow in the opposite direction and finally meet at the RS near the weld center. At the bottom of the weld, plastic metal flows from RS to AS, and the plastic metal flow basically stops near the center of the weld. Based on the above flow behavior, the plastic metal at the bottom of the weld will eventually form the "S line" weak connection defect and lack penetration at the center of the bottom of the weld.

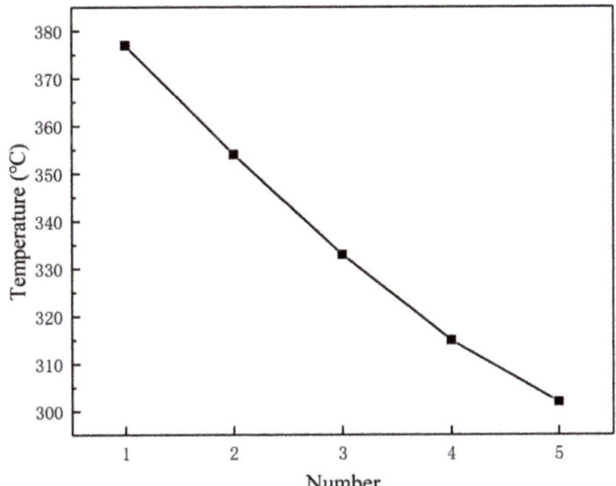

Figure 18. Weld bottom temperature curve.

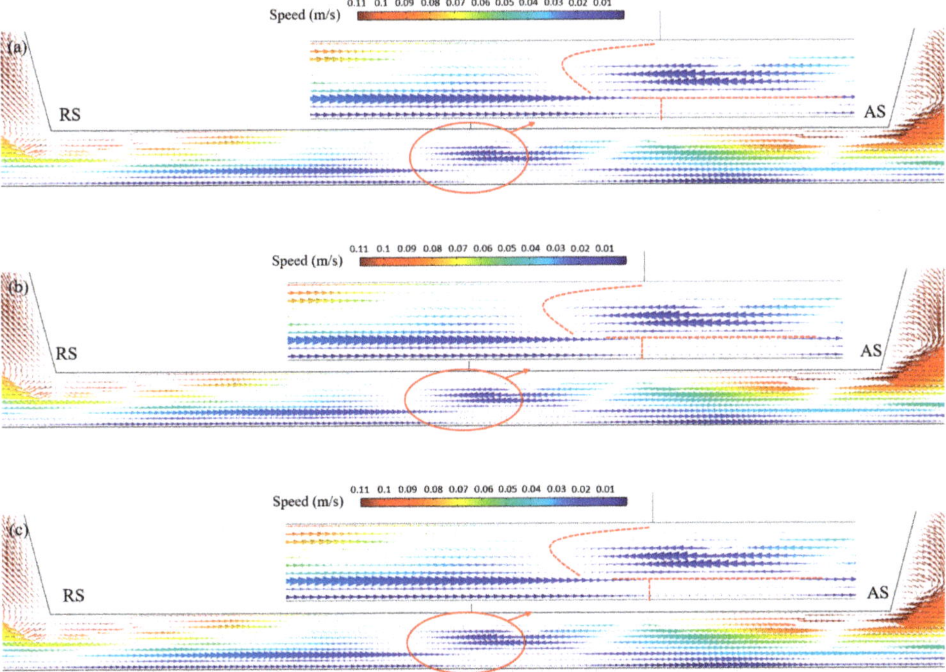

Figure 19. *Cont.*

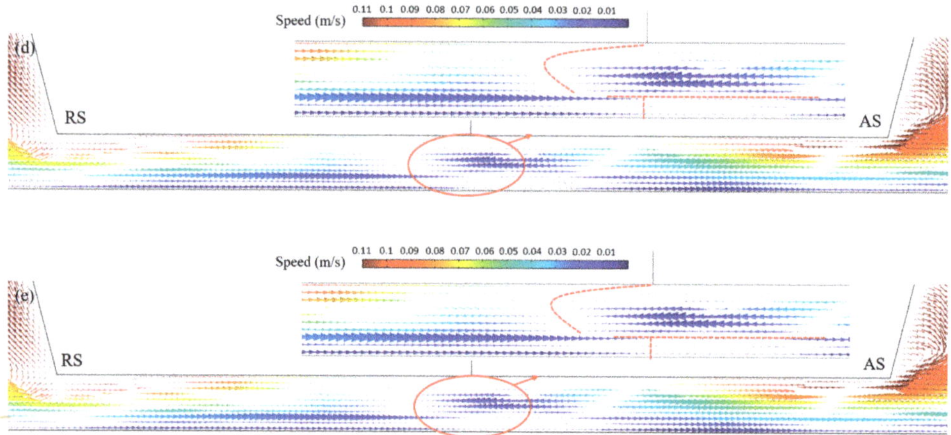

Figure 19. Flow behavior of plastic metal at different temperatures (**a**) No. 1; (**b**) No. 2; (**c**) No. 3; (**d**) No. 4; (**e**) No. 5.

The flow behavior at the bottom of the weld is due to the lack of shear stress on the plastic metal by agitation. The distribution of shear stress at the bottom of the weld is illustrated in Figure 20. When the temperature at the bottom of weld decreases gradually, the shear stress of the plastic metal at the bottom of weld seam increases gradually. In the position of the weld center, the shear stress is the smallest—that is, plastic metal at the bottom of the weld is the weakest. When the temperature at the bottom of the weld decreases, the viscosity of the plastic metal increases, and the critical driving force required for its flow increases. Therefore, when the temperature at the bottom of the weld decreases, although the shear effect of agitation on the plastic metal at the bottom is enhanced, the flow of the plastic metal is not improved.

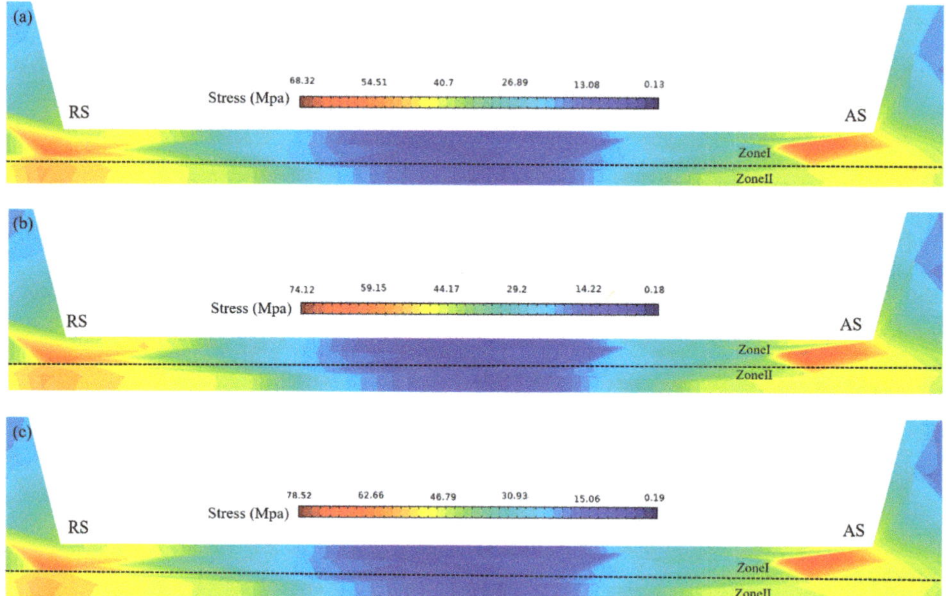

Figure 20. *Cont.*

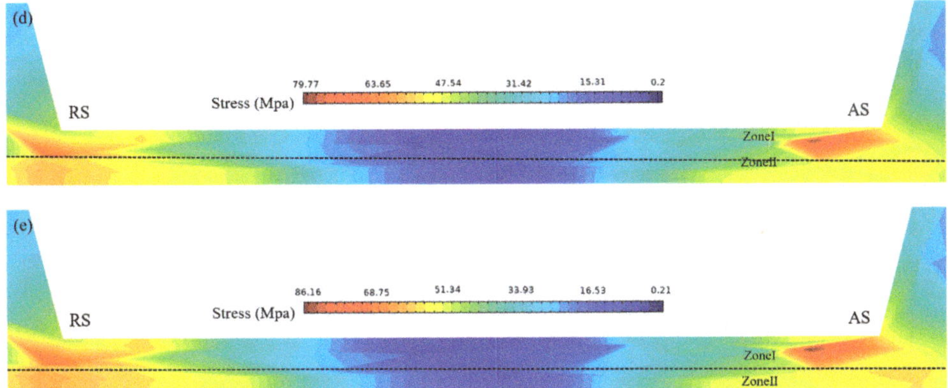

Figure 20. Distribution of shear stress at the root of the joint at different temperatures. (**a**) No. 1; (**b**) No. 2; (**c**) No. 3; (**d**) No. 4; (**e**) No. 5.

4. Conclusions

In this paper, material flow behavior at the root of the FSW weld is studied through an established 3D thermal–mechanical coupled FSW model. The results are as follows:

(1) The material at the root of the FSW welding seam can be divided into two different zones according to flow behavior. The material near the bottom surface of the pin is zone I, where material from AS and RS flow in opposite directions and mix on the RS side near the weld center. The material near the bottom surface of the workpiece is zone II, where the material flows in the same direction near the weld center.

(2) Owing to the low pressure and shear stress at the weld center, the fluidity of the material is low in the central area of zones I and II. The "S line" defect will appear at the RS side of zone I when the material is not fully mixed owing to low fluidity. Furthermore, the non-penetration defect will appear at the center of zone II when the material in zone II cannot flow through the weld center.

(3) With the increase of the pin length, the material flow behavior changes significantly on the margin of the pin of the RS side, and the shape of the "s line" defect also changes to the "right-tilting line" shape. If the pin length further increases, the tilting angle of the "right-tilting line" reduces, and the sizes of the "s line" defect and non-penetration decrease. The "s line" defect and non-penetration defect are finally eliminated when the pin length increases to a certain extend.

(4) In this paper, the optimal pin length is 5.95 mm under a 6 mm butt FSW welded joint of aluminum alloy, which is determined from two aspects. On the one hand, if the pin length is shorter than 5.95 mm, the thickness of zone II is not negligible, which will result in the insufficient fluidity of the material near the bottom surface of workpiece so that non-penetration defect is prone to occur. On the other hand, if the pin length is longer than 5.95 mm, the temperature at the root of the weld will increase to a relatively high level, which will coarsen the recrystallization tissue and degrade the mechanical properties of the FSW joint or even make an intense wear of stirring pin, reducing the stability of FSW process.

Author Contributions: J.L.: Supervision, writing, review and editing. J.W.: Formal analysis, methodology, data curation, writing, and original draft. H.L.: Conceptualization, data curation, writing, and reviewing. L.Y.: Software, validation. J.G.: Investigation, software, writing, and reviewing. H.G.: Investigation, writing, editing, and formal analysis. All authors have read and agreed to the published version of the manuscript.

Funding: Support Program of State Key Lab of Advanced Welding and Joining, Harbin Institute of Technology (AWJ-20-Z02), the State Key Laboratory of Solidification Processing in Northwestern Polytechnical University (SKLSP201903), and the Minjiang Scholar Support Program of Fujian, China (2016).

Acknowledgments: The authors acknowledge the financial support from the State Key Lab of Advanced Welding and Joining, Harbin Institute of Technology (AWJ-20-Z02), the State Key Laboratory of Solidification Processing in Northwestern Polytechnical University (SKLSP201903), and the Minjiang Scholar Support Program of Fujian, China. The authors are also thankful to Qingzhong Liu, Associate Professor in the Department of Computer Science at Sam Houston State University, USA, for his help with some revision suggestions.

Conflicts of Interest: The authors declare that they have no known competing financial interests or personal relationships that could have appeared to influence the work reported in this paper.

References

1. Thomas, W.M.; Nicholas, E.D.; Needham, J.C. Friction Stir Butt Welding: Great. Britain Patent Application. No. 9125978.8, 8 December 1991.
2. Mishra, R.S.; Ma, Z.Y. Friction stir welding and processing. *Mater. Sci. Eng. R-Rep.* **2005**, *50*, 1–78. [CrossRef]
3. Luo, J.; Li, F.; Chen, W. Experimental researches on resistance heat aided friction stir welding of Mg alloy. *Q. J. Jpn. Weld. Soc.* **2013**, *31*, 65–68. [CrossRef]
4. Song, Z.; Nakata, K.; Wu, A.; Liao, J.; Zhou, L. Influence of probe offset distance on interfacial microstructure and mechanical properties of friction stir butt welded joint of Ti6Al4V and A6061 dissimilar alloys. *Mater. Des.* **2014**, *57*, 269–278. [CrossRef]
5. Salih, O.S.; Ou, H.; Sun, W.; Mccartney, D.G. A review of friction stir welding of aluminium matrix composites. *Mater. Des.* **2015**, *86*, 61–71. [CrossRef]
6. Jia, Y.; Lin, S.C.; Liu, J.Z.; Qin, Y.G.; Wang, K.H. The influence of pre- and post-heat treatment on mechanical properties and microstructures in friction stir welding of dissimilar age-hardenable aluminum alloys. *Metals* **2019**, *9*, 1162. [CrossRef]
7. Oliveira, J.P.; Duarte, J.F.; Inácio, P.; Schell, N.; Miranda, R.M.; Santos, T.G. Production of Al/NiTi composites by friction stir welding assisted by electrical current. *Mater. Des.* **2017**, *113*, 311–318. [CrossRef]
8. Wan, L.; Huang, Y.X. Microstructure and mechanical properties of Al/Steel friction stir lap weld. *Metals* **2017**, *7*, 542. [CrossRef]
9. Zhou, C.; Yang, X.; Luan, G. Effect of kissing bond on fatigue behavior of friction stir welds on Al 5083 alloy. *J. Mater. Sci.* **2006**, *41*, 2771–2777. [CrossRef]
10. Sato, Y.S.; Takauchi, H.; Park, S.H.C.; Kokawa, H. Characteristics of the kissing-bond in friction stir welded Al alloy 1050. *Mater. Sci. Eng. A-Struct. Mater. Prop. Microstruct. Process.* **2005**, *405*, 333–338. [CrossRef]
11. Khan, N.Z.; Siddiquee, A.N.; Khan, Z.A.; Shihab, S.K. Investigations on tunneling and kissing bond defects in FSW joints for dissimilar aluminum alloys. *J. Alloys. Compd.* **2015**, *648*, 360–367. [CrossRef]
12. Le Jolu, T.; Morgeneyer, T.F.; Denquin, A.; Sennour, M.; Laurent, A.; Besson, J.; Gourgues, L. Microstructural Characterization of Internal Welding Defects and Their Effect on the Tensile Behavior of FSW Joints of AA2198 Al-Cu-Li Alloy. *Metall. Mater. Trans. A-Phys. Metall. Mater. Sci.* **2014**, *45A*, 5531–5544. [CrossRef]
13. Le Jolu, T.; Morgeneyer, T.F.; Denquin, A.; Gourgues, L. Fatigue lifetime and tearing resistance of AA2198 Al-Cu-Li alloy friction stir welds: Effect of defects. *Int. J. Fatigue* **2015**, *70*, 463–472. [CrossRef]
14. Le Jolu, T.; Morgeneyer, T.F.; Gourgues, L. Effect of joint line remnant on fatigue lifetime of friction stir welded Al-Cu-Li alloy. *Sci. Technol. Weld. Join.* **2010**, *15*, 694–698. [CrossRef]
15. Zhou, C.Z.; Yang, X.Q.; Luan, G.H. Effect of root flaws on the fatigue property of friction stir welds in 2024-T3 aluminum alloys. *Mater. Sci. Eng. A-Struct. Mater. Prop. Microstruct. Process.* **2006**, *418*, 155–160. [CrossRef]
16. Kadlec, M.; Růžek, R.; Nováková, L. Mechanical behaviour of AA 7475 friction stir welds with the kissing bond defect. *Int. J. Fatigue* **2015**, *74*, 7–19. [CrossRef]
17. Sato, Y.S.; Yamashita, F.; Sugiura, Y.; Park, S.H.C.; Kokawa, H. FIB-assisted TEM study of an oxide array in the root of a friction stir welded aluminium alloy. *Scr. Mater.* **2004**, *50*, 365–369. [CrossRef]
18. Okamura, H.; Aota, K.; Sakamoto, M.; Ezumi, M.; Ikeuchi, K. Behavior of oxides during friction stir welding of aluminium alloy and their effect on its mechanical properties. *Weld. Int.* **2002**, *16*, 266–275. [CrossRef]
19. Chen, G.; Li, H.; Shi, Q. On the material bonding behaviors in friction stir welding. In *Friction Stir Welding and Processing*; X (The Minerals, Metals & Materials Series); Springer: Cham, Switzerland, 2019; pp. 99–108. Available online: https://link.springer.com/chapter/10.1007/978-3-030-05752-7_10#citeas (accessed on 12 February 2019).
20. Luo, J.; Li, S.X.; Chen, W.; Xiang, J.F. Wang Hong. Simulation of aluminum alloy flowing in friction stir welding with a multiphysics field model. *Int. J. Adv. Manuf. Technol.* **2015**, *81*, 349–360. [CrossRef]

21. Al-Moussawi, M.; Smith, A.J. Defects in Friction Stir Welding of Steel. *Metallogr. Microstruct. Anal.* **2018**, *7*, 194–202. [CrossRef]
22. Hou, X.; Yang, X.; Cui, L.; Zhou, G. Influences of joint geometry on defects and mechanical properties of friction stir welded AA6061-T4 T-joints. *Mater. Des.* **2014**, *53*, 112–123. [CrossRef]
23. Zhou, N.; Song, D.; Qi, W.; Li, X.; Zou, J.; Attallah, M.M. Influence of the kissing bond on the mechanical properties and fracture behavior of AA5083-H112 friction stir welds. *Mater. Sci. Eng. A-Struct. Mater. Prop. Microstruct. Process.* **2018**, *719*, 12–20. [CrossRef]
24. Santos, T.G.; Miranda, R.M.; Vilaça, P. Friction Stir Welding assisted by electrical Joule effect. *J. Mater. Process. Technol.* **2014**, *214*, 2127–2133. [CrossRef]
25. Luo, J.; Chen, W.; Fu, G. Hybrid-heat effects on electrical-current aided friction stir welding of steel, and Al and Mg alloys. *J. Mater. Process. Technol.* **2014**, *214*, 3002–3012. [CrossRef]
26. Luo, J.; Wang, X.J.; Wang, J.X. New technological methods and designs of stir head in resistance friction stir welding. *Sci. Technol. Weld. Join.* **2009**, *14*, 650–654. [CrossRef]
27. Fuse, K.; Badheka, V. Bobbin tool friction stir welding: A review. *Sci. Technol. Weld. Join.* **2019**, *24*, 277–304. [CrossRef]
28. Wang, F.F.; Li, W.Y.; Shen, J.; Wen, Q.; Dos Santos, J.F. Improving weld formability by a novel dual-rotation bobbin tool friction stir welding. *J. Mater. Sci. Technol.* **2018**, *34*, 135–139. [CrossRef]
29. Liu, H.; Hu, Y.; Wang, H.; Du, S.; Sekulic, D.P. Stationary shoulder supporting and tilting pin penetrating friction stir welding. *J. Mater. Process. Technol.* **2018**, *255*, 596–604. [CrossRef]
30. Kumar, S. Ultrasonic assisted friction stir processing of 6063 aluminum alloy. *Arch. Civ. Mech. Eng.* **2016**, *16*, 473–484. [CrossRef]
31. Tarasov, S.Y.; Rubtsov, V.Y.; Kolubaev, E.A.; Ivanov, A.N.; Fortuna, S.V.; Eliseev, A.A. Ultrasonic-Assisted Friction Stir Welding on V95AT1 (7075) Aluminum Alloy. *AIP Conf. Proc.* **2015**, *1683*, 020231.
32. Salari, E.; Jahazi, M.; Khodabandeh, A.; Ghasemi-Nanesa, H. Influence of tool geometry and rotational speed on mechanical properties and defect formation in friction stir lap welded 5456 aluminum alloy sheets. *Mater. Des.* **2014**, *58*, 381–389. [CrossRef]
33. Do Vale, N.L.; Torres, E.A.; Santos, T.F.D.A.; Urtiga Filho, S.L.; Dos Santos, J.F. Effect of the energy input on the microstructure and mechanical behavior of AA2024-T351 joint produced by friction stir welding. *J. Braz. Soc. Mech. Sci. Eng.* **2018**, *40*, 467–481. [CrossRef]
34. Mandache, C.; Levesque, D.; Dubourg, L.; Gougeon, P. Non-destructive detection of lack of penetration defects in friction stir welds. *Sci. Technol. Weld. Join.* **2012**, *17*, 295–303. [CrossRef]
35. Riahi, M.; Nazari, H. Analysis of transient temperature and residual thermal stresses in friction stir welding of aluminum alloy 6061-T6 via numerical simulation. *Int. J. Adv. Manuf. Technol.* **2011**, *55*, 143–152. [CrossRef]
36. Ren, J.G.; Wang, L.; Xu, D.K.; Xie, L.Y.; Zhang, Z.C. Analysis and Modeling of Friction Stir Processing-Based Crack Repairing in 2024 Aluminum Alloy. *Acta Metall. Sin. (Engl. Lett.)* **2017**, *30*, 228–237. [CrossRef]
37. Zeng, X.H.; Xue, P.; Wang, D.; Ni, D.R.; Xiao, B.L.; Ma, Z.Y. Effect of Processing Parameters on Plastic Flow and Defect Formation in Friction-Stir-Welded Aluminum Alloy. *Met. Mater. Trans. A* **2018**, *49*, 2673–2683. [CrossRef]
38. Hernández, C.A.; Ferrer, V.H.; Mancilla, J.E.; Mart'ınez, C. Three-dimensional numerical modeling of the friction stir welding of dissimilar steels. *Int. J. Adv. Manuf. Technol.* **2017**, *93*, 1–15. [CrossRef]

Article

Real-Time Temperature Measurement Using Infrared Thermography Camera and Effects on Tensile Strength and Microhardness of Hot Wire Plasma Arc Welding

Nirut Naksuk *, Jiradech Nakngoenthong, Waravut Printrakoon and Rattanapon Yuttawiriya

National Metal and Materials Technology Center (MTEC), National Science and Technology Development Agency, 114 Thailand Science Park, Khlong Luang, Pathum Thani 12120, Thailand; jiradecn@mtec.or.th (J.N.); waravutp@mtec.or.th (W.P.); rattanapon.yut@mtec.or.th (R.Y.)

* Correspondence: nirutn@mtec.or.th; Tel.: +66-2564-6500 (ext. 4149)

Received: 8 July 2020; Accepted: 26 July 2020; Published: 3 August 2020

Abstract: The hot wire plasma arc welding process, a hybrid process between the plasma arc welding (PAW) process and hot wire process, is used to weld 316 stainless steel sheets, in which the temperature generated during welding is recorded in real time with a high-speed infrared thermography camera. Therefore, this research studies the factors in the hot wire process, of which there are two: (1) wire feed rate and (2) wire current; this study investigated the tensile strength, microhardness, and relationship of cooling rate per tensile strength and microhardness. The study found that the hot wire current plays an important role in cooling rates and tensile strength. The temperature results from high-speed infrared thermography camera show that the maximum welding temperature is around 1300 °C. The weld pool has a temperature between 900 and 1300 °C and the temperature profile of the weld pool will look like an "M" shaped, which is caused by the hot wire process. Finally, the appropriate hot wire parameters are 1.5 m/min for wire feed rate and 40A for wire current, which will give the workpiece cooling rate of 800–500 °C as 13.42 °C/s, tensile strength of 610.95 MPa, and the average Vickers microhardness of 195 HV.

Keywords: real time; infrared thermography; camera; tensile strength; microhardness; plasma arc welding; hot wire

1. Introduction

Hot wire plasma arc welding is relatively a new welding process, a hybrid process between the plasma arc welding (PAW) process and the hot wire process. The hot wire process can be used as a hybrid with other welding processes such as gas tungsten arc welding (GTAW) process or laser welding process [1–3]. The basic principle of the hot wire process is to heat the consumable filler wire by a separate power source until close to the melting point and feed to the weld pool [4]. There are many advantages over the cold wire process, mainly high deposition rates [5,6], which will assist to increase welding speed and productivity. The work of [7] has been conducted to study mechanical properties to compare the results between the use of hot wire and cold wire of GTAW. The results showed that the mechanical properties obtained were not significantly different between the use of hot wire and cold wire. In general, hot wire process parameters (such as wire current, wire feed rate, wire feed angle, wire size, wire contact length, gas shield) have a significant effect on the quality of the obtained workpiece [8–10]. The selection of inappropriate parameters can result in weld defects [11]. Therefore, choosing the appropriate parameters will increase the mechanical properties and achieve quality welding workpieces. The optimization of hot wire laser welding parameters has been studied [12] based on ensemble metamodels and non-dominated sorting genetic algorithm. This research studied the three parameters, that is, laser power, welding speed, and hot wire current. The results showed that the optimal process parameters gave results consistent with the experiment. Although research articles related to the hot wire process are available [13–15], research on hot wire plasma arc welding is limited. Therefore, the study of the basic parameters of the hot wire process is interesting to obtain good mechanical properties of the welding workpiece. Recently, cameras have been widely used to study the real-time welding process [16–18], which are non-contact measure and give results with high accuracy and precision. The information obtained from the camera gives insight into various forms of weld results such as temperature, geometric shape, and defects. The temperature data recorded from the camera can be used to calculate the cooling rates that occur in the range of 800–500 °C, which is a material phase transformation that affects the microstructure and mechanical properties. The effects of cooling rate in a super duplex stainless steel on pulsed GTAW welding have been studied [19] using three process parameters (heat input, wire feed rate, wire feed technique). The results found that the heat input plays an important role in the cooling rate, which affects microstructure in the heat-affected zone (HAZ) and weld zone. From the literature review [20–22], we found that the main factors affecting the cooling rate were heat input, type of material, and the thickness of the material.

This paper presents the hot wire plasma arc welding process using a high-speed infrared thermography camera to measure the temperature in real time. The resulting temperature profile allows to calculate the cooling rate of the workpiece, which will obtain the relationship between the cooling rate and the mechanical properties (tensile strength, microhardness). Furthermore, the appropriate parameters for the hot wire plasma arc welding process are obtained that allow the welding workpiece to have tensile strength and microhardness within the specified standards.

2. Materials and Methods

The material used in this research is a stainless steel sheet in accordance with AISI 316 (equivalent to DIN 1.4401 and JIS SUS 316), which has a chemical composition measured from the machine brand, Thermo Scientific, model: ARL 3460 Metals Analyzer by optical emission spectrometry (OES) compared with the standard chemical composition [23] and the mechanical property values as in Tables 1 and 2, respectively.

Table 1. Chemical composition of AISI 316 by optical emission spectrometry.

Elements	Standard (% *w/w*)	Measured (% *w/w*)
C (Max)	0.08	0.019
Cr	16–18	15.697
Fe	61.8–72	70.354
Mn (Max)	2.0	1.377
Mo	2.0–3.0	2.140
Ni	10–14	10.039
P (Max)	0.045	0.029
Si (Max)	1.0	0.341
S (Max)	0.03	0.004

Table 2. The mechanical property of AISI 316.

Mechanical Properties	Standard Value	Measured Value
Hardness, Vickers	146 HV	197.96 HV
Ultimate Tension Strength	580 MPa	598.52 MPa
Yield Strength	290 MPa	360.76 MPa
Elongation at Break	30–50%	35.36%
Modulus of Elasticity	193 GPa	198.42 GPa

The workpiece used was 100 mm in width, 150 mm in length, and 3 mm in thickness using two sheets per experiment and using a closed square butt joint. The workpiece will be clamped by jigs to prepare for hot wire plasma arc welding by machine brand, Cebora, model: EVO 450T (Bologna, Italy), which is used as the current source. The plasma welding controller from the machine brand, Cebora, model: digital Console PW30 (Bologna, Italy), was employed, which is used for controlling various parameters for plasma welding. The hot wire machine used the brand, MAC, model: Power Assist IV-642 (Osaka, Japan), with a wire feed system to the welding torch, which is attached to the ABB robotic arm, model: IRB4400 (Västerås, Sweden) used for controlling the hot wire plasma arc welding path, as shown in Figure 1. The hot wire plasma arc welding parameters used in this research are shown in Table 3. The welding process starts from the right-hand side of the workpiece and ends at the left-hand side. Here, the plasma torch was at the front, while the hot wire was feed behind the plasma welding torch, with a total welding distance of 130 mm.

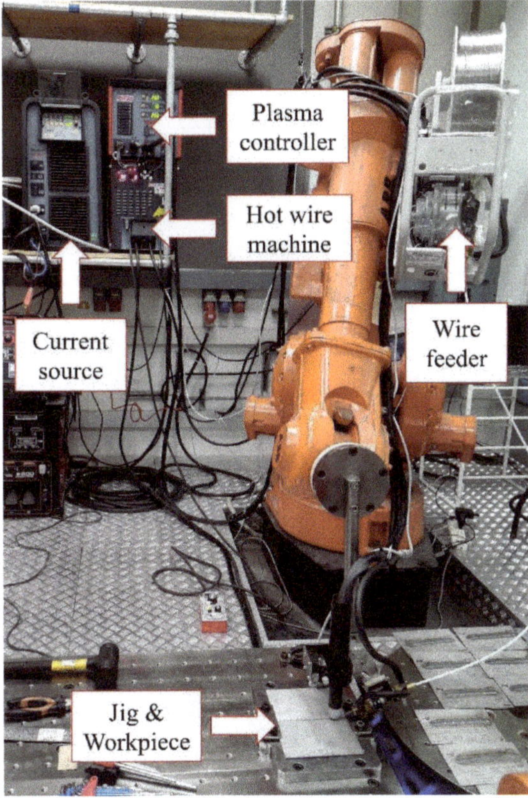

Figure 1. The hot wire plasma arc welding system.

For real-time temperature measurement during hot wire plasma arc welding, the high-speed infrared thermography camera brand, Infratec, model: ImageIR 8300 (Dresden, Germany), was used, with a temperature measurement accuracy of ±1%. The principle of temperature measurement was that all objects with temperatures above zero kelvin (−273.15 °C) emit infrared radiation that is invisible to the human eye. The camera with an Indium antimonide (InSb) sensor will measure the mediumwave infrared (MWIR) radiation, wavelength between 2 and 5 μm. The conversion of infrared radiation detected by sensors into temperature units, based on Planck's law and Stefan–Boltzmann's law according to the behavior of infrared pyrometer. Infrared emission of the object to be measured may be more or less depending on the wavelength. The parameters such as material composition, surface roughness, and measurement angle have some influence on wavelength [24]. Thus, the installation of the camera must control the environment to be constant throughout the experiment to not affect the results. This camera is equipped with an ImageIR® standard lens with a focal length of 25 mm and a motorized filter wheel with spectral filters. This camera will be installed at a distance of 80 cm from the welding workpiece. The camera mounting point will be on the side of the workpiece so that the temperature can be seen throughout the welding length, as shown in Figure 2. This research sets the camera to record the temperature every 0.5 s (f = 2 Hz), using the recording time of 240 s. Using a high-speed infrared thermography camera will make it possible to be aware of the temperature occurring throughout the welding process, and to know the temperature in every position such as weldment position, HAZ, and base material, including the temperature that occurs on hot wire feed to the weld pool, from the beginning to the cooling rate after the welding is done. Although the plasma light is brighter, the infrared radiation is always emitted, which passes through the spectral filters into the camera's sensor. This camera uses an Indium antimonide (InSb) sensor that is narrow-gap sensitive at wavelengths between 2 and 5 μm. Therefore, other wavelengths will not affect the measurement of the temperature.

Table 3. Parameters for hot wire plasma arc welding.

Parameters	Value
Current of arc	120 A
Current of pilot arc	20 A
Flow of gas plasma	0.5 L/min
Flow of gas shield	20 L/min
Plasma and shield gas	Argon
Nozzle tip diameter	2.5 mm
Torch angle	90 degree
Arc distance	6 mm
Welding speed	180 mm/min
Wire diameter	1.2 mm
Wire type	316 LSi
Wire feed angle	40 degree
Room temperature	24 °C

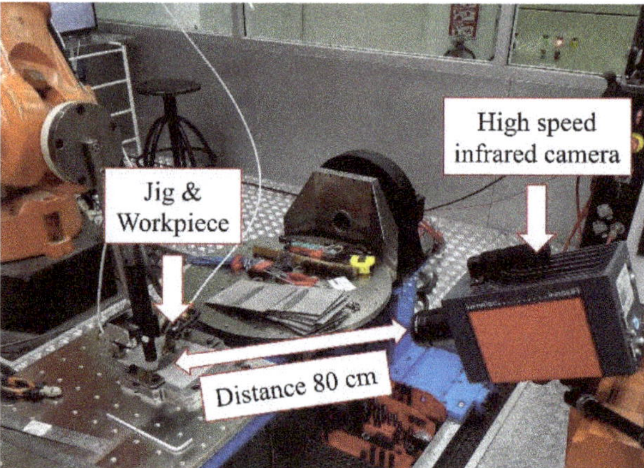

Figure 2. Installed high-speed infrared thermography camera.

The cooling rate at the temperature range 800–500 °C will be studied because the temperature range is the material phase transformation, which has a significant impact on the properties of the workpiece. The formula for calculating the cooling rate range 800–500 °C is calculated from a simple equation as shown in Equation (1):

$$Cooling\ rate_{800-500^{\circ}C} = \frac{\Delta Temp_{800-500^{\circ}C}}{\Delta time_{800-500^{\circ}C}} \tag{1}$$

where $\Delta Temp_{800-500^{\circ}C}$ = the temperature difference at 800 °C and 500 °C, $\Delta time_{800-500^{\circ}C}$ = the time difference at 800 °C and 500 °C.

Each workpiece uses three positions to measure the resulting temperature, as shown in Figure 3. Here, position 1 (P1) is the center of the weldment; position 2 (P2) is at a distance of 5 mm away from the center of the weldment, which is the area of the HAZ; and position 3 (P3) is at a distance of 10 mm from the center of the weldment, which is the base material. Besides, the temperature of the hot wire at a height of 7.5 mm from the workpiece surface is measured.

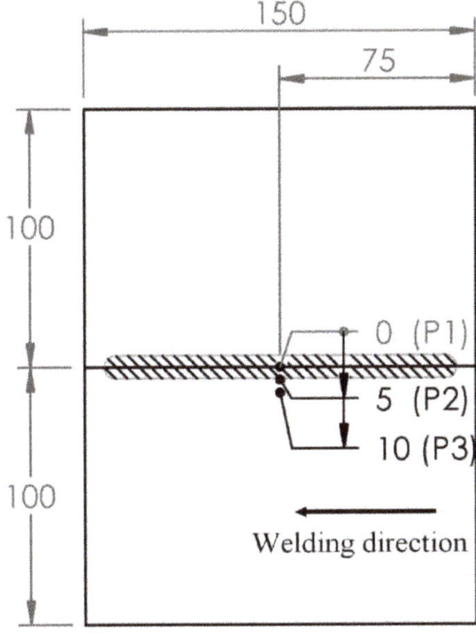

Figure 3. Position for measuring temperature.

Once the installation of the hot wire plasma arc welding system is finished, the next step will be to design the experiment, in which this research will use the design of a full factorial experiment. The factors used in this study will be the factors of the hot wire process, which has two factors, wire feed rate and wire current, where each factor is assigned to have two levels caused by the trial and error of the preliminary study, whereby such parameters and ranges do not cause defect workpieces and obtain complete welds, such as no sputtering during welding, good penetration, no cracking of welds, uniform hot wire during welding, and the experimental design table, as shown in Tables 4 and 5, respectively.

Table 4. Determining the level of factors used in the experiment.

Factors	Symbol	Unit	Low Level	High Level
Wire feed rate	A	m/min	1.5	1.7
Wire current	B	A	30	40

Table 5. Full factorial experimental design table.

Run Order	A (Wire Feed Rate) (m/min)	B (Wire Current) (A)
1	1.7	40
2	1.7	30
3	1.5	40
4	1.5	30

When the hot wire plasma arc welding process is finished according to the experimental design. The workpieces in each experiment are divided into three parts, as shown in Figure 4. The parts number 1 and number 3 of each welded workpiece will be tested for ultimate tensile strength in accordance with ASTM-E8 to find the average ultimate tensile strength value and the dimension of the tensile test workpiece, as shown in Figure 5. Tensile testing uses the machine brand, Instron, model: 8801 (Norwood, MA, USA), equipped with a strain gauge to find an elongation at break.

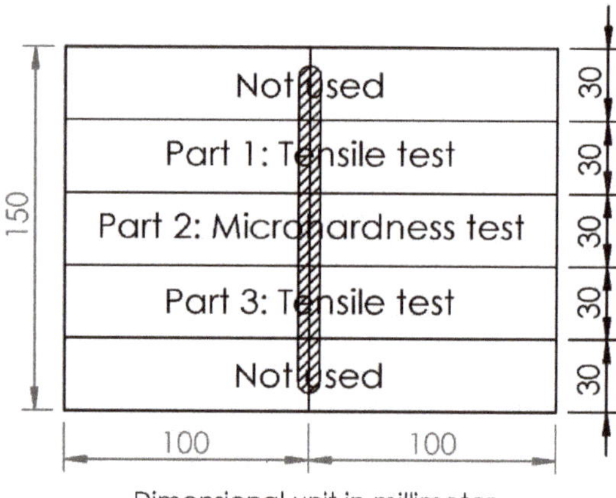

Figure 4. Separation of welded workpieces.

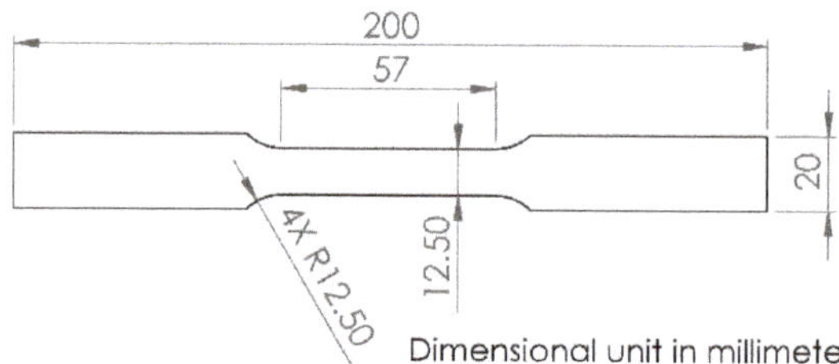

Figure 5. The dimension of the test workpiece in accordance with ASTM-E8.

Part number 2 of each welded workpiece after going through the grinding and polishing processes will be tested for Vickers microhardness by machine brand, Anton-Paar, model: MHT-10 (Graz, Austria). The parameters used are 200 g of testing force or HV0.2; time of withstanding the load of 15 s; loading speed of 25 g/s, which will test the microhardness at the center of the cross-sectional area (measured at a distance below the surface about 1.5 mm); a total of 15 points; and two times per point to find the average, as shown in Figure 6. The results will be displayed in a graph that will show the microhardness value in each position.

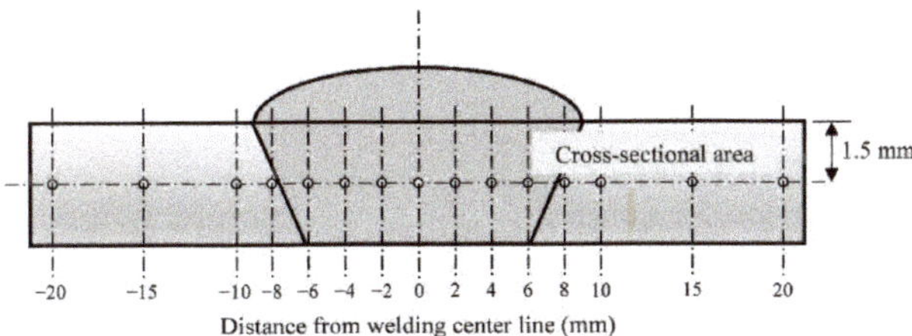

Figure 6. Position for measuring Vickers microhardness.

3. Results and Discussion

The program used for high-speed infrared thermography camera for temperature recording and data retrieval was the IRBIS® 3.1 Professional program. Figure 7 shows an example of an image captured from a high-speed infrared thermography camera. Each pixel of the image can show the resulting temperature and, therefore, allows knowing the exact location of the temperature. Real-time temperature measurement results using a high-speed infrared thermography camera according to the design of the experimental design could be graphed as shown in Figures 8–11, which are shown, at position 1 (position of weldment), position 2 (position of HAZ), and position 3 (position of base material), to have a graph in the shape like an "M". Position 1 has the highest measured temperature of around 1300 °C and a minimum of about 900 °C. It was shown that the weld pool will have a temperature between 900 and 1300 °C and then gradually cool down. The occurrence of the graph in the shape like an "M" owing to the area in front of the weld pool has the area of plasma charge that will cause the workpiece to form a keyhole and this area has the highest temperature, as shown in the graph. Behind the plasma area, caused by the hot wire feeding, causing the temperature to decrease and the tail of the weld pool, the temperature will drop slightly close to the temperature of the plasma area, as shown in Figure 12. Therefore, using a high-speed infrared thermography camera to record the real-time temperature will result in the graph in the shape of an "M". Whereas, if the recording frequency is not high or using a thermocouple or if the hot wire is not fed, the "M" graph will not be produced. The weldment position will have a maximum temperature of about 1300 °C higher than the HAZ position, which has a maximum temperature of about 1000 °C, higher than the base material position, which has a maximum temperature of about 700 °C, respectively. The temperature decreases in this order owing to the increasing distance from the center of the weld, in which the highest temperature corresponds to Figure 13. For peak temperature at different distances from the welding center line, as shown in Figure 13, we found that the resulting graph has very little temperature differences owing to the same type and size of material used, even when using different parameters of welding. The maximum temperature at the weldment position of run order 1 is 1324.64 °C, run order 2 is 1324.67 °C, run order 3 is 1324.69 °C, and run order 4 is 1324.62 °C.

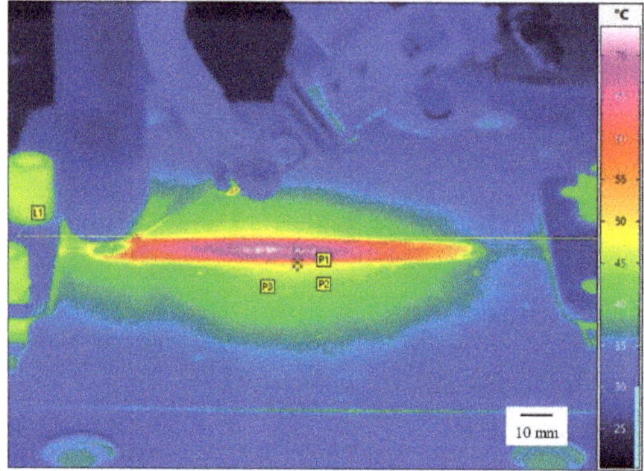

Figure 7. The captured image from a high-speed infrared thermography camera.

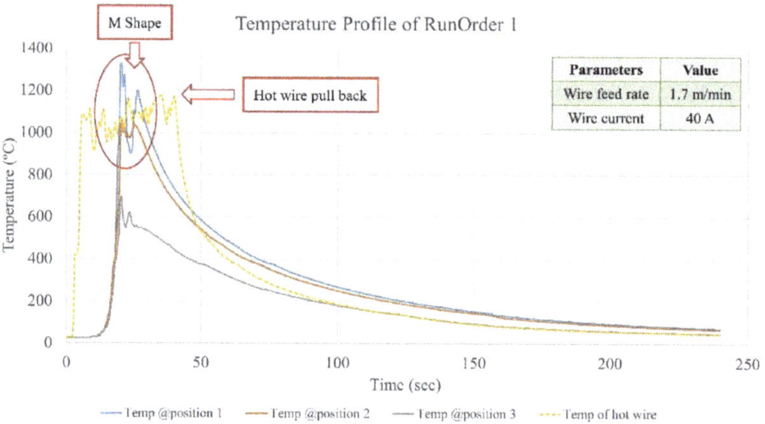

Figure 8. The temperature profile of run order 1.

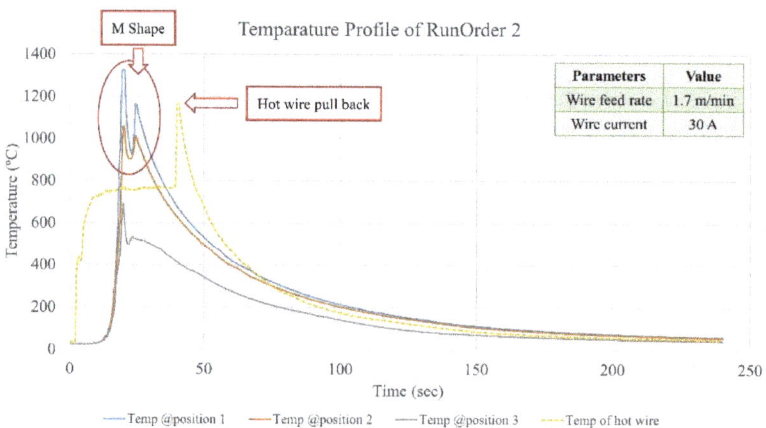

Figure 9. The temperature profile of run order 2.

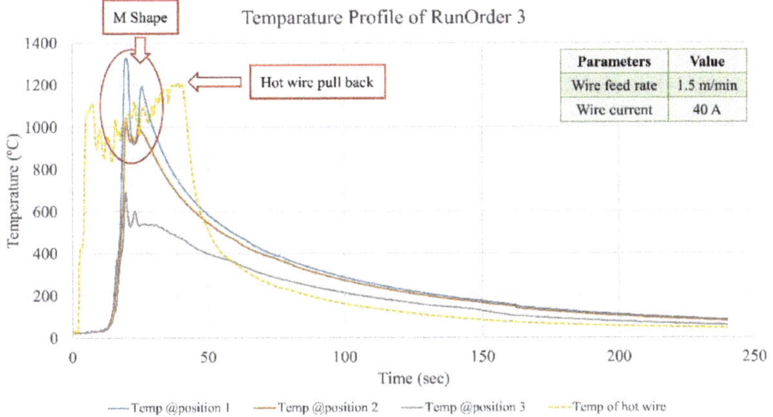

Figure 10. The temperature profile of run order 3.

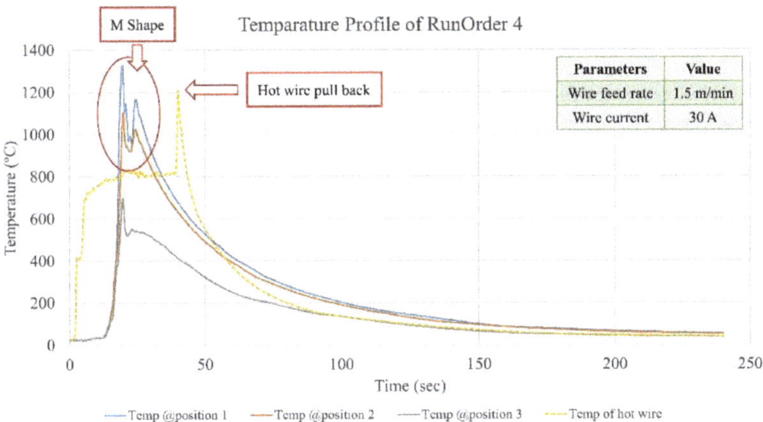

Figure 11. The temperature profile of run order 4.

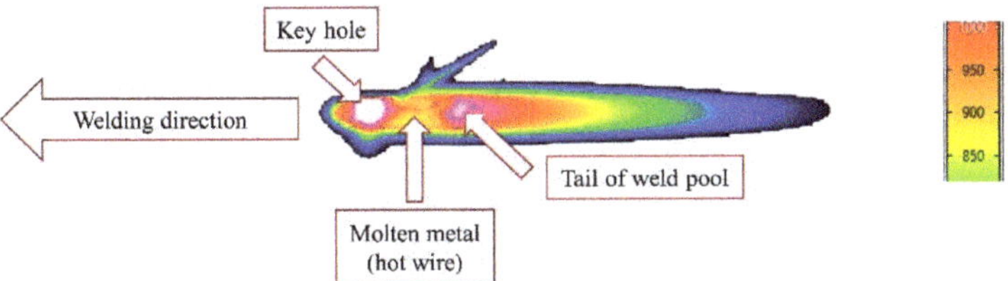

Figure 12. The characteristics of the weld pool results in an M-shaped graph.

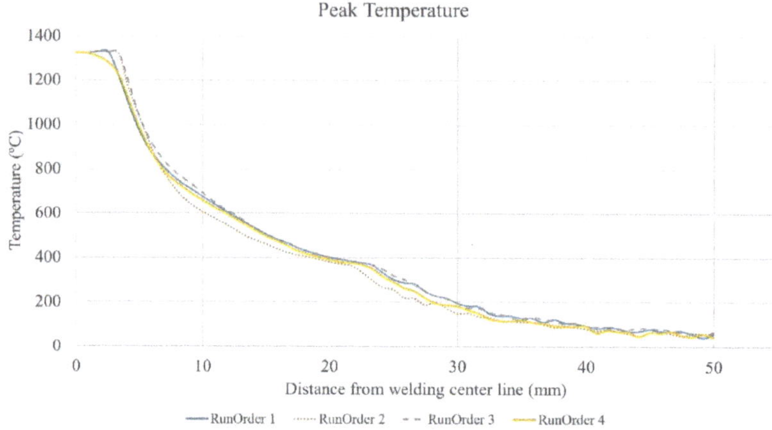

Figure 13. Peak temperature by the distance of the workpiece.

In the graph temperature of the hot wire from Figures 9 and 11, which use hot wire current at 30 A, the hot wire has an average temperature of about 800 °C throughout the welding process, and the final position when the hot wire is pulled back, as shown in Figure 14, at the tip of the hot wire, will have a temperature of around 1200 °C. It was found that the hot wire during the welding process is still lower than the weld pool, resulting in the hot wire possibly not melting homogeneously with the workpiece, resulting in the workpiece having lower strength according to the tensile strength results obtained from Table 7. In Figures 8 and 10, which use a hot wire current at 40 A, the hot wire has an average temperature of about 1000 °C throughout the welding process, and the final position when the hot wire is pulled back at the tip of the hot wire will have the temperature around 1200 °C. During the welding process, the hot wire has the temperature close to the weld pool, causing the hot wire to melt homogeneously with the workpiece, resulting in the workpiece being stronger, such as in the results obtained from Table 7. Therefore, if the workpiece needs to be strong as per the standards, it is necessary to control the hot wire current so that the hot wire has an average temperature during welding around 900–1300 °C to make the workpiece as strong as possible.

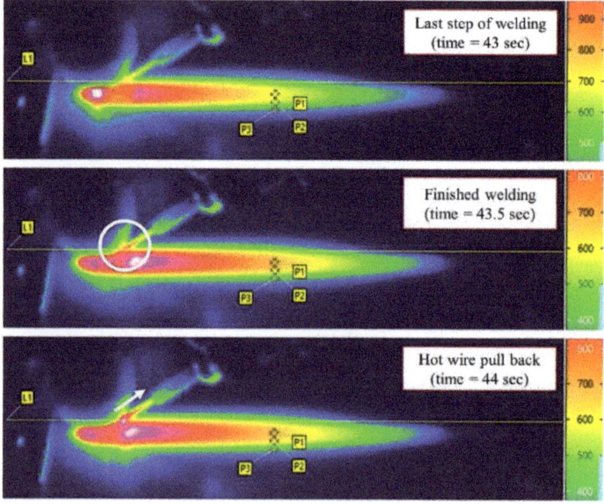

Figure 14. The captured image while the hot wire pulls back.

From the recorded temperature, the cooling rate can be calculated according to Equation (1); the values used to calculate the cooling rate are shown in Table 6, which found that the cooling rate of 800–500 °C is between 13.42 and 17.31 °C/s. Here, position 1 (position of weldment) has a slightly lower cooling rate than position 2 (position of HAZ) in the same workpiece. This is because position 1 has a higher temperature than position 2. Moreover, it was found that the parameters of the hot wire process affect the cooling rate. If the wire current increased, the cooling rate tends to be slower owing to the hot wire temperature.

Table 6. Cooling rate calculation at position 1 and position 2.

Run Order	Position	Temperature @800 °C (°C)	Temperature @500 °C (°C)	Time @800 °C (s)	Time @500 °C (s)	Cooling Rate (°C/s)
1	1	810.62	497.88	36.5	58.5	14.22
1	2	809.49	497.39	33.5	54.0	15.22
2	1	796.04	495.32	34.5	53.0	16.26
2	2	804.68	494.07	31.0	50.0	16.35
3	1	800.79	498.91	36.5	59.0	13.42
3	2	809.94	497.66	32.5	55.0	13.88
4	1	810.62	497.77	34.0	52.5	16.91
4	2	809.32	497.77	31.5	49.5	17.31

The tensile test results found that run order 3 gave the best results, followed by run order 1, 2, and 4 respectively. Run order 3 and 1 gave the ultimate tensile strength, yield strength, and elongation close to the reference workpiece, which shows that the parameters of run order 3 and 1 are suitable for use. Run order 4 gave the least value, by visual inspection of a broken workpiece, it was found that the fracture was in the position of the workpiece's joint (center of the welds). The breakage characteristics are brittle, resulting in lower ultimate tensile strength, yield strength, and elongation compared with other workpieces, where other workpieces fracture at the base material near the HAZ. The results of the tensile strength given in Table 7 show the consistency of the hot wire parameters, the temperature of the hot wire, and the cooling rate, as shown in Figure 8 to Figure 11. If the hot wire current was 30 A, the temperature of hot wire was about 800 °C during welding, which was lower than the temperature of the weld pool (around 900–1300 °C). On the other hand, if the hot wire current was 40 A, the hot wire will have a temperature of around 1000 °C, which was in the temperature range of the weld pool, causing the hot wire to melt homogeneously, making the workpiece stronger. Furthermore, it was found that the relationship between the cooling rate and tensile strength was inversely proportional; the workpiece with a lower cooling rate will be stronger than the fast cooling rate workpiece, as shown in Figure 15.

Table 7. Tensile strength results according to the experimental design.

Run Order	A (Wire Feed Rate) (m/min)	B (Wire Current)(A)	Ultimate Tensile Strength (MPa) Mean ± SD	Yield Strength (MPa) Mean ± SD	Elongation at Break (%) Mean ± SD
0 *	-	-	598.52 ± 0.91	360.76 ± 2.45	35.36 ± 0.06
1	1.7	40	607.23 ± 7.42	361.40 ± 6.38	30.60 ± 5.98
2	1.7	30	515.22 ± 142.81	351.09 ± 32.15	17.63 ± 20.62
3	1.5	40	610.95 ± 2.76	363.40 ± 2.53	30.28 ± 3.56
4	1.5	30	428.70 ± 61.45	294.39 ± 86.92	4.78 ± 1.99

* As-received material.

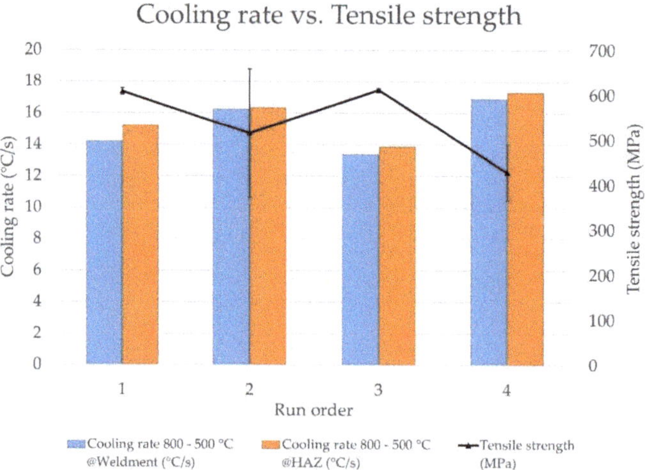

Figure 15. The relationship between cooling rate and tensile strength.

For Vickers microhardness values measured at each specified position shown in Figure 16, the microhardness characteristics were shaped like a "W", with all the smallest microhardness values at the HAZ area that was away from the welding center line about 5 mm, the measured microhardness was in the range of 180–210 HV, which was more than the standard microhardness of AISI 316 specified at 146 HV. It can be concluded that this parameter was suitable for use. The relationship between cooling rate and microhardness is shown in Figure 17. It was found that, if the cooling rate of 800–500 °C was between 13 and 16 °C/s, the microhardness will increase as the cooling rate increases. Although the cooling rate of 800–500 °C greater than 16 °C/s will result in a significant decrease in microhardness, the microhardness value is still higher than the standard value. However, the cooling rate was very important, thus the cooling rate of 800–500 °C should be controlled between 13 and 16 °C/s for the suitable results. Therefore, to control the cooling rate appropriately for this research, the hot wire current should be used at 40 A and the hot wire feed rate of 1.5 m/min or 1.7 m/min will give the most appropriate tensile strength and microhardness results.

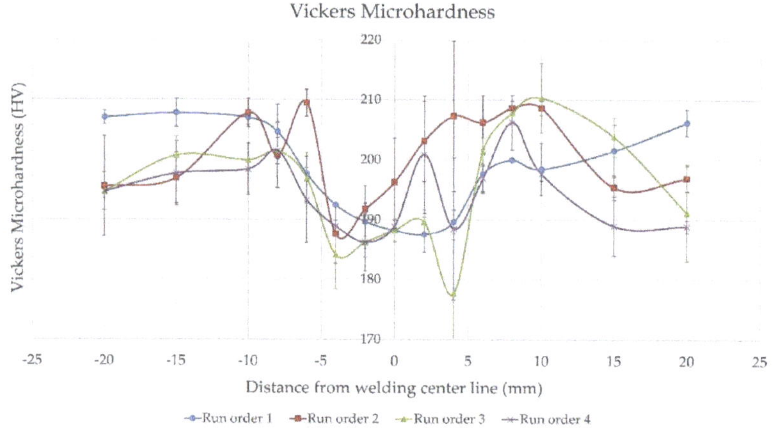

Figure 16. Vickers microhardness results according to the experimental design.

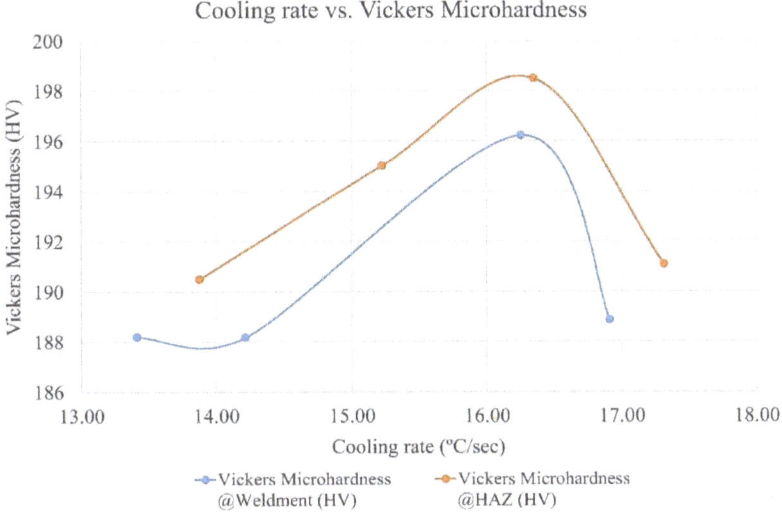

Figure 17. The relationship between cooling rate and Vickers microhardness.

4. Conclusions

This experiment can lead to the following conclusions:

1. The high-speed infrared thermography camera can record the temperature throughout the hot wire plasma arc welding process, allowing the maximum temperature to be known around 1300 °C and an "M" shaped graph form is generated. In the graph, the weld pool temperature is between 900 and 1300 °C.
2. The factors studied by the hot wire current at 40 A will give the hot wire temperature around 1000 °C, which is the temperature in the weld pool range, causing the hot wire to melt homogeneous with the base material.
3. The cooling rate of 800–500 °C is between 13.42 to 17.31 °C/s, which affects the tensile strength, with the low cooling rate getting the maximum tensile strength at 610.95 MPa and the tensile strength decreasing according to the high cooling rate. The cooling rate also affects the hardness, which should be controlled to between 13 and 16 °C/s to get the most suitable hardness. This cooling rate can be controlled by controlling the hot wire current.
4. The appropriate hot wire parameters are the hot wire current at 40A and the wire feed rate of 1.5 m/min, which result in the workpiece having a cooling rate at 800–500 °C of 13.42 °C/s, tensile strength 610.95 MPa, and Vickers microhardness average at 195 HV—all such results are higher than standard materials values.

This study found that the hot wire current plays an important role in cooling rates and tensile strength. Therefore, the hot wire process parameters should be used appropriately to maximize the benefits of the hot wire plasma arc welding workpiece. In the future, the optimization of the hot wire process parameters will be performed to obtain the most suitable parameters and study the phenomena that affect the results. This manuscript only reports appropriate parameters of the hot wire process effects on tensile strength and microhardness.

Author Contributions: Conceptualization, N.N.; methodology, N.N.; data curation, J.N., W.P., and R.Y.; validation, N.N. and J.N.; writing—original draft preparation, N.N.; writing—review and editing, N.N.; supervision, N.N.; project administration, J.N. and W.P.; funding acquisition, N.N. All authors have read and agreed to the published version of the manuscript.

Funding: This research received no external funding.

Acknowledgments: The authors would like to thank the National Metal and Materials Technology Center (MTEC), National Science and Technology Development Agency, Thailand for financial support.

Conflicts of Interest: The authors declare no conflict of interest.

References

1. Wen, P.; Cai, Z.; Feng, Z.; Wang, G. Microstructure and mechanical properties of hot wire laser clad layers for repairing precipitation hardening martensitic stainless steel. *Opt. Laser Technol.* **2015**, *75*, 207–213. [CrossRef]
2. Oliveira, J.P.; Crispim, B.; Zeng, Z.; Omori, T.; Fernandes, F.M.B.; Miranda, R.M. Microstructure and mechanical properties of gas tungsten arc welded Cu-Al-Mn shape memory alloy rods. *J. Mater. Process. Technol.* **2019**, *271*, 93–100. [CrossRef]
3. Oliveira, J.P.; Curado, T.M.; Zeng, Z.; Lopes, J.G.; Rossinyol, E.; Park, J.M.; Schell, N.; Fernandes, F.M.B.; Kim, H.S. Gas tungsten arc welding of as-rolled CrMnFeCoNi high entropy alloy. *Mater. Des.* **2020**, *189*, 108505. [CrossRef]
4. Hori, K.; Watanabe, H.; Myoga, T.; Kusano, K. Development of hot wire TIG welding methods using pulsed current to heat filler wire – research on pulse heated hot wire TIG welding processes. *Weld. Int.* **2004**, *18*, 456–468. [CrossRef]
5. Olivares, E.A.G.; Díaz, V.M.V. Study of the hot-wire TIG process with AISI-316L filler material, analysing the effect of magnetic arc blow on the dilution of the weld bead. *Weld. Int.* **2018**, *32*, 139–148. [CrossRef]
6. Bambach, M.; Sizova, I.; Silze, F.; Schnick, M. Comparison of laser metal deposition of Inconel 718 from powder, hot and cold wire. *Procedia CIRP* **2018**, *74*, 206–209. [CrossRef]
7. Pai, A.; Sogalad, I.; Basavarajappa, S.; Kumar, P. Results of tensile, hardness and bend tests of modified 9Cr 1Mo steel welds: Comparison between cold wire and hot wire gas tungsten arc welding (GTAW) processes. *Int. J. Press. Vessel. Pip.* **2019**, *169*, 125–141. [CrossRef]
8. Benyounis, K.Y.; Olabi, A.G.; Hashmi, M.S.J. Effect of laser welding parameters on the heat input and weld-bead profile. *J. Mater. Process. Technol.* **2005**, *164*, 978–985. [CrossRef]
9. Peng, W.; Jiguo, S.; Shiqing, Z.; Gang, W. Control of wire transfer behaviors in hot wire laser welding. *Int. J. Adv. Manuf. Technol.* **2016**, *83*, 2091–2100. [CrossRef]
10. Liu, W.; Ma, J.; Liu, S.; Kovacevic, R. Experimental and numerical investigation of laser hot wire welding. *Int. J. Adv. Manuf. Technol.* **2015**, *78*, 1485–1499. [CrossRef]
11. Zhou, Q.; Rong, Y.; Shao, X.; Jiang, P.; Gao, Z.; Cao, L. Optimization of laser brazing onto galvanized steel based on ensemble of metamodels. *J. Intell. Manuf.* **2018**, *29*, 1417–1431. [CrossRef]
12. Yang, Y.; Longchao, C.; Chaochao, W.; Qi, Z.; Ping, J. Multi-objective process parameters optimization of hot-wire laser welding using ensemble of metamodels and NSGA-II. *Robot. Comput. Integr. Manuf.* **2018**, *53*, 141–152. [CrossRef]
13. Li, J.; Sun, Q.; Kang, K.; Zhen, Z.; Liu, Y.; Feng, J. Process stability and parameters optimization of narrow-gap laser vertical welding with hot wire for thick stainless steel in nuclear power plant. *Opt. Laser Technol.* **2020**, *123*, 105921. [CrossRef]
14. Liu, W.; Liu, S.; Ma, J.; Kovacevic, R. Real-time monitoring of the laser hot-wire welding process. *Opt. Laser Technol.* **2014**, *57*, 66–76. [CrossRef]
15. Spaniol, E.; Ungethüm, T.; Trautmann, M.; Andrusch, K.; Hertel, M.; Füssel, U. Development of a novel TIG hot-wire process for wire and arc additive manufacturing. *Weld. World* **2020**, *3*, 1–12. [CrossRef]
16. Liu, X.F.; Jia, C.B.; Wua, C.S.; Zhang, G.K.; Gao, J.Q. Measurement of the keyhole entrance and topside weld pool geometries in keyhole plasma arc welding with dual CCD cameras. *J. Mater. Process. Technol.* **2017**, *248*, 39–48. [CrossRef]
17. Pinto-Lopera, J.E.; Motta, J.M.S.T.; Alfaro, S.C.A. Real-Time Measurement of Width and Height of Weld Beads in GMAW Processes. *Sensors* **2016**, *16*, 1500. [CrossRef]
18. Silwal, B.; Santangelo, M. Effect of vibration and hot-wire gas tungsten arc (GTA) on the geometric shape. *J. Mater. Process. Technol.* **2018**, *251*, 138–145. [CrossRef]
19. Wang, H. Effect of Welding Variables on Cooling Rate and Pitting Corrosion Resistance in Super Duplex Stainless Weldments. *Mater. Trans.* **2005**, *46*, 593–601. [CrossRef]

20. Mohammed, G.R.; Ishak, M.; Aqida, S.N.; Abdulhadi, H.A. Effects of Heat Input on Microstructure, Corrosion and Mechanical Characteristics of Welded Austenitic and Duplex Stainless Steels: A Review. *Metals* **2017**, *7*, 39. [CrossRef]
21. Arora, A.; Roy, G.G.; DebRoy, T. Cooling rate in 800 to 500 °C range from dimensional analysis. *Sci. Technol. Weld. Join.* **2010**, *15*, 423–427. [CrossRef]
22. Poorhaydari, K.; Patchett, B.M.; Ivey, D.G. Estimation of Cooling Rate in the Welding of Plates with Intermediate Thickness. *Weld. J.* **2005**, *10*, 149–155.
23. 316 Stainless Steel, Annealed Sheet. Available online: http://www.matweb.com/ (accessed on 5 June 2020).
24. Thermography Theory. Available online: https://www.infratec.eu/thermography/service-support/ (accessed on 18 July 2020).

Article

Laser Oscillating Welding of TC31 High-Temperature Titanium Alloy

Zhimin Wang [1], Lulu Sun [2], Wenchao Ke [3], Zhi Zeng [3], Wei Yao [2] and Chunming Wang [4,*]

1 School of Mechanical Science and Engineering, Huazhong University of Science and Technology, Wuhan 430074, China; wangzhimin2020@yeah.net
2 Special Machining Department, Beijing Hangxing Machine Manufacture Co., Ltd., Beijing 100013, China; sunll@live.cn (L.S.); yaowei239@sina.com (W.Y.)
3 School of Mechanical and Electrical Engineering, University of Electronic Science and Technology of China, Chengdu 611731, China; ke@std.uestc.edu.cn (W.K.); zhizeng@uestc.edu.cn (Z.Z.)
4 School of Materials Science and Engineering, Huazhong University of Science and Technology, Wuhan 430074, China
* Correspondence: cmwang@hust.edu.cn; Tel./Fax: +86-027-875-438-94

Received: 15 July 2020; Accepted: 20 August 2020; Published: 3 September 2020

Abstract: The joining of high-temperature titanium alloy is attracting much attention in aerospace applications. However, the defects are easily formed during laser welding of titanium alloys, which weakens the joint mechanical properties. In this work, laser oscillating welding was applied to join TC31 high-temperature titanium alloy. The weld appearance, microstructure and mechanical properties of the laser welds were investigated. The results show that sound joints were formed by using laser oscillating welding method, and a large amount of martensite was presented in the welds. High mechanical properties were achieved, which was approaching to (or even equaled) the strength of the base material. The joints exhibited a tensile strength of up to 1200 ± 10 MPa at room temperature and 638 ± 6 MPa at 923 K. Laser oscillating welding is beneficial to the repression of porosity for welding high-temperature titanium alloy.

Keywords: laser oscillating welding; high temperature titanium alloy; microstructure; mechanical properties

1. Introduction

Titanium alloys exhibit high specific strength, excellent corrosion resistance and high thermal strength, which have been widely used in aerospace field to reduce weight [1–4]. In recent years, as the flight speed and distance of air vehicles increase significantly, traditional titanium alloys such as TC4, TA15 cannot cover the operating requirements. High-temperature titanium alloys possess excellent mechanical properties at high temperature, and the service temperature of these alloys is increased to above 873 K [5–7]. Due to these excellent performances, high-temperature titanium alloys have become advanced materials in aerospace applications, such as the components of aircraft engines, wings and rudders [8]. TC31 alloy is a high-temperature double-phase titanium alloy of the Ti-Al-Sn-Zr-Mo-Nb-W-Si system with high aluminum content, which can be used in a temperature range of 923–973 K [9–11]. This alloy also has good durability and creep properties under high load and high temperature [12]. Therefore, it has shown a prefect application in the aerospace field. To obtain a sound joint with good mechanical performance is a key factor to achieve structural integrity, light weight and low-cost manufacturing for its applications.

As an important part of laser manufacturing technology, laser welding is currently one of the most notable and most promising welding technologies. Due to its advantages of high quality, low deformation, high precision, high efficiency and high speed, laser welding technology has become a key joining method for airplanes and automobiles—with improved safety and structural weight reduction, clean and efficient shipbuilding—and nuclear power plant construction [13,14]. It is an effective way to achieve the upgrading of traditional industrial structures and achieve energy saving and emission reduction. Laser welding has been able to achieve the joining of many types of materials, and has many unmatched advantages of fusion welding processes [15–17]. For example, laser welding is used to weld aircraft skin and stringers [18,19]. Compared with riveted structure, the weight and cost of airplane is reduced dramatically. However, due to high degree of alloying and low plasticity-reserve for high-temperature titanium alloy, the possibility of cracking of the joint is higher than that of traditional titanium alloy, and it is easy to generate weld defects during welding [20–22]. The fewer the weld defects, the better the weld mechanical performances. In addition, the mechanical performances have a relationship with the laser welding parameters, such as laser power and welding speed, as verified by finite element method (FEM) and artificial neural network (ANN) simulation techniques [23–25]. Recently, laser oscillating welding has been promising in reducing the weld porosities and increasing the weld ductility [26]. However, most of the studies about laser oscillating welding purely focused on the mechanism of porosity suppression; there is a need for an exhaustive description of the relationship between the weld mechanical performances and the process parameters.

The paper gives the comparison of the weld profiles, microstructures and mechanical properties under different laser powers, welding speeds, laser beam weaving frequencies and amplitudes to obtain the sound welding parameters. In the study, high-temperature titanium alloy TC31 was welded by a laser oscillating welding method. The appearance, internal quality and microstructure of the weld under different welding parameters were observed. The mechanical properties of the joints at room temperature and high temperature were investigated. The mechanism for improved joint strength is also discussed.

2. Experimental

The base metal is high-temperature titanium alloy TC31 in this study, and its composition and mechanical properties are listed in Tables 1 and 2, respectively. TC31 alloy was machined into plates with a dimension of 200 mm × 100 mm × 3 mm. The plates were polished with SiC sandpaper, acid pickled in HF solution and ultrasonically cleaned with acetone and ethyl alcohol.

Table 1. The composition of high-temperature titanium alloy TC31 (wt%).

Element	Ti	Al	Sn	Zr	Mo	Nb	W	Si
Composition	Balance	6.0–7.2	2.5–3.5	2.5–3.5	1.0–3.2	1.0–3.2	0.3–1.2	0.1–0.5

Table 2. Mechanical properties of alloy TC31 with the thickness of 3 mm at room temperature and high temperature 923 K.

Temperature	UTS/MPa	YS/MPa	Elongation/%
RM	1205	1091	15.2
650 °C	648	-	17.3

Butt welding experiments were carried out on the base material using laser oscillating welding method. The laser welding parameters for two welding methods are listed in Table 3. In laser oscillating welding experiments, the laser beam oscillated in a figure-eight manner with the weaving frequency of 200–400 Hz and the weaving amplitude of 0.1–0.5 mm. The laser power was in the range of 1900–2100 W, and the welding speed is in the range of 1200–1800 mm/min. Since titanium alloy is easily reacted with ambient gases, argon was used as shielding gas with the flow rate of 15 L/min during welding process. The laser focused on the workpiece surface with a diameter of 0.45 mm, defocus amount of 0 and a working distance of 247 mm. A D50 wobble seam oscillating head (IPG Laser GmbH, Burbach, Germany) was utilized to adjust the oscillating frequency and amplitude.

Table 3. Laser oscillating welding parameters.

Sample	Weaving Frequency of Laser Beam (Hz)	Weaving Amplitude of Laser Beam (mm)	Laser Power (W)	Welding Speed (mm/min)
W1	200	0.1	1900	1200
W2	200	0.3	2000	1500
W3	200	0.5	2100	1800
W4	300	0.1	2000	1800
W5	300	0.3	2100	1200
W6	300	0.5	1900	1500
W7	400	0.1	2100	1500
W8	400	0.3	1900	1800
W9	400	0.5	2000	1200

After welding, the internal quality of the welds was examined by X-ray nondestructive testing instrument (NDT, MU2000-D) (YXLON International GmbH, Hamburg, Germany) using a tube voltage of 120 kV and a tube current of 3.4 mA. Due to the limitation of capability of equipment, the nondestructive testing was conducted twice on different parts for each sample. The microstructure of the weld was studied by a GX53 metallurgical microscope (Olympus Corporation, Tokyo, Japan) after being etched with Kroll's reagent: 2 mL HF, 6 mL HNO_3 and 92 mL H_2O. X-ray diffraction (XRD, D/MAX-2500) (Rigaku Corporation, Tokyo, Japan) using Cu-K_α radiation was employed to examine the structure of the welds. For X-ray diffraction measurements, a diffractometer with the X-ray tube operating at 40 kV and 200 mA target current was used. The microstructure of the joints was also studied using an electron probe microanalyzer (EPMA, JXA-8100) (JEOL Ltd., Tokyo, Japan). To determine the mechanical properties of the joints at room temperature and high temperature 923 K, the weld plate samples were machined into joint samples with dimensions depicted in Figure 1 for measurement (as GB/T 228.1-2010 standard and GB/T 228.2-2015 standard). The samples were cut along the direction perpendicular to the weld. Tensile tests were performed, respectively, on three joint samples by using a universal testing machine at a tensile speed of 0.5 mm/min. After the tensile test, fracture surfaces of the weld joints were observed by a scanning electron microscope (SEM, JSM-7001F) (JEOL Ltd., Tokyo, Japan) with an energy dispersive spectrometer (EDS, Pegasus XM2 EDS) (EDAX Inc., Mahwah, NJ, USA) system.

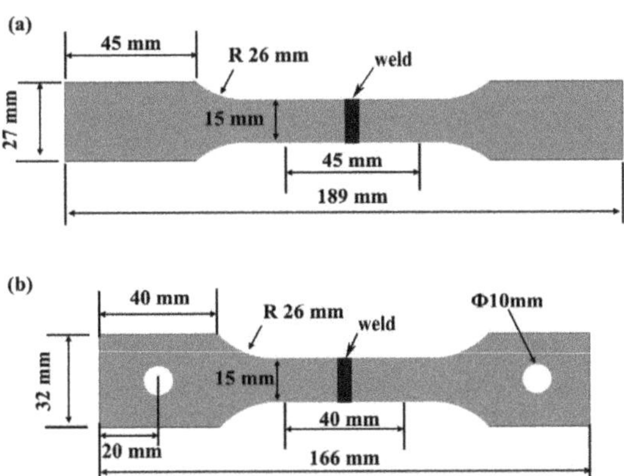

Figure 1. Dimensions of the specimens in tensile testing: (**a**) sample design for room temperature testing; (**b**) sample design for high temperature testing.

3. Results and Discussion

3.1. Weld Appearance and Internal Quality

High-temperature titanium alloy TC31 plate samples were weld under different laser welding conditions listed in Table 3. The front and back appearances of the butt weld joints are shown in Figure 2. As shown in Figure 2, the welds are continuous in the surfaces without cracks, and uniform fish-scale patterns are observed. The samples except samples W5 and W9 display the front appearance as yellow or silvery white. With increasing the laser power up to 2100 W or decreasing welding speed to 1200 mm/min, the front appearances of the welds present a color of green, purple or blue, which indicates that the oxidation of the front appearances is further deepened [27]. By contrast, the back appearance of all the welds is silvery-white in color, indicating an effective protection for the joints through gas shielding. Besides, as can be seen from Figure 2e,i, the back weld width of W5 and W9 samples are obviously larger than those of other samples, indicating the higher heat input.

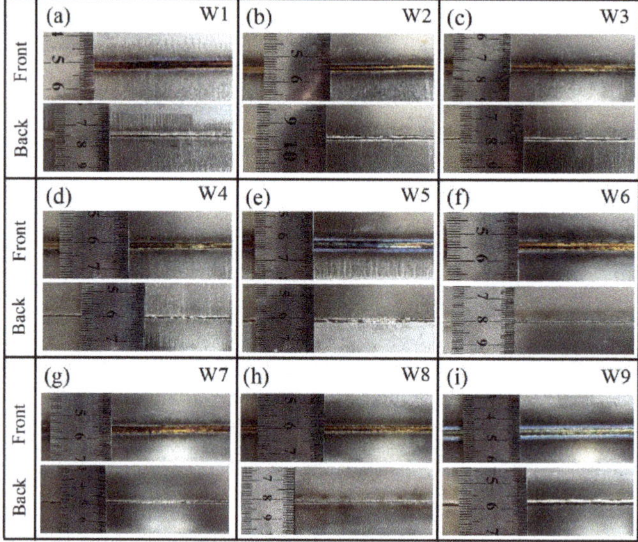

Figure 2. The front and back appearances of the laser-welded TC31 joints: (**a–i**) represent W1 to W9 samples.

For further investigating the internal quality of the welds, the X-ray non-destructive testing was conducted and the result is shown in Figure 3. Except that porosity defects are observed in W8, other joints exhibit sound bonding without lack of fusion, slag or hot cracks.

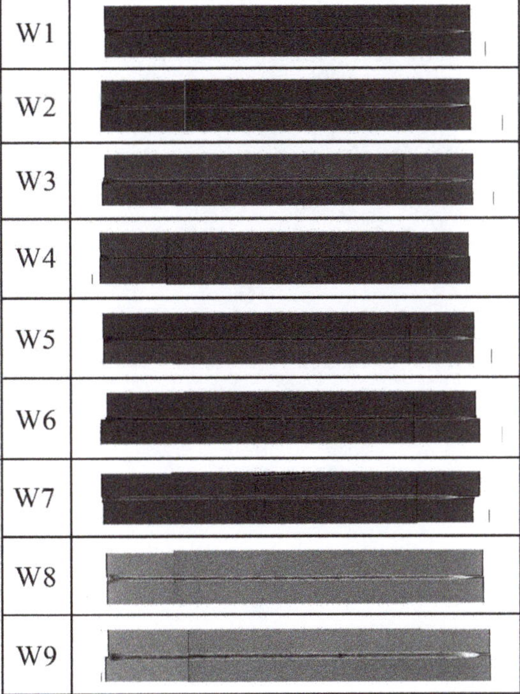

Figure 3. Images of the laser-welded TC31 joints in X-ray detection.

3.2. Microstructure Characterization

Figure 4 shows the optical micrographs of the cross-sections of W1 to W9 laser welded joints. The joints exhibit slight defect of undercut. The undercut with a maximum depth (≈0.165 mm) was occurred in W9 sample, and the undercut imperfection is 0.055t (t is the thickness of the alloy plate), which is lower than the limit of for quality C level (0.1t) according to ISO 5817 standard. Thus, the laser-welded TC31 joints reach the weld quality C level (medium quality requirements). In addition, no misalignment defect was observed in these joints. Three distinct regions in the joints can be observed, i.e., the fusion zone (FZ), the heat-affect zone (HAZ) and the base material (BM). The FZ mainly consists of coarse prior β columnar grains. There are a few equiaxed grains at the junction of the FZ and the HAZ. The columnar grains in the middle of cross section grow horizontally and towards the weld centerline, while the columnar grains closed to the weld front and back surfaces grow from the weld fusion line to the weld surface. Figure 5 shows the magnification view of the fusion zone and heat-affected zone of the W5 joint. It can be seen that there is a large amount of staggered needle-shaped α' phases inside the columnar crystal, which is the typical structure of martensite [28].

The phases in the FZ of the welded joints are examined by XRD. The examination is carried out on W2. As shown in Figure 6, the XRD pattern consists of the peaks corresponding to α'-Ti phase, and no peak corresponding to β-Ti phase is observed. This result implies that the major phase in the weld is α'-Ti phase. For further analyzing the microstructure of the joints, EPMA analysis is also conducted on the weld of W2. The backscattered electron image of the cross-section of the joint is shown in Figure 7. It is seen that the laser welded joint is free from obvious defects, such as voids and cracks. The BM consists of elongated α grain and intergranular β grain (Figure 7a). A large amount of acicular Ti and columnar prior β grain boundaries can be observed in the FZ as shown in Figure 7b,c, which implies that the prior β phase gradually transform into the secondary α phase [1,29].

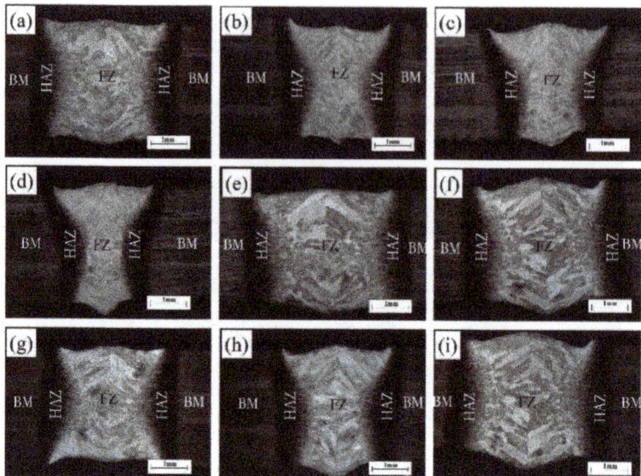

Figure 4. Optical micrographs of cross sections of as the laser-welded TC31 joints: (**a**–**i**) represent W1 to W9 samples.

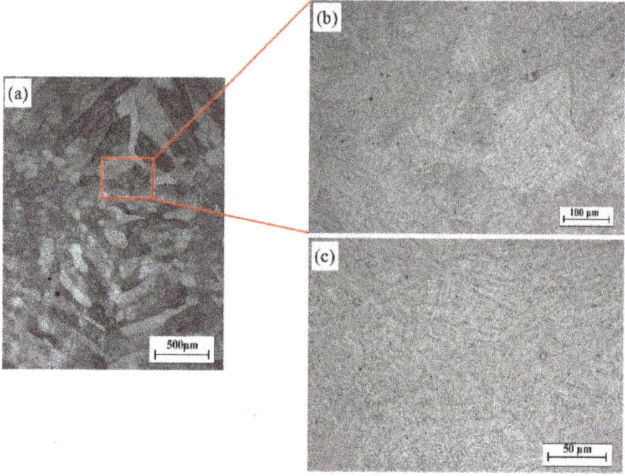

Figure 5. Magnification view of fusion zone and heat-affected zone of W5 joint: (**a**) fusion zone and heat-affected zone; (**b**) high-magnification scanning electron microscope image of the rectangular region in (**a**); (**c**) high-magnification scanning electron microscope image of (**b**).

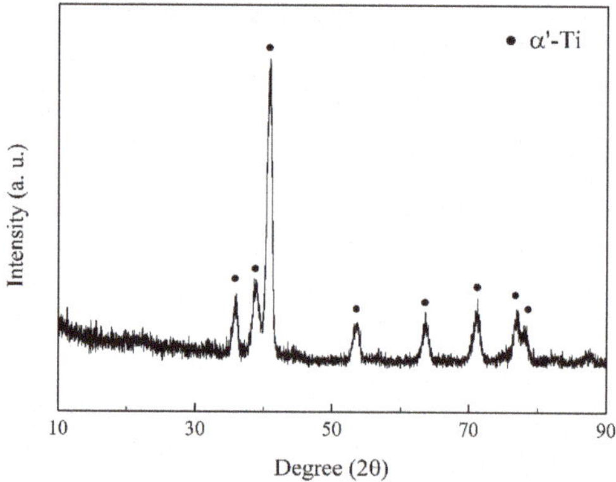

Figure 6. X-ray diffraction pattern of W2 weld.

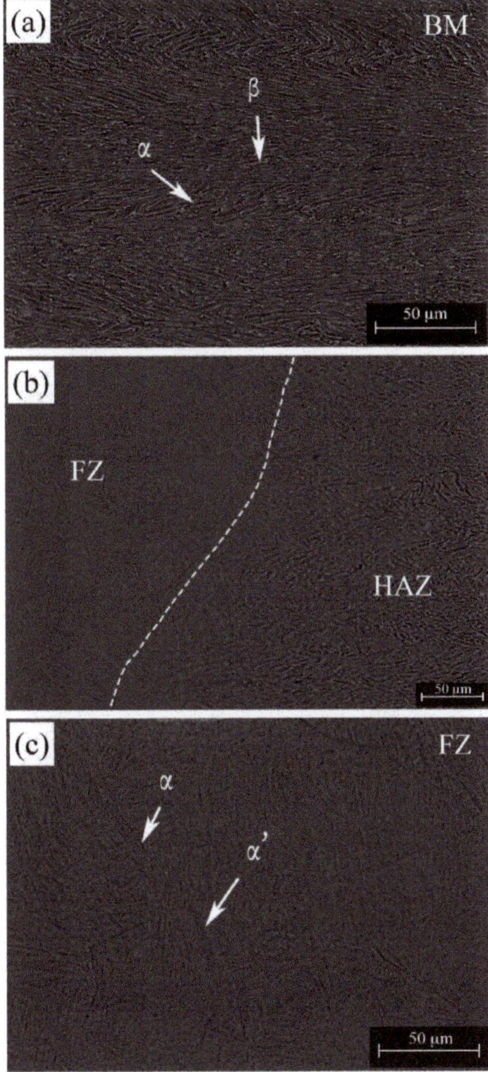

Figure 7. Backscattered electron images of the cross-section of W2 joint: (**a**) Base material (BM); (**b**) Heat-affect zone/Fusion zone (HAZ/FZ) transition zone; (**c**) Fusion zone (FZ).

The formation mechanism of the joint is discussed as follows. The microstructural transformation of the welds for TC31 laser-welded joints are influenced by the welding thermal cycle and the ingredient distribution of the bonding zone [13]. As illustrated in Figure 4, the size of prior grain boundaries significantly increased. This may be due to the growth of grain is sensitive to overheating. During the laser welding process, the unmelt BM closed to the bottom of welding pool and the top of heat-affecting zone is under the condition of ultra-high temperature reaching or exceeding the liquidus temperature of phase. The grain nucleates at the surface of partially molten base material, and then grow quickly towards the center of the weld. With the growth of columnar grain, the temperature gradient of the melt gradually flattens, and the solute concentration in melt increases, resulting in the constitutional undercooling in the front of the solid-liquid interface increasing and the equiaxed crystals forming in the center of the weld. Due to the solidification behavior of weld center occur during the final stage of solidification, the growth of the equiaxed crystals is limited by the solidified columnar grains. Moreover, the columnar grains have an insulating effect on the equiaxed grains, which results in the coarse grains in the weld center. Near the surface of the weld, due to the change in heat dissipation conditions in the middle solidification of the weld, the columnar crystal grows toward the surface of the weld, and there is a large angle between the growth direction and the growth direction of the central columnar crystal. Laser welding is a rapid heating and cooling process, rapid quenching will cause martensite transformation of titanium alloy [1]. During the cooling process, the initial β columnar crystals of the weld metal generate α′ phase via shear transformation. It is due to the rapid cooling rate frustrating the atomic diffusion of the β phase. The α′ martensite grows and form one or several primary needle-like martensite parallel to each other, and then form a series of relatively fine secondary needle-like martensite. These secondary martensitic grains stop growing when encountering grain boundary or primary martensite, resulting in the formation of a typical staggered needle-like structure in the welding seam of the laser oscillating welds [29,30].

3.3. Mechanical Properties

The tensile strength of the welded specimens at room temperature are listed in Figure 8, and the photographs of W1 and W2 joints after the tensile test are shown in the insets of Figure 8. It is observed that the average tensile strength of W2 reaches 1200 ± 10 MPa, which is equal to the tensile strength of TC31 alloy. Meanwhile, the average tensile strength of W1, W6, W7 and W9 samples also exhibit high tensile strength, reaching 1183 ± 4 MPa, 1185 ± 31 MPa, 1184 ± 63 MPa and 1190 ± 1 MPa, respectively. The fracture of two specimens of W1 joint occurred at the base material, while the ruptures of W2 specimens occurred in the HAZ. As seen in Figure 8, the W8 sample exhibits relatively low mechanical properties, which is caused by the porosity defects in the weld [31,32].

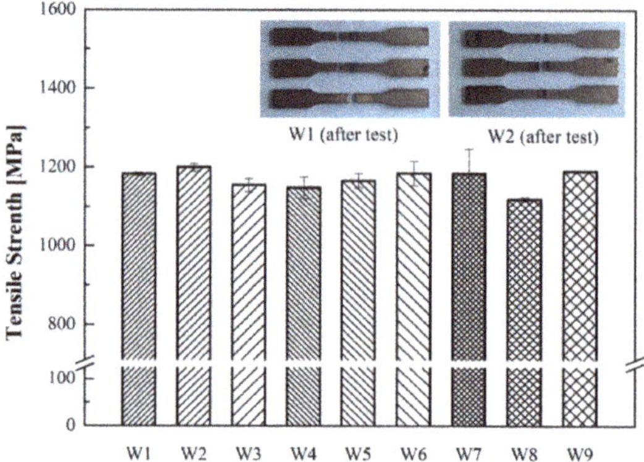

Figure 8. Tensile strength of welded TC31 specimens by wave laser oscillating welding method at room temperature. (The insets show the photographs of W1 and W2 joints after the tensile test.).

Given that the working temperature of TC31 alloy can reach from 923 K to 973 K, the mechanical properties of the joints at high temperature conditions is also essential. The tensile strength of the welded specimens at 923 K are also tested and listed in Figure 9. As seen in Figure 9, the tensile strength of W2 joint reaches 635 ± 3 MPa. Moreover, the highest tensile strength for TC31 specimens by laser oscillating welding at 923K can reach up to 638 ± 6 MPa for W1 joint. The fracture of W1 joint specimens at 923 K occurred at the BM away from the weld seams.

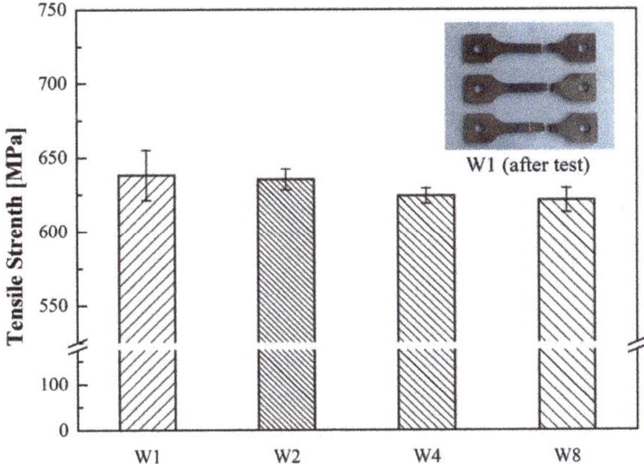

Figure 9. Tensile strength of welded TC31 specimens by laser oscillating welding method at 923 K. (The inset show the photograph of the W1 joint after the high-temperature tensile test.).

In order to investigate the mechanism for the high strength of the joints by laser oscillating welding, the fracture surfaces of the welded joints were examined by SEM. Figure 10 presents the magnified view of the fracture surface of W2 after the tensile test performed at room temperature. A large number of dimples densely distribute on the fracture zone, which is a typical feature of ductile failure. The formation of the dimples implies plastic deformation in the micro regions, which is suggested to be favorable for improving mechanical properties of the joint. Figure 11 presents the magnified view of the fracture surface of W1 joint after the tensile test performed at 923 K. It can be seen from Figure 11 that the area where the fracture is close to the surface of the weld (Figure 11b) and the middle part of the welded joint (Figure 11c,d) show distinct features. Region i fracture morphology shows the characteristics of tough dimple fractures. The fracture surface is rock sugar-like. There are a large number of small dimples on the grain boundaries of the fracture surface and show slip characteristics. This results from nucleus, growth, and connection. Meanwhile, there are also a small number of dimples concentrated in some areas (red dotted area in Figure 11b). As can be seen from Figure 11c,d, there are a large number of dimple structures in Region ii in the middle part of the welded joint. The above results indicate that the joint undergoes ductile fracture under high temperature conditions.

The TC31 joint fabricated by the laser oscillating welding method exhibits high tensile mechanical properties at room temperature, which is exceeding or approaching the mechanical properties of the base alloy. On the one hand, as can be seen from Figures 4 and 7, no defects, such as cracks, inclusions, or unwelded joints, are generated except W8 sample. Laser welding is a rapid melting and solidification process, and it is difficult for the pores formed in the molten pool to escape, which becomes the stress concentration point of the joint. Compared with pulsed laser welding method [12], the content of pores in the weld seam obtained by wave laser oscillating welding is significantly reduced, and the depth-to-width ratio of the weld seam is relatively small. This is due to the adoption of wave laser oscillating welding method, the reciprocating swing of the laser beam on the weld seam causes part of the weld to remelt repeatedly, prolonging the residence time of the molten metal in the weld pool. The deflection of the laser beam also increases the heat input per unit area, reduce the depth-to-width ratio of the weld, which is favorable for the escape of bubbles. Besides, the oscillation of the laser beam causes the air holes to oscillate, which can also provide the stirring force for the welding molten pool, increase the convection and stirring of the molten pool and then eliminate the air holes. This is a favorable factor for improving the mechanical properties of the joint [33–35].

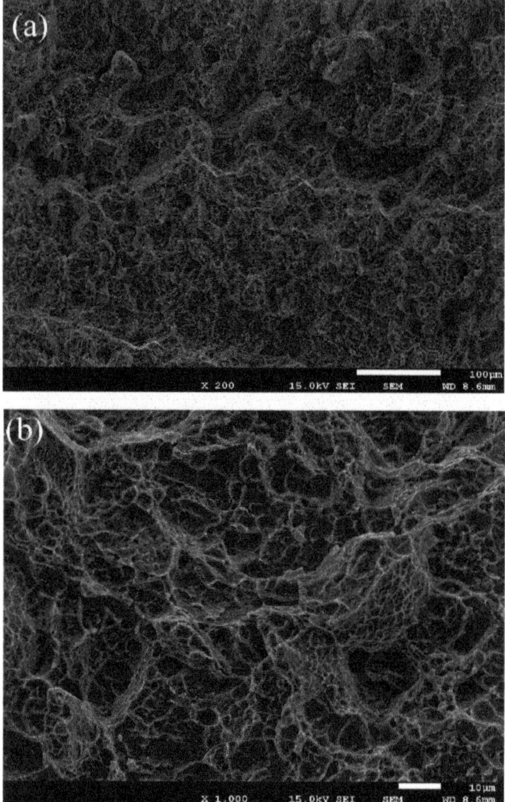

Figure 10. Fracture surface of W2 welded joint after tensile test performed at room temperature, (**b**) is the magnified view of (**a**).

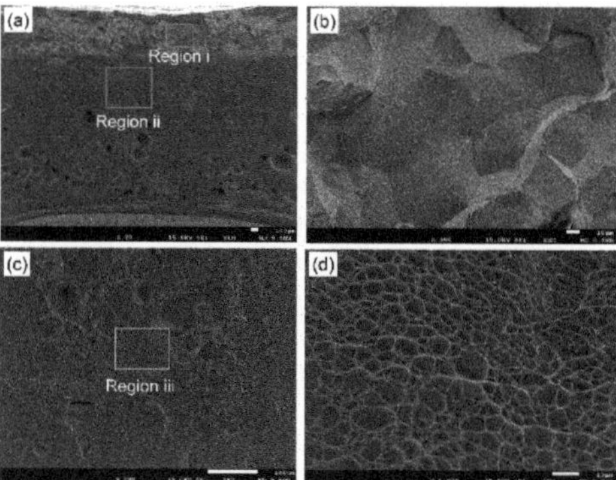

Figure 11. Fracture surface of W1 welded joint after tensile test performed at 923 K: (**a**) macroscopic appearance of fracture; (**b**) high magnification magnification of Region i; (**c**) high magnification magnification of Region ii; (**d**) high magnification magnification of Region iii.

On the other hand, the TC31 high temperature titanium alloy forms a martensite structure in the weld after laser welding. Martensite α' is a supersaturated solid solution of alloying elements inα phase. During the cooling of the weld from the β-phase region to room temperature at a rapid cooling rate, the atoms have no time to diffuse, and only a small needle-shaped and unevenly distributed martensite structure can be precipitated by shearing. α' grows inside the initial β columnar crystals, forming one or several parallel primary α' firstly, and extending through the entire grain over a long distance, stopping at the grain boundary. Then a series of relatively fine secondary needles α' are formed, which stops at the grain boundaries or primary martensite, resulting in the formation of a typical basket structure in the weld. This structure has good comprehensive properties, such as plasticity, creep resistance and high-temperature endurance strength [1].

4. Conclusions

In this study, high-temperature titanium alloy TC31 was joined by the laser oscillating welding method and good mechanical properties were achieved. The appearance, microstructure and mechanical properties of the welded joints were investigated. High-temperature titanium alloy TC31 formed sound joints by the laser oscillating welding method under the optimized process parameters with the weaving frequency of 200 Hz, weaving amplitude of 0.3 mm, laser power of 2000 W and welding speed of 1500 mm/min. The joints exhibit sound bonding without lack of fusion, slag or hot cracks. The welded region was composed of acicular α' phase resulting from the high degree of supercooling during the laser welding process. The welded joints exhibit the highest tensile strength of 1200 ± 10 MPa at room temperature, which is approaching that of the base metal. The tensile strength of the joints at the high temperature of 923 K reaches 638 ± 6 MPa. The welded joints exhibit ductile fracture with dimples at both room temperature and high temperature. Considering the mechanical properties of TC31 laser-weld joints at different temperatures and the potential usage environment of this high-temperature titanium alloy, W1 joint with 200 Hz weaving frequency, 0.1 mm weaving amplitude, 1900 W laser power and 1200 mm/min welding speed, exhibit outstanding mechanical properties in this study. Furthermore, laser oscillating welding is beneficial to the repression of porosity for welding high temperature titanium alloy.

Author Contributions: Conceptualization, data curation, investigation and resources, Z.W.; writing—original draft preparation, L.S.; methodology, W.K.; validation, Z.Z.; supervision, W.Y.; funding acquisition, writing—review and editing, C.W. All authors have read and agreed to the published version of the manuscript.

Funding: This research was funded by the National Key R&D Program of China, grant number 2017YFB1301603.

Conflicts of Interest: The authors declare no conflict of interest.

References

1. Leyens, C.; Peters, M. *Titanium and Titanium Alloys*; Wiley-VCH: Weinheim, Germany, 2003.
2. Boyer, R.R.; Briggs, R.D. The use of β titanium alloys in the aerospace industry. *J. Mater. Eng. Perform.* **2005**, *14*, 681–685. [CrossRef]
3. Qiao, Y.; Xu, D.; Wang, S.; Ma, Y.; Chen, J.; Wang, Y.; Zhou, H. Corrosion and tensile behaviors of Ti-4Al-2V-1Mo-1Fe and Ti-6Al-4V titanium alloys. *Metals* **2019**, *9*, 1213. [CrossRef]
4. Singh, P.; Pungotra, H.; Kalsi, N.S. On the characteristics of titanium alloys for the aircraft applications. *Mater. Today Proc.* **2017**, *4*, 8971–8982. [CrossRef]
5. Gogia, A.K. High-temperature titanium alloys. *Def. Sci. J.* **2005**, *55*, 149–173. [CrossRef]
6. Evans, R.W.; Hull, R.J.; Wilshire, B. The effects of alpha-case formation on the creep fracture properties of the high-temperature titanium alloy IMI834. *J. Mater. Process. Technol.* **1996**, *56*, 492–501. [CrossRef]
7. Narayana, P.L.; Kim, S.W.; Hong, J.K.; Reddy, N.S.; Yeom, J.T. Tensile properties of a newly developed high-temperature titanium alloy at room temperature and 650 °C. *Mat. Sci. Eng. A* **2018**, *718*, 287–291. [CrossRef]
8. Zhao, Z.L.; Li, H.; Fu, M.W.; Guo, H.Z.; Yao, Z.K. Effect of the initial microstructure on the deformation behavior of Ti60 titanium alloy at high temperature processing. *J. Alloy Compd.* **2014**, *617*, 525–533. [CrossRef]
9. Su, Y.; Kong, F.T.; You, F.H.; Wang, X.P.; Chen, Y.Y. The high-temperature deformation behavior of a novel near-α titanium alloy and hot-forging based on the processing map. *Vacuum* **2019**, *173*, 109135. [CrossRef]

10. Song, X.Y.; Zhang, W.J.; Ma, T.; Ye, W.J.; Hui, S.X. Effect of heat treatment on the microstructure evolution of Ti-6Al-3Sn-3Zr-3Mo-3Nb-1W-0.2Si titanium alloy. *Mater. Sci. Forum* **2016**, *879*, 1828–1833. [CrossRef]
11. Zhang, W.J.; Song, X.Y.; Hui, S.X.; Ye, W.J.; Wang, Y.L.; Wang, W.Q. Tensile behavior at 700 °C in Ti–Al–Sn–Zr–Mo–Nb–W–Si alloy with a bi-modal microstructure. *Mat. Sci. Eng. A* **2014**, *595*, 159–164. [CrossRef]
12. Wu, D.; Wu, Y.; Chen, M.; Xie, L.; Wang, B. High Temperature Flow Behavior and Microstructure Evolution of TC31 Titanium Alloy Sheets. *Rare Met. Mater. Eng.* **2019**, *48*, 3901–3910.
13. Hong, K.M.; Shin, Y.C. Prospects of laser welding technology in the automotive industry: A review. *J. Mater. Process. Technol.* **2017**, *245*, 46–69. [CrossRef]
14. Lopes, J.G.; Oliveira, J.P. A short review on welding and joining of high entropy alloys. *Metals* **2020**, *10*, 212. [CrossRef]
15. Meijer, J. Laser beam machining (LBM), state of the art and new opportunities. *J. Mater. Process. Technol.* **2004**, *149*, 2–17. [CrossRef]
16. Mehrpouya, M.; Gisario, A.; Elahinia, M. Laser welding of NiTi shape memory alloy: A review. *J. Mater. Process. Technol.* **2018**, *31*, 162–186. [CrossRef]
17. Cao, X.; Jahazi, M.; Immarigeon, J.P.; Wallace, W. A review of laser welding techniques for magnesium alloys. *J. Mater. Process. Technol.* **2006**, *171*, 188–204. [CrossRef]
18. Grbović, A.; Sedmak, A.; Kastratović, G.; Petrašinović, D.; Vidanović, N.; Sghayer, A. Effect of laser beam welded reinforcement on integral skin panel fatigue life. *Eng. Fail. Anal.* **2019**, *101*, 383–393. [CrossRef]
19. Reitemeyer, D.; Schultz, V.; Syassen, F.; Seefeld, T.; Vollertsen, F. Laser welding of large scale stainless steel aircraft structures. *Phys. Procedia* **2013**, *41*, 106–111. [CrossRef]
20. Nakai, M.; Niinomi, M.; Akahori, T.; Hayashi, K.; Itsumi, Y.; Murakami, S.; Oyama, H. Microstructural factors determining mechanical properties of laser-welded Ti–4.5Al–2.5Cr–1.2Fe–0.1C alloy for use in next-generation aircraft. *Mat. Sci. Eng. A* **2012**, *550*, 55–65. [CrossRef]
21. Gursel, A. Crack risk in Nd: YAG laser welding of Ti-6Al-4V alloy. *Mater. Lett.* **2017**, *197*, 233–235. [CrossRef]
22. Quazi, M.M.; Ishak, M.; Fazal, M.A.; Arslan, A.; Rubaiee, S.; Qaban, A.; Manladan, S.M. Current research and development status of dissimilar materials laser welding of titanium and its alloys. *Opt. Laser Technol.* **2020**, *126*, 106090. [CrossRef]
23. Casalino, G.; Losacco, A.M.; Arnesano, A.; Facchini, F.; Pierangeli, M.; Bonserio, C. Statistical analysis and modelling of an Yb: KGW femtosecond laser micro-drilling process. *Procedia CIRP* **2017**, *62*, 275–280. [CrossRef]
24. Su, X.; Tao, W.; Chen, Y.B.; Fu, J.Y. Microstructure and tensile property of the joint of laser-MIG hybrid welded thick-section TC4 alloy. *Metals* **2018**, *8*, 1002. [CrossRef]
25. Casalino, G.; Facchini, F.; Mortello, M.; Mummolo, G. ANN modelling to optimize manufacturing processes: The case of laser welding. *IFAC-PapersOnLine* **2016**, *49*, 378–383. [CrossRef]
26. Wang, L.; Gao, M.; Zhang, C.; Zeng, X.Y. Effect of beam oscillating pattern on weld characterization of laser welding of AA6061-T6 aluminum alloy. *Mater. Des.* **2016**, *108*, 707–717. [CrossRef]
27. Li, X.; Xie, J.; Zhou, Y. Effects of oxygen contamination in the argon shielding gas in laser welding of commercially pure titanium thin sheet. *J. Mater. Sci.* **2005**, *40*, 3437–3443. [CrossRef]
28. Zhang, H.; Hu, S.S.; Shen, J.Q.; Li, D.L.; Bu, X.Z. Effect of laser beam offset on microstructure and mechanical properties of pulsed laser welded BTi-6431S/TA15 dissimilar titanium alloys. *Opt. Laser Technol.* **2015**, *74*, 158–166. [CrossRef]
29. Zeng, Z.; Oliveira, J.P.; Bu, X.; Yang, M.; Li, R.; Wang, Z. Laser Welding of BTi-6431S High Temperature Titanium Alloy. *Metals* **2017**, *7*, 504. [CrossRef]
30. Junaid, M.; Baig, M.N.; Shamir, M.; Khan, F.N.; Rehman, K.; Haider, J. A comparative study of pulsed laser and pulsed TIG welding of Ti-5Al-2.5Sn titanium alloy sheet. *Mater. Process. Technol.* **2016**, *242*, 24–38. [CrossRef]
31. Martínez, C.; Guerra, C.; Silva, D.; Cubillos, M.; Briones, F.; Muñoz, L.; Sancy, M. Effect of porosity on mechanical and electrochemical properties of Ti–6Al–4V alloy. *Electrochim. Acta* **2020**, *338*, 135858. [CrossRef]
32. Zhang, W.F.; Liu, X.P.; Wang, H.X.; Dai, W.; Fu, G.C. Quantitative analysis of weld-pore size and depth and effect on fatigue life of Ti-6Al-2Zr-1Mo-1V alloy weldments. *Metals* **2017**, *7*, 417. [CrossRef]

33. Zhao, L.; Zhang, X.D.; Chen, W.Z.; Bao, G. Repression of porosity with beam weaving laser welding. *Trans. China Weld. Inst.* **2004**, *251*, 29–32.
34. Panwisawas, C.; Perumal, B.; Mark Ward, R.; Turner, N.; Turner, R.P.; Brooks, J.W.; Basoalto, H.C. Keyhole formation and thermal fluid flow-induced porosity during laser fusion welding in titanium alloys: Experimental and modelling. *Acta Mater.* **2017**, *126*, 251–263. [CrossRef]
35. Assuncao, E.; Williams, S. Comparison of continuous wave and pulsed wave laser welding effects. *Opt. Laser Eng.* **2013**, *51*, 674–680. [CrossRef]

Review

Research Status and Prospect of Laser Impact Welding

Kangnian Wang [1], Huimin Wang [1,*], Hongyu Zhou [1], Wenyue Zheng [1,*] and Aijun Xu [2]

[1] National Center for Materials Service Safety, University of Science and Technology Beijing, Beijing 100083, China; g20189092@xs.ustb.edu.cn (K.W.); hyzhou@ustb.edu.cn (H.Z.)

[2] Beijing Satellite Manufacturing Co. LTD, China Academy of Space Technology, Beijing 100090, China; hitxuaijun@163.com

* Correspondence: wanghuimin@ustb.edu.cn (H.W.); zheng_wenyue@ustb.edu.cn (W.Z.)

Received: 17 September 2020; Accepted: 25 October 2020; Published: 29 October 2020

Abstract: The demands for the connection between thin dissimilar and similar materials in the fields of microelectronics and medical devices has promoted the development of laser impact welding. It is a new solid-state metallurgical bonding technology developed in recent years. This paper reviews the research progress of the laser impact welding in many aspects, including welding principle, welding process, weld interface microstructure and performance. The theoretical welding principle is the atomic force between materials. However, the metallurgical combination of two materials in the solid state by atomic force but almost no diffusion has not been confirmed by microstructure observation. The main theories used to explain the wave formation in impact welding were compared to conclude that caved mechanism and the Helmholz instability mechanism were accepted by researchers. The rebound of the flyer is still a critical problem for its application. With proper control of the welding parameters, the weld failure occurs on the base materials, indicating that the weld strength is higher than that of the base materials. Laser impact welding has been successfully applied in joining many dissimilar materials. There are issues still remained unresolved, such as surface damage of the flyer. The problems faced by laser impact welding were summaried, and its future applications were proposed. This review will provide a reference for the studies in laser impact welding, aiming process optimization and industrial application.

Keywords: laser impact welding; interfacial bonding mechanism; interface wave; diffusion

---

1. Introduction

In the hush service environment, such as the nuclear power plant's primary water reactor, the performance of single metal materials is difficult to meet the requirement. Composite materials can ensure that the components have two or more of the properties of lightweight, high strength, good toughness, corrosion resistance, human body compatibility, and low cost, which has become a direction of current material development [1,2]. Welding is an important process for joining materials, for example, welding is indispensable for the assembly of high-temperature shape memory alloys [3], and it can also perform metallurgical bonding of high-entropy alloy workpieces [4]. However, the huge differences of dissimilar materials, especially in the microstructure, physical and chemical properties, which leads to the production of intermetallic compounds and the large residual stresses during the fusion welding process. Thus, the performance of the bonding area will be reduced [5–7]. To obtain a good metallurgical bond between dissimilar materials, researchers consider using solid-state welding processes for welding dissimilar materials, such as diffusion welding, friction stir welding [8], ultrasonic welding [9], and impact welding (explosive welding [10], magnetic pulse welding [11], etc.). The impact welding temperature is relatively low, and the thermal cycle time is extremely short. In theory, it can be widely used for welding between any metals. Metal cladding is already a common process conducted with explosive welding.

In recent years, there has been an increasing demand for connections between similar or dissimilar metal thin foils, especially in the fields of battery electrodes, special medical materials, metal anti-corrosion coatings, and the coating of heat dissipation layers of semiconductor materials. For example, the electrodes of the heart-pace maker need to be optimized with a good weld between aluminium and titanium. However, the solid-state welding technologies have certain limitations in the connection of dissimilar metal foils. Ultrasonic welding, as shown in Figure 1, produces surface indentation, and friction welding leads to layered intermetallic compounds [12,13]. In 2009, Daehn and Lippold [14] from The Ohio State University in the United States proposed a non-contact welding method with laser as the energy source—Laser impact welding. This process could realize solid metallurgical bonding between dissimilar metal foils with a thickness of millimeters/micrometers, with precise positioning and flexible as well as adjustable welding area size.

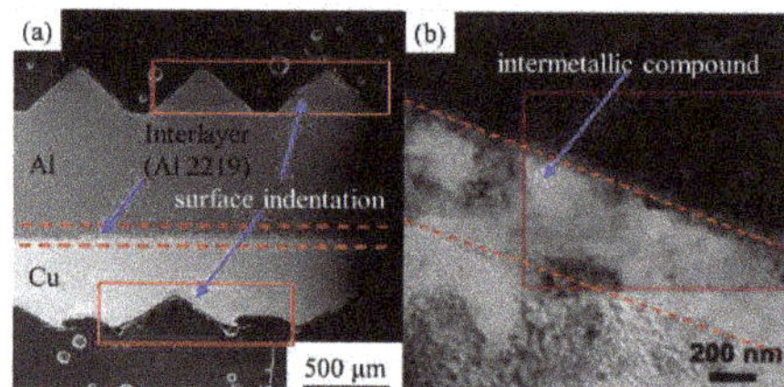

Figure 1. (**a**) Surface indentation of ultrasonic welding (reproduced from [9], with permission of Elsevier 2020); (**b**) Intermetallic compound layer with friction stir welding (reproduced from [15], with permission of Elsevier 2015).

As a new type of solid-state welding process, laser impact welding is still in the laboratory process exploration stage. This review summarizes the progress of the laser impact welding process and the results achieved at this stage from the mechanism of impact welding, the process of impact welding, the structure, and performance of the impact welded joint. It also proposes the development direction of the laser impact welding process and provides its maturity reference.

2. Laser Impact Welding Process

Laser irradiation on the surface of the material will cause temperature and force effects. According to the order of magnitude of energy input from small to large, the phenomenon of temperature rise, melting, vaporization, and plasma excitation will occur in sequence. While vaporizing and exciting the plasma, an instantaneous stress action is formed on the surface.

In the decades since the 1970s, the stability and high speed of the laser-driven flyer flying were verified, and the Gurney mathematical model of flyer flying speed and its influencing factors were established [16–18]:

$$\rho x_d E = (\rho/2)(x_0 - x_d)v_0^2 + (\rho/2)\int_0^{x_d} (v_0 x/x_d)^2 dx \quad (1)$$

ρ is the density of ablated material, x_d is the thickness ablated away, E is called Gurney energy, x_0 is the original thickness, and v_0 is the final velocity.

The formula is based on the principle of conservation of energy. The left side is the energy released by the ablation layer, and the right side is the kinetic energy. Equation (1) was simplified to obtain the final velocity:

$$v_0 = \sqrt{\frac{3E}{3x_0/2x_d - 1}} \quad (2)$$

In the 1940s, Carl first proposed the use of explosives to drive metal and metal collisions for metallurgical bonding, named explosive welding [19]. Nowadays, people use chemical energy [20], electromagnetic field energy [21], high-energy-density light energy [22], high-pressure gas [23], etc. as driving sources to achieve various forms of impact welding by the transient release of high energy and drive high-speed collision of welding parts.

The principle of atomic bonding at the impact welding interface is shown in Figure 2. When atoms reach a certain position, interatomic bonding will occur. However, the obstacles on the surface of the flyer and the target such as oxides, oil stains, and surface impurities prevented the atoms from the flyer and the target to get close within the atomic distance. The basic principle of high-speed impact welding is to remove the bonding obstacles by jet flow, and with the help of the transient and huge impact force of the high-speed impact to make the atoms reach a close enough distance to achieve the bond between the atoms [23].

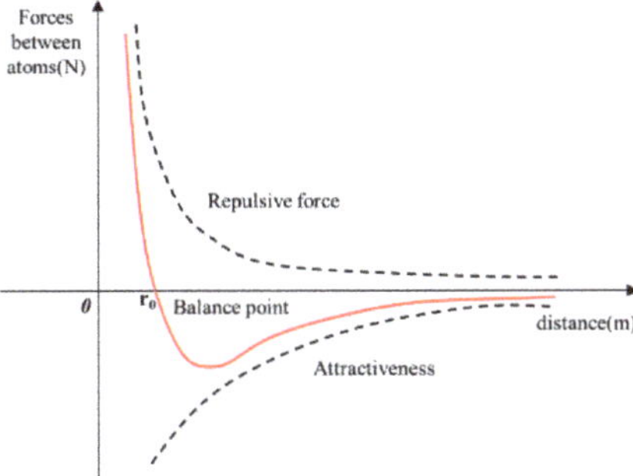

Figure 2. Atomic force-distance curve.

Laser impact welding is also an impact welding process with laser as the driving source, which is mainly used in spot welding of millimeter/micron-scale [24,25]. As shown in Figure 3a, the laser impact welding system is divided into two parts: the energy source, namely the laser system, and the weldment support system [26,27]. To prevent non-uniform force caused by continuous energy input and to ensure energy transfer efficiency, a pulsed laser with a pulse width of about 10 ns and a wavelength of 1064 nm is usually used. Since the laser can achieve 0–100% capacity adjustment, the greater the energy that the laser can achieve, the wider the applicability, but generally the minimum energy required to achieve millimeter-level spot welding is about 1 J. The commonly used laser types are flat-top laser and Gaussian laser. The energy distribution of the laser beam is shown in Figure 3b,c respectively. We can change the laser beam diameter by adjusting the position of the convex lens to determine the energy density and solder joint size. The arrangement sequence of the support system from left to right is confinement layer, ablation layer, flyer, standoff, and target. The whole set of equipment is fixed on the stander.

During the impact welding process, due to collision and extrusion, a jet along the welding direction is generated at the collision point to clean the surface, which is a necessary condition for the metallurgical bonding of impact welding. For the flat-top laser-driven flyer, the flat-top light action area on the flyer first collides with the target in parallel, that is, the impact angle is zero degrees, there is no metallurgical bonding in this area, and rebound occurs for a large area after the collision. As the impact collision progresses, the impact angle gradually increases, enters the welding window, and the interface metallurgical bond is formed. Therefore, the solid-state metallurgical bonding area produced by the flat-top laser is a narrow ring shape. However, the metallurgical bonding area produced using Gaussian laser to drive the flyer is a wider circular ring shape.

The parameters of laser impact welding are also categorized into two groups: laser system parameters and weldment combination parameters. The parameters of the laser system are the laser energy and laser spot size, that is, the laser energy and laser spot size acting on the ablation layer. The combination parameters of weldment are complex, as shown in Table 1, including various indexes of the weldment support system, such as confinement layer, binder, ablative layer, the thickness of the flyer, preset flying distance, and so on [26].

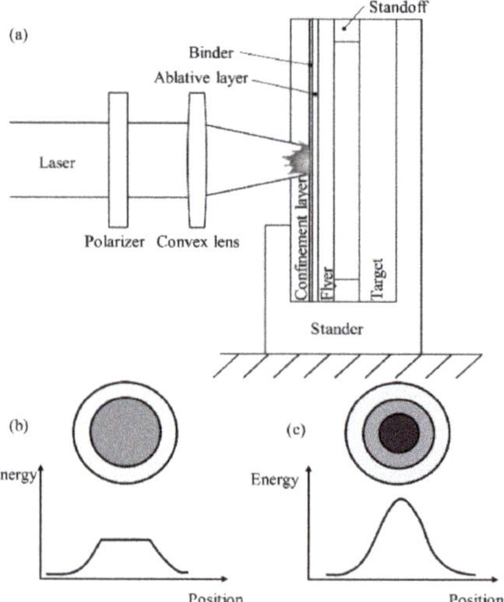

Figure 3. (**a**) Schematic diagram of laser impact welding system; (**b**) flat-top laser energy distribution; (**c**) Gaussian laser energy distribution.

After starting the laser, laser impact welding can be divided into three stages, as shown in Figure 4 [28]:

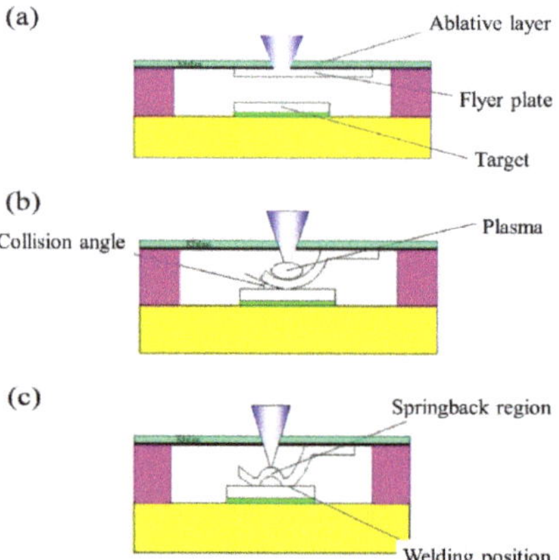

Figure 4. Schematic diagram of laser impact welding process (reproduced from [28], with permission of Elsevier 2019). (**a**) Excitation stage: The laser irradiates the ablation layer through the confinement layer, and the ablation layer is vaporized into plasma. Due to the limitation of the confinement layer, the reaction force of the plasma drives the flyer to emit; (**b**) Flight phase: the flyer passes the preset flight distance (standoff) and collides with the target at a certain speed and angle; (**c**) Welding stage: The behavior of the impact point is shown in Figure 5 (the welded area on the left). The flyer and the target collide at a certain angle from the starting position of the metallurgical bonding to the end position and the welding is over.

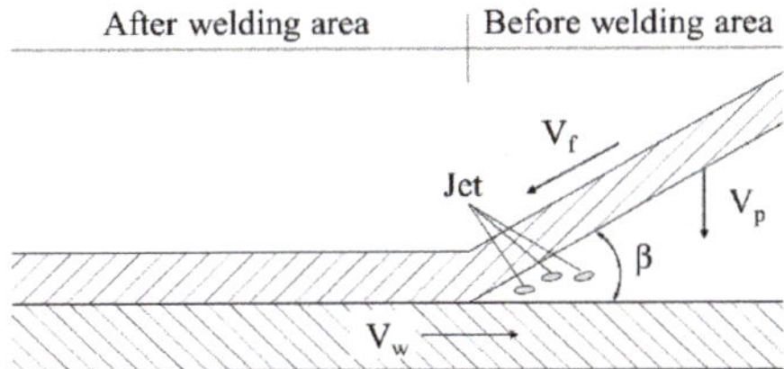

Figure 5. Schematic diagram of the impact welding process.

Table 1. Influence of Weldment Combination Parameters.

Weldment Combination Parameters	Influence of Impact Welding
Confinement layer	The higher the hardness, the greater the reaction force on the flyer, and the higher the impact speed under the same energy; but when the hardness of the confinement layer is high, the toughness is poor, and the service life is shorter in the continuous high-speed impact service environment. Currently, the commonly used material is polycarbonate and high light transmission glass
Binder	Liquid super glue has high performance and can reach the highest speed, but it is difficult to spread evenly; solid double-sided glue spreads evenly, but it will lose a certain amount of energy due to self-adhesion. Therefore, its impact speed is lower than that of liquid super glue.
Ablative layer	The ablation layer is excited by the laser to form plasma to accelerate the flyer. The stronger its ability to absorb laser energy, the more plasma formed, and the higher the conversion efficiency of light energy to kinetic energy. Currently, black spray paint is commonly used
Thickness of flyer	The thicker the flyer, the larger the mass and the smaller the impact speed; however, as the thickness increases, the thickness of the welded joint becomes larger, thereby improving the joint strength
Preset flying distance	The laser-driven flyer flight is a variable speed process. First, it accelerates and then decelerates. Under certain conditions, there is an optimal position for the highest impact velocity.

The laser system parameters and weldment combination parameters affect the metallurgical bonding process by controlling the impact velocity V_p and impact angle β in Figure 5 at the collision point. The jet only starts in the shaded area process window as shown in Figure 6. Therefore, studying the effects of various parameters on the impact velocity and angle is very important for optimizing the laser impact welding process. It is worth noting that the jet is also affected by the welding metal itself, and the jet process window of different metals is different [24,29].

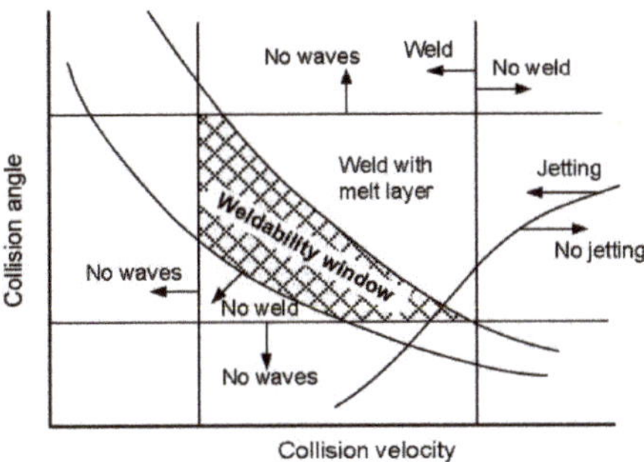

Figure 6. Generic welding window (reproduced from [10], with permission of Elsevier 2019).

3. Microstructure of Laser Impact Welding Interface

3.1. Macroscopic Morphology of Laser Impact Welding Interface

The typical macroscopic morphology of the laser impact welding interface is shown in Figure 7a. The welding joint area has a ring shape, and the middle is the collision rebound area. As shown in Figure 7c,e, a certain amount of damage was produced on the top surface of the flyer-copper foil irradiated by the laser, and a convex area was produced on the back of the target. For thicker targets, this problem does not exist. Additionally, the generation, propagation, rebound and superposition of stress waves may cause tearing between the flyer and the target [27,28,30]. The center rebound zone of the flat-top laser is larger than that of the Gaussian laser, which may be affected by the impact angle factor mentioned above. The huge rebound zone seriously affects the industrial application and joint performance of laser impact welding (LIW), so eliminating the rebound zone is the primary task of current process optimization.

Liu et al. [31] performed Cu-Al-Cu three-layer impact welding of weldments, that is, using Cu as the flyer, first impact the middle layer Al, and finally, under the impact of the impact, the middle layer accelerates the impact to the Cu of the bottom target, completing three-layer impact welding. Due to the first impact welding of the flyer and the middle layer, impact velocity and impact angle are adjusted, thus reducing the springback of the middle layer Al and the bottom target Cu during the second impact welding process, but the springback of the first flyer and the middle layer still not be controlled. Sadeh et al. [32] found in experiments that the use of black tape between the target and the fixed plate reduces the rebound of the flyer. They used a black tape buffer layer to eliminate the center spring back phenomenon, greatly increasing the area of the weldment area. Convert the weldment area from a ring to a dot and outer ring shape.

Three-layer impact welding and the use of black tape have better eliminated the center spring back and obtained an ideal circular solder joint. They confirmed the possibility of laser impact welding to eliminate the center spring back and made a great contribution to the application of the process. These two experiments jointly pointed out that "buffering" is the key factor for laser impact welding to eliminate center spring back.

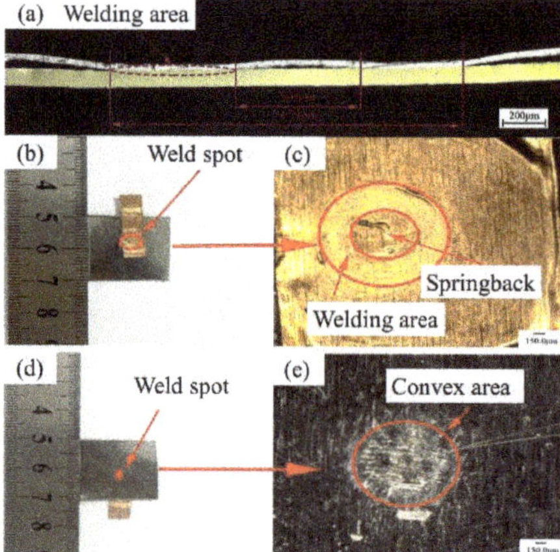

Figure 7. (**a**) Cross-section of weld interface at 1550 mJ energy (reprodueced from [27], with permission of Elsevier 2019); (**b**–**e**) Weld spot between molybdenum and copper (reproduced from [28], with permission of Elsevier 2019).

3.2. Laser Impact Welding Interface Wave

A periodic wave-like interface is a typical interface morphology in impact welding. On the one hand, the interface wave can increase the area of metallurgical bonding in a limited welding area and increase the welding bonding strength. On the other hand, it can also be mechanically interlocked to improve the strength of the interface. Accordingly, the interface wave characteristics are related to welding parameters such as input energy and impact angle. Figure 8a–c show the interface wave characteristics of explosive welding, magnetic pulse welding and laser impact welding [33]. The wavelength and peak of the interface wave increase with the increase in energy. Due to the low energy input in laser impact welding, the interface wave presents irregular characteristics. It is almost a straight line under the ordinary optical microscope and low-magnification scanning electron microscope, and the undulations of tiny waves can only be seen under the high-magnification electron microscope. Wang et al. [24] studied the relationship between the characteristics of the laser impact welding interface wave and the laser energy density, as shown in Figure 8d, which further confirmed the irregularity of the laser impact welding interface wave.

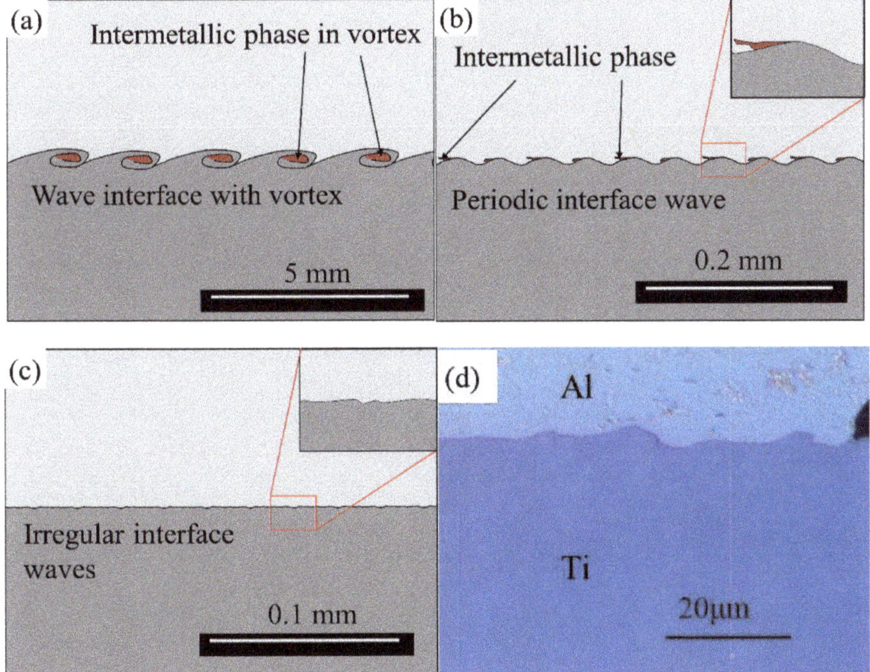

Figure 8. Weld interface morphology of three typical impact welded joints with (**a**) Explosive welding; (**b**) Magnetic pulse welding; (**c**) Laser impact welding; (**d**) Weld wave interface morphology with laser impact welding (reproduced from [24], with permission of Laser Institute of America 2016).

The ideal interface wave can make the welded joint get excellent performance, but the formation mechanism of the interface wave is still controversial. At present, there are mainly the following four theories regarding the formation mechanism of interface waves:

1. Bahrani and Black [34,35] first proposed the flyer flow penetration mechanism (caved mechanism), as shown in Figure 9. Since the stress generated by the impact is much greater than the yield strength of the material, this mechanism regards the flying stream as a fluid with a certain viscosity, and the target is a non-fluid ductile metal. It is believed that the initial impact of

the flyer on the target will cause the target to sink and bulge with plastic deformation. The work hardening caused by deformation makes it more and more difficult for the target to deform, reaching the limit. After periodic action, a wavy interface formed.

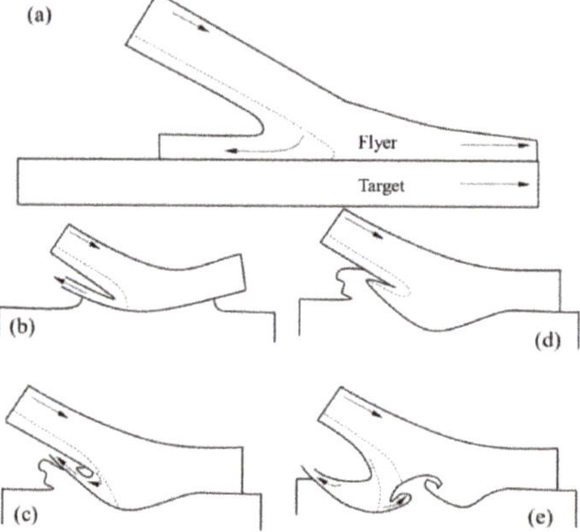

Figure 9. (**a–e**) the process of Bahrani caved mechanism (reproduce from [34], with permission of Royal Society 1967).

2. The Helmholz instability mechanism was proposed by Hunt et al. [36]. This mechanism regards the two metals under high-speed impact as fluids. During the impact and collision, the flyer and the target will have their own characteristics at the interface between the two. The tangential velocities u_1 and u_2 parallel to the interface, due to that the different properties of the two metals, the different driving forces they receive, and the reflection from the fixed surface of the target, cause the tangential velocities u_1 and u_2 to be inconsistent. As shown in Figure 10a, the speed difference between the two fluids at the interface position will cause small disturbances. This disturbance will cause the interface to instability and produce a wave-shaped interface. This wave-shaped cloud commonly found in nature is Kelvin–Helmholtz instability. They believe that a similar situation will also occur during the impact, so a wavy interface will be formed.

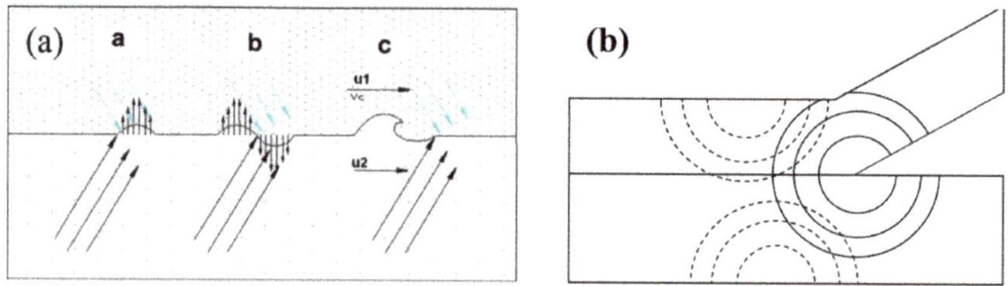

Figure 10. (**a**) Helmholz instability mechanism (reproduced from [37], with permission of Elsevier 2010); (**b**) stress wave mechanism.

3. The stress wave mechanism says that the impact wave generated by the release of energy and the various stress waves reflected from the target superimpose on the interface to produce interface waves as shown in Figure 10b [38,39]. At the collision between the flyer and the target, stress waves generated at the interface and propagate into both the flyer and the target. The higher the input laser energy, the stronger the stress waves are. The waves are reflected when they meet an interface/surface. Until now, no quantitative relationship was built between the wave

characteristics and the stress waves. According to this mechanism, the size of the interface waveform is only related to the thickness of the weldment. However, the size of the waveform will be significantly different under different energy [24]. Therefore, this mechanism is not the main factor affecting the formation of the interface waveform.

4. The vortex street mechanism is also called the vortex flow mechanism. Sherif [40] and Hay [41] drew on the principles of fluid mechanics and regarded the flyer and target as fluids. As shown in Figure 11, according to the vortex mechanism, when the fluid vortex encounters an obstacle, they will rotate in opposite directions from both sides of the obstacle to form a vortex line. Therefore, the flow of the flyer and the target will flow out during the impact welding process. Separation and convergence eventually form a wavy interface. However, in fact, the impact process is not blocked by obstacles.

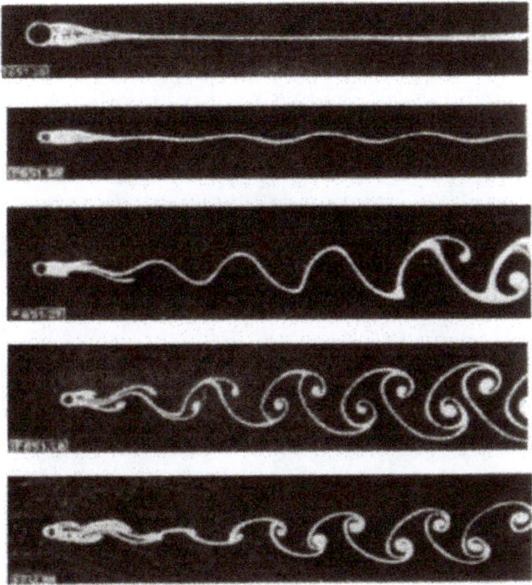

Figure 11. Vortex Street mechanism (reproduced from [42], with permission of Elsevier 2018).

At present, the caved mechanism and the Helmholz instability mechanism are interface wave formation mechanisms accepted by most scholars, especially to explain the periodic interface waves in explosive welding and magnetic pulse welding. The energy input by the two is large, and the interface metal can be approximated as fluid during the impact. The laser impact welding interface presents irregular interface waves or flat interfaces. Whether the Helmholz instability mechanism can accurately predict the shape of the laser impact welding interface still needs to be explored.

At present, the simulation of the wave-shaped interface of impact welding by researchers often regards the material as a fluid and applies the penetration model. The material models adopted by most researchers are the Johnson–Cook materials model [43]. The Johnson–Cook materials model has the following formula.

$$\sigma = \left(A + B\varepsilon_{eff}^{n}\right)\left(1 + C\ln\dot{\varepsilon}\right)\left(1 - T^{*m}\right) \tag{3}$$

σ is flow stress; ε_{eff} is effective plastic strain; $\dot{\varepsilon} = \frac{\varepsilon_{eff}}{\dot{\varepsilon}_0}$ is plastic strain rate; $T^* = \frac{T - T_{room}}{T_{melt} - T_{room}}$ is homologous temperature; A, B, C, n, m are materials parameters.

3.3. Microstructure of Laser Impact Welding Interface

3.3.1. Interface with and without Transition Layer

Figure 12 shows two typical interface structures in impact welding revealed by electron microscopy: welding interface with transition layer and welding interface without transition layer [44,45]. For the welding interface with a transition layer, the transition layer is a new phase different from the base material produced after melting and solidification of the interface metals, which belongs to the "rapid melting-solidification" interface bonding mechanism [46]. For the interface without a transition layer, there are currently two opposing views: on the one hand, scholars represented by Stern [47] believe that the interface combination is attributed to the "mechanical mixing" effect and there is no melting. The high plastic deformation of the interface leads to the rapid refinement of crystals and the formation of an intermediate thin layer. On the other hand, scholars represented by Marya [48] believe that under high-speed impact conditions, the temperature increase in the impact interface is inevitable. This type of interface is also formed by the "rapid melting-solidification" of the thin metal layer. The formation of the transition layer is related to the input energy.

For the foil vaporization welding with higher input energy than LIW, Sridharan N et al. [49] used the latest TEM and APT techniques to observe the structure of the weld interface without transition layers from the nanoscale. As shown in Figure 12a, they found that there is a nano-scale amorphous layer at the interface. The amorphous layer contains elements of the two weldments, confirming the diffusion of interface atoms during the impact. In this regard, they proposed a "liquid film" diffusion mechanism. That is, the metal atoms with the lower melting point are first melted into a liquid film during the impact welding process, and the higher melting point atoms on the other side will enter the liquid film for diffusion. Their study indicated that the interface reaction in impact welding is complex and exhibited different phenomena at a different scale.

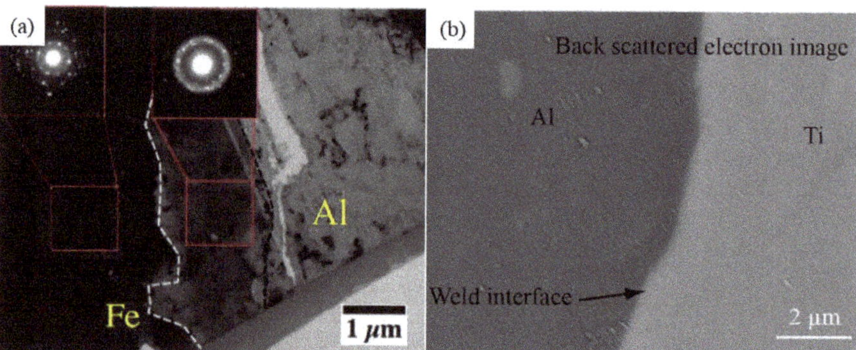

Figure 12. Two welding interfaces of impact welding (**a**) Transition zone interface (reproduced from [49], with permission of Elsevier 2019); (**b**) Interface without transition zone (reproduced from [24], with permission of Laser Institute of America 2016).

Under SEM observation, the laser impact welding interface mostly belongs to the interface without the transition layer. Wang H. et al. [50] used EBSD to confirm that the grains in the vicinity of the joint are significantly refined, and there are nanocrystalline ribbons like the vaporized foil, but no obvious continuous transition layer is found on the welded joints of dissimilar metals. However, the impact is a rapid process of energy accumulation and release. Especially in the second half of the cycle, as shown below, the impact will abnormally increase the energy to form a discontinuous intermetallic compound, and finally form a mixed interface without a transition layer and a transition layer. At present, the mixed interface formed by laser impact welding with such low energy input has not been explored. Exploring the formation and distribution of these two interface structures is of great significance to reveal the order of the formation of the impact welding interface. In particular, the influence of discontinuous intermetallic compounds on the brittleness of the bonding surface has a certain significance for the improvement of process performance.

3.3.2. Laser Impact Welding Interface Diffusion and Intermetallic Compounds

At the impact welding interface with the corresponding occurring obvious melting phenomenon, the atom diffusion mechanism is the same as that in the melting and welding process, and intermetallic compounds forms. This kind of regional melting phenomenon is common in explosive welding and magnetic pulse welding/foil vaporization welding interface. Wang X et al. [27] found that laser impact welding also showed a local melting lump as shown in Figure 13 at high welding input energy, and a platform also appeared on the EDS (Energy Dispersive Spectrometer) concentration curve. According to previous studies by Akbari and Behnagh [51] and Zhang et al. [33], it is an intermetallic compound. Therefore, in laser impact welding, with high laser energy, melting at the interface also occurs. EDS can only qualitatively analyze the existing problems of intermetallic compounds. In-depth analysis is still needed for the composition and shape of intermetallic compounds.

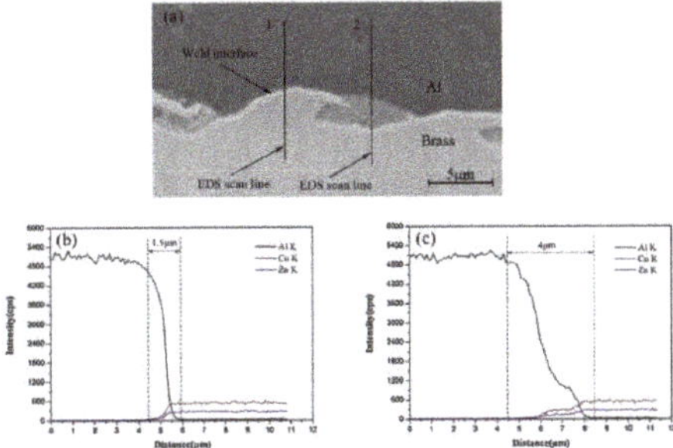

Figure 13. EDS (Energy Dispersive Spectrometer) curves of impact weld melting block (reproduced from [27], with permission of Elsevier 2019) (**a**) SEM image of aluminum/copper weld interface with fusion block; (**b**,**c**) EDS curves at positions 1 and 2.

For laser impact welding, the interface is generally no obvious melting phenomenon and the intermetallic compounds [24]. For example, Wang et al. [52] studied the laser impact welding of amorphous and crystalline materials. They used XRD to compare and analyze the changes in amorphous materials before and after laser impact welding. As shown in Figure 14, the EDS curve is the same as other materials. It is a continuous curve and no intermetallic compounds were found. In addition, they found that LIW could not cause structural changes in the amorphous matrix.

Chen S et al. [53] and Ning L et al. [54] found that the crystal structure at the interface was destroyed during an impact, resulting in a disordered organization. The two elements diffused and interacted with each other in a disordered structure, then, followed into the interior of the ordered crystal structure. Although the calculated diffusion layer thickness is slightly lower than the experimentally measured diffusion layer thickness, it provides an effective way to study atomic diffusion at the impact welding interface. This numerical simulation calculation result is consistent with "liquid film", that is, the key to the inter-diffusion of LIW across the bonding interface is the disordered layer of atoms produced by the impact. In the future, high-precision characterization technology and numerical simulation technology are expected to reveal the atomic diffusion mechanism of laser impact welding interface.

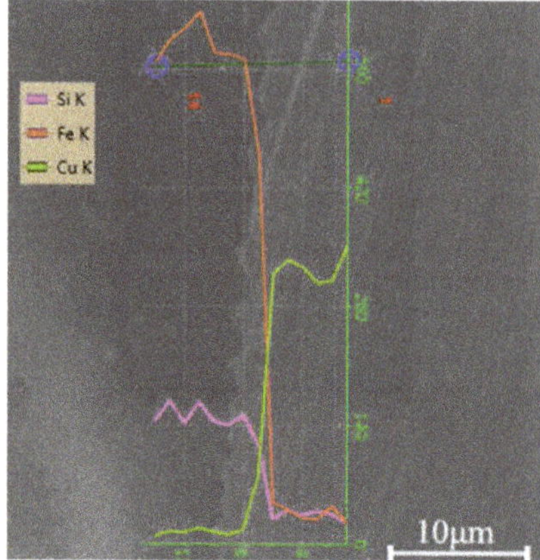

Figure 14. EDS concentration curve for laser impact welding of crystal and amorphous materials [52].

4. Mechanical Properties of Laser Impact Welding Interface

4.1. Laser Impact Welding Interface Strength

The welding area of laser impact welding is a millimeter-sized ring, and it is difficult to measure its area under mechanical testing in real-time. Therefore, the maximum force (N) that can be achieved in the tensile fracture of the welded joint is usually used to characterize the bonding performance of the welded joint. The test methods mainly include the peeling test and shearing test [24,28], as shown in Figure 15.

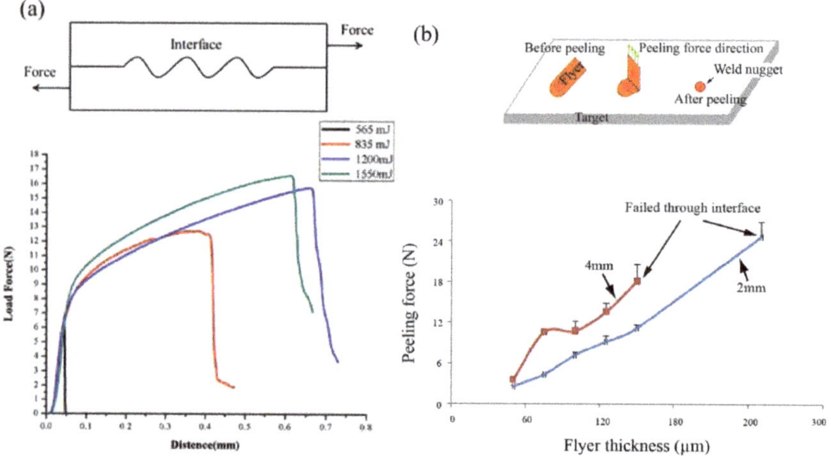

Figure 15. Test method for bonding force of welded joints (**a**) shearing test and result (reproduced from [28], with permission of Elsevier 2019); (**b**) peeling test and result (reproduced from [24] with permission of Laser Institute of America 2016).

The mechanical properties of laser impact welded joints are generally considered to be related to energy density, welding area, and flyer thickness. The improvement of bonding force is affected by the higher the energy density, the larger the welding area, and the thicker the flying piece. As mentioned above, the higher the energy density, which leads to more generation of interface waves. It can be seen as increasing the bonding force by increasing the welding area and "mechanical interlocking".

According to the failure location, laser impact welding can be divided into joint damage failures under low energy state and matrix damage failures under high energy state. The joint failure under low input energy is shown in Figure 16a,c,e. The joint bonding force is lower than the strength of the matrix, and the failure location is located in the joint, which is mostly brittle fracture; the matrix fails under high input energy. As shown in Figure 16b,d,f, the strength of the metallurgical joint is higher than the strength of the matrix, and the failure location is in the matrix, which is generally ductile fracture. Therefore, when the input energy is low, the input energy can be increased to increase the joint bonding force, and when the input energy is high, the thickness of the weldment can be increased to directly strengthen the target material and improve the damage resistance [27].

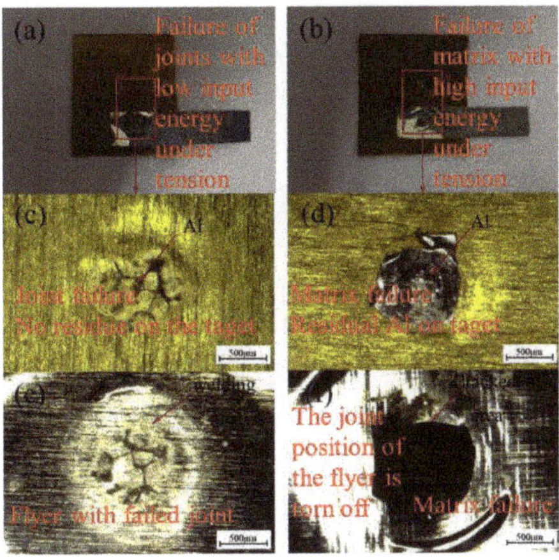

Figure 16. Two forms of failure (reproduced from [27], with permission of Elsevier 2019) (**a**,**c**,**e**) Low-energy joint failure; (**b**,**d**,**f**) High-energy base metal failure.

4.2. Interface Hardness of Laser Impact Welding

In addition to the bonding force, the nanoindentation experiment is used to test the hardness of the tiny interface area to characterize its strengthening under the high-speed impact. As shown in Figure 17, the laser impact welding joint is impact-strengthened, and the hardness is higher. The hardness of the matrix on both sides of the joint gradually decreases from the interface. However, the hardening effect is different for LIW (Laser Impact Welding) and MPW (Magnetic Pulse Welding) due to the different input energy. The higher hardness at the same position is affected by the more energy input [33].

Laser impact welding is a transient, high-temperature, and high-pressure process. The microstructure of the material undergoes abrupt changes, such as the increase in dislocations, the refinement of grains, and the formation of cellular structures, which change the properties of the material. On the one hand, the welded joint produces high-rate strain plastic deformation strengthening under impact; on the other hand, the rapid plastic deformation during the impact process will generate a lot of heat and cool down in a short period, which can be regarded as a quenching process [28,33,55,56].

However, it should be noted that the formation of brittle intermetallic compounds will also significantly increase its hardness. Therefore, to evaluate its performance, the bonding force of the welded joint and its hardening degree should be considered together.

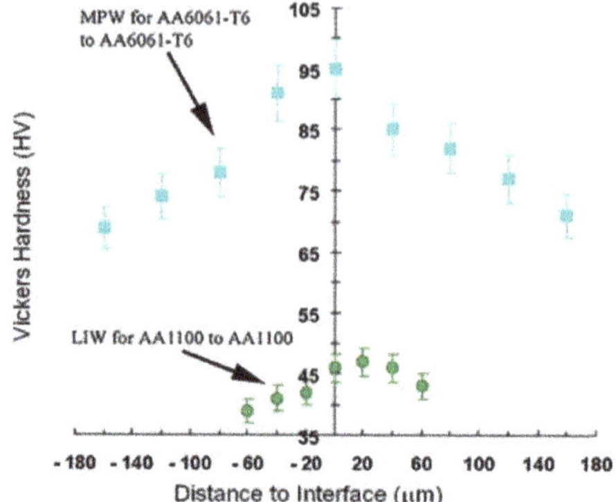

Figure 17. Hardness of weld interface with MPW (Magnetic Pulse Welding) and LIW (Laser Impact Welding) (reproduced from [33], with permission of Elsevier 2011).

5. Conclusions and Prospects

The principle and process of impact welding are described, and the difference between laser impact welding and another impact welding is illustrated above in this work. It also describes the research progress in recent years, and summarizes its application prospects, the difficulties in industrial applications and the mechanism problems currently to be studied as follows:

1. In the laboratory research stage, the development of laser impact welding technology has made great progress, realizing the joining between a variety of dissimilar materials, including the joining between amorphous and crystalline and multilayer composite materials. It shows that laser impact welding has a good application prospect in the joining of dissimilar materials.
2. Laser impact welding still has problems such as the rebound of the flyer, the surface damage of the flyer, and the small welding area. It is necessary to use a higher energy laser to try a larger welding area and use a "buffer layer" to solve the problem of center spring back. Before industrial applications, it is necessary to further optimize the laser impact welding process and design evaluation standards for the performance of welded joints.
3. The evolution of the microstructure of the laser impact welding interface is complex, the mechanism of interface atom diffusion is not clear, and the existence of micro-domain melting at the interface has not been confirmed. The discovery of the interface bonding mechanism requires more in-depth research on the atomic scale.

Author Contributions: Conceptualization, H.W.; Investigation, K.W., H.W., H.Z. and A.X.; Writing—Original Draft Preparation, K.W., H.Z. and H.W.; Writing—Review & Editing, K.W., H.W. and W.Z. All authors have read and agreed to the published version of the manuscript.

Funding: This work was supported by Beijing Municipal Science and Technology (Z201100004520011), National Natural Science Foundation of China (61409220124), Fundamental Research Funds for the Central Universities (06500107).

Acknowledgments: The authors thank all the lab members for the discussion and help. Particularly, the authors appreciate the excellent research work from the researchers included in this review.

Conflicts of Interest: The authors declare no conflict of interest.

References

1. Liu, H.; Gong, J.; Ma, Y.; Cui, J.; Li, M.; Wang, X. Investigation of novel laser shock hydroforming method on micro tube bulging. *Opt. Lasers Eng.* **2020**, *129*, 106073. [CrossRef]
2. Chen, Y.; Nakata, K. Microstructural characterization and mechanical properties in friction stir welding of aluminum and titanium dissimilar alloys. *Mater. Des.* **2009**, *30*, 469–474. [CrossRef]

3. Oliveira, J.; Schell, N.; Zhou, N.; Wood, L.; Benafan, O. Laser welding of precipitation strengthened Ni-rich NiTiHf high temperature shape memory alloys: Microstructure and mechanical properties. *Mater. Des.* **2019**, *162*, 229–234. [CrossRef]
4. Oliveira, J.; Curado, T.; Zeng, Z.; Lopes, J.; Rossinyol, E.; Park, J.M.; Schell, N.; Fernandes, F.B.; Kim, H.S. Gas tungsten arc welding of as-rolled CrMnFeCoNi high entropy alloy. *Mater. Des.* **2020**, *189*, 108505. [CrossRef]
5. Fronczek, D.M.; Wojewoda-Budka, J.; Chulist, R.; Sypien, A.; Korneva, A.; Szulc, Z.; Schell, N.; Zieba, P. Structural properties of Ti/Al clads manufactured by explosive welding and annealing. *Mater. Des.* **2016**, *91*, 80–89. [CrossRef]
6. Arya, H.K.; Saxena, R.K.; Kumar, R. *Effect of Heat Input on Residual Stress in Submerged Arc Welding*; LAMBERT Academic Publishing: Rīgā, Latvia, 2014; ISBN 13: 978-3-659-66032-0.
7. Ishigami, A.; Roy, M.; Walsh, J.N.; Withers, P.J. The effect of the weld fusion zone shape on residual stress in submerged arc welding. *Int. J. Adv. Manuf. Technol.* **2016**, *90*, 3451–3464. [CrossRef]
8. Mehta, K.P.; Carlone, P.; Astarita, A.; Scherillo, F.; Rubino, F.; Vora, P. Conventional and cooling assisted friction stir welding of AA6061 and AZ31B alloys. *Mater. Sci. Eng. A* **2019**, *759*, 252–261. [CrossRef]
9. Ni, Z.; Yang, J.; Gao, Z.; Hao, Y.; Chen, L.; Ye, F. Joint formation in ultrasonic spot welding of aluminum to copper and the effect of particle interlayer. *J. Manuf. Process.* **2020**, *50*, 57–67. [CrossRef]
10. Mousavi, S.A.; Sartangi, P.F. Experimental investigation of explosive welding of cp-titanium/AISI 304 stainless steel. *Mater. Des.* **2009**, *30*, 459–468. [CrossRef]
11. Pereira, D.; Oliveira, J.; Santos, T.; Miranda, R.; Lourenço, F.; Gumpinger, J.; Bellarosa, R. Aluminium to Carbon Fibre Reinforced Polymer tubes joints produced by magnetic pulse welding. *Compos. Struct.* **2019**, *230*, 111512. [CrossRef]
12. Watanabe, T.; Takayama, H.; Yanagisawa, A. Joining of aluminum alloy to steel by friction stir welding. *J. Mater. Process. Technol.* **2006**, *178*, 342–349. [CrossRef]
13. Wang, K.; Shriver, D.; Li, Y.; Banu, M.; Hu, S.; Xiao, G.; Arinez, J.; Fan, H.-T. Characterization of weld attributes in ultrasonic welding of short carbon fiber reinforced thermoplastic composites. *J. Manuf. Process.* **2017**, *29*, 124–132. [CrossRef]
14. Daehn, G.S.; Lippold, J.C. Low-Temperature Spot Impact Welding Driven without Contact. WO/2009/111,774, 11 September 2009.
15. Wu, A.; Song, Z.; Nakata, K.; Liao, J.; Zhou, L. Interface and properties of the friction stir welded joints of titanium alloy Ti6Al4V with aluminum alloy 6061. *Mater. Des.* **2015**, *71*, 85–92. [CrossRef]
16. Lawrence, R.; Trott, W.M. Theoretical analysis of a pulsed-laser-driven hypervelocity flyer launcher. *Int. J. Impact Eng.* **1993**, *14*, 439–449. [CrossRef]
17. Paisley, D.L. Laser-Driven Miniature Flyer Plates For Shock Initiation Of Secondary Explosives. In Proceedings of the American Physical Society Topical Conference on Shock Compression of Condensed Matter, Albuquerque, NM, USA, 14–17 August 1989.
18. Wang, H.; Wang, Y. Characteristics of Flyer Velocity in Laser Impact Welding. *Metals* **2019**, *9*, 281. [CrossRef]
19. Crossland, B. *Explosive Welding of Metals and Its Application*; Clarendon Press: Oxford, UK, 1982.
20. Blazynski, T. *Explosive Welding, Forming and Compaction*; Springer: Dordrecht, The Netherland, 1983.
21. Katzenstein, J. System and Method for Impact Welding by Magnetic Implosion. U.S. Patent 4,513,188, 23 April 1985.
22. Daehn, G.S.; Lippold, J.; Liu, D.; Taber, G.; Wang, H. Laser impact welding—Process introduction and key variables. In Proceedings of the International Conference on High Speed Forming, Dortmund, Germany, 24–26 April 2012; Volume 2012.
23. Szecket, A. Impact Welding. U.S. Patent 4,842,182, 27 June 1989.
24. Wang, H.; Vivek, A.; Taber, G.; Daehn, G. Laser impact welding application in joining aluminum to titanium. *J. Laser Appl.* **2016**, *28*, 32002. [CrossRef]
25. Wang, H.; Wang, Y. High-Velocity Impact Welding Process: A Review. *Metals* **2019**, *9*, 144. [CrossRef]
26. Wang, H.; Taber, G.; Liu, D.; Hansen, S.; Chowdhury, E.; Terry, S.; Lippold, J.C.; Daehn, G.; Taber, G.A.; Hansen, S.R. Laser impact welding: Design of apparatus and parametric optimization. *J. Manuf. Process.* **2015**, *19*, 118–124. [CrossRef]
27. Wang, X.; Shao, M.; Jin, H.; Tang, H.; Liu, H. Laser impact welding of aluminum to brass. *J. Mater. Process. Technol.* **2019**, *269*, 190–199. [CrossRef]

28. Wang, X.; Tang, H.; Shao, M.; Jin, H.; Liu, H. Laser impact welding: Investigation on microstructure and mechanical properties of molybdenum-copper welding joint. *Int. J. Refract. Met. Hard Mater.* **2019**, *80*, 1–10. [CrossRef]
29. Zhang, Z.; Liu, M. Numerical studies on explosive welding with ANFO by using a density adaptive SPH method. *J. Manuf. Process.* **2019**, *41*, 208–220. [CrossRef]
30. Wang, X.; Li, F.; Huang, T.; Wang, X.; Liu, H. Experimental and numerical study on the laser shock welding of aluminum to stainless steel. *Opt. Lasers Eng.* **2019**, *115*, 74–85. [CrossRef]
31. Liu, H.; Jin, H.; Shao, M.; Tang, H.; Wang, X. Investigation on Interface Morphology and Mechanical Properties of Three-Layer Laser Impact Welding of Cu/Al/Cu. *Met. Mater. Trans. A* **2018**, *50*, 1273–1282. [CrossRef]
32. Sadeh, S.; Gleason, G.H.; Hatamleh, M.I.; Sunny, S.F.; Yu, H.; Malik, A.; Qian, D. Simulation and Experimental Comparison of Laser Impact Welding with a Plasma Pressure Model. *Metals* **2019**, *9*, 1196. [CrossRef]
33. Zhang, Y.; Babu, S.S.; Prothe, C.; Blakely, M.; Kwasegroch, J.; Laha, M.; Daehn, G.S. Application of high velocity impact welding at varied different length scales. *J. Mater. Process. Technol.* **2011**, *211*, 944–952. [CrossRef]
34. Bahrani, A.S.; Black, T.J.; Crossland, B. The mechanics of wave formation in explosive welding. *Proc. R. Soc. London. Ser. A Math. Phys. Sci.* **1967**, *296*, 123–136. [CrossRef]
35. Abrahamson, G.R. Permanent Periodic Surface Deformations Due to a Traveling Jet. *J. Appl. Mech.* **1961**, *28*, 519–528. [CrossRef]
36. Hunt, J.N. Wave formation in explosive welding. *Philos. Mag.* **1968**, *17*, 669–680. [CrossRef]
37. Ben-Artzy, A.; Stern, A.; Frage, N.; Shribman, V.; Sadot, O. Wave formation mechanism in magnetic pulse welding. *Int. J. Impact Eng.* **2010**, *37*, 397–404. [CrossRef]
38. Godunov, S.; Deribas, A.; Zabrodin, A.; Kozin, N. Hydrodynamic effects in colliding solids. *J. Comput. Phys.* **1970**, *5*, 517–539. [CrossRef]
39. Godunov, S.K.; Deribas, A.A.; Kozin, N.S. Wave formation in explosive welding. *J. Appl. Mech. Tech. Phys.* **1973**, *12*, 398–406. [CrossRef]
40. Reid, S.R.; Sherif, N.H.S. Prediction of the Wavelength of Interface Waves in Symmetric Explosive Welding. *J. Mech. Eng. Sci.* **1976**, *18*, 87–94. [CrossRef]
41. Kowalick, J.; Hay, D. A mechanism of explosive bonding. *Metall. Trans.* **1971**, *2*, 1953–1958. [CrossRef]
42. Zhang, Z.; Feng, D.; Liu, M. Investigation of explosive welding through whole process modeling using a density adaptive SPH method. *J. Manuf. Process.* **2018**, *35*, 169–189. [CrossRef]
43. Nassiri, A.; Abke, T.; Daehn, G. Investigation of melting phenomena in solid-state welding processes. *Scr. Mater.* **2019**, *168*, 61–66. [CrossRef]
44. Göbel, G.; Kaspar, J.; Herrmannsdörfer, T.; Brenner, B.; Beyer, E. Insights into intermetallic phases on pulse welded dissimilar metal joints. In Proceedings of the 4th International Conference High Speed Form, Columbus, OH, USA, 9–10 March 2010. [CrossRef]
45. Kore, S.D.; Imbert, J.; Worswick, M.J.; Zhou, Y. Electromagnetic impact welding of Mg to Al sheets. *Sci. Technol. Weld. Join.* **2009**, *14*, 549–553. [CrossRef]
46. Stern, A.; Aizenshtein, M.; Moshe, G.; Cohen, S.R.; Frage, N. The Nature of Interfaces in Al-1050/Al-1050 and Al-1050/Mg-AZ31 Couples Joined by Magnetic Pulse Welding (MPW). *J. Mater. Eng. Perform.* **2013**, *22*, 2098–2103. [CrossRef]
47. Stern, A.; Shribman, V.; Ben-Artzy, A.; Aizenshtein, M. Interface Phenomena and Bonding Mechanism in Magnetic Pulse Welding. *J. Mater. Eng. Perform.* **2014**, *23*, 3449–3458. [CrossRef]
48. Marya, M.; Marya, S.; Priem, D. On The Characteristics of Electromagnetic Welds Between Aluminium and other Metals and Alloys. *Weld. World* **2005**, *49*, 74–84. [CrossRef]
49. Sridharan, N.; Poplawsky, J.D.; Vivek, A.; Bhattacharya, A.; Guo, W.; Meyer, H.; Mao, Y.; Lee, T.; Daehn, G. Cascading microstructures in aluminum-steel interfaces created by impact welding. *Mater. Charact.* **2019**, *151*, 119–128. [CrossRef]
50. Wang, H.; Liu, D.; Lippold, J.C.; Daehn, G.S. Laser impact welding for joining similar and dissimilar metal combinations with various target configurations. *J. Mater. Process. Technol.* **2020**, *278*, 116498. [CrossRef]
51. Akbari, M.; Behnagh, R.A. Dissimilar Friction-Stir Lap Joining of 5083 Aluminum Alloy to CuZn34 Brass. *Met. Mater. Trans. A* **2012**, *43*, 1177–1186. [CrossRef]

52. Wang, X.; Luo, Y.; Huang, T.; Liu, H. Experimental Investigation on Laser Impact Welding of Fe-Based Amorphous Alloys to Crystalline Copper. *Materials* **2017**, *10*, 523. [CrossRef]
53. Chen, S.Y.; Wu, Z.W.; Liu, K.X.; Li, X.J.; Luo, N.; Lu, G.X. Atomic diffusion behavior in Cu-Al explosive welding process. *J. Appl. Phys.* **2013**, *113*, 044901. [CrossRef]
54. Luo, N.; Shen, T.; Junxiang, X. Diffusion Mechanism of Explosive Welding Interface Between Memory Alloy Ni 50 Ti 50 and Cu. *Xiyou Jinshu Cailiao Yu Gongcheng Rare Met. Mater. Eng.* **2018**, *47*, 3238–3242.
55. Zhang, Y.; Babu, S.; Daehn, G. Interfacial ultrafine-grained structures on aluminum alloy 6061 joint and copper alloy 110 joint fabricated by magnetic pulse welding. *J. Mater. Sci.* **2010**, *45*, 4645–4651. [CrossRef]
56. Lee, K.-J.; Kumai, S.; Arai, T.; Aizawa, T. Interfacial microstructure and strength of steel/aluminum alloy lap joint fabricated by magnetic pressure seam welding. *Mater. Sci. Eng. A* **2007**, *471*, 95–101. [CrossRef]

Publisher's Note: MDPI stays neutral with regard to jurisdictional claims in published maps and institutional affiliations.

Article

Influence of Rolling Temperatures on Interface Microstructure and Mechanical Properties of Multi-Pass Rolling TA1/Q235B Explosive Welded Sheets

Huizhong Li [1,2,3], Liangming Cao [1], Xiaopeng Liang [1,2,3,*], Wending Zhang [1], Chunping Wu [1], Zhiheng Zeng [4] and Chengshang Zhou [2]

1 School of Materials Science and Engineering, Central South University, Changsha 410083, China; lhz606@csu.edu.cn (H.L.); caoliangming@csu.edu.cn (L.C.); zwd960123@csu.edu.cn (W.Z.); chunpingwu@csu.edu.cn (C.W.)

2 State Key Laboratory of Powder Metallurgy, Central South University, Changsha 410083, China; chengshang.zhou@csu.edu.cn

3 Key Laboratory of Nonferrous Metal Materials Science and Engineering, Ministry of Education, Central South University, Changsha 410083, China

4 Hunan Phohom New Material Technology Co., Ltd., Changsha 410083, China; zengzh@phohom.com

* Correspondence: liangxp@csu.edu.cn; Tel.: +86-731-888-7794

Received: 2 November 2020; Accepted: 6 December 2020; Published: 9 December 2020

Abstract: The effect of rolling temperatures on the interface microstructure and mechanical properties is investigated using 2-mm-thick TA1/Q235B composite sheets, which were prepared after nine passes of hot rolling of explosive welded plates. The results show that the vortex region and the transition layer exist in the interface at the explosive welded plate, while only the transition layer exists in the interface after hot rolling. The transition layer is composed of α-Ti, TiC, Fe, and FeTi, and the thickness increases with the increasing rolling temperature. The microhardness of the explosive welded plate is higher than that of the hot-rolling sheet, and the microhardness of interface are higher than that of matrix metals. The interface shear strength and tensile elongation of the hot-rolled sheet increase with the increasing hot rolling temperature, while the ultimate tensile strength (UTS), yield strength (YS) and Young modulus decrease with the increase of hot rolling temperature. The shear strength of sheets is related to the interfacial compounds, and the tensile strength is mainly affected by the grain morphology of the matrix.

Keywords: TA1/Q235B composite sheets; rolling temperature; explosive welding; microstructures; mechanical properties

1. Introduction

As an important metal-based material, bimetallic composite sheets have been widely used in petroleum, chemical, metallurgy, light industry, electric power, seawater desalination, shipbuilding, marine engineering, and other industries due to their unique physical chemical and mechanical properties [1–7].

As an efficient welding method, explosive welding has been widely used in preparing heterogeneous metal composite plates [8–10]. During explosive welding, the high pressure forces the different metals to achieve close contact at atomic level, promoting an excellent metallurgical combination. At the bonding interface of the two metal plates, the high pressure caused by the explosion provides energy for the joint process, and causes the flyer plate to impact with the base plate at a high speed. At high pressure, the thin flyer plate presents a dual solid–liquid state at the joint, and forms a jet to remove the contaminants and improves the quality of joint. Besides, high pressure can reduce the residual stress of the structure, thus reducing the deformation during explosive welding [11–13]. In addition, explosive welding is not limited by the shape and area of materials. This method can combine the excellent properties of different materials to obtain high bonding strength and good machining performance.

Titanium and titanium alloys are characterized by high specific strength, high temperature resistance, low density and good corrosion resistance, but their production costs are relatively high [14,15]. Carbon steel not only has weldability, formability, and thermal conductivity, it is also very cheap [16,17]. When titanium/steel bimetallic composite material is used as corrosion-resistant structural material in chemical equipment and marine engineering, it makes full use of the advantages of the two metal materials, especially the high specific strength and corrosion resistance of titanium, and also significantly reduces the material cost compared with pure titanium material.

Titanium and steel are difficult to combine by traditional welding methods due to their different lattice types, large difference in atomic radius and weak mutual solubility. In addition, a large number of intermetallic compounds (such as FeTi and Fe_2Ti) forming in the weld after welding will worsen the performance of the welded joint. As a strong carbide forming element, titanium will combine with carbon to form brittle TiC, which further increases the brittleness of the welding joint [18,19]. In addition, due to the different linear expansion coefficients of the two metals and a large internal stress, the welding joint is prone to crack. In order to overcome the above shortcomings, explosive welding has become a common method to prepare titanium/steel bimetallic composite.

In previous literature, many studies reported on the titanium/steel bimetallic composite. Jiang et al. [18] found that the heat treatment process results in a significant enhancement of diffusion and microstructural transformation in explosive-rolled Ti-Steel clad plate. Besides, TiC has formed near the carbon steel side and the bonding surface was fractured at the TiC first through tensile tests. Chu et al. [20] combined experimental and numerical approaches to quantitatively investigate microstructure evolution and mechanical properties of Ti/Fe explosive-bonded interfaces, finding that the Ti/Fe interface features a wavy structure with melted zones embedded in the crests. The relationship between microstructure and mechanical properties was established. Li et al. [21] found that the σ phase was continuously distributed at the bonding interface, when the rolling temperature was 1223K, reducing the yield strength and ultimate tensile strength of the titanium/steel composite plate. Arisova et al. [22] investigated the influence of explosion welding and hot rolling on the explosive welding-rolled five-layer titanium-steel composite micromechanical properties, structure, and phase composition, founding that local melted zones formed by solid solutions based on titanium and iron because of explosive welding and the diffusion zones on all boundaries as a result of rolling.

In this paper, in order to obtain thin titanium/steel composite sheets, the explosive welded TA1/Q235B thick composite plate was hot-rolled at different temperatures. After nine passes of rolling, the 20-mm-thick explosive composite plate became a 2-mm-thick explosive-rolling composite sheet. The explosive welding and hot rolling can combine the advantages of the two separate methods. Explosive welding realizes the metallurgical bonding of TA1/Q235B, and hot rolling eliminates the defects of explosive welding and optimizes the structure. The study on the effect of rolling temperature on the interface microstructure and properties of the TA1/Q235B composite sheets can provide reference for the preparation of the titanium/steel composite sheets.

2. Materials and Methods

2.1. *Raw Materials and Multi-Pass Rolling*

The 17-mm-thick carbon steel plate (Q235B: C-0.18%; Mn-0.17%; Si-0.13%; S-0.017%; P-0.018%; Fe-Bal.) and 3-mm-thick commercial pure titanium sheet (TA1: Fe-0.023%; C-0.0024%; Ti-Bal.) were used as the raw materials for explosion welding. The explosion welded TA1/Q235B composite plates with a 2000 mm × 1000 mm × 20 mm size were prepared by Hunan Phohom New Material Technology

Co., Ltd. (Changsha, China). The flyer plate (TA1) and base plate (Q235B) were placed parallel to the ground and the explosive was spread over the flyer plate. An emulsion explosive was selected and the explosive process follows the Jones–Wilkins–Lee (JWL) equation as Equation (1), as shown [23].

$$p = A(1 - \frac{\omega}{R_1 V})e^{-R_1 V} + B(1 - \frac{\omega}{R_2 V})e^{-R_2 V} + \frac{\omega E}{V} \quad (1)$$

where A, B, R_1, R_2, ω are constants relating to JWL equation, E is the energy per volume, V is relative specific capacity and p is pressure. Moreover, D (the detonation velocity) and ρ_0 (the density of explosive) are physical parameters relating to emulsion explosive. Before the explosive welding experiment, according to the physical parameters of the explosive and the characteristics of the flyer plate and base plate, the JWL equation was simulated and optimized by computer. By adjusting the constants of the JWL equation, the appropriate explosion equation was obtained, and then the explosive welding experiment was carried out. The relevant parameters of the JWL equation and the physical parameters are shown in Table 1.

Table 1. The parameters of the Jones–Wilkins–Lee (JWL) equation and physical parameters.

Parameters	A (GPa)	B (GPa)	R_1	R_2	ω	D (m/s)	V	E (kJ/m^3)	ρ_0 (g/cm^3)	p (GPa)
Value	8.615	0.818	3.754	0.807	0.01	2100	1	5.252×10^5	0.8	1.278

The preparation process of the TA1/Q235B sheets is shown in Figure 1. The size of explosive welded TA1/Q235B plates is 1000 mm × 2000 mm × 20 mm (3-mm-thick TA1 plate and 17-mm-thick Q235B plate). The sample with a size of 40 mm × 50 mm × 20 mm was cut from the center of the explosive welded TA1/Q235B composite plate with wire-electrode cutting. Then the samples were hot rolled at the temperatures of 1003 K, 1053 K and 1103 K, respectively. The rolling direction was consistent with the explosive direction. After nine passes of hot rolling, the TA1/Q235B sheets were obtained. The thicknesses and reduction of TA1 and Q235B before and after hot rolling are shown in Table 2.

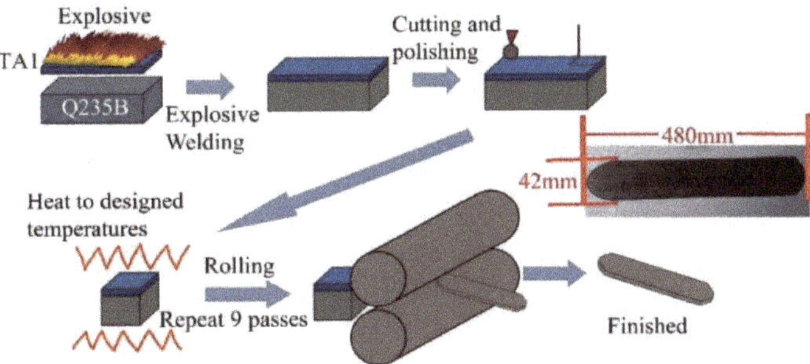

Figure 1. Schematic diagram of the preparation process of 2-mm-TA1/Q235B composite sheets.

Table 2. The thickness and reduction of TA1/Q235B Sheets.

Materials	Thickness Before Rolling (mm)	Thickness After Rolling (mm)	Reduction (%)
TA1	3	0.4	86.7
Q235B	17	1.6	90.6

2.2. Microstructure and Mechanical Properties

The metallographic samples were parallel to the rolling direction (as Figure 2a shown), and examined with an optical microscope (OM, 4XC-II, Shanghai Optics Instrument Factory Inc., Shanghai, China). The interface microstructure and tensile fracture morphology are observed on scanning electron microscopy (SEM, QUANTA-200, FEI Inc., Hillsboro, OR, USA). The EBSD analysis is performed on a scanning electron microscope (SEM, SIRION 200, FEI Inc., Hillsboro, OR, USA) equipped with an electron backscatter diffraction (EBSD) analyzer (XM4-Hikari, EDAX Inc., Mahwah, NJ, USA). The accelerating voltage was 20 kV, working distance was 10 mm, and the step size of data acquisition was 100 nm. The element distributions near the interface are analyzed with energy dispersive spectroscopy (EDS, GENESIS60S, EDAX Inc., Mahwah, NJ, USA) and electron probe micro-analysis (EPMA, JXA-8230, JEOL Ltd., Tokyo, Japan), respectively. The surfaces in both the TA1 and Q235B were peeled with a mechanical method and characterized by X-ray diffraction (XRD, D/Max 2500, Rigaku Inc., Tokyo, Japan), using Cu Kα radiation scanning range from 30° to 90° at a step size of 0.02°.

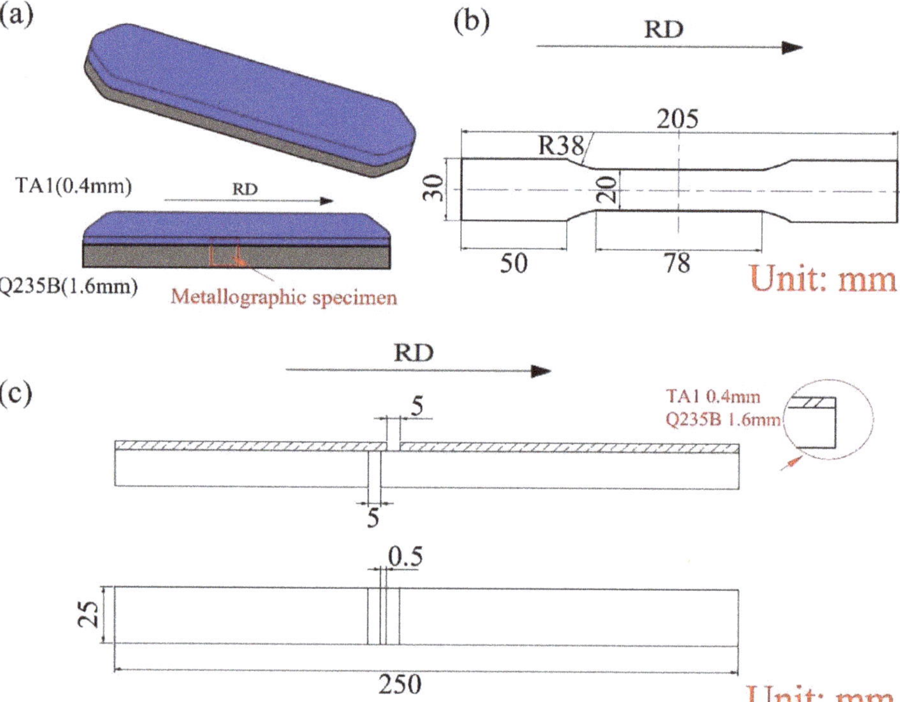

Figure 2. Schematic diagrams of specimen for (**a**) metallographic specimen, (**b**) tensile test, and (**c**) tensile shear test.

The microhardness tests was performed at a Vickers hardness tester (200HVS-5, Huayin Inc., Shandong, China), the load was 19.61 N and the holding time was 10 s. The tensile shear tests and tensile tests were repeated three times on the MTS-810 test machine (MTS Inc., Eden Prairie, MN, USA) at a constant strain rate of 2.5×10^{-3} s^{-1} at room temperature (298 K). Moreover, the tensile direction was parallel to the rolling direction. The size of tensile specimen and the shear specimens (GB/T6396-2008 standard) are shown in Figure 2b,c and the shear strength was calculated by Equation (2).

$$\tau = \frac{F}{bd} \tag{2}$$

where F is the applied peak force, b is the notches distance and d is the bonding width of the specimen.

3. Results and Discussion

3.1. The Interface Microstructure of the Explosive Welded Plate

The typical interface microstructure of explosive welded TA1/Q235B composite plate is shown in Figure 3. It exhibits a typical periodic wave and the wavelength is about 400 μm (Figure 3a). The wave interfaces of the TA1/Q235B composite plate are divided into two categories: The vortex region is shown in Figure 3b and the transition layer as shown in Figure 3c. The wave expands along the effective bonding area of titanium and steel hinders crack propagation, which is an ideal microstructure of a welded joint [24]. The vortex region is composed of titanium and steel, and the transition layer is composed of compounds [25,26].

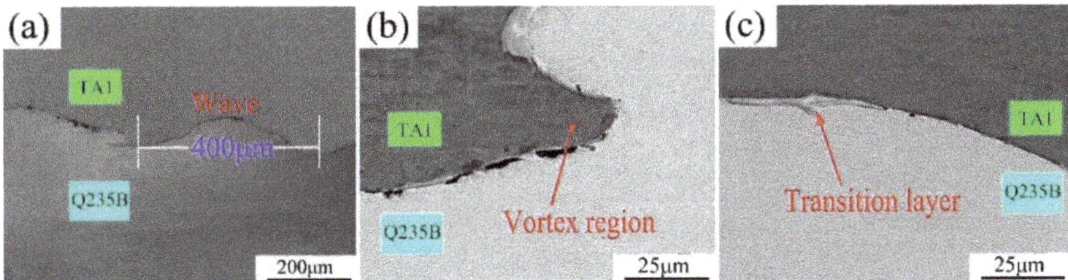

Figure 3. (**a**) SEM of the TA1/Q235B interface: (**b**) the vortex region and (**c**) transition layer.

Figure 4a shows the details of vortex region at higher magnifications. The Ti and Fe element distributions measured by EDS mapping are displayed in Figure 4b,c, respectively. The Ti element is marked in green, and the red represents Fe element. It can be seen that the vortex region is a mixture of the two melted elements. The segregation of Fe (red) can be observed in the center of the melted vortex region, while Ti (green) is uniformly distributed in the red areas (Fe element). In addition, a wide red layer can be found in the crest, revealing the liquid steel flow from the high-pressure part to the low-pressure part. When titanium and carbon steel collide at high temperatures, pressures and speed, the accumulated heat does not have enough time to transfer outwards, causing the carbon steel and part of the titanium to melt at 1811 K. The most deformed steel melts first at the bonding interface, while the titanium does not directly melt. Only when more liquid steel flows to the titanium part and is mixed with the molten titanium can the vortex region be formed; this can explain why Fe dominates the central vortex region. Finally, under the action of relatively low wave-front pressure, the mixture of liquid steel and titanium accumulates gradually and flows towards the vortex region along the direction of the wave crest.

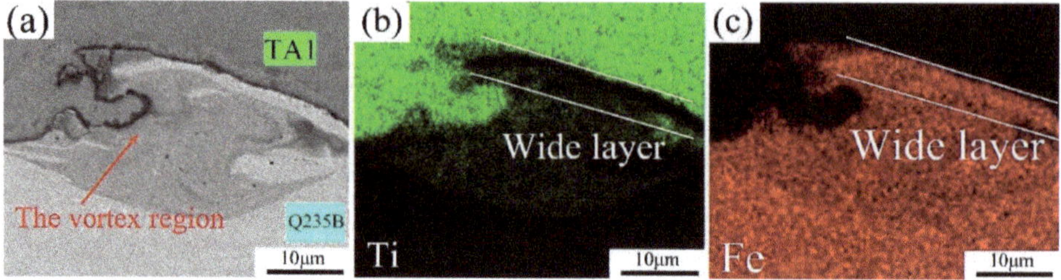

Figure 4. (**a**) SEM microstructure of melted vortex regions of TA1/Q235B welded plate (**b**) EDS map for Ti (Green) and (**c**) Fe (Red).

Another interface is a transition layer. As seen from Figure 5a, the large proportion of microscopic interface is widely distributed at the joint surface of the two metals. They tend to present behind the waveform. The energy dispersive spectroscopy (EDS) element line distribution in Figure 5b also shows the concentration gradient distributions of the two elements in the transition layer, indicating that the diffusion of titanium and carbon steel occurred in the transition layer. Additionally, because the pressure behind the wave is lower than at the peak, the metal in the transition layer is less likely to melt compared to the vortex region. When the two metals explosive welding, the heat generated by the collision is difficult to spread and accumulated at the back of the wave [27]. Although the metal in the transition layer did not melt, high temperature and friction region during the explosion were caused by collision, more importantly, which also caused the atoms of two metals to diffuse rapidly, creating a concentration gradient.

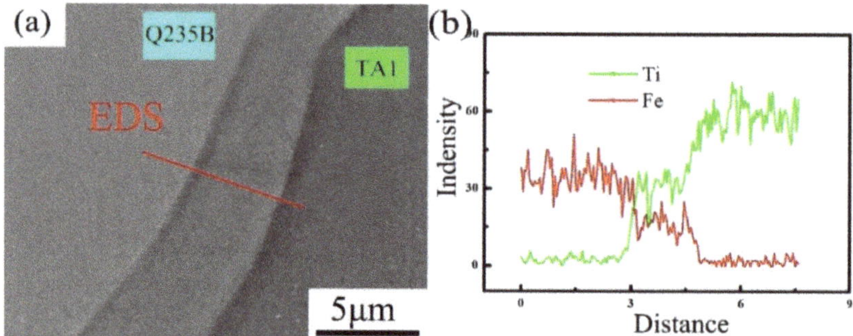

Figure 5. (**a**) SEM microstructure of transition layer, and (**b**) EDS element line distribution for Ti (Green) and Fe (Red) in transition layer of explosive welded TA1/Q235B plate.

3.2. The Interfacial Microstructure of Hot-Rolled Sheets at Different Temperatures

The interface SEM microstructure of TA1/Q235B hot rolled sheets at different temperatures (1003 K, 1053 K and 1103 K) is shown in Figure 6. The vortex region disappears after hot rolling, and the interfacial is straight linearity. The transition layer is obvious. With the increases of rolling temperature, the width of transition layer increases from 3 μm (1003 K) to 4.5 μm (1053 K), and finally reaches 5 μm at 1103 K.

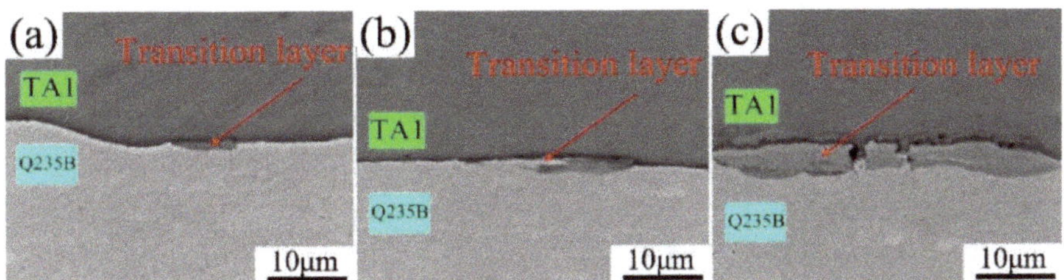

Figure 6. Interface SEM microstructure of TA1/Q235B sheets hot-rolled at (**a**) 1003 K, (**b**) 1053 K, (**c**) 1103 K.

The TA1 layer and Q235B layer of the 1103 K hot-rolled composite sheet were separated by a mechanical method, and each of the two separated surfaces were tested by XRD. Figure 7 shows the XRD patterns of the two surfaces. It indicates that the interfacial compounds are primarily TiC and FeTi, and Ti mainly exists in the form of α-Ti rather than β-Ti, because the rolling temperature is not high enough to reach the phase transition temperature. The TiC and FeTi compounds are high-temperatures phases, which have relatively high hardness and low ductility. In addition, the FeTi distributes mainly in the carbon steel side, while TiC distributes on both sides. When the rolling temperature is lower than the phase transition temperature (1155 K) of titanium, the solid solubility of Fe in α-Ti is lower than that of Ti in Fe; therefore, Ti is easier to diffuse into Fe. During cooling after rolling, the FeTi intermetallic compound is formed in the Q235B layer. Due to the strong combining capacity of Ti and C, TiC compounds are easily formed at the interface.

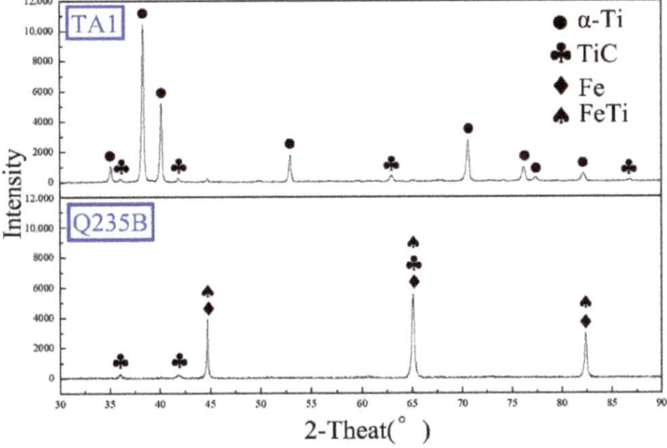

Figure 7. XRD patterns of both separation surfaces of TA1/Q235B hot rolled sheet.

The Figure 8 shows the distribution of C element in the TA1/Q235B interface. The thickness of the C element gathering region decreases from 5 μm at 1003 K to 2 μm at 1103 K. With the increase of rolling temperature, the distribution of the C element gathering region shows a substantial decrease along with the transition layer. At 1003 K, the diffusion of Ti in Q235B is very slow, and the TiC is formed preferentially at the interface, which hinders the grain boundary diffusion of Ti element and delays the formation of FeTi intermetallic compounds. Due to the high temperature and inhibition effect on carbides TiC [19], the amount of FeTi increases and that of TiC decreases, when the rolling temperature rises to 1103 K.

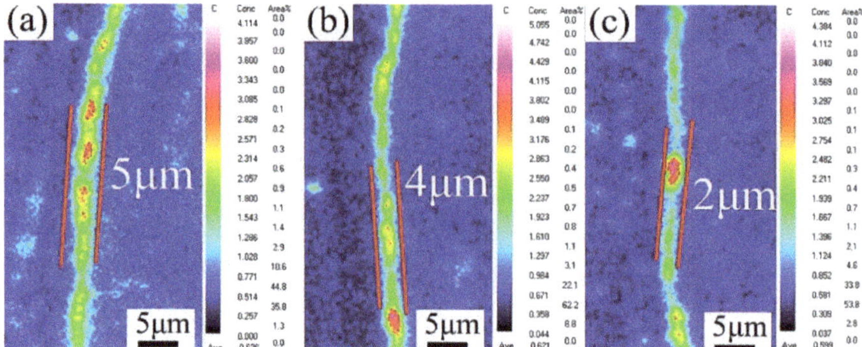

Figure 8. Electron probe micro-analysis (EPMA) for C element distribution in the interface of TA1/Q235B hot-rolled sheets at the temperatures of (a) 1003 K (b) 1053 K (c) 1103 K.

Figure 9 shows the EPMA element distribution of Ti and Fe in the interface of hot-rolling TA1/Q235B sheets at different temperatures. It can be seen that the interface of the two metals indicates the element concentration gradient of Ti and Fe. From the Ti distribution maps in Figure 9b,e,h, it can be found that titanium diffuses to 3 μm, 4.5 μm, and 5 μm to the Q235B when the rolling temperature is 1003 K, 1053 K, and 1103 K, respectively, while in the Fe element distribution map from Figure 9c,f,i, Fe diffuses less than 3 μm to the titanium owing to the titanium diffusion coefficient in steel is much larger than that of Fe in Ti [28,29].

The composition of diffusion layers are two different interfacial compounds, the carbide TiC forms earlier, and FeTi forms later. The main factor affecting the composition of TA1/Q235B composite plate is the carbide TiC and the interface intermetallic FeTi. In addition, the growth rate of interfacial compounds varies from the rolling conditions. Figures 8 and 9 show that TiC layer is always thicker than the FeTi layer, which is attributed to the segregation of C element in the transition layer and it distributes linearly along the bonding interface of the Q235B. Atoms can spread along the grain interface, and the interface of Q235B provides a rapid diffusion channel for C and Ti atoms [30], so TiC generates more rapidly than FeTi.

The EBSD IPF mapping of TA1/Q235B sheets hot rolled at different temperatures is shown in Figure 10. The size of the grain gradually increases with the increasing rolling temperatures both in the TA1 and Q235B. In the Q235B, the grain is pulled into strips because of hot rolling, and the recrystallization of the grain occurs at a relatively high temperature [31]. While the grain grows from about 10 μm in 1003 K to about 30 μm in 1153 K in the TA1. In addition, there are some new grains generated due to the breakage and refinement of the original grains.

3.3. Mechanical Properties of TA1/Q235B Sheets Hot-Rolled at Different Temperatures

Figure 11 shows the microhardness and test schematic of explosive welding plates and hot-rolled sheets at different temperatures. Since the thickness of the sheet is only 2 mm after rolling, it is difficult to test the hardness gradient along a straight line, and to ensure that the adjacent test points are far enough so the test method as shown in Figure 11a,b is adopted. The microhardness of matrix metals is measured at the distance of 225 μm near the bonding interface, as shown in Figure 11b. Assuming that the interface bonding is the origin, the direction from the interface to Q235B side is positive. TA1, Q235B and the interface are tested five times, and the results are averaged. The interface microhardness distribution is shown in Figure 11c. The microhardness of the explosive welded plate is higher than that of the rolled sheets and the hardness of the matrix, and the interface decreases with the increase of the rolling temperature. The microhardness of the interface is higher than that of the matrix. The maximum interface microhardness is 217.8 HV at the explosive welded plate, with an increase of rolling temperature, which is gradually decreases to 184.2 HV. When the rolling temperature rises, the change of interface microhardness is related to the amount of TiC (as shown in Figure 8), while the decrease of TA1 and Q235B matrix microhardness is related to the recrystallization grain size (as shown in Figure 10).

Shear strengths, yield strength (YS), ultimate tensile strength (UTS), maximum elongation and Young's modulus of rolled sheets under different temperatures are shown in Figure 12. With the increase of the rolling temperature from 1003 K to 1103 K, the shear strength of the composite sheet increases from 142.1 MPa to 171.4 MPa. The interface shear strength of the TA1/Q235 composite sheet is affected mainly by the interface brittle compounds. It can be seen from Figure 8 that the amount of the C element in the transition layer decreases when rolling temperatures rises, it means the amount of brittle TiC decreases with the increase of rolling temperature. Hence, the formation of brittle carbides in TiC may suppress by the increasing rolling temperature, and the shear strength improved with the disappearance of brittle compounds TiC at a relatively high rolling temperature.

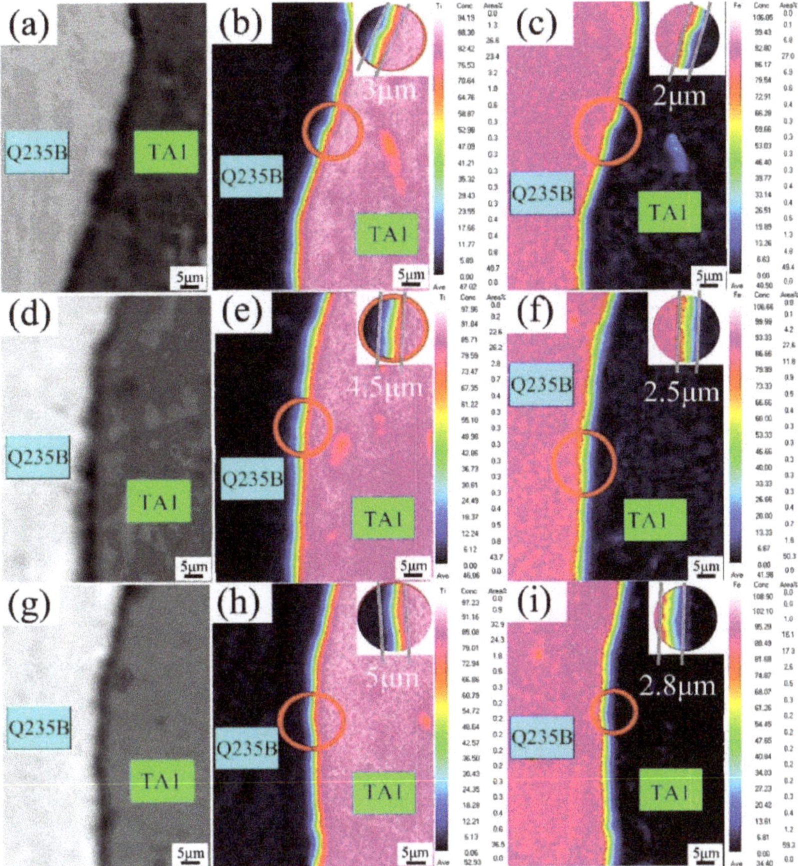

Figure 9. Electron probe micro-analysis (EPMA) element distribution in the interface of TA1/Q235B sheets hot rolled at (**a**–**c**) 1003 K, (**d**–**f**) 1053 K and (**g**–**i**) 1103 K, (**a**,**d**,**g**) are microstructures, (**b**,**e**,**h**) are Ti elements and (**c**,**f**,**i**) are Fe elements.

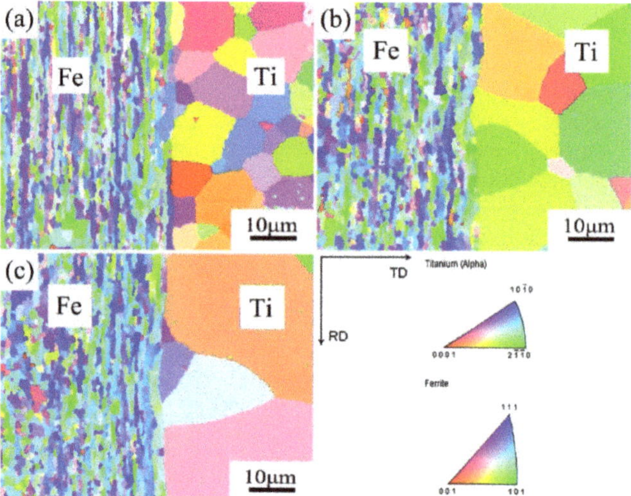

Figure 10. EBSD Inverse pole figure (IPF) mapping of TA1/Q235B sheets hot rolled at the temperature of (**a**) 1003 K (**b**) 1053 K (**c**) 1103 K.

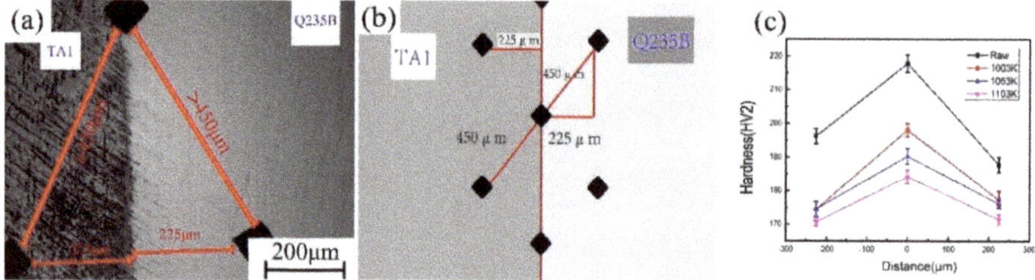

Figure 11. (**a**) Micrograph of microhardness test (**b**) schematic of microhardness test (**c**) microhardness of explosive welding plate and different temperatures rolled sheets.

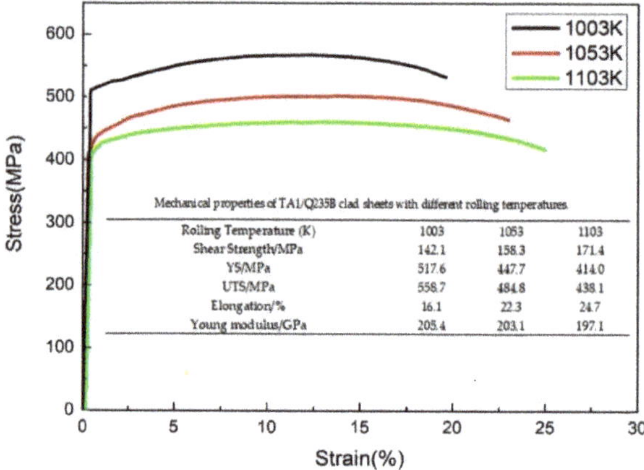

Rolling Temperature (K)	1003	1053	1103
Shear Strength/MPa	142.1	158.3	171.4
YS/MPa	517.6	447.7	414.0
UTS/MPa	558.7	484.8	438.1
Elongation/%	16.1	22.3	24.7
Young modulus/GPa	205.4	203.1	197.1

Figure 12. The tensile stress-strain curves and mechanical properties of TA1/Q235B composite sheets at different rolling temperatures.

The YS and UTS of hot rolled TA1/Q235 sheets decrease with the increase of rolling temperature. After rolling at 1003 K, the maximum YS and UTS of the hot rolled sheet are 517.6 MPa and 558.7 MPa, respectively. The sheet rolled at 1103 K shows lower yield strength (414.0 MPa) and ultimate tensile strength (438.1 MPa) than that rolled at 1003 K and 1053 K. The lower strength of 1103 K rolled sheet is due to the increase of recrystallization in titanium and carbon steel matrix at higher temperature (as shown in Figure 10). The increase of recrystallization also leads to a gradual increase in elongation. The Figure 12 also shows that the rolling temperature has little influence on the Young's modulus of the sheet, and when rolling temperature rises from 1003 K to 1103 K, while the Young's modulus of the sheet does not change significantly and only decreases slightly.

Figure 13 shows the macroscopic morphology of sheets rolled at different temperatures after tensile fractured. It can be seen that titanium and carbon steel are separated near the fracture surface. During the tensile test, the brittle compounds TiC and FeTi in the bonding interface fractured firstly. Subsequently, the fracture occurred in the matrix metals. Figure 14 shows the SEM fracture morphology after tensile fracture of samples rolled at different temperatures. It is obvious that interfacial fractures of the composite sheet rolled at 1003 K is small, and the extent of interfacial fractures increases with the increasing rolling temperature. It illustrates that the interfacial compounds layer causes the cracks to initiate in the composite sheets, and crack propagation is more serious with the thicker interfacial compound layer. It can be seen from Figure 14d–f that the fracture profiles of titanium at different rolling temperatures are similar in fracture characteristics. The fracture profile of the TA1 plate includes cleavage fractures, while the fracture profile of carbon steel is a dimple fracture. As shown in Figure 14g–i, the depth, size, and proportion of the dimple increased with the increase in rolling temperature, and the plasticity of the composite sheets increased simultaneously [32,33]. The more in-depth, more extensive, as well as larger proportion of dimple fractures exist on the carbon steel side of the composite sheet at 1103 K, which is related to the increase in plasticity after proper heating [34]. Because of the appearance of both dimples and cleavage planes in the rolled composite sheet, it infers that the fractures of TA1/Q235B composite sheets are a mixed-mode of ductile and brittle fractures.

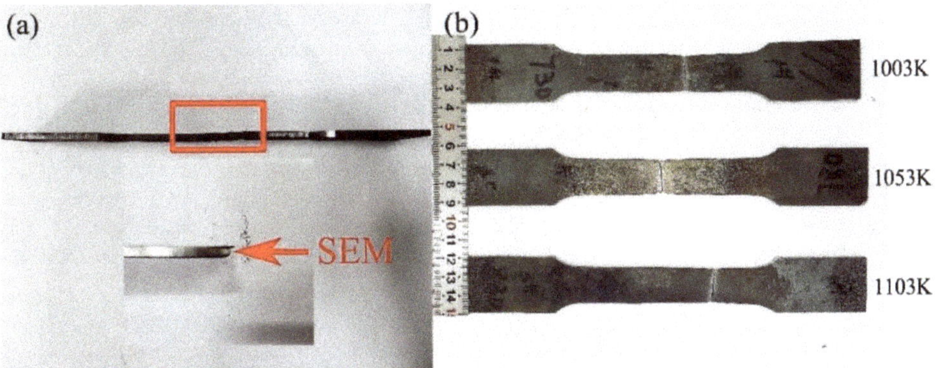

Figure 13. (**a**) The area of SEM of fracture profiles, (**b**) Macroscopic morphology of sheets rolled at different temperatures after tensile fracture.

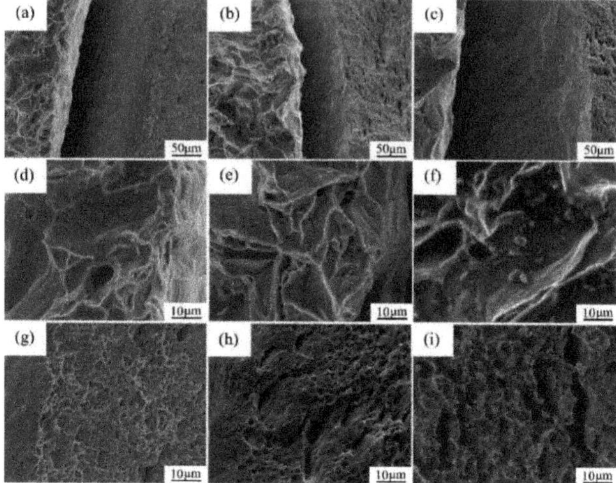

Figure 14. SEM fracture morphology after tensile fracture of samples rolled at different temperatures: (**a**) 1003 K, (**b**) 1053 K, (**c**) 1103 K, (**d**,**g**) Ti and steel matrices in (a), (**e**,**h**) Ti and steel matrices in (**b**), (**f**,**i**) Ti and Steel matrices in (**c**).

4. Conclusions

The 2-mm TA1/Q235B composite sheets (TA1 is 0.4 mm and Q235B is 1.6 mm) are fabricated through explosive welding and nine passes of hot rolling at different temperatures. The microstructure and mechanical properties were studied and the following conclusions were drawn:

(1) The vortex region and the transition layer microstructures are formed in the TA1/Q235B explosive welded plate. After hot rolling, only the transition layer exists in the bonding surfaces.

(2) The thickness of transition layer increases with the increase of rolling temperatures. The component of the transition layer is α-Ti, TiC, Fe, and FeTi. In addition, the fracture is a mixed-mode of ductile and brittle fractures, which mainly occurred in interface layer.

(3) Due to fragile and hard interfacial compounds, the microhardness of the explosive plate is higher than that of hot rolled sheets. High temperatures reduce the proportion of brittle TiC compounds, resulting in an increment of interface shear strength. However, due to the increase of recrystallization degree of TA1 and Q235B at a high temperature, the yield strength and tensile strength of the sheet decreased. Furthermore, when the rolling temperature rises, the elongation increases, while the Young's modulus shows little change.

Author Contributions: H.L. and X.L. conceived and designed the experiments; L.C. performed the experiments; W.Z. and C.W. analyzed the data; Z.Z. contributed reagents/materials/analysis tools; C.Z. wrote the paper. All authors have read and agreed to the published version of the manuscript.

Funding: This research was funded by the Fundamental Research Fund for the Central Universities of Central South University (Grant Number: 502041002) and the Open Sharing Fund for the Large-scale Instruments and Equipment of Central South University.

Acknowledgments: This research was financially supported by the Fundamental Research Fund for the Central Universities (Changsha, China) and the Open Sharing Fund for the Large-scale Instruments and Equipment of Central South University. The authors are grateful to the Hunan Phohom New Material Technology Co., Ltd for the support in explosive welding experiment.

Conflicts of Interest: The authors declare no conflict of interest.

References

1. Nishida, M.; Chiba, A.; Honda, Y.; Hirazumi, J.-I.; Horikiri, K. Electron Microscopy Studies of Bonding Interface in Explosively Welded Ti/Steel Clads. *ISIJ Int.* **1995**, *35*, 217–219. [CrossRef]
2. Wachowski, M.; Sniezek, L.; Szachogluchowicz, I.; Kosturek, R.; Plocinski, T. Microstructure and fatigue life of Cp-Ti/316L bimetallic joints obtained by means of explosive welding. *Bull. Pol. Acad. of Sci. Tech. Sci.* **2018**, *66*, 925–933.
3. Li, Y.; Liu, C.; Yu, H.; Zhao, F.; Wu, Z. Numerical Simulation of Ti/Al Bimetal Composite Fabricated by Explosive Welding. *Metals* **2017**, *7*, 407. [CrossRef]
4. Ning, J.; Zhang, L.; Xie, M.-X.; Yang, H.-X.; Yin, X.-Q.; Zhang, J.-X. Microstructure and property inhomogeneity investigations of bonded Zr/Ti/steel trimetallic sheet fabricated by explosive welding. *J. Alloys Compd.* **2017**, *698*, 835–851. [CrossRef]
5. Carvalho, G.; Galvão, I.; Mendes, R.; Leal, R.; Loureiro, A. Aluminum-to-Steel Cladding by Explosive Welding. *Metals* **2020**, *10*, 1062. [CrossRef]
6. Yu, C.; Xiao, H.; Yu, H.; Qi, Z.; Xu, C. Mechanical properties and interfacial structure of hot-roll bonding TA2/Q235B plate using DT4 interlayer. *Mater. Sci. Eng. A* **2017**, *695*, 120–125. [CrossRef]
7. Sun, Z.; Shi, C.; Wu, X.; Shi, H. Comprehensive investigation of effect of the charge thickness and stand-off gap on interface characteristics of explosively welded TA2 and Q235B. *Compos. Interfaces* **2020**, *27*, 977–993. [CrossRef]
8. Paul, H.; Chulist, R.; Bobrowski, P.; Perzynski, K.; Madej, L.; Mania, I.; Miszczyk, M.M.; Cios, G. Microstructure and properties of the interfacial region in explosively welded and post-annealed titanium-copper sheets. *Mater. Charact.* **2020**, *167*, 110520. [CrossRef]
9. Bina, M.H.; Dehghani, F.; Salimi, M. Effect of heat treatment on bonding interface in explosive welded copper/stainless steel. *Mater. Des.* **2013**, *45*, 504–509. [CrossRef]
10. Chen, Z.; Wang, D.; Cao, X.; Yang, W.; Wang, W. Influence of multi-pass rolling and subsequent annealing on the interface microstructure and mechanical properties of the explosive welding Mg/Al composite plates. *Mater. Sci. Eng. A* **2018**, *723*, 97–108. [CrossRef]
11. Lysak, V.; Kuzmin, S.V. Energy balance during explosive welding. *J. Mater. Process. Technol.* **2015**, *222*, 356–364. [CrossRef]
12. Vieille, B.; Casado, V.M.; Bouvet, C. About the impact behavior of woven-ply carbon fiber-reinforced thermoplastic- and thermosetting-composites: A comparative study. *Compos. Struct.* **2013**, *101*, 9–21. [CrossRef]
13. Pereira, D.; Oliveira, J.P.; Pardal, T.; Miranda, R.M.; Santos, T.G. Magnetic pulse welding: Machine optimisation for aluminium tubular joints production. *Sci. Technol. Weld. Join.* **2018**, *23*, 172–179. [CrossRef]
14. Ha, J.S.; Hong, S. Deformation and fracture of Ti/439 stainless steel clad composite at intermediate temperatures. *Mater. Sci. Eng. A* **2016**, *651*, 805–809. [CrossRef]
15. Liu, J.; Cai, W.; Liu, L.; Han, J.; Liu, J. Investigation of interfacial structure and mechanical properties of titanium clad steel sheets prepared by a brazing-rolling process. *Mater. Sci. Eng. A* **2017**, *703*, 386–398. [CrossRef]
16. Cui, Y.; Liu, D.; Fan, M.Y.; Deng, G.P.; Sun, L.X.; Zhang, Y.; Chen, D.; Zhang, Z.W. Microstructure and mechanical properties of TA1/3A21 composite plate fabricated via explosive welding. *Mater. Sci. Technol.* **2020**, *36*, 425–433. [CrossRef]
17. Song, J.; Kostka, A.; Veehmayer, M.; Raabe, D. Hierarchical microstructure of explosive joints: Example of titanium to steel cladding. *Mater. Sci. Eng. A* **2011**, *528*, 2641–2647. [CrossRef]
18. Jiang, H.-T.; Yan, X.-Q.; Liu, J.-X.; Duan, X.-G. Effect of heat treatment on microstructure and mechanical property of Ti–steel explosive-rolling clad plate. *Trans. Nonferrous Met. Soc. China* **2014**, *24*, 697–704. [CrossRef]
19. Jin, Y.X.; Zeng, S.Y.; Wang, H.W. Changes of carbides morphology in the alloy Ti15Al7C during heat treatment. *Rare Metal Mater. Eng.* **2002**, *31*, 358–362.
20. Chu, Q.; Zhang, M.; Li, J.; Yan, C. Experimental and numerical investigation of microstructure and mechanical behavior of titanium/steel interfaces prepared by explosive welding. *Mater. Sci. Eng. A* **2017**, *689*, 323–331. [CrossRef]

21. Li, B.; Chen, Z.; He, W.; Wang, P.; Lin, J.; Wang, Y.; Peng, L.; Li, J.; Liu, Q. Effect of interlayer material and rolling temperature on microstructures and mechanical properties of titanium/steel clad plates. *Mater. Sci. Eng. A* **2019**, *749*, 241–248. [CrossRef]
22. Arisova, V.N.; Gurevich, L.M.; Trudov, A.F.; Serov, A.G.; Kharlamov, V.G. Structure Formation in the Zones of Joints Obtained by Explosion Welding with Subsequent Rolling of a Five-Layer Titanium-Steel Composite. *Metals* **2019**, *63*, 96–104. [CrossRef]
23. Sun, Z.; Shi, C.-G.; Shi, H.; Li, F.; Gao, L.; Wang, G. Comparative study of energy distribution and interface morphology in parallel and double vertical explosive welding by numerical simulations and experiments. *Mater. Des.* **2020**, *195*, 109027. [CrossRef]
24. Acarer, M.; Gülenç, B.; Fındık, F. Investigation of explosive welding parameters and their effects on microhardness and shear strength. *Mater. Des.* **2003**, *24*, 659–664. [CrossRef]
25. Zhang, H.; Jiao, K.X.; Zhang, J.L.; Liu, J. Comparisons of the microstructures and micro-mechanical properties of copper/steel explosive-bonded wave interfaces. *Mater. Sci. Eng. A* **2019**, *756*, 430–441. [CrossRef]
26. Wronka, B. Testing of explosive welding and welded joints: Joint mechanism and properties of explosive welded joints. *J. Mater. Sci.* **2010**, *45*, 4078–4083. [CrossRef]
27. Qin, L.; Wang, J.; Wu, Q.; Guo, X.; Tao, J. In-situ observation of crack initiation and propagation in Ti/Al composite laminates during tensile test. *J. Alloys Compd.* **2017**, *712*, 69–75. [CrossRef]
28. Kundu, S.; Chatterjee, S. Effect of bonding temperature on interface microstructure and properties of titanium–304 stainless steel diffusion bonded joints with Ni interlayer. *Mater. Sci. Technol.* **2006**, *22*, 1201–1207. [CrossRef]
29. Klugkist, P.; Herzig, C. Tracer diffusion of titanium in α-iron. *Phys. Status Solidi* **2010**, *148*, 413–421. [CrossRef]
30. Sam, S.; Kundu, S.; Chatterjee, S. Diffusion bonding of titanium alloy to micro-duplex stainless steel using a nickel alloy interlayer: Interface microstructure and strength properties. *Mater. Des.* **2012**, *40*, 237–244. [CrossRef]
31. Kahraman, N.; Gülenç, B.; Findik, F. Joining of titanium/stainless steel by explosive welding and effect on interface. *J. Mater. Process. Technol.* **2005**, *169*, 127–133. [CrossRef]
32. Blach, J.; Falat, L.; Ševc, P. Fracture characteristics of thermally exposed 9Cr–1Mo steel after tensile and impact testing at room temperature. *Eng. Fail. Anal.* **2009**, *16*, 1397–1403. [CrossRef]
33. Xie, M.-X.; Zhang, L.-J.; Zhang, G.; Zhang, J.-X.; Bi, Z.-Y.; Li, P.-C. Microstructure and mechanical properties of CP-Ti/X65 bimetallic sheets fabricated by explosive welding and hot rolling. *Mater. Des.* **2015**, *87*, 181–197. [CrossRef]
34. Rozumek, D.; Bański, R. Crack growth rate under cyclic bending in the explosively welded steel/titanium bimetals. *Mater. Des.* **2012**, *38*, 139–146. [CrossRef]

Publisher's Note: MDPI stays neutral with regard to jurisdictional claims in published maps and institutional affiliations.

Article

Heat Input and Mechanical Properties Investigation of Friction Stir Welded AA5083/AA5754 and AA5083/AA7020

Mohamed M. Z. Ahmed [1,2,*], Sabbah Ataya [2,3], Mohamed M. El-Sayed Seleman [2], Abdalla M. A. Mahdy [4], Naser A. Alsaleh [3] and Essam Ahmed [2]

1 Department of Mechanical Engineering, College of Engineering at Al-Kharj, Prince Sattam Bin Abdulaziz University, Al Kharj 11942, Saudi Arabia
2 Department of Metallurgical and Materials Engineering, Faculty of Petroleum and Mining Engineering, Suez University, Suez 43512, Egypt; sabbah.ataya@suezuniv.edu.eg (S.A.); mohamed.elnagar@suezuniv.edu.eg (M.M.E.-S.S.); essam.ahmed@suezuniv.edu.eg (E.A.)
3 Department of Mechanical Engineering, College of Engineering, Al-Imam Mohammad Ibn Saud Islamic University, Riyadh 11432, Saudi Arabia; alsaleh@engineer.com
4 Nasr Petroleum Company, Suez 43511, Egypt; abdalla_mahdy@yahoo.com
* Correspondence: moh.ahmed@psau.edu.sa; Tel.: +966-011-588-1200

Abstract: The current work presents a detailed investigation for the effect of a wide range friction stir welding (FSW) parameters on the dissimilar joints' quality of aluminum alloys. Two groups of dissimilar weldments have been produced between AA5083/AA5754 and A5083/AA7020 using tool rotational rates range from 300 to 600 rpm, and tool traverse speeds range from 20 to 80 mm/min. In addition, the effect of reversing the position of the high strength alloy at the advancing side and at retreating side has been investigated. The produced joints have been investigated using macro examination, hardness testing and tensile testing. The results showed that sound joints are obtained at the low heat input FSW parameters investigated while increasing the heat input results in tunnel defects. The hardness profile obtained in the dissimilar AA5083/AA5754 joints is the typical FSW hardness profile of these alloys in which the hardness reduced in the nugget zone due to the loss of the cold deformation strengthening. However, the profile of the dissimilar AA5083/AA7020 showed increase in the hardness in the nugget due to the intimate mixing the high strength alloy with the low strength alloy. The sound joints in both groups of the dissimilar joints showed very high joint strength with efficiency up to 97 and 98%. Having the high strength alloy at the advancing side gives high joint strength and efficiency. Furthermore, the sound joints showed ductile fracture mechanism with clear dimple features mainly and significant plastic deformation occurred before fracture. Moreover, the fracture in these joints occurred in the base materials. On the other, the joints with tunnel defect showed some features of brittle fracture due to the acceleration of the existing crack propagation upon tensile loading.

Keywords: friction stir welding; dissimilar welding; aluminum; mechanical properties; fracture

Citation: Ahmed, M.M.Z.; Ataya, S.; El-Sayed Seleman, M.M.; Mahdy, A.M.A.; Alsaleh, N.A.; Ahmed, E. Heat Input and Mechanical Properties Investigation of Friction Stir Welded AA5083/AA5754 and AA5083/AA7020. *Metals* **2021**, *11*, 68. https://doi.org/10.3390/met11010068

Received: 12 December 2020
Accepted: 26 December 2020
Published: 31 December 2020

Publisher's Note: MDPI stays neutral with regard to jurisdictional claims in published maps and institutional affiliations.

1. Introduction

AA5754 and AA5083 are aluminum magnesium alloys, and their most prominent features are the high corrosion resistance and good formability. Thus, they have been extensively used in pressure vessels, tanks, trucks and shipbuilding [1,2]. AA7020 is a precipitation-hardened aluminum alloy, demonstrating high strength per weight ratio [3]. The use of the dissimilar alloys leads to the sustainable advantages such as overall cost reduction and hybrid properties that are available in the two different alloys. Appropriate joining process and its parameters optimization plays a vital role in the service performance of these alloys. Challenges like solidification cracking, porosity, intermetallic formation and so on are present due to the difference in the chemical and physical properties of the dissimilar alloy's combinations. Recently, the friction stir welding (FSW) of dissimilar aluminum alloys combinations has been studied extensively, which proved the potential of the process to join these alloy combinations [4–6]. However, improper FSW parameters give rise to the formation of intermetallic compounds and internal and external defects (e.g., tunnel formation, voids, surface grooves and flash) [6–10]. Therefore, the investigation of FSW parameters is very important for obtaining defect-free joints with good mechanical properties. The placement of the higher strength aluminum alloys at the advancing side (AS) or at the retreating side (RS) affects material flow as it strongly influences material the stirring and flow behavior [10,11]. This can be a crucial parameter affecting the final joint microstructure, particularly when the selected combinations of base material (BM) have significant differences in their mechanical properties, microstructure and texture [4,12–16]. Some researchers studied the effect of the placement of BM on the material flow and the resulting FSWed microstructure and the mechanical properties [17,18]. Palanivel et al. [15] revealed that the tool rotational rate and tool pin profiles affected the AA5083-H111/AA6351-T6 joint strength because of the loss of cold work in the heat affected zone (HAZ) of AA5083 side, dissolution and over-aging of precipitates of AA6351 side and macroscopic defects formation in the weld zone. Jannet and Mathews [17] concluded that the AA6061 T6/AA5083 O joints fabricated at tool rotational rate of 900 rpm yielded a higher tensile strength than those fabricated at 750 rpm contributed by the thorough plastic flow and dissolving of dissimilar alloys and due to reduction in heat generated from plastic flow of the metal at 750 rpm. Park et al. [18] showed that the materials were more properly mixed when the AA5052-H32 was in the AS and the AA6061-T6 was in the RS than the reverse case on RS. Leitao et al. [19] reported that the global mechanical behavior of the AA6016-T4/AA5182-H111 welds was a 10–20% strength reduction relative to the base materials and important losses in ductility. Khanna et al. [20] concluded that softer alloy should be placed on AS with tool offset towards it for better FSWed AA6061-T6/AA 8011-H14 qualities. Kailainathan et al. [16] showed that the tensile strength of the 6-mm-thick AA6063/AA8011 joints was increased with the increase in the tool rotational speed due to the uniform temperature distribution at the weld region. However, beyond 1200 rpm, an adverse effect was noticed due to the distortion in the weld region. Abd Elnabi et al. [21] reported that the traverse speed has the highest contribution to the process for ultimate tensile strength of AA5454/AA7075 joints. Cole et al. [22] estimated that the AA6061/AA7075 joint strength was improved with decreasing the power input to the weld because of the sensitivity of alloy to heat input and weld temperature. The work of Ouyang and Kovacevic [23] suggested that the lower-strength alloy should be placed on AS for obtaining a better weld quality. Gerard and Ehrstrom [24] mentioned that the material with the higher solidus temperature should be on the AS not only for joint quality improvement but also for internal defects/porosity elimination. Guo et al. [25] revealed that the material mixing is much more effective when AA6061 alloy was located on the AS for AA6061/AA7075 joints. The ultimate tensile strength of the joints increases with the decrease of the heat input induced by friction. Kim et al. [26] demonstrated that excessive agglomerations and defects generated by joints when the high strength Al alloy on the AS of AA5052/AA5J32 are placed due to limited flow of material. Lee et al. [27] concluded that the mechanical properties of the stir zone showed higher values when AA6061 were positioned at the RS due to the complex microstructure of the stir zone. On the other hand, Jonckheere et al. [28] showed that material flow and joint quality are more dependent on the FSW conditions and their effects on heat input and temperature distribution in weld nugget, regardless of BM placement. Due to the material plastic flow during FSW, the heat generation is controlled by tool rotation and welding speed [29–31]. However, very high rotation speeds lead to macroscopic defects because of the excessive heat input [1,32–34]. To the author's knowledge, the FSW of AA5083/AA5754 and AA5083/AA7020 have not been reported in the open literature. The present work focuses on the influences of FSW parameters more deeply including the traverse speeds (20–80 mm/min) and AS/RS positions of base materials on the quality and the mechanical properties of the dissimilar AA5083/AA5754 and AA5083/AA7020 joints.

2. Experimental Procedure

2.1. Materials

Three commercial aluminum alloys AA5083-O, AA5754-H14 and AA7020-T6 were chosen for producing dissimilar friction stir butt welds. The alloys were purchased in the form of rolled plates of 10 mm thick. The butt welds were designed to be 200 mm total width which composed of two plates; each plate was 100 mm wide and 200 mm long. The nominal chemical compositions of the parent materials are listed in Table 1. Moreover, the tensile strength, temper condition and hardness of the parent materials are summarized in Table 2.

Table 1. Nominal chemical composition of aluminum alloys AA5083, AA5754 and AA7020.

Alloy	Elements in wt.%								
	Si	Fe	Cu	Mn	Mg	Zn	Cr	Ti	Al
AA5083	0.40	0.40	0.10	0.4–1.0	4.0–4.9	0.25	0.05–0.25	0.15	Bal.
AA5754	0.40	0.40	0.10	0.50	2.6–3.6	0.20	0.30	<0.15	Bal.
AA7020	0.35	0.40	0.20	0.05–0.50	1.0–1.4	4.50	0.1–0.35	<0.35	Bal.

Table 2. Mechanical properties of the aluminum alloys AA5083, AA5754 and AA7020.

Alloy	Condition	Tensile Strength, MPa	Hardness, HV
AA5083-O	Annealed	233	68
AA5754-H14	Strain hardened-1/2 hard	251	74
AA7020-T6	Solution heat treated and artificially aged	364	117

2.2. Friction Stir Welding Procedure

The welding process was performed on the friction stir welding machine (EG-FSW-M1) at Suez University. This machine has been locally designed and manufactured in Egypt. The main motor power of this machine is 30 HP (22 kW) and can deliver torque up to 100 N·m, rotational speed up to 3000 rpm and tilt angel up to ±5°. The travel speed of the table up to 1000 mm/min. The tool design is an important parameter in FSW processes, which influences the heat generation, plastic flow, the resulting microstructure and mechanical properties of the welded material. The used rotating tool was of a cylindrical threaded pin with scrolled shoulder made of H13 tool steel that heat treated to obtain hardness of 50 HRC. The shoulder diameter was 25 mm, the pin (probe) diameter was 8 mm, and pin height was 9.8 mm, which is slightly less than the material thickness (10 mm). The angle between the edge of shoulder and the pin was 3°. The configuration of the tool used in this study is shown in Figure 1.

The hardness of the as-received tool steel was 25.3 HRC. After manufacturing the FSW tool, it has been hardened by heating to 950 °C and holding for 30 min then oil quenched, then tempered by heating to 550 °C and holding for one hour then air-cooled to room temperature. The heat treatment process was carried out using an electric resistant furnace of type Nabertherm-1200 °C. The hardness of the hardened tool steel was measured as 61 HRC. The tempering process has decreased the tool hardness to 54 HRC. Al alloy plates were prepared to obtain the required dimensions of 200 mm length and 100 mm width. The plates were clamped properly on the FSW machine table as shown in Figure 2a, b shows the butt joint after completing the FSW process.

For the system AA5083/AA5754, the plate of the alloy AA5083 was positioned in the AS, while the AA5754 plate was in the RS as illustrated in Figure 2a. Workpieces were rigidly clamped, to prevent the plates from lifting apart during the welding process. For the system AA5083/AA7020, the plate of the alloy AA5083 was positioned in the AS, while the AA7020 plate was in the RS for a set of welding conditions. For the same set of the welding conditions, the plate of the AA7020 was reversed to be positioned in the AS and the plate of the AA5083 was in the RS, as shown in Table 3. The welding process progressed as follows: The tool was rotated and slowly plunged into the workpiece with speed of 0.1 mm/s until the shoulder of the tool forcibly contacts the upper surface of the material. After that, the tool was traversed along the weld line for a single pass weld. The tool was tilted by a constant angle of 3° against the vertical axis, so that the rear of the tool is lower than the front. This has been found to assist the forging process and the material flow during FSW. Table 3 summarizes the different combinations of operating conditions parameters investigated in this work during FSW. The desired welding parameters are based on the ongoing research at the authors laboratory in FSW of the different aluminum alloys of 10 mm thick.

2.3. Macrostructural Investigation

Cross sections of welded joints were prepared for metallographic analysis using standard metallographic procedure [32]. The samples were etched using Keller's reagent for a period of 40–50 s. at room temperature to reveal the macrostructure of the welded samples and then washed with water and acetone, and then air-dried.

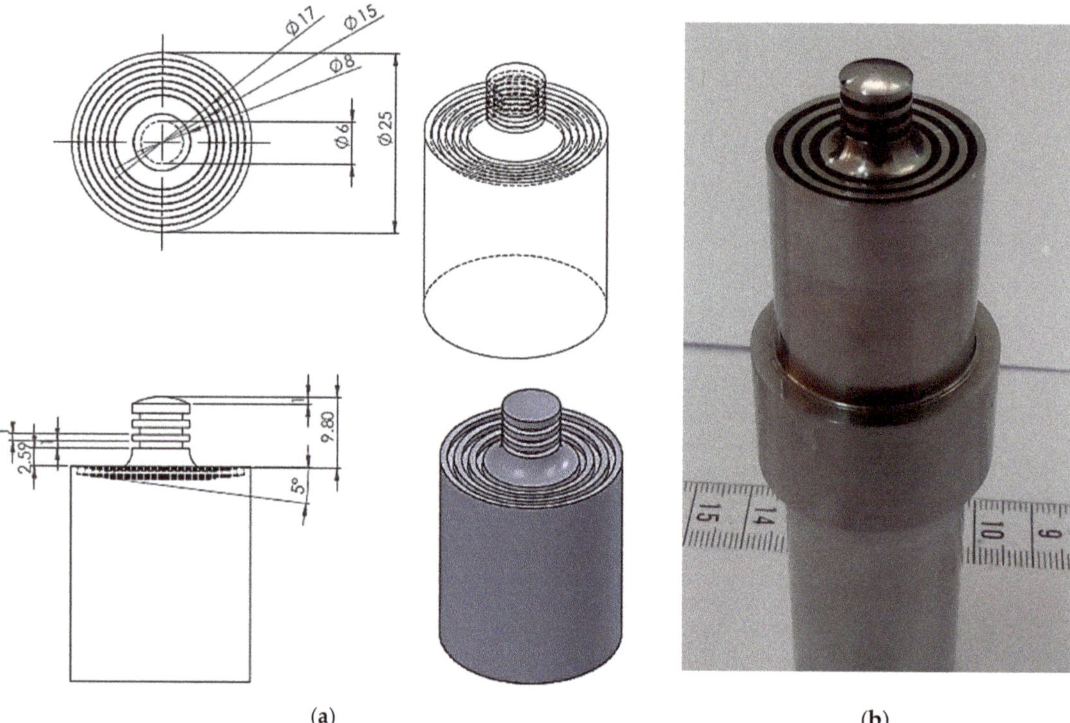

Figure 1. Friction stir welding (FSW) tool used in the FSW experiments (**a**) Computer aided design (CAD) drawings with detailed dimensions in mm and (**b**) image of the used FSW tool.

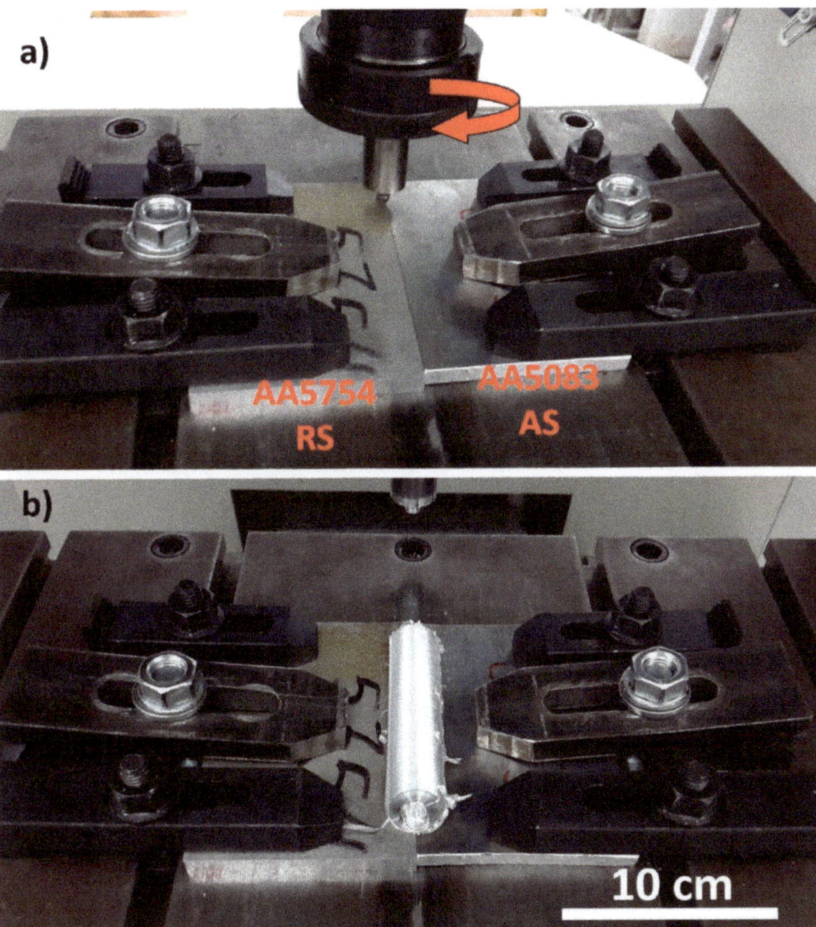

Figure 2. (**a**) The clamping of the plates on the FSW machine table with the advancing side (AS) and retreating side (RS) indicated and the tool rotation direction indicated also by the red arrow and (**b**) Aluminum alloys butt joint on the table after FSW.

Table 3. FSW welding parameters and position of alloys at the AS and RS.

AA5083/AA5754			AA5083/AA7020		
Rotation Speed (rpm)	Travel Speed (mm/min)	Position	Rotation Speed (rpm)	Travel Speed (mm/min)	Position
400	20	AA5083 AS	500	20	AA5083 AS
	40	AA5083 AS		40	AA5083 AS
	60	AA5083 AS		80	AA5083 AS
600	20	AA5083 AS	500	20	AA7020 AS
	40	AA5083 AS		40	AA7020 AS
	60	AA5083 AS		80	AA7020 AS

2.4. Mechanical Properties

The materials were mechanically tested before and after FSW for comparison. To have an insight into the mechanical properties, hardness measurements and tensile testing were carried out. Vickers macro-hardness tests were performed on the transverse cross-sections perpendicular to the welding direction with an interspacing distance of 2 mm using a test load of 1000 g force and dwell time of a 15 s. To evaluate the tensile properties of the welded stir zone, transverse flat tensile specimens were used. Specimens were machined perpendicular to the FSW direction to the dimensions: length of 80 mm, width of 15 mm, and thickness of 8.5 mm. The specimen's dimensions agree with the DIN EN10002-1 2001(D) standards. After machining, both surfaces of the samples were flushed to avoid any dimensional irregularity. Figure 3 shows the dimensions of the tensile specimen and an image of the sample after tensile testing. Tensile tests were carried out at room temperature with an initial crosshead speed of 0.1 mm/s using the universal testing machine Instron 4210, Norwood, MA, USA. The tensile data acquired were analyzed to determine tensile properties and joint efficiency. The fracture surface of the tension tested samples was examined using the Scanning Electron Microscope Type: Quanta 250 with a Field Emission Gun, FEI company (Hillsboro, OR, USA) to determine the failure mode of the welded samples.

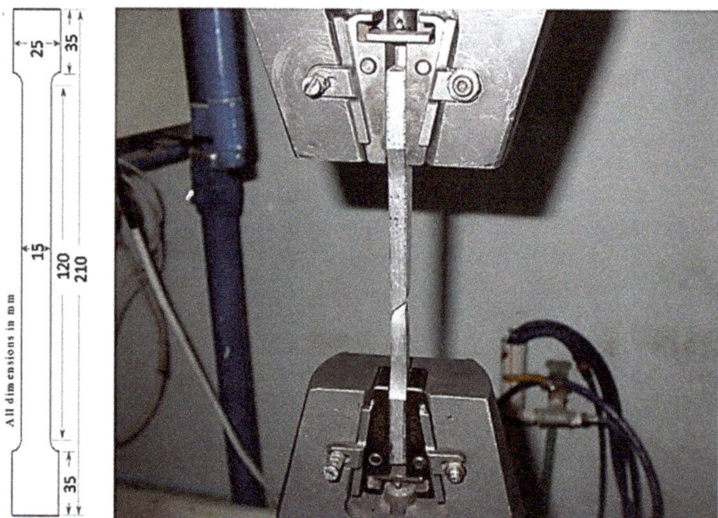

Figure 3. The dimensions of the tensile specimen and an image of the sample after tensile testing.

3. Results and Discussion

3.1. *Effect of FSW Parameters on the Heat Input*

Heat input is one of the important parameters associated with all welding processes and affects the weld quality and properties. Although, FSW is characterized by low heat input relative to the fusion welding processes, still heat input plays a significant role in controlling the joints properties and quality [33]. In this work, the control system in the FSW machine used allows the recording of the spindle torque T (N·m) that can be used with the other FSW parameters such as rotational speed ω (rpm) and the welding speed v (mm/min) to calculate the heat input. Heat input is defined as the heat energy applied to the workpiece per unit length in the unit of (J/mm). The source of heat generated during FSW is mainly from the friction between the tool and the stirred material and the heat input during FSW can be calculated using Equation (1) [33–35]:

$$\text{Heat Input (J/mm)} = \frac{power}{speed} = \eta\left(\frac{\omega T}{v}\right) \quad (1)$$

$$\text{Where } \omega = \left(\frac{2\pi r}{60}\right) \quad (2)$$

where T is the torque (N·m), ω is the rotational speed (rpm), v is the linear speed (mm/min) and η is the efficiency of heat transfer, (η = 0.9) [36,37]. The pseudo heat index is represented by the ratio of the square of the rotational speed to travel speed (ω^2/v). As a function of FSW parameters, it can be considered a simple heat input metric and a well-known method to predict the heat generated during FSW. The maximum temperature highly depends on the rotation tool speed while the heating rate depends on the welding speed at a given tool geometry and plunge depth. The rotation tool speed term is squared because of its significant effect on the heat generated during the process [38]. The pseudo-steady-state welding parameters, calculated heat input and heat index are presented in Table 4.

Table 4. Key pseudo-steady-state welding parameters.

Joint AS-RS	ω (rpm)	v (mm/min)	ω/v	Torque (N·m)	HI (J/mm)	Heat Index ω^2/v
AA5083-AA5754	400	20	20.0	91	171	8000
	400	40	10.0	117	110	4000
	400	60	6.7	116	73	2666
AA5083-AA5754	600	20	30.0	73	206	18,000
	600	40	15.0	87	123	9000
	600	60	10.0	85	80	6000
AA5083-AA7020	500	20	25.0	87	205	12,500
	500	40	12.5	91	107	6250
	500	80	6.3	65	39	3125
AA7020-AA5083	500	20	25.0	101	238	12,500
	500	40	12.5	85	100	6250
	500	80	6.3	104	62	3125

For the joints AA5083/AA5754, Figure 4a shows that the relatively high travel speed (60 mm/min) with low rotational speed (400 rpm) resulted in low ω/v value (6.66) and consequently low heat input value. Decreasing the welding speed to 40 mm/min for the same rotational speed of 400 rpm in Figure 4a increased the ω/v value to 10, leading to an increase in the heat input level. In Figure 4a, although the value of is the same (ω/v = 10) for a travel speed of 60 mm/min and rotational speed of 600 rpm as that in Figure 4a, the increased travel speed of 60 mm/min has showed a more dominant effect than the rotational speed (600 rpm) and resulted in decreasing the HI level. For the other system of joint (AA5083/AA7020; Figure 4b), the heat input can be interpreted in the same manner as explained in Figure 4a. The increased level of ω/v value (25) in Figure 4b (500 rpm and 20 mm/min) has resulted in obvious increase in the heat input level which reach the value of 261 J/mm. Changing the arrangement of the plates from AA5083/AA7020 to AA7020/AA5083 for the same ω/v value (6.25) has showed no difference in the power and heat input values, as shown in Figure 4b.

3.2. Joint Appearance and Internal Quality

To investigate the joint appearance, the top surfaces of all joints have been visually investigated and pictured. Figures 5 and 6 show the top view of the FSWed AA5083/AA5754 and AA5083/AA7020, respectively. It should be mentioned here that the alloys position at the AS and RS has been ignored in case of the alloys of the same series AA5083/AA5754, while this parameter has been taken into consideration in case of the different series alloys AA5083/AA7020. Figure 5 clearly shows top surfaces free of any surface defects almost at all FSW conditions investigated for this group of alloys except the tool pin breakage at the 600-rpm rotation rate and both 40 and 60 mm/min traverse speeds. The position of the tool breakage is indicated in each top surface by a black arrow. This breakage of the tool pin at the high tool rotation rate and the high tool traverse speed can be attributed to the increase in the applied pressure at the high welding speeds to keep the plunge depth constant. In terms of the flash at the top surface, it is almost minimum under all conditions.

Figure 6 shows the top surface of the FSWed AA5083/AA7020 and AA7020/AA5083 at the same welding conditions for each combination. The surfaces are clearly free of any surface defects expect little flash at the AS especially in case of AA5083/AA7020 that have reduced by reversing the alloys position. Moreover, the reversing alloys position has resulted in tool pin breakage at the highest welding speed condition in the combination AA7020/AA5083. This can be attributed to the resistance of the high strength alloy at the AS especially at the high welding speed of 60 mm/min.

Similar surface features can be visualized in the FSWed joints AA5083/AA7020 and AA7020/AA5083 shown in Figure 6. In some samples, the plasticized flash became clear thick as shown in samples welded at low travel speed (20 and 40 mm/min) where the heat input is higher, and the material is more ductile. This thick flash could be also related to the high applied pressure by the shoulder which leads to excessive penetration of the shoulder in the hot stirred material. At higher travel speed (80 mm/min, Figure 6) the formed flash is thin, discontinuous, and easily dethatched from the FSWed samples. One additional defect is the keyhole formed at the exit of the pin from the material at the end of the welding pass, which is a characteristic defect in the FSWed samples. Finally, it can be said that the welded surface showed a relatively minimum amount of flash which consider as materials loss due to either a higher plunging force or a hotter condition, e.g., a higher rotational speed and/or a lower traverse speed as it will be discussed in studying the FSW heat input.

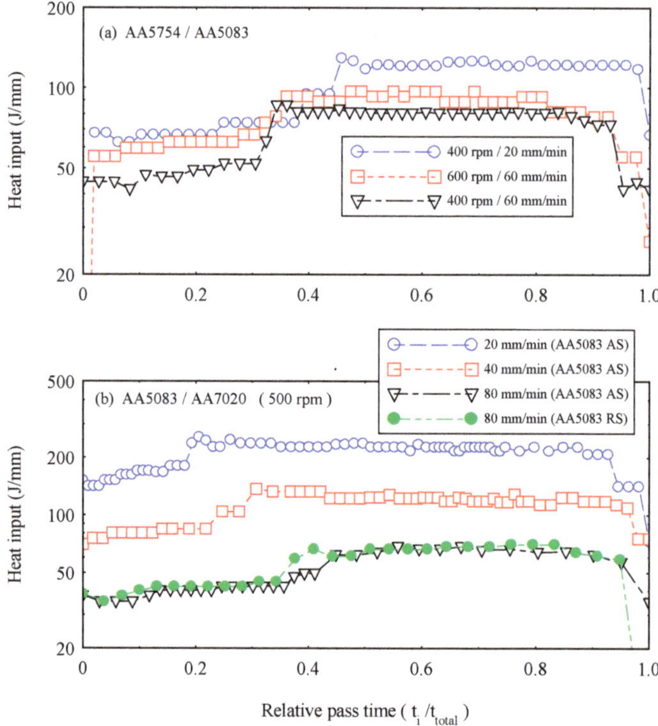

Figure 4. Calculated heat input of friction stir welded joints at different rotation and traverse speeds (i.e., at different ω/v ratios) versus the relative pass time (t_i/t_{total}) for FSW for the joints (**a**) AA5083/AA5754 and (**b**) AA5754/AA7020.

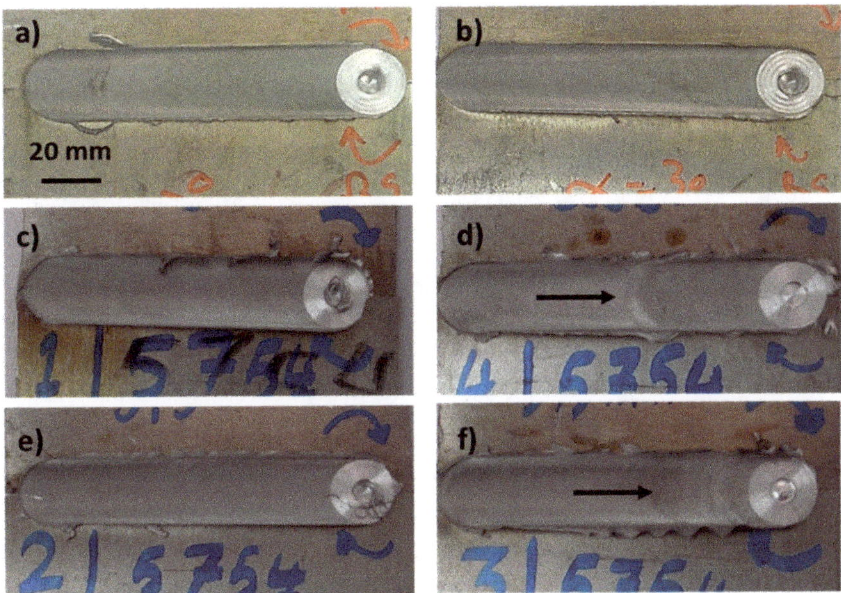

Figure 5. Surface appearance of friction stir welded AA5083/AA5754 joints at different rotation and traverse speeds: (**a**) 400 rpm, 20 mm/min, (**b**) 600 rpm, 20 mm/min, (**c**) 400 rpm, 40 mm/min, (**d**) 600 rpm, 40 mm/min, (**e**) 400 rpm, 60 mm/min and (**f**) 600 rpm, 60 mm/min.

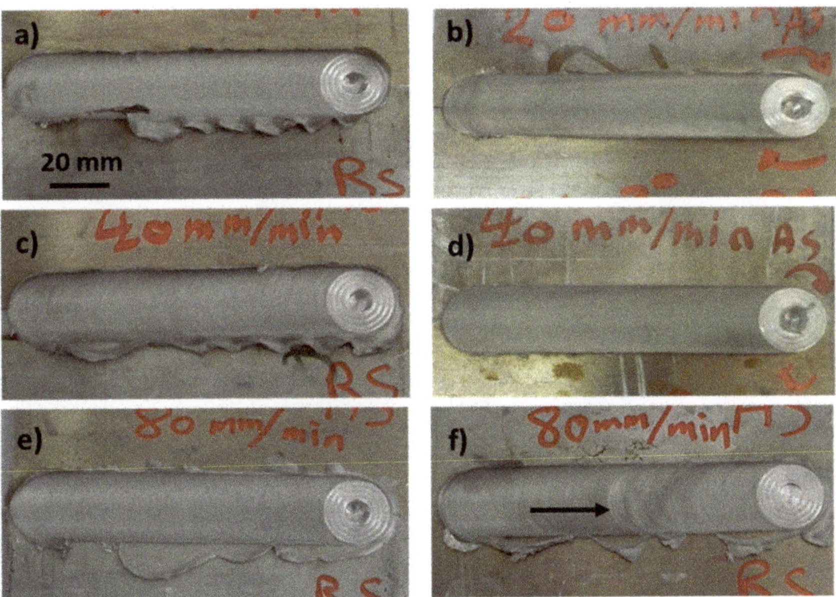

Figure 6. Surface appearance of friction stir welded joints AA5083/AA7020 at different rotation and traverse speeds: (**a**) AA5083/AA7020, 500 rpm, 20 mm/min, (**b**) AA7020/AA5083, 500 rpm, 20 mm/min, (**c**) AA5083/AA7020, 500 rpm, 40 mm/min, (**d**) AA7020/AA5083, 500 rpm, 40 mm/min, (**e**) AA5083/AA7020, 500 rpm, 80 mm/min and (**f**) AA7020/AA5083, 500 rpm, 80 mm/min.

Figure 7 shows the transverse cross section macrographs of the polished and etched FSW joints AA5083/AA5754 at rotational speeds of 400 and 600 rpm using traverse speeds of 20, 40 and 60 mm/min. Although several Keller's' reagents have been used at different concentrations to etch the polished sections; the boundaries separating the stirred zone (SZ) and base material are difficult to identify due to the difficulties of etching the AA5XXX Al-alloy group. However, the presented transverse cross section macrographs show defect free joints in two joints out of six made for this combination. The two of the joints welded at 400 rpm and traverse speeds of 40 mm/min and 60 mm/min are completely sound and defect-free, while the joint made at welding speed of 20 mm/min contains a tiny tunnel defect indicated by arrow on the macrograph. This implies that at 400 rpm increasing the welding speed from 20 mm/min to 40 mm/min and 60 mm/min eliminates the tiny tunnel defect. The calculated heat input data above indicates that increasing the welding speed at constant rotation rate results in a decrease of the heat input. The three joints welded at 600 rpm for the same base materials with same arrangement (AS & RS) contain different sizes of tunnel defect from tiny or small to medium size. This indicates that the high heat input will result in tunnel defect, and this can be attributed to the change in the friction condition during the FSW process. There are two friction conditions reported to occur during FSW based on the FSW conditions or based on the heat input namely sticking friction and sliding friction [38]. This would result in some frictional slippage at the shoulder. There could also be instances where the FSW process may alternate between plastic flow and frictional slippage or a stick-slip mode operating at the shoulder. Alternating boundary conditions at the interface may act to destabilize the temperature, which may cause stick-slip oscillations [38]. The AA5083 is reported display poor weldability during FSW due to the strong influence of the plastic properties at high temperatures, on material flow during welding, as well as on contact conditions at the tool workpiece interface [39].

Figure 8 shows the optical macrographs of the transverse cross sections of the FSW joints AA5083/AA7020 and AA7020/AA5083 produced at rotational speed of 500 rpm and traverse speeds of 20, 40 and 80 mm/min. Etching shows the deformation lines of the alloy AA7020 (as it is well known in the AA7XXX Al-alloy series) and the welding zone can be distinguished from the two base plates. The optical macrographs in Figure 8 clearly show that the boundaries between the nugget zone (NG) and the base materials are well defined through the whole thickness. The shape of the NG is wide conical near the top surface due to the large shoulder diameter dominating the stirring and deformation at the top surface. While it is narrow cylindrical near the lower surface due to the small pin diameter dominating the stirring and deformation at the lower surface. A transition can be noted with the conical shape narrowing towards the base. In this transition zone both the shoulder and the pin are contributing to the stirring and deformation. It can be observed that the interface near to the AA7020 is clearer and more distinguished regardless of the AS or RS. This can be due to the effective etching in revealing the micro and macro-features for this alloy in contrast to AA5083. In addition, it can be observed that the interface near to the AA7020 is always free of any defects regardless of the position of the alloy in the AS or RS. Having the AA5083 at the AS has resulted in defect free joint at the welding speed of 20 mm/min, while by increasing the welding speed a very small tunnel defect has occurred at the AS at the welding speed of 40 mm/min and increased in size at 80 mm/min. These tunnel defects can be formed due to the insufficient down force applied during the FSW. This can be attributed to the high resistance of the AA7020 that does not allow the required pressure for the complete consolidation at the applied constant plunge depth. The level of the NG region at the top surface is slightly higher than the level of the base material which supports the scenario of the lack in the applied pressure that causes the tunnel defects. On the other hand, having the AA5083 at the RS has resulted in two defect free joints at the welding speeds of 20 and 80 mm/min. At the welding speed of 40 mm/min, a tunnel defect occurred at the center of the NG near the lower base. This can be attributed to the low applied pressure during the FSW process.

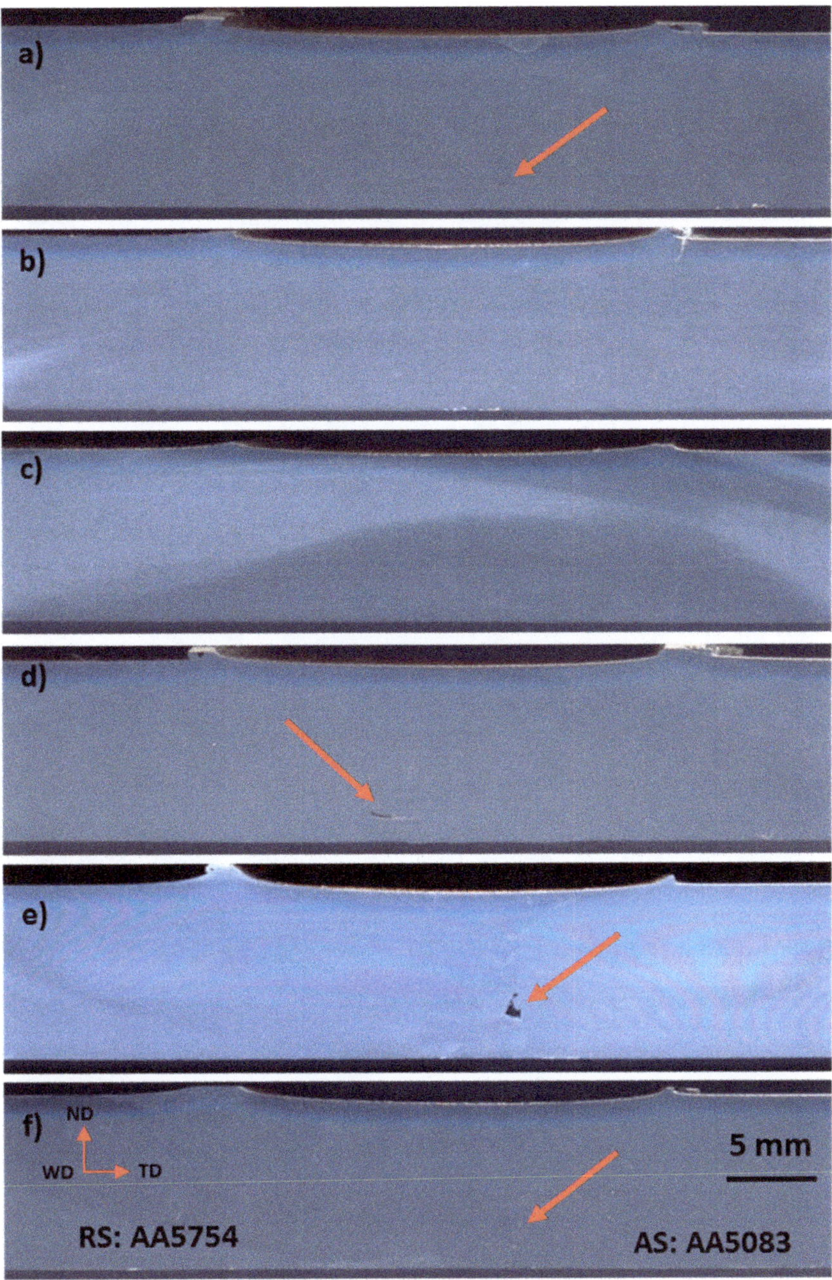

Figure 7. Polishing cross-sections of friction stir welded joints AA5083/AA5754 at different rotational and traverse speeds. Arrows refer to the tunnel defects: (**a**) 400 rpm, 20 mm/min, (**b**) 400 rpm, 40 mm/min, (**c**) 400 rpm, 60 mm/min, (**d**) 600 rpm, 20 mm/min, (**e**) 600 rpm, 40 mm/min and (**f**) 600 rpm, 60 mm/min.

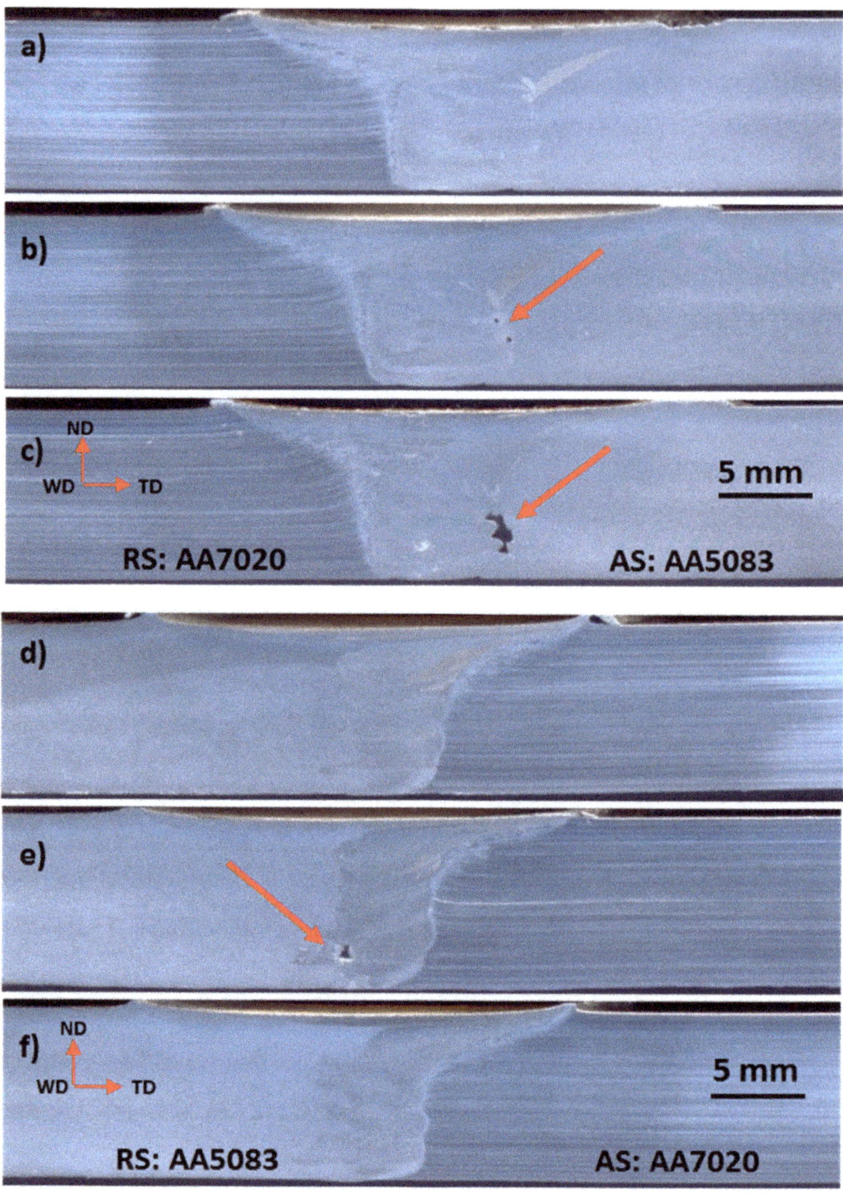

Figure 8. Cross sections of friction stir welded joints at different rotational and traverse speeds for AA5083/AA7020 and AA7020/AA5083 after etching: (**a**,**d**) 500 rpm, 20 mm/min, (**b**,**e**) 500 rpm, 40 mm/min and (**c**,**f**) 500 rpm, 80 mm/min.

Table 5 summarizes the defects formed in the produced FSWed joints and the welding conditions. For instance, it is difficult to relate the formed defect to the welding conditions such as the rotational speed (ω), travel speed (v) or their combination (ω/v). The general observation is that the internal defects (pin hole or tunnel) are shifted to the side of the softer plate (AA5083) and formed in the lower half of the joint away for the part of the SZ produced by the rotation of the shoulder. This reflects the positive effect of the pressure excreted by the shoulder on the SZ.

Table 5. Visual inspection of the macrostructure of cross section of the FSWed joints and the welding conditions (ω and v).

Joint	ω (rpm)	v (mm/min)	ω/v	Surface Thin Flash and Track Lines	Internal Defect
AA5083/AA5754	400	20	20	-	-
	400	40	10	thin flash	-
	400	60	6.66	-	-
	600	20	30	-	pin hole
	600	40	15	track lines, thin flash	tunnel
	600	60	10	thin flash	-
AA5083/AA7020	500	20	25	thick flash	-
	500	40	12.5	thick flash	-
	500	80	6.25	thin flash	tunnel
AA7020/AA5083	500	20	25	thin flash	-
	500	40	12.5	-	pin hole
	500	80	6.25	track lines, thin flash	-

3.3. *Mechanical Properties*

3.3.1. Macro-Hardness Distribution

Figure 9 shows the hardness profile measured at the midsection of the transverse cross sections of the FSWed joints AA5083/AA5754 at 400 rpm with the different welding speeds in Figure 9a and at 600 rpm with the different welding speeds in Figure 9b. It can be observed that the hardness is reduced in the weld zone with more reduction by increasing the rotation rate from 400 rpm to 600 rpm at each welding speed, and this is mainly due to the increase in the heat input. Moreover, the reduction in the NG hardness is also affected by the increase in the welding speed at the constant rotation rate. It can be noted that the NG hardness is reduced more by decreasing the welding speed from 60 mm/min up to 20 mm/min at the constant rotation rate. In terms of the width of the heat affected zone at each rotation rate, it is reduced by increasing the welding speed. Generally, this hardness reduction in the weld zone is mainly because of the thermal cycle on the strain hardened alloys. The thermal cycle leads to softening of the strain hardened material through the recovery and recrystallization processes that take place during FSW of the aluminum alloys [40].

Figure 10 shows the hardness profile measured at the midsection of the transverse cross sections of the FSWed joints AA7020/AA5083 at 500 rpm with AA5083 at the AS and different welding speeds in Figure 10a and with the AA7020 at the AS and different welding speeds in Figure 10b. It can be observed that both profiles show almost no reduction in the base materials hardness in all the heat affected zone of the two alloys; however, a slight increase in the hardness of the AA7020 alloy can be observed towards the NG center regardless of its position at the AS or RS. This can be attributed to the solid solution strengthening that can occur due to the stirring of the dissimilar alloys [41]. On the other hand, there is a slight decrease in the hardness of the AA5083 towards the NG center regardless of its position at the RS or the AS. This is mainly because of thermal cycle on the softening of the FSWed alloy.

3.3.2. Tensile Properties Analysis

The tensile strength properties of the FSW joints are usually compared with those of the alloy having lower tensile strength [41–43]; here AA5083. For design purposes, the yield stress is used more frequently than the ultimate tensile strength so that the yield stress is determined ($\sigma_{0.2\%}$) for the tested base materials and FSW joints. Relative ultimate tensile strength (σ_{UTS} joint/σ_{UTS} 5083) and relative yield stress ($\sigma_{0.2\%}$ joint/$\sigma_{0.2\%}$ 5083) of the produced FSW joints were determined from the tensile stress–strain curves. These relative values (ultimate tensile strength and yield stress) were presented as a function of the welding speeds as shown in Figure 11. Figure 11a summarizes the tensile properties of FSWed joints AA5083/AA5754 at different welding speeds of 20, 40 and 60 mm/min and at rotational speeds of 400 and 600 rpm obtained from the tensile stress strain curves of the FSW samples and related to the tensile properties of the base alloy.

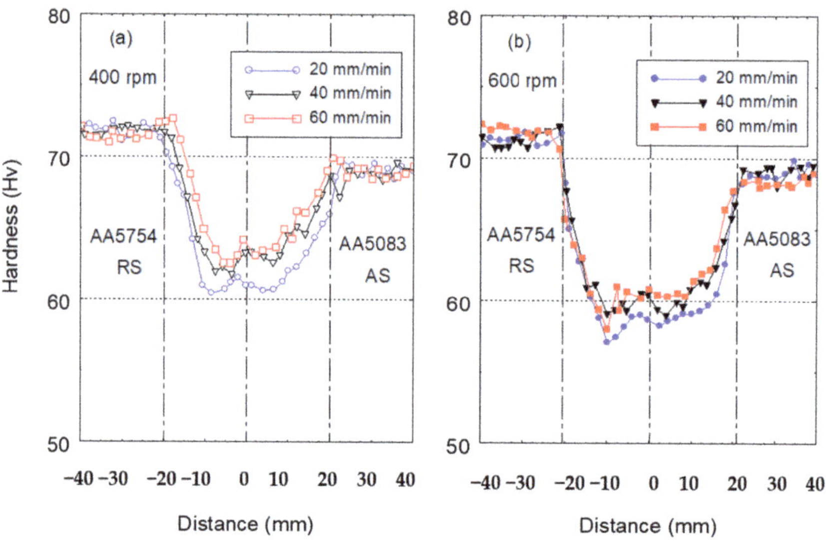

Figure 9. Hardness values along FSWed joints AA5083/AA5754 at travel speeds of 20, 40 and 60 mm/min and rotation speeds of (**a**) 400 rpm and (**b**) 600 rpm.

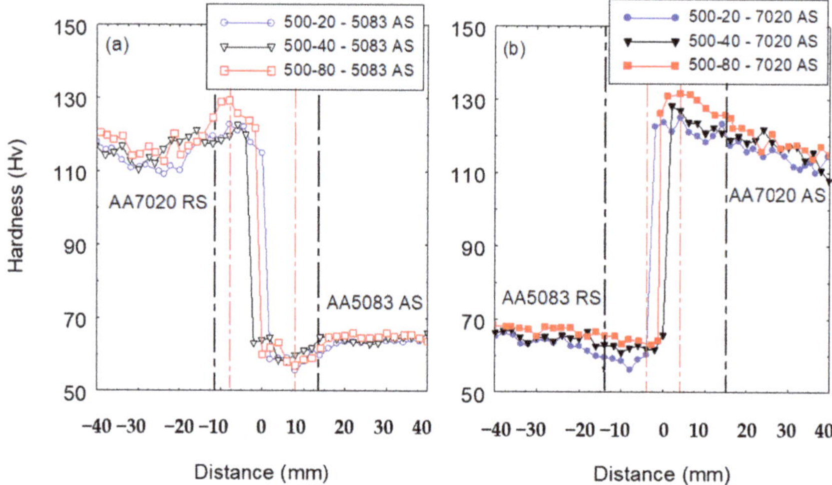

Figure 10. Hardness distribution across the joints (**a**) AA5083/AA7020 and (**b**) AA7020/AA5083 welded at travel speeds of 20, 40 and 80 mm/min showing the effect of the joint arrangement.

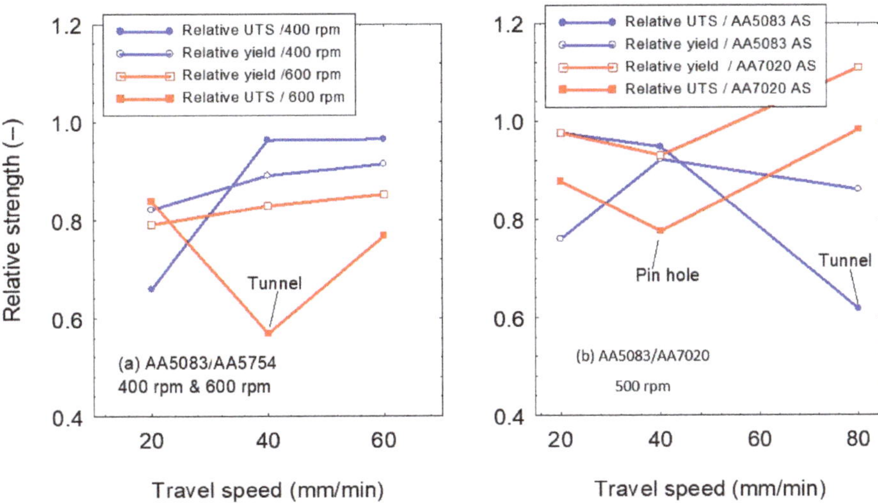

Figure 11. Relative tensile strength (σUTS joint/σUTS 5083) and relative yield stress (σ0.2% joint/σ0.2% 5083) of the produced FSW joints against the welding speeds.

As a usual trend, the tensile strength increases with increasing the travel speed and with decreasing the rotational speed, because of the lower generated heat input which permits materials softening. This statement can be supported by the relative yield stress presented in Figure 11a. The relative ultimate tensile strength of FSWed joints is greatly affected by the internal defects in some samples causing early fracture. FSW of the joints AA7020/AA5083 at rotational speed of 500 rpm and travel speed of 80 mm/min has produced joint with ultimate tensile strength comparable with the base alloy AA5083, while the yield stress is even higher than that of the base alloy Figure 11b. Regardless the defect happened at a travel speed of 40 mm/min, clamping the higher strength plate as an AS in friction stir welding of materials with high differences in strength increases the joint strength.

Figure 12 shows macrographs for the fracture positions of the tensile samples of AA5083/AA5754 joints at different welding parameters: (a) 400 rpm–20 mm/min, (b) 400 rpm–40 mm/min, (c) 400 rpm–60 mm/min. and fracture locations of AA5083/AA7020 joints at 500 rpm but different travel speeds and positions; (d) AA7020 AS-20 mm/min, (e) AA7020 AS-40 mm/min and (f) AA7020 AS-80 mm/min. It can be observed that the fracture occurred at the nugget zone in one joint of AA5083/AA5754 (Figure 12a) mainly due to the defect while the fracture occurred away from the NG zone in two joints (Figure 12b,c). In terms of the AA7020/AA5083 joints, it can be observed that the low-speed joint fracture occurred away from the NG (Figure 12d) and the high-speed joints the fracture occurred inside the NG (Figure 12e,f) mainly due to the defects noted. The fracture surface of two samples indicated in Figure 12 are investigated using SEM and EDS analysis. Clearly, it can be observed that from Figure 13 that the fracture mechanism of the dissimilar AA AA5083/AA5754 is ductile mode with very clear dimple features as can be seen in the enlarged micrographs of Figure 13b,c. The inclusions shown in Figure 13d are mostly aluminum and magnesium oxides as detected by the EDX analysis. The lack of adherence of such oxide inclusions with the matrix has accelerated the formation of pores around the inclusions through decohesion between the inclusions and the surrounding material which make the microcracks initiation by the coalescence of the neighboring pores possible. With increasing the applied load, the microcracks grow faster and propagate until the reaching the complete fracture. Figure 14a–d shows the fracture surface SEM micrographs of the dissimilar joint AA5083/AA7020. Clearly the dimple features are dominating, which confirms the ductile fracture mechanism.

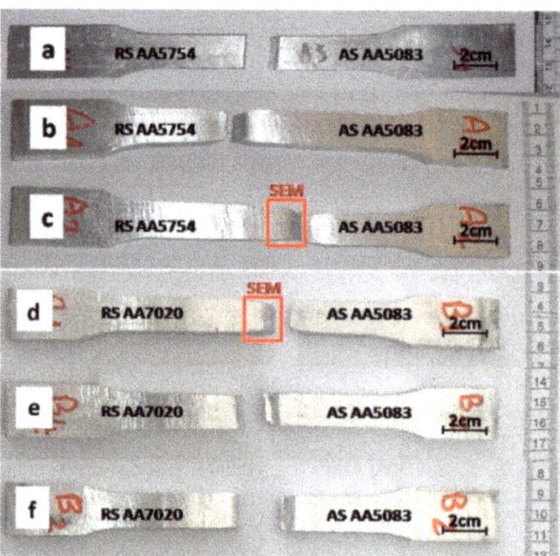

Figure 12. Fracture locations of tensile test specimens of AA5083/AA5754 joints at different welding parameters: (**a**) 400 rpm–20 mm/min, (**b**) 400 rpm–40 mm/min and (**c**) 400 rpm–60 mm/min and fracture locations of AA5083/AA7020 joints at 500 rpm but different travel speeds and positions; (**d**) AA7020 AS-20 mm/min, (**e**) AA7020 AS-40 mm/min and (**f**) AA7020 AS-80 mm/min.

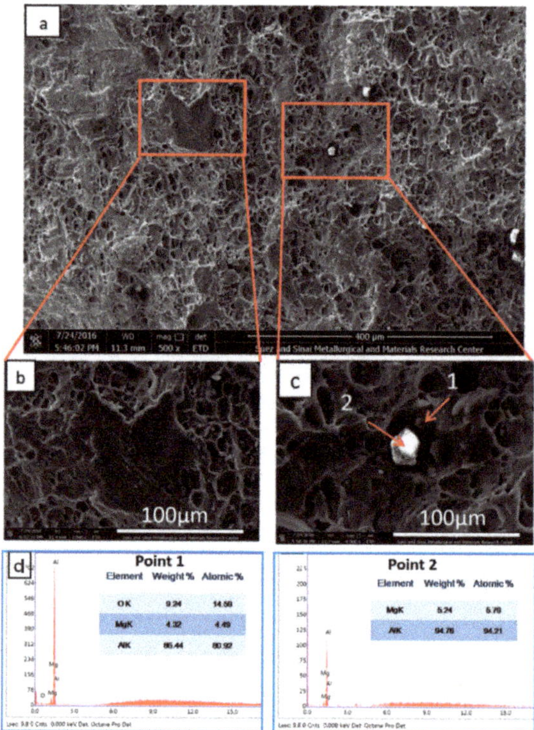

Figure 13. SEM micrographs of the fracture surface of tensile test sample of AA5083/AA5754 joint at 400 rpm–60 mm/min, (**a**) mixed fracture mode (**b**) brittle fracture features, (**c**) ductile fracture features, (**d**) EDX spot analysis of the points 1 and 2.

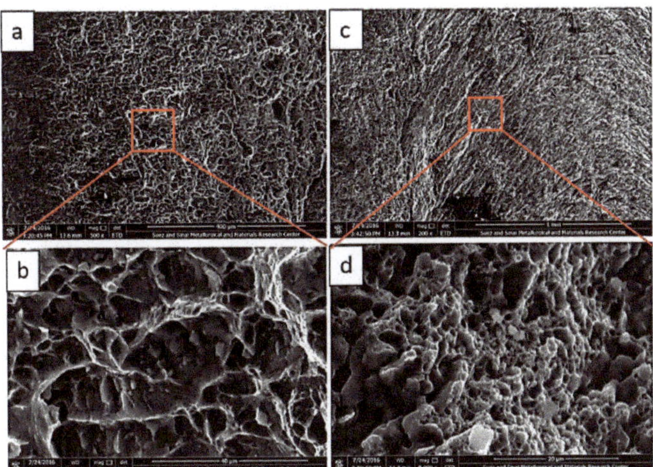

Figure 14. (**a**,**c**) SEM images of the fracture surface of tensile test samples of AA7020/AA5083 joint of AA7020 AS, 500 rpm–20 mm/min; (**b**,**d**) show the shape deep dimples.

4. Conclusions

In the present study, dissimilar aluminum alloys (AA5083/AA5754 and AA5083/AA7020) were successfully joined by FSW at a wide range tool rotation speed of 300–600 rpm, a traverse welding speed range of 20–80 mm/min and reversing the alloys between the AS and the RS. From the obtained results the following conclusions can be drawn:

- Sound joints are obtained at the low heat input FSW parameters investigated while increasing the heat input results in tunnel defects.
- The hardness profile obtained in the dissimilar AA5083/AA5754 joints is the typical FSW hardness profile of these alloys that reduced in the NG zone due to the loss of the cold deformation strengthening. However, the profile of the dissimilar AA5083/AA7020 showed increase in the hardness in the NG due to the intimate mixing the high strength alloy with the low strength alloy.
- The sound joints in both groups of the dissimilar joints showed very high joint strength with efficiency up to 97 and 98%. Having the high strength alloy at the advancing sides gives high joint strength and efficiency.
- The sound joints showed ductile fracture mechanism with clear dimple features, and significant plastic deformation occurred before fracture. Moreover, the fracture in these joints occurred in the base materials. On the other hand, the joints with tunnel defect showed some features of brittle fracture due the acceleration of the existing crack propagation upon tensile loading.

Author Contributions: Conceptualization, S.A., E.A. and M.M.Z.A.; methodology, A.M.A.M., E.A. and M.M.E.-S.S.; validation, S.A., N.A.A. and E.A.; formal analysis, S.A. and M.M.Z.A.; investigation, M.M.E.-S.S., S.A. and E.A.; writing—original draft preparation, S.A. and E.A.; writing—review and editing, M.M.Z.A. and N.A.A.; project administration, M.M.Z.A. and M.M.E.-S.S. All authors have read and agreed to the published version of the manuscript.

Funding: This research received no external funding.

Institutional Review Board Statement: Not applicable.

Informed Consent Statement: Not applicable.

Data Availability Statement: The data presented in this study are available on request from the corresponding author. The data are not publicly available due to the extremely large size.

Acknowledgments: The authors acknowledge the financial support rendered by the Science and Technology Development Fund (STDF), Ministry of Higher Education and Scientific Research, Egypt.

Conflicts of Interest: The authors declare no conflict of interest.

Abbreviations

ω:	rotational speed, rpm
η:	efficiency of heat transfer, %
$\sigma_{0.2\%}$:	0.2 offset yield stress, MPa
σ_{UTS}:	Ultimate tensile strength, MPa
AA:	Aluminum alloy
AS:	advancing side
BM:	Base Material
EDX:	Energy Dispersive X-Ray
FSW:	Friction stir welding
FSWed:	Friction Stir Welded
HAZ:	Heat affected zone
HI:	Heat Input, J/mm
HP:	Horsepower
HRC:	Hardness Rockwell C
HV:	Hardness Vickers
NG:	Nugget zone
Rpm:	Revolution per minute
RS:	retreating side
SEM:	Scanning electron microscope
SZ:	stirred zone
T:	Torque, N·m
TMAZ:	Thermomechanical affected zone
v:	welding speed, mm/min
WN:	Welding nugget

References

1. Kasman, Ş.; Yenier, Z. Analyzing dissimilar friction stir welding of AA5754/AA7075. *Int. J. Adv. Manuf. Technol.* **2014**, *70*, 145–156. [CrossRef]
2. El Rayes, M.M.; Soliman, M.S.; Abbas, A.T.; Pimenov, D.Y.; Erdakov, I.N.; Abdel-Mawla, M.M. Effect of Feed Rate in FSW on the Mechanical and Microstructural Properties of AA5754 Joints. *Adv. Mater. Sci. Eng.* **2019**, *2109*, 4156176. [CrossRef]
3. Chen, Y.; Ding, H.; Cai, Z.; Zhao, J.; Li, J. Microstructural and Mechanical Characterization of a Dissimilar Friction Stir-Welded AA5083-AA7B04 Butt Joint. *J. Mater. Eng. Perform.* **2017**, *26*, 530–539. [CrossRef]
4. Patel, V.; Li, W.; Wang, G.; Wang, F.; Vairis, A.; Niu, P. Friction stir welding of dissimilar aluminum alloy combinations: State-of-the-art. *Metals* **2019**, *9*, 270. [CrossRef]
5. El-Sayed, M.M.; Shash, A.Y.; Abd-Rabou, M. Finite element modeling of aluminum alloy AA5083-O friction stir welding process. *J. Mater. Process. Technol.* **2018**, *252*, 13–24. [CrossRef]
6. Oliveira, J.P.; Duarte, J.F.; Inácio, P.; Schell, N.; Miranda, R.M.; Santos, T.G. Production of Al/NiTi composites by friction stir welding assisted by electrical current. *Mater. Des.* **2017**, *113*, 311–318. [CrossRef]
7. Safeen, M.W.; Spena, P.R. Main issues in quality of friction stir welding joints of aluminum alloy and steel sheets. *Metals* **2019**, *9*, 610. [CrossRef]
8. Threadgill, P.L.; Ahmed, M.M.Z.; Martin, J.P.; Perrett, J.G.; Wynne, B.P. The use of bobbin tools for friction stir welding of aluminium alloys. *Mater. Sci. Forum* **2010**, *638*, 1179–1184. [CrossRef]
9. Ahmed, M.M.Z.; Wynne, B.P.; El-Sayed Seleman, M.M.; Rainforth, W.M. A comparison of crystallographic texture and grain structure development in aluminum generated by friction stir welding and high strain torsion. *Mater. Des.* **2016**, *103*, 259–267. [CrossRef]
10. Kalemba-Rec, I.; Hamilton, C.; Kopyściański, M.; Miara, D.; Krasnowski, K. Microstructure and Mechanical Properties of Friction Stir Welded 5083 and 7075 Aluminum Alloys. *J. Mater. Eng. Perform.* **2017**, *26*, 1032–1043. [CrossRef]
11. Vilaa, P.; Santos, T.G. Non-Destructive Testing Techniques for Detecting Imperfections in Friction Stir Welds of Aluminium Alloys. In *Aluminium Alloys, Theory and Applications*; IntechOpen Limited: London, UK, 2011.
12. Ahmed, M.M.Z.; Wynne, B.P.; Rainforth, W.M.; Martin, J. Crystallographic texture investigation of thick section friction stir welded AA6082 and AA5083 using EBSD. *Key Eng. Mater.* **2019**, *786*, 44–51. [CrossRef]
13. Ahmed, M.M.Z.; El-Sayed Seleman, M.M.; Shazly, M.; Attallah, M.M.; Ahmed, E. Microstructural Development and Mechanical Properties of Friction Stir Welded Ferritic Stainless Steel AISI 409. *J. Mater. Eng. Perform.* **2019**, *28*. [CrossRef]
14. Ahmed, M.M.Z.; Ahmed, E.; Hamada, A.S.; Khodir, S.A.; El-Sayed Seleman, M.M.; Wynne, B.P. Microstructure and mechanical properties evolution of friction stir spot welded high-Mn twinning-induced plasticity steel. *Mater. Des.* **2016**, *91*. [CrossRef]

15. Palanivel, R.; Koshy Mathews, P.; Murugan, N.; Dinaharan, I. Effect of tool rotational speed and pin profile on microstructure and tensile strength of dissimilar friction stir welded AA5083-H111 and AA6351-T6 aluminum alloys. *Mater. Des.* **2012**, *40*, 7–16. [CrossRef]
16. Kailainathan, S.; Sundaram, S.K.; Nijanthan, K. Influence of Friction Stir Welding Parameter on Mechanical Properties in Dissimilar (AA6063-AA8011) Aluminium Alloys. *Int. J. Innov. Res. Sci. Eng. Technol.* **2014**, *3*, 15691–15695. [CrossRef]
17. Jannet, S.; Mathews, P.K. Effect of welding parameters on mechanical and microstructural properties of dissimilar aluminum alloy joints produced by friction stir welding. In *Applied Mechanics and Materials*; Trans Tech Publications: Zurich, Switzerland, March 2014.
18. Park, S.K.; Hong, S.T.; Park, J.H.; Park, K.Y.; Kwon, Y.J.; Son, H.J. Effect of material locations on properties of friction stir welding joints of dissimilar aluminium alloys. *Sci. Technol. Weld. Join.* **2010**, *15*, 331–336. [CrossRef]
19. Leitao, C.; Leal, R.M.; Rodrigues, D.M.; Loureiro, A.; Vilaça, P. Mechanical behaviour of similar and dissimilar AA5182-H111 and AA6016-T4 thin friction stir welds. *Mater. Des.* **2009**, *30*, 101–108. [CrossRef]
20. Khanna, N.; Sharma, P.; Bharati, M.; Badheka, V.J. Friction stir welding of dissimilar aluminium alloys AA 6061-T6 and AA 8011-h14: A novel study. *J. Braz. Soc. Mech. Sci. Eng.* **2020**, *42*, 7. [CrossRef]
21. Abd Elnabi, M.M.; Elshalakany, A.B.; Abdel-Mottaleb, M.M.; Osman, T.A.; El Mokadem, A. Influence of friction stir welding parameters on metallurgical and mechanical properties of dissimilar AA5454-AA7075 aluminum alloys. *J. Mater. Res. Technol.* **2019**, *8*, 1684–1693. [CrossRef]
22. Cole, E.G.; Fehrenbacher, A.; Duffie, N.A.; Zinn, M.R.; Pfefferkorn, F.E.; Ferrier, N.J. Weld temperature effects during friction stir welding of dissimilar aluminum alloys 6061-t6 and 7075-t6. *Int. J. Adv. Manuf. Technol.* **2014**, *71*, 643–652. [CrossRef]
23. Ouyang, J.H.; Kovacevic, R. Material flow and microstructure in the friction stir butt welds of the same and dissimilar aluminum alloys. *J. Mater. Eng. Perform.* **2002**, *11*, 51–63. [CrossRef]
24. Gérard, H.; Ehrström, J.C. Friction stir welding of dissimilar alloys for aircraft. In Proceedings of the 5th International Friction Stir Welding Symposium, Metz, France, 14–16 September 2004.
25. Guo, J.F.; Chen, H.C.; Sun, C.N.; Bi, G.; Sun, Z.; Wei, J. Friction stir welding of dissimilar materials between AA6061 and AA7075 Al alloys effects of process parameters. *Mater. Des.* **2014**, *56*, 185–192. [CrossRef]
26. Kim, N.K.; Kim, B.C.; An, Y.G.; Jung, B.H.; Song, S.W.; Kang, C.Y. The effect of material arrangement on mechanical properties in friction stir welded dissimilar A5052/A5J32 aluminum alloys. *Met. Mater. Int.* **2009**, *15*, 671–675. [CrossRef]
27. Lee, W.B.; Yeon, Y.M.; Jung, S.B. The joint properties of dissimilar formed Al alloys by friction stir welding according to the fixed location of materials. *Scr. Mater.* **2003**, *49*, 423–428. [CrossRef]
28. Jonckheere, C.; De Meester, B.; Denquin, A.; Simar, A. Dissimilar friction stir welding of 2014 to 6061 aluminum alloys. *Advanced Mater. Res. Trans Tech Publ. Switz.* **2012**, *409*, 269–274. [CrossRef]
29. Kalemba-Rec, I.; Kopyściański, M.; Miara, D.; Krasnowski, K. Effect of process parameters on mechanical properties of friction stir welded dissimilar 7075-T651 and 5083-H111 aluminum alloys. *Int. J. Adv. Manuf. Technol.* **2018**, *97*, 2767–2779. [CrossRef]
30. Mastanaiah, P.; Sharma, A.; Reddy, G.M. Dissimilar Friction Stir Welds in AA2219-AA5083 Aluminium Alloys: Effect of Process Parameters on Material Inter-Mixing, Defect Formation, and Mechanical Properties. *Trans. Indian Inst. Met.* **2016**, *69*, 1397–1415. [CrossRef]
31. Essa, A.R.S.; Ahmed, M.M.Z.; Mohamed, A.K.Y.A.; El-Nikhaily, A.E. An analytical model of heat generation for eccentric cylindrical pin in friction stir welding. *J. Mater. Res. Technol.* **2016**, *5*, 234–240. [CrossRef]
32. ASTM International. *ASTM E3-11(2017) Preparation of Metallographic Specimens 1*; ASTM: West Conshohocken, PA, USA, 2017.
33. Ahmed, M.M.Z.; Wynne, B.P.; Rainforth, W.M.; Addison, A.; Martin, J.P.; Threadgill, P.L. Effect of Tool Geometry and Heat Input on the Hardness, Grain Structure, and Crystallographic Texture of Thick-Section Friction Stir-Welded Aluminium. *Metall. Mater. Trans. A* **2018**, *50*, 271–284. [CrossRef]
34. Kumar, R.; Singh, K.; Pandey, S. Process forces and heat input as function of process parameters in AA5083 friction stir welds. *Trans. Nonferrous Met. Soc. China* **2012**, *22*, 288–298. [CrossRef]
35. Costa, A.M.S.; Oliveira, J.P.; Pereira, V.F.; Nunes, C.A.; Ramirez, A.J.; Tschiptschin, A.P. Ni-based Mar-M247 superalloy as a friction stir processing tool. *J. Mater. Process. Technol.* **2018**, *262*, 605–614. [CrossRef]
36. Akinlabi, E.T.; Akinlabi, S.A. Effect of Heat Input on the Properties of Dissimilar Friction Stir Welds of Aluminium and Copper. *Am. J. Mater. Sci.* **2012**, *2*, 147–152. [CrossRef]
37. Schneider, J.; Beshears, R.; Nunes, A.C. Interfacial sticking and slipping in the friction stir welding process. *Mater. Sci. Eng. A* **2006**, *435*, 297–304. [CrossRef]
38. Fall, A.; Jahazi, M.; Khdabandeh, A.R.; Fesharaki, M.H. Effect of process parameters on microstructure and mechanical properties of friction stir-welded ti–6al–4v joints. *Int. J. Adv. Manuf. Technol.* **2017**, *91*, 2919–2931. [CrossRef]
39. Leitão, C.; Louro, R.; Rodrigues, D.M. Analysis of high temperature plastic behaviour and its relation with weldability in friction stir welding for aluminium alloys AA5083-H111 and AA6082-T6. *Mater. Des.* **2012**, *37*, 402–409. [CrossRef]
40. Etter, A.L.; Baudin, T.; Fredj, N.; Penelle, R. Recrystallization mechanisms in 5251 H14 and 5251 O aluminum friction stir welds. *Mater. Sci. Eng. A* **2007**, *445*, 94–99. [CrossRef]
41. Jia, Y.; Lin, S.; Liu, J.; Qin, Y.; Wang, K. The Influence of Pre- and Post-Heat Treatment on Mechanical Properties and Microstructures in Friction Stir Welding of Dissimilar Age-Hardenable Aluminum Alloys. *Metals* **2019**, *9*, 1162. [CrossRef]

42. Ahmed, M.M.Z.; Ataya, S.; El-Sayed Seleman, M.M.; Ammar, H.R.; Ahmed, E. Friction stir welding of similar and dissimilar AA7075 and AA5083. *J. Mater. Process. Technol.* **2017**, *242*. [CrossRef]
43. Ilangovan, M.; Rajendra Boopathy, S.; Balasubramanian, V. Effect of tool pin profile on microstructure and tensile properties of friction stir welded dissimilar AA 6061–AA 5086 aluminium alloy joints. *Def. Technol.* **2015**, *11*, 174–184. [CrossRef]

Article

Microstructure and Mechanical Properties of Dissimilar Friction Welding Ti-6Al-4V Alloy to Nitinol

Ateekh Ur Rehman [1,*], Nagumothu Kishore Babu [2], Mahesh Kumar Talari [2], Yusuf Siraj Usmani [1] and Hisham Al-Khalefah [3]

1 Department of Industrial Engineering, College of Engineering, King Saud University, Riyadh 11421, Saudi Arabia; yusmani@ksu.edu.sa
2 Department of Metallurgical and Materials Engineering, National Institute of Technology, Warangal 506004, India; kishorebabu@nitw.ac.in (N.K.B.); talari@nitw.ac.in (M.K.T.)
3 Advanced Manufacturing Institute, King Saud University, Riyadh 11421, Saudi Arabia; halkhalefah@ksu.edu.sa
* Correspondence: arehman@ksu.edu.sa

Abstract: In the present study, a friction welding process was adopted to join dissimilar alloys of Ti-Al-4V to Nitinol. The effect of friction welding on the evolution of welded macro and microstructures and their hardnesses and tensile properties were studied and discussed in detail. The macrostructure of Ti-6Al-4V and Nitinol dissimilar joints revealed flash formation on the Ti-6Al-4V side due to a reduction in flow stress at high temperatures during friction welding. The optical microstructures revealed fine grains near the Ti-6Al-4V interface due to dynamic recrystallization and strain hardening effects. In contrast, the area nearer to the nitinol interface did not show any grain refinement. This study reveals that the formation of an intermetallic compound (Ti_2Ni) at the weld interface resulted in poor ultimate tensile strength (UTS) and elongation values. All tensile specimens failed at the weld interface due to the formation of intermetallic compounds.

Keywords: friction welding; Ti-6Al-4V; nitinol; intermetallic compound; fractography; dissimilar metal joining

Citation: Rehman, A.U.; Babu, N.K.; Talari, M.K.; Usmani, Y.S.; Al-Khalefah, H. Microstructure and Mechanical Properties of Dissimilar Friction Welding Ti-6Al-4V Alloy to Nitinol. *Metals* **2021**, *11*, 109. https://doi.org/10.3390/met11010109

Received: 9 December 2020
Accepted: 2 January 2021
Published: 7 January 2021

1. Introduction

Titanium and titanium alloys are being widely used in several industrial applications because of their attractive properties, with a low density being one of them [1]. The most widely used $\alpha+\beta$ titanium alloy is Ti-6Al-4V (the workhorse grade in the titanium alloy group). This alloy finds extensive use in the medical industry and aerospace applications, due to its high specific strength, corrosion resistance, good fracture toughness, fatigue resistance, elevated temperature strength up to 500 °C, biocompatibility and weldability [1–4]. Nitinol, or titanium nickelide, belongs to the category of shape memory alloys, and it consists of an equiatomic alloy of nickel (Ni) and titanium (Ti). Nitinol is applied in the biomedical, aerospace, sensorics, fashion and automotive industries, as well as in structural elements and actuators, owing to its shape memory, biocompatibility and pseudoelasticity properties [5,6].

The joining of Ti-Al-4V to Nitinol would be of great interest for many applications, including the hybrid welded structure of an adaptive gas turbine engine's toothed nozzle [7]. A joint made of Ti-Al-4V joined to Nitinol adds the superelastic behavior of Nitinol to the excellent biocompatibility and corrosion resistance properties of these alloys. However, welding Ti-Al-4V to Nitinol is challenging because of differences in their chemical and physical properties [8–10]. The functional behavior of NiTi alloys is strongly influenced by the chemical compositions and heat input of welding processes. Fusion welding readily forms stable intermetallic compounds (Ti_2Ni, Ni_3Ti) when titanium alloy is welded to Nitinol. The migration of Ni from Nitinol to the liquid titanium leads to the formation of

Ti_2Ni. These intermetallic compounds contribute to degradation of the mechanical properties of the welded joints [9–12]. These problems can be avoided by using an interlayer in between the Ti-Al-4V and Nitinol that inhibits or decreases the formation of intermetallic compounds in the weld zone [13].

NiTi was joined to another shape memory alloy, CuAlMn, by laser welding. A complex microstructure with islands of base metal deep inside the fusion zone were seen [14]. Datta et al. [15] did a feasibility study on the dissimilar joining of NiTi to Ti by laser welding. They reported that dissimilar welds exhibited poor strength and ductility due to the formation of Ti_2Ni phases and transverse cracks in the weld [15]. Similar problems occurred in Ti-6Al-4V-NiTiNb welds, too [9,16]. Though intermetallics in general reduce weld ductility in dissimilar metal combinations, some types of intermetallics seem to be less harmful. In NiTi-Cu laser welds, a larger Cu dilution means a greater amount of Cu-based intermetallics in the weld metal, and these Cu-based intermetallics prove beneficial for the weld ductility when compared to Ti-based intermetallics [17,18].

The problem of brittle phases was overcome by placing an Nb interlayer with a thickness of 50 μm in between the Ti-6Al-4V and NiTi in laser welding [19]. The interlayer acted as a diffusion barrier between the two base metals. By varying the laser power and thereby controlling the Nb dilution, Zhou et al. could control the overall amount of intermetallics within the weld [20]. When the Nb melted completely, the joint strength shot up to as high as 82% of the Ti-6Al-4V base metal strength. Even electron beam welding produced similar effects [21].

Another possibility is to explore the solid state welding process to join Ti-Al-4V to Nitinol, in which the joining of materials takes place without melting and without the use of any filler. Solid state welding, like friction welding, can alleviate to some extent the fusion welding problems described above. Friction welding is a solid state welding process in which joining is achieved by one placing work piece in the rotating fixture and the other in a stationary fixture, and frictional heat is generated by the pressure and relative motion between these work pieces. Senkevich et al. [22] studied the joints of Ti-54.2% Ni with VT6 titanium alloy by the diffusion bonding technique. They observed a transition zone between the connected alloys, and this zone exhibited a higher hardness compared with the two base metals due to enrichment of the titanium. It was reported that a maximum shear strength of 170 MPa was achieved at 950 °C and a 30 min holding time. Wei Zhang et al. [23] investigated the microstructure and mechanical properties of Nitinol-Nitinol joints by ultrasonic welding, using Cu as an interlayer. They found that no intermetallic layer was observed at the joint interface. It was reported that the ultimate shear load increased with the increasing weld energy from 500 J to 1000 J due to metallurgical bonding at the interface.

There were quite a few successful attempts at joining Nitinol to Nitinol, whether it was through fusion welding [24,25] or solid-state welding [26,27]. However, when it came to joining Nitinol to Ti-Al-4V, the reports were few and far between, and even then, the welds turned out to be substantially poor in quality. All these published works dealt with fusion welding only [9–21]. This knowledge gap threw open the opportunity for us to attempt the same weld combination through a solid state joining technique like friction welding. Friction welding had already proved itself as a promising technique in several other dissimilar metal combinations. To the authors' best knowledge, this is one of the first attempts at joining Nitinol to Ti-Al-4V. This study demonstrates how, by choosing weld parameters carefully, we could obtain significantly strong welds between Nitinol and Ti-Al-4V. At the same time, some of the challenges that still need to be overcome before this promising combination of dissimilar metals could be used in wide-ranging applications have been identified.

2. Materials and Methods

Ti-6Al-4V and Nitinol rods 100 mm in length with a cross-sectional diameter of 10 mm were used in the present investigation. The chemical compositions of these alloys are listed in Table 1. The rods of Ti-6Al-4V and Nitinol alloys were face turned and cleaned with acetone before friction welding. Continuous drive friction welding with a capacity of 150 kN was employed for the friction welding of dissimilar materials (Figure 1). Ti-6Al-4V was held in the non-rotating vice and Nitinol in the rotating chuck. At the start of the friction welding process, the rotating spindle quickly reached a set speed (spindle speed). The non-rotating specimen was then pushed toward the rotating one under high pressure (friction pressure). Due to the relative motion between the two specimens under pressure, frictional heat was generated at the interface. This heat very quickly plasticized both of the base metals, resulting in a flash. The flash, in fact, helps by ejecting out impurities and oxide layers from the surface of the specimens. The loss in the overall length of the rods was monitored by a sensor, and when the set burn-off length was achieved, the rotating chuck was suddenly brought to rest using a brake. This ended the first stage of the weld cycle, called the friction stage. In the next stage (the upset stage), the pressure was further increased (upset pressure) and held constant for a certain length of time (upset time). The bond consolidation between the two specimens was thus completed.

Table 1. Composition of base metals (wt%).

Elements	Al	V	Fe	Cr	C	N	O	H	Ti
Ti-6Al-4V	6.01	4.0	0.04	0.14	0.01	0.004	0.114	0.008	balance
Elements	**Ni**	**Co**	**Cr**	**Fe**	**Nb**	**C**	**O**	**H**	**Ti**
Nitinol	55.7	0.005	0.003	0.015	0.005	0.04	0.036	<0.001	balance

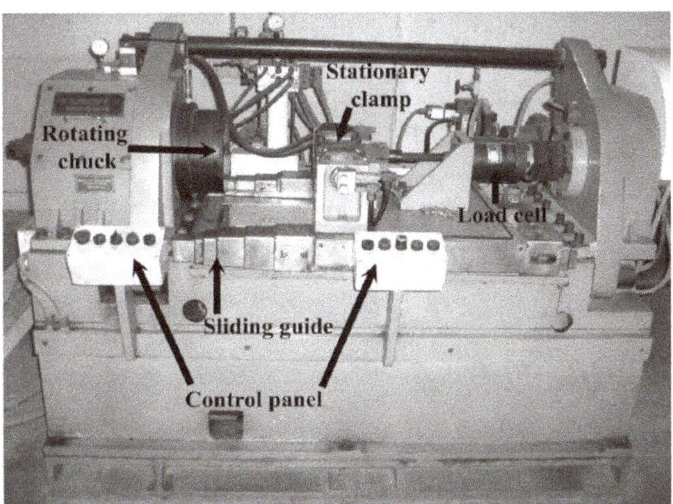

Figure 1. Friction welding machine used in the current study.

The initial parameter window considered is listed in Table 2. At first, one parameter was varied at a time to see how it affected the quality of the welds. Here, quality was assessed by two simple and quick methods: visual inspection of the flash and how the welds survived a drop test. The upset pressure, spindle speed and upset time did not have a significant effect, whereas the friction pressure and burn-off length noticeably influenced the weld quality. In the second stage, just these parameters (friction pressure and burn-off length) were changed by two levels each. Again, based on visual inspection, a final set of parameters was selected, and the same set was used for all the subsequent welds in the current work (Table 2).

Table 2. Welding parameters.

	Explored Parameter Range	Final Parameters Chosen
Friction Pressure	50–250 MPa	60 MPa
Upset Pressure	100–400 MPa	150 MPa
Spindle Speed	1000–2000 rev/min	1400 rev/min
Burn-off Length	1–7 mm	5 mm
Upset Time	4–10 s	5 s

A solution containing 2% HF and 3% HNO_3 in 95% distilled water was used to etch the Ti-6Al-4V alloy weld, and 40% HNO_3 and 10% HF in 50% distilled water was used to etch the Nitinol weld. The chemical compositions of the Ti-6Al-4V alloy to the Nitinol rods were analyzed by employing a LECO TCH 400 instrument (LECO Corporation, St. Joseph, MI, USA). The low-magnification macrostructure of the friction welds was observed using a Nikon SMZ745T stereo microscope (Nikon Instruments Inc., New York, NY, USA). The microstructural features of the friction welded samples were observed using a Leitz optical microscope. The microstructure and energy dispersive spectroscopy (EDS) line scan were investigated using a VEGA 3LMV, TESCAN scanning electron microscope and oxford instruments, respectively (TESCAN ORSAY HOLDING, Brno, Czech Republic). X-ray diffraction (XRD) analysis (PANalytical, Malvern, UK, X'pert powder XRD) was used to identify phases in the base metal and weld. Vickers microhardness measurements were carried out as per the ASTM E384 standard (West Conshohocken, PA, USA) across the weld region by using a diamond pyramid indenter under a load of 500 g for 15 s (MMT-X Matsuzawa, Akita Prefecture, Kawabetoshima, Japan). Transverse weld specimens of a 25 mm gauge length and 4 mm diameter were machined from friction welded samples. Tensile tests were carried out according to the ASTM E8 standard on the base metal, as well as dissimilar friction welded samples using a servo hydraulic testing machine at a constant displacement rate of 0.5 mm/min.

3. Results and Discussion

3.1. Base Metal Microstructures

The optical microstructures of the Ti-6Al-4V and Nitinol alloys are shown in Figure 2a,b respectively. Both microstructures showed equiaxed grains, and the average grain sizes of the Ti-6Al-4V and the Nitinol were 7 ± 1 µm and 50 ± 8 µm, respectively. The optical microstructure of the Ti-6Al-4V revealed alpha grains (brighter phase) in a transformed beta matrix (darker phase). Predominantly, a single-phase equiaxed microstructure was seen, with distinct twin lines in the Nitinol. Nitinol transforms to an austenitic phase when heated to a point above its austinite start (A_s) temperatures, but reverts back to its stable state of a martensitic phase when cooled [5]. Depending on the Ni content, the start and finish temperatures of the phase transformation varies. However, it is safe to assume that the present alloy remained predominantly austenitic at room temperature, with possible tiny volumes of a martensitic phase.

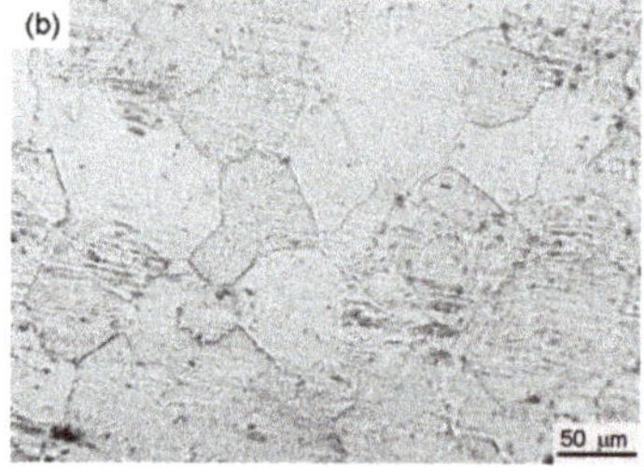

Figure 2. Optical microstructures of the base metals (**a**) Ti-6Al-4V and (**b**) Nitinol.

3.2. Macro and Microstructures of Dissimilar Friction Welds

An actual Ti-6Al-4V and Nitinol joint is shown in Figure 3. Visual examination revealed no obvious macroscopic defects. The low-magnification friction welded joint made between the Ti-6Al-4V and Nitinol alloys showed a flash formation, which is a typical characteristic of the friction welding process (Figure 3). The flash predominantly occurred on the Ti-6Al-4V side, but for the Nitinol it was notably absent. From Figure 4, it is worthwhile to note that a precipitous drop in flow stresses in Ti-6Al-4V at high temperatures during friction welding makes the alloy softer compared with the Nitinol side, even though the yield strength (YS) of Ti-6Al-4V is higher than Nitinol to begin with at room temperature [28]. In addition to this, the poor heat conducting properties of Ti-6Al-4V cause the temperature to rise quickly on its side of the joint.

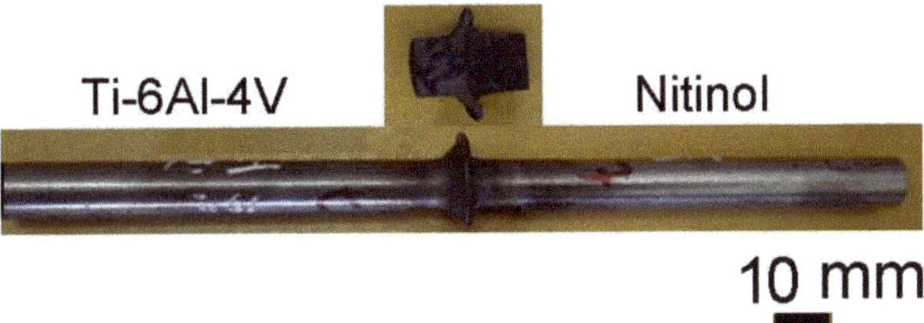

Figure 3. The visual view of a Ti-6Al-4V and Nitinol dissimilar friction welded joint.

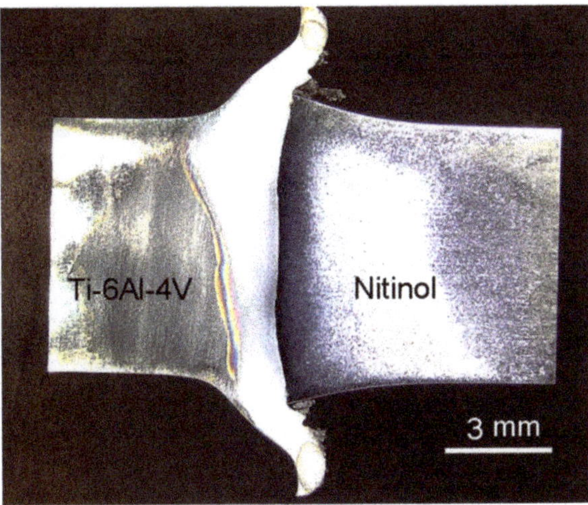

Figure 4. Dissimilar friction welded Ti-6Al-4V and Nitinol sample showing the flash on the Ti-6Al-4V side.

The optical microstructure of the Ti-6Al-4V/Nitinol joint interface is shown in Figure 5a. The grains nearer to the interface on the Ti-6Al-4V side underwent refinement, whereas the Nitinol microstructure largely remained unaffected. Friction welding introduced a lot of dislocations into the materials because of the heavy plastic deformation that occurred during the process and, at the same time, it generated high temperatures close to the melting point of the base metals. The subgrain structure could also be seen if the dislocation density increased. These low-angle grain boundaries rotated to form stain-free grains, which are fine grains referred to as dynamic recrystallization (DRX) [29]. DRX was observed on the Ti-6Al-4V side, adjacent to the joint interface when compared with the Nitinol side. This means that the deformation was more on the Ti-6Al-4V side due to the heavy plastic deformation and high temperatures encountered in friction welding. Due to dynamic recrystallization, nucleation and growth of the grains occurred; hence, a large amount of fine equiaxed grains were observed adjacent to the Ti-6Al-4V interface. A darker region (intermixed zone) was noticed between the Ti-6Al-4V/Nitinol joint interface. This may be due to the formation of intermetallics at the interface. The grains along the Ti-6Al-4V interface aligned in the rotating direction due to deformation of the material, as shown in Figure 5 b. The dark spots observed on the welded samples were referred to as etch pits, developed during etching of the sample. There was no significant change in grain size adjacent to the Nitinol interface, and the microstructure consisted mainly of austenite rather than martensite due to the heating of the material during welding.

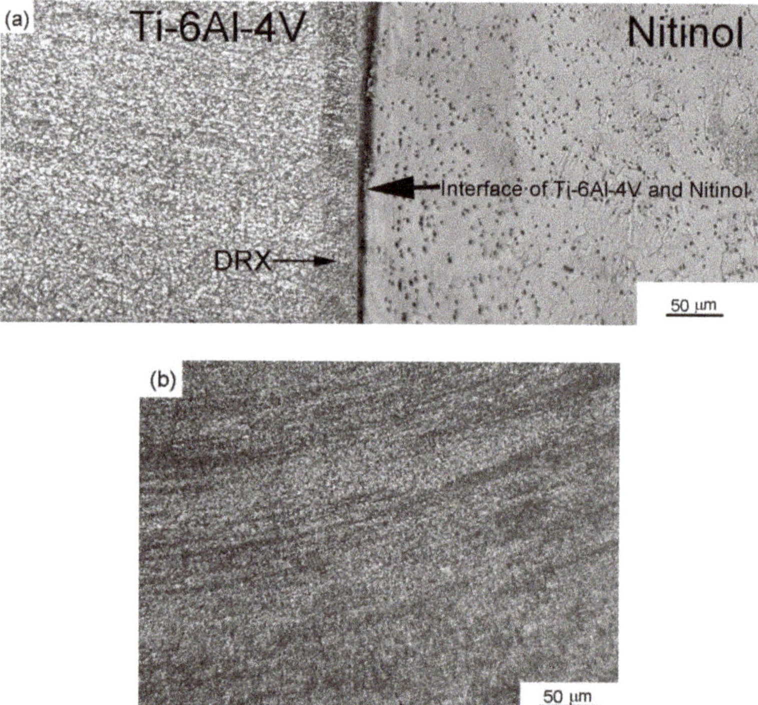

Figure 5. (**a**) The optical microstructure of the Ti-6Al-4V/Nitinol joint interface. (**b**) The grains along the titanium interface, aligned in the rotating direction.

Figure 6 shows a scanning electron microscopy (SEM) image of the Ti-6Al-4V/Nitinol joint and the corresponding SEM EDS line scan, which shows the distribution of different elements across the joint. The EDS line scan revealed an interface between the Ti-6Al-4V/Nitinol joint, and here, a ~10 µm wide intermixed zone could be seen. The intermixed zone was the result of the huge amount of plastic strains and high temperatures seen in this region. It is also reasonable to expect that this intermixed zone would not be of the same width and composition between the center and the periphery [30]. This is because the intensity of rubbing differs in these zones. Ni has low solubility in Ti in its solid state and forms the brittle intermetallic compound Ti_2Ni. These hard intermetallic compounds play a major role in the poor ductility of Ti-6Al-4V/Nitinol welds [9–12]. It is worth noting that even though the initial diffusion rates depend upon the alloying elements present in each of the base metals, the newly formed intermetallic phases would also start to influence the diffusion phenomenon very quickly [30]. There is also the possibility of formation of Kirkendall voids when a particular element undergoes mass transport while it forms the reaction products in dissimilar material joining [31].

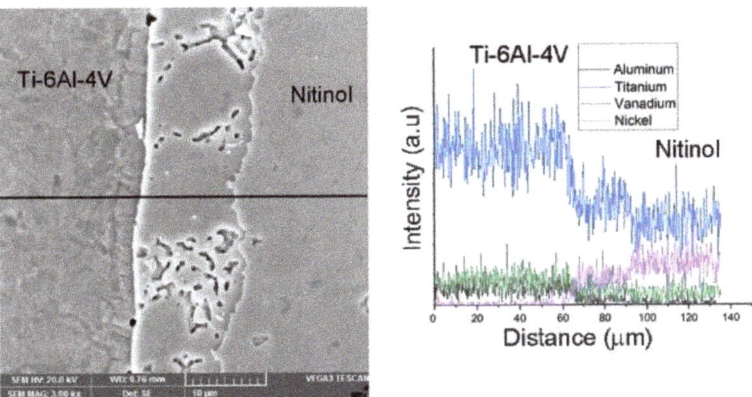

Figure 6. SEM image of the Ti-6Al-4V/Nitinol joint and the corresponding SEM energy dispersive spectroscopy (EDS) line scan.

The diffusion of Ni into Ti-6Al-4V was relatively higher. Figure 7 shows the X-ray diffraction analysis of the Ti-6Al-4V base metal, Nitinol base metal and fracture surface of the weld. The Nitinol base metal exhibited a B2 austenitic phase, and the Ti-6Al-4V base metal exhibited a hexagonal α phase and a weak β phase. In contrast, the fracture surface of the weld revealed a hexagonal α phase, β phase and the formation of a Ti_2Ni brittle intermetallic phase.

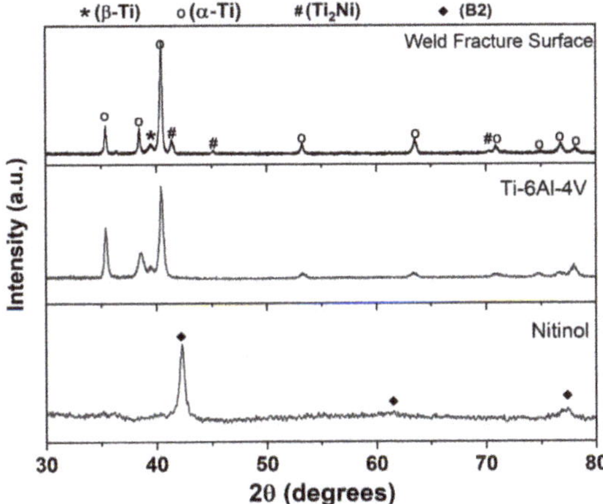

Figure 7. X-ray diffraction (XRD) profiles of the Ti-6Al-4V base metal, Nitinol base metal and fracture surface of the Ti-6Al-4V/Nitinol friction welds.

3.3. Mechanical Properties

3.3.1. Hardness

Figure 8 shows the hardness distribution across the interfaces of dissimilar Ti-6Al-4V/Nitinol welds. It is observed that there was an increase in hardness on the Ti-6Al-4V side due to stain hardening that occurred during the friction welding process. However, no significant increase in hardness is observed on the Nitinol side, which confirms the previous observation that no there was no significant deformation on the Nitinol side. Similar results were observed in previous studies involving friction welding of commercially pure Ti to 304L stainless steel [28,32]. Dissimilar friction welds prepared from Ti to 304L stainless steel have shown an increase in hardness on the Ti side due to the stain hardening effect when compared with the stainless steel side. However, a decrease in hardness on the stainless steel side was attributed to a limited amount of deformation or the lack of a strain hardening effect [28]. The Ti-6Al-4V and Nitinol base materials had average hardnesses of 367 ± 18 HV and 260 ± 16 HV, respectively. The highest hardness observed at the Ti-6Al-4V/Nitinol interface may be attributed to the formation of intermetallics of Ti and Ni, such as Ti_2Ni, a result conformed by XRD analysis.

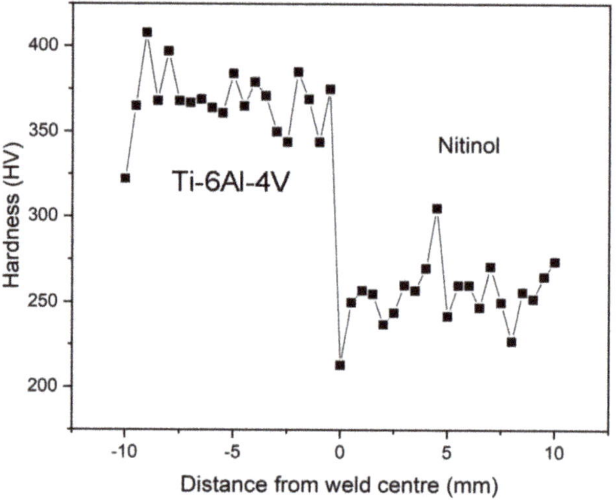

Figure 8. The hardness distribution across the interface of dissimilar Ti-6Al-4V/Nitinol welds.

3.3.2. Tensile Properties

The tensile properties of dissimilar Ti-6Al-4V/Nitinol friction welds are shown in Figure 9. The important objective of the present investigation was to determine whether the joint strength of a dissimilar weld shows any improvement in its weld zone tensile properties. The data of two base materials' tensile properties have also been included in Figure 8 for comparison. It is revealed from Figure 9 that the dissimilar Ti-6Al-4V/Nitinol friction weld exhibited a lower ultimate tensile strength (UTS) and ductility (UTS = 589 MPa, 3.9% elongation) compared with the base metals (Ti-6Al-4V: UTS = 1073 MPa, 15% elongation; Nitinol: UTS = 980 MPa, 16% elongation). This could be because of intermetallic formations (Ti_2Ni) at the interface which are brittle in nature. Similar results were observed in laser welding of dissimilar Ti-6Al-4V to NiTi joints [10]. These dissimilar welds exhibited a UTS value of 250 MPa and 3% elongation. The authors attributed the reason for the inferior properties to the formation of a Ti_2Ni phase and transverse cracks in the weld metal. Electron beam welding was better in that the UTS was 480 MPa with 2.3 % elongation [21]. Three tensile tests were conducted for each condition, and the average of these specimens was taken. The dissimilar weld failed in the intermixed zone, except for the Ti-6Al-4V and Nitinol base metals. Failure in the intermixed zone indicates that the weld region was weaker than the base metals in the dissimilar welded joint. The tensile fracture surfaces of the Ti-6Al-4V base metal, Nitinol base metal and dissimilar Ti-6Al-4V/Nitinol friction welds are shown in Figure 10. Very fine and ductile fracture features were observed for the Ti-6Al-4V and Nitinol base metals in Figure 10a,b respectively. It is well established that the fine equiaxed dimples are observed in the base metal samples because of microvoid formation and joining. Figure 10c shows the cleavage fracture (brittle) features of dissimilar Ti-6Al-4V/Nitinol friction welds.

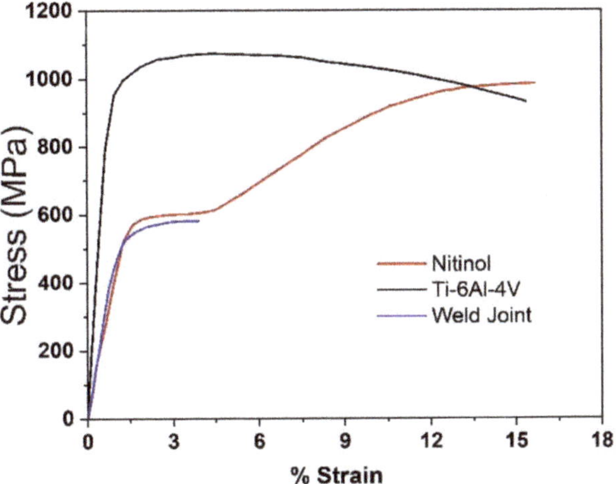

Figure 9. The tensile properties of base metals and dissimilar Ti-6Al-4V/Nitinol friction welds.

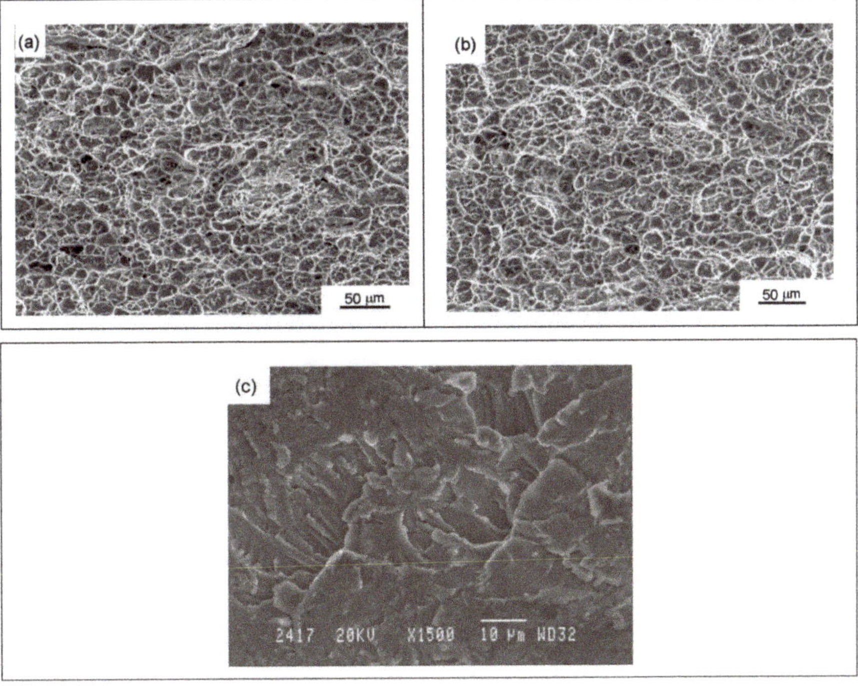

Figure 10. Fracture surfaces of the base metals and dissimilar welded joint for (**a**) Ti-6Al-4V, (**b**) Nitinol and (**c**) dissimilar Ti-6Al-4V/Nitinol friction welded joint.

4. Conclusions

Dissimilar Ti-6Al-4V/Nitinol friction welds have been analyzed for weld macro and microstructures, grain sizes, hardnesses and tensile properties. The following conclusions can be drawn:

1. A defect-free dissimilar friction weld could be obtained by a continuous drive friction welding machine;
2. The macrostructure of dissimilar Ti-6Al-4V/Nitinol friction welds revealed flash formation only on the Ti-6Al-4V side due to the reduction in flow stress at high temperatures experienced during friction welding;
3. XRD studies revealed the formation of an intermetallic compound (Ti_2Ni) on the fracture surface of dissimilar welds;
4. The dissimilar Ti-6Al-4V/Nitinol friction weld exhibited low strength and ductility (UTS = 589 MPa, 3.9% elongation) compared with the base metals, and this may be attributed to the formation of Ti_2Ni intermetallics at the interface, which are brittle in nature;
5. The tensile fracture surfaces observed in dissimilar Ti-6Al-4V/Nitinol friction welds had cleavage (brittle) fracture features due to the formation of intermetallics at the intermixed zone when compared with the fine dimples noticed in the base metal samples.

Author Contributions: Conceptualization, A.U.R., Y.S.U. and H.A.-K.; methodology, A.U.R., N.K.B., M.K.T. and Y.S.U.; formal analysis, A.U.R., N.K.B., M.K.T., Y.S.U. and H.A.-K.; investigation, A.U.R., N.K.B., M.K.T. and Y.S.U.; resources, A.U.R., Y.S.U. and H.A.-K.; data curation, A.U.R., N.K.B. and M.K.T.; writing—original draft preparation, A.U.R., N.K.B. and M.K.T.; writing—review and editing, A.U.R., N.K.B., M.K.T., Y.S.U. and H.A.-K.; visualization, A.U.R., N.K.B. and M.K.T.; supervision, A.U.R., N.K.B. and M.K.T.; project administration, A.U.R., and Y.S.U.; funding acquisition, A.U.R., Y.S.U. and H.A.-K. All authors have read and agreed to the published version of the manuscript.

Funding: This research was funded by the National Plan for Science, Technology and Innovation (MAARIFAH), King Abdulaziz City for Science and Technology, Kingdom of Saudi Arabia, Award Number (14-ADV110-02).

Institutional Review Board Statement: Not applicable.

Informed Consent Statement: Not applicable.

Data Availability Statement: Data is contained within the article "Microstructure and Mechanical Properties of Dissimilar Friction Welded Ti-6Al-4V to Nitinol Joints".

Acknowledgments: This Project was funded by the National Plan for Science, Technology and Innovation (MAARIFAH), King Abdulaziz City for Science and Technology, Kingdom of Saudi Arabia, Award Number (14-ADV110-02). The authors acknowledge Ashfaq Mohammad at the National Manufacturing Institute Scotland for his technical inputs during the experimental and writeup stages of the present work.

Conflicts of Interest: The authors declare no conflict of interest.

References

1. Boyen, R.R. Titanium and titanium alloys. In *Metals Handbook*; Davis, J.R., Ed.; ASM International: Materials Park, OH, USA, 1998; pp. 575–588. ISBN 978-0-87170-654-6.
2. Sidambe, A. Biocompatibility of Advanced Manufactured Titanium Implants—A Review. *Materials* **2014**, *7*, 8168–8188. [CrossRef] [PubMed]
3. Niinomi, M. Mechanical properties of biomedical titanium alloys. *Mater. Sci. Eng. A* **1998**, *243*, 231–236. [CrossRef]
4. Gurrappa, I. Characterization of titanium alloy Ti-6Al-4V for chemical, marine and industrial applications. *Mater. Charact.* **2003**, *51*, 131–139. [CrossRef]
5. Jani, J.M.; Leary, M.; Subic, A.; Gibson, M.A. A Review of Shape Memory Alloy Research, Applications and Opportunities. *Mater. Des.* **2013**. [CrossRef]
6. Saedi, S.; Turabi, A.S.; Andani, M.T.; Moghaddam, N.S.; Elahinia, M.; Karaca, H.E. Texture, aging, and superelasticity of selective laser melting fabricated Ni-rich NiTi alloys. *Mater. Sci. Eng. A* **2017**, *686*, 1–10. [CrossRef]

7. Chau, E.T.F. Comparative Study of Joining Methods for a SMART Aerospace Application. Ph.D. Thesis, Cranfield University, Cranfield, UK, 2007.
8. Sun, Z.; Ion, J.C. Laser welding of dissimilar metal combinations. *J. Mater. Sci.* **1995**, *30*, 4205–4214. [CrossRef]
9. Miranda, R.M.; Assunção, E.; Silva, R.J.C.; Oliveira, J.P.; Quintino, L. Fiber laser welding of NiTi to Ti-6Al-4V. *Int. J. Adv. Manuf. Technol.* **2015**, *81*, 1533–1538. [CrossRef]
10. Shojaei Zoeram, A.; Akbari Mousavi, S.A.A. Laser welding of Ti–6Al–4V to Nitinol. *Mater. Des.* **2014**, *61*, 185–190. [CrossRef]
11. Falvo, A.; Furgiuele, F.M.; Maletta, C. Functional behaviour of a NiTi-welded joint: Two-way shape memory effect. *Mater. Sci. Eng. A* **2008**, *481*, 647–650. [CrossRef]
12. Otsuka, K.; Wayman, C.M. (Eds.) *Shape Memory Materials*, 1st ed.; Cambridge Univ. Press: Cambridge, UK, 1999; ISBN 978-0-521-44487-3.
13. Oliveira, J.P.; Miranda, R.M.; Braz Fernandes, F.M. Welding and Joining of NiTi Shape Memory Alloys: A Review. *Prog. Mater. Sci.* **2017**, *88*, 412–466. [CrossRef]
14. Oliveira, J.P.; Zeng, Z.; Andrei, C.; Braz Fernandes, F.M.; Miranda, R.M.; Ramirez, A.J.; Omori, T.; Zhou, N. Dissimilar laser welding of superelastic NiTi and CuAlMn shape memory alloys. *Mater. Des.* **2017**, *128*, 166–175. [CrossRef]
15. Datta, S.; Raza, M.S.; Kumar, S.; Saha, P. Exploring the possibility of dissimilar welding of NiTi to Ti using Yb-fiber laser. *Adv. Mater. Process. Technol.* **2018**, *4*, 614–625. [CrossRef]
16. Yuhua, C.; Yuqing, M.; Weiwei, L.; Peng, H. Investigation of welding crack in micro laser welded NiTiNb shape memory alloy and Ti6Al4V alloy dissimilar metals joint. *Opt. Laser Technol.* **2017**, *91*, 197–202. [CrossRef]
17. Zeng, Z.; Panton, B.; Oliveira, J.P.; Han, A.; Zhou, Y.N. Dissimilar laser welding of NiTi shape memory alloy and copper. *Smart Mater. Struct.* **2015**, *24*, 1–8. [CrossRef]
18. Zeng, Z.; Oliveira, J.P.; Yang, M.; Song, D.; Peng, B. Functional fatigue behavior of NiTi-Cu dissimilar laser welds. *Mater. Des.* **2017**, *114*, 282–287. [CrossRef]
19. Oliveira, J.P.; Panton, B.; Zeng, Z.; Andrei, C.M.; Zhou, Y.; Miranda, R.M.; Braz Fernandes, F.M. Laser joining of NiTi to Ti6Al4V using a Niobium interlayer. *Acta Mater.* **2016**, *105*, 9–15. [CrossRef]
20. Zhou, X.; Chen, Y.; Huang, Y.; Mao, Y.; Yu, Y. Effects of niobium addition on the microstructure and mechanical properties of laser-welded joints of NiTiNb and Ti6Al4V alloy. *J. Alloys Compounds* **2018**, *735*, 2616–2624. [CrossRef]
21. Zhan, Z.; Chen, Y.; Wang, S.; Huang, Y.; Mao, Y. Prevention of crack formation in electron-beam welded joints of dissimilar metal compounds (TiNi/Ti-6Al-4V). *Metal Sci. Heat Treat.* **2019**, *61*, 373–378. [CrossRef]
22. Senkevich, K.S.; Knyazev, M.I.; Runova, Y.E.; Shlyapin, S.D. Special Features of Formation of a TiNi–VT6 Diffusion Joint. *Met. Sci. Heat Treat.* **2013**, *55*, 419–422. [CrossRef]
23. Zhang, W.; Ao, S.; Oliveira, J.; Zeng, Z.; Huang, Y.; Luo, Z. Microstructural Characterization and Mechanical Behavior of NiTi Shape Memory Alloys Ultrasonic Joints Using Cu Interlayer. *Materials* **2018**, *11*, 1830. [CrossRef]
24. Gugel, H.; Schuermann, A.; Theisen, W. Laser welding of NiTi wires. *Mater. Sci. Eng. A* **2008**, *481*, 668–671. [CrossRef]
25. Yang, D.; Jiang, H.C.; Zhao, M.J.; Rong, L.J. Microstructure and mechanical behaviors of electron beam welded NiTi shape memory alloys. *Mater. Des.* **2014**, *57*, 21–25. [CrossRef]
26. London, B.; Fino, J.; Pelton, A.R.; Mahoney, M. Friction stir processing of Nitinol. In Proceedings of the Friction Stir wielding and Processing III, TMS Annual Meeting, San Francisco, CA, USA, 13–17 February 2005; Jata, K.V., Mahoney, M.W., Mishra, R.S., Lienert, T.J., Eds.; TMS (The Minerals, Metals and Materials Society): San Francisco, CA, USA, 2005; pp. 67–74.
27. Shinoda, T.; Tsuchiya, T.; Takahashi, H. Friction welding of shape memory alloy. *Weld. Int.* **1992**, *6*, 20–25. [CrossRef]
28. Muralimohan, C.H.; Muthupandi, V.; Sivaprasad, K. Properties of Friction Welding Titanium-stainless Steel Joints with a Nickel Interlayer. *Procedia Mater. Sci.* **2014**, *5*, 1120–1129. [CrossRef]
29. Mishra, R.S.; Ma, Z.Y. Friction stir welding and processing. *Mater. Sci. Eng. R Rep.* **2005**, *50*, 1–78. [CrossRef]
30. Zhang, Y.; Sun, D.Q.; Gu, X.Y.; Li, H.M. Nd:YAG pulsed laser welding of dissimilar metals of titanium alloy to stainless steel. *Int. J. Adv. Manuf. Technol.* **2018**, *94*, 1073–1085. [CrossRef]
31. Wang, Y.; Prangnell, P.B. The significance of intermetallic compounds formed during interdiffusion in aluminum and magnesium dissimilar welds. *Mater. Charact.* **2017**, *134*, 84–95. [CrossRef]
32. Dey, H.C.; Ashfaq, M.; Bhaduri, A.K.; Rao, K.P. Joining of titanium to 304L stainless steel by friction welding. *J. Mater. Process. Technol.* **2009**, *209*, 5862–5870. [CrossRef]

MDPI

Article

Microstructure and Mechanical Properties of Dissimilar Friction Stir Welded AA2024-T4/AA7075-T6 T-Butt Joints

Mohamed M.Z. Ahmed [1,2,*], Mohamed M. El-Sayed Seleman [2], Zeinab A. Zidan [2], Rashad M. Ramadan [2], Sabbah Ataya [2,3] and Naser A. Alsaleh [3]

[1] Department of Mechanical Engineering, College of Engineering at Al Kharj, Prince Sattam Bin Abdulaziz University, Al Kharj 11942, Saudi Arabia

[2] Department of Metallurgical and Materials Engineering, Faculty of Petroleum and Mining Engineering, Suez University, Suez 43512, Egypt; mohamed.elnagar@suezuniv.edu.eg (M.M.E.-S.S.); Eng.zeinab12@yahoo.com (Z.A.Z.); rashadrmdn@yahoo.co.uk (R.M.R.); sabbah.ataya@suezuniv.edu.eg (S.A.)

[3] Department of Mechanical Engineering, College of Engineering, Al Imam Mohammad Ibn Saud Islamic University, Riyadh 11432, Saudi Arabia; alsaleh@engineer.com

* Correspondence: moh.ahmed@psau.edu.sa; Tel.: +966-011-588-1200

Abstract: Aircraft skin and stringer elements are typically fabricated from 2xxx and 7xxx series high strength aluminum alloys. A single friction stir welding (FSW) pass using a specially designed tool with shoulder/pin diameter ratio (D/d) of 3.20 is used to produce dissimilar T-butt welds between AA2024-T4 and AA7075-T6 aluminum alloys at a constant travel speed of 50 mm/min and different rotational speeds of 400, 600 and 800 rpm. The AA2024-T4 is the skin and the AA7075-T6 is the stringer. Sound joints are produced without macro defects in both the weld top surfaces and the joint corners at all rotational speeds used (400, 600, and 800 rpm). The hardness value of the nugget zone increases by increasing the rotational speed from 150 ± 4 Hv at 400 rpm to 167 ± 3 Hv at 600 rpm, while decreases to reach the as-received AA2024-T4 hardness value (132 ± 3 Hv) at 800 rpm. Joint efficiency along the skin exhibits higher values than that along the stringer. Four morphologies of precipitates were detected in the stir zone (SZ); irregular, almost-spherical, spherical and rod-like. Investigations by electron back scattered diffraction (EBSD) technique showed significant grain refinement in the sir zone of the T-welds compared with the as-received aluminum alloys at 600 rpm due to dynamic recrystallization. The grain size reduction percentages reach 85 and 90 % for AA2024 and AA7075 regions in the mixed zone, respectively. Fracture surfaces along the skin and stringer of T-welds indicate that the joints failed through mixed modes of fracture.

Keywords: dissimilar friction stir welding; AA2024-T4/AA7075-T6Al alloys; t-butt joints; microstructure evaluation; EBSD; fracture surfaces

Citation: Ahmed, M.M.Z.; El-Sayed Seleman, M.M.; Zidan, Z.A.; Ramadan, R.M.; Ataya, S.; Alsaleh, N.A. Microstructure and Mechanical Properties of Dissimilar Friction Stir Welded AA2024-T4/AA7075-T6 T-Butt Joints. *Metals* **2021**, *11*, 128. https://doi.org/doi:10.3390/met11010128

Received: 28 December 2020
Accepted: 6 January 2021
Published: 10 January 2021

Publisher's Note: MDPI stays neutral with regard to jurisdictional claims in published maps and institutional affiliations.

1. Introduction

Friction Stir Welding (FSW) is a solid-state welding technique, where similar [1–3] and dissimilar metals [2] are welded without reaching their melting point. FSW has gained significant attention due to its numerous advantages [4–6] over the traditional arc welding methods [7]. However, FSW has also some drawbacks [8–10] which are very little when compared with the traditional welding processes. Various joining configurations in terms of spot [3,11], butt [12–14], corner [15], lap [16–18] and fillet [19] joints have been studied and established for different metals and alloys using FSW technology through optimizing their welding process parameters. However, FSW of high strength to weight ratio aluminum alloys to produce T-joints are still demanding more studies and efforts to evaluate the joint efficiency, and to optimize the welding parameters. Arora et al. [20] reported that the hardness and tensile strength of the friction stir similar butt joint of AA2219 were lower than the base metal. However, a toughness value of the nugget zone enhanced compared to the base metal. Babu et al. [18] studied the effect of pin profiles on the friction stir welded (FSWed) lap joint quality of the AA2014 aluminum alloy and reported that the strength and the failure mode of joints are related to the hook geometry. Sadeesh et al. [21] studied the influence of using five different tool designs in FSW of dissimilar AA2024 and AA6061aluminum alloys in butt joints and concluded that the joints efficiency is related to the varying welding process parameters and the shoulder to pin diameter ratio is the main dominant parameter. Manuel et al. [22] investigated the behavior of the friction stir welds of three dissimilar aluminum alloys in a T-joint using three base materials namely, AA2017-T4, AA5083-H111, and AA6082-T6. They reported that the arrangement of the skin materials with respect to the pin rotation direction influence the morphology and the mechanical properties of the joints [22]. In fact, the 2xxx and 7xxx aluminum alloys before FSW were known to be non-weldable by traditional fusion welding processes since number of problems such as porosity, solidification cracks and residual internal stresses are found in the weld. Moreover, the formed dendritic structure during solidification in the fusion zone is significantly deteriorating the mechanical properties of the joints [5].

AA2024 and AA7075 aluminum alloys are typically representing the 2xxx and 7xxx alloys series, respectively, which are usually used intensively to fabricate structural components of automotive and aircraft industries [23–25]. In many applications, especially the aircraft fuselage panels in the form of T-joints; where, the aluminum alloys AA2024 and AA7075 are used as skins and stingers, respectively. The stinger is the stiffening element which reinforces the section of the load carrying skin to avoid buckling and failure. In FSW of T-joints, Cui et al. [26] studied the effect of different rotational speeds on the microstructure and mechanical properties of the T-lap, T-butt-lap, and T-butt joints. They ascribed the formed defects such as tunnel defect, kissing bond, original joint line defect, and zigzag line to the initial matching modes of the blanks, material flow patterns, and insufficient heat input. Acerra et al. [15] investigated the dissimilar AA2024-T4 and AA7075-T6 T-joints and evaluated the design rules for the setup of FSW operations. They concluded that a large shoulder diameter is required to generate a sufficient heat input to fulfill the weld joint. Moreover, the presence of coating layers negatively affects the mechanical properties by a generation of macro and micro defects. In T-butt joints, especially in aerospace industry, the tool pin has been penetrated vertically into the materials setup surface to join three separate parts, two skin parts and vertical stringer element, to produce one-piece T-joint [15].

This work aims to conduct single pass FSW of T-butt joints between AA2024-T4 and AA7075-T6 aluminum alloys as skin and stringer; respectively, at different tool rotational rates of 400, 600 and 800 rpm and a constant welding speed of 50 mm/min. The role of FSW parameters and the welded materials on the microstructure, the strength and the fracture mode will be examined. Attention will be given to investigate the evolved texture using EBSD.

2. Experimental Procedure

Plates of the dissimilar aluminum alloys of AA2024-T4 and AA7075-T6 were friction stir welded in T-butt joints. The dimensions for FSW samples were 4 mm × 50 mm × 200 mm for AA2024-T4 and 5 mm × 100 mm × 200 mm for AA7075-T6 alloys. The T-butt joint setup was designed to be two plates of AA2024-T4 alloy as a skin and one plate of AA7075-T6 as a stringer. The chemical composition of the aluminum alloys was analyzed using Q2 ION-OES -Optical Emission Spectrometry (Bruker, Billerica, MA, USA). The chemical composition of the investigated alloys is listed in Table 1.

Table 1. Chemical analysis (wt. %) of the AA2024-T4 and AA7075-T6 alloys.

Alloy	Elements in wt. %								
	Si	**Fe**	**Cu**	**Mn**	**Mg**	**Cr**	**Zn**	**Ti**	**Al**
AA2024-T4	0.20	0.40	3.90	0.40	1.50	0.02	0.17	0.02	Bal
AA7075-T6	0.05	0.22	1.94	0.05	2.66	0.21	5.97	0.01	Bal

The T-butt joints were friction stir welded using gantry type FSW (EG-FSW-M1, made by Suez University, Suez, Egypt) machine [2,6]. Adaptive fixture was designed [15] and fabricated from steel for T-butt joint configuration setup, Figure 1a,b. A constant travel speed of 50 mm/min and three various rotational speeds of 400, 600 and 800 rpm were used in this work. The shoulder tilting angle and the shoulder plunge depth were 3° and 0.20 mm, respectively. The other welding conditions were kept constant. A concave tool shoulder with a diameter of 32 mm and a threaded taper pin with a length of 3.8 mm was used (Figure 1c). The bottom and top diameters of the pin were 12 mm and 10 mm, respectively. The tool was machined from H13 tool steel and heat treated to obtain a hardness value of 58 HRC. Figure 1c shows top view appearance of a friction stir welded joint.

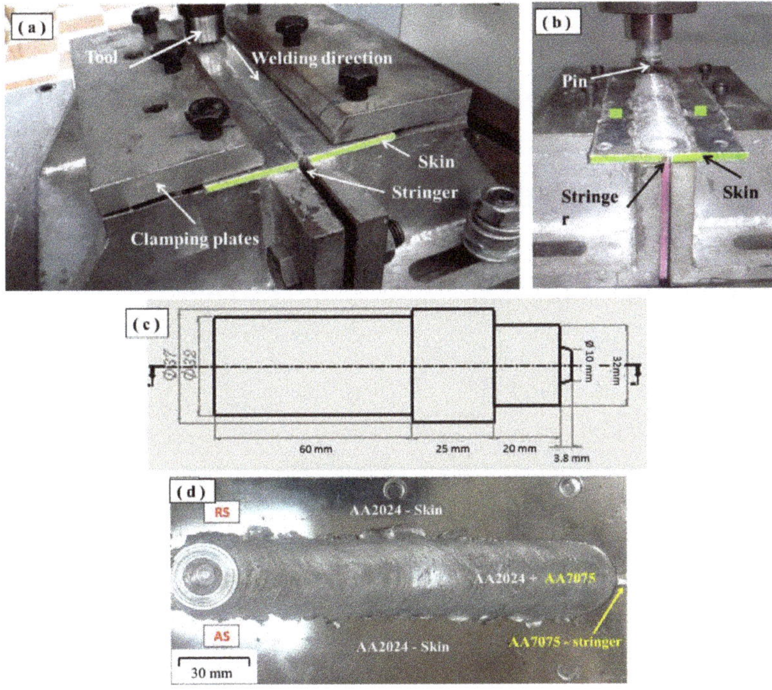

Figure 1. (**a**,**b**) FSW fixture setup configuration of AA2024 and AA7075 T-butt joint, (**c**) Engineering drawing of the used tool (**d**) top view appearance of a friction stir welded joint.

In order to investigate the developed T-joints microstructure and mechanical properties, the samples were cross-sectioned perpendicular to the welding direction (WD). For microstructural analysis, the sectioned samples were ground using SiC papers with different grit up to 2400 and polished on felt cloth with alumina 0.05 μm paste and then etched with Keller's reagent (6 mL hydrofluoric acid, 6 mL hydrochloric acid, 5 mL nitric acid and 150 mL water). Stereo microscope (Optica SZR 10, Optica, Ponteranica (BG), Itaty) was used to examine macrostructures of the joints. Microstructural analyses were carried out using optical microscope (OM) Olympus -BX41M-LED, Olympus, Tokyo, Japan. And scanning electron microscopy (SEM) Type Quanta FEG-250 (FEI company, Hillsboro, OR, USA) equipped with electron back scatter diffraction (EBSD) and energy dispersive spectroscopy (EDS) were also used to examine the grain structure, texture and precipitates composition.

Vickers hardness tester machine (HWDV-75, TTS Unlimited, Osaka, Japan) was used to evaluate the hardness profiles along the width of the weld samples using a load of 2.0 kg and dwell time of 15 s, where the free space between any two indentations was 0.5 mm, which represents at least 2.5 times the maximum indentation diameter. The hardness maps were also analyzed and drawn by collecting three lines measurements across the stir zone (SZ) and the stringer. Two tensile tests were carried out along skin and stringer axes using tensile testing machine (Instron-4208-300 kN capacity, Norwood, MA, USA) at a room temperature and a cross-head speed of 1.0 mm/min, Figures 2 and 3, respectively. At least three tensile samples were prepared from each T-butt joint and the average value of tensile strength was considered. The fracture surfaces of tensile test samples were investigated using the SEM. For EBSD investigation, the polished samples were subjected to electro-polish for 60 sec at −15 °C and 14 V. The electro-polishing electrolyte consists of 30 vol.% nitric acid in 70 vol.% methanol. EBSD was performed for the base materials (BMs) and stir zone (SZ) of T-butt joints at 20 kV and 0.5 µm step size. The collected EBSD data were analyzed by OIM DC 7.3 software, developed by EDAX, AMETEK, Draper, UT, USA.

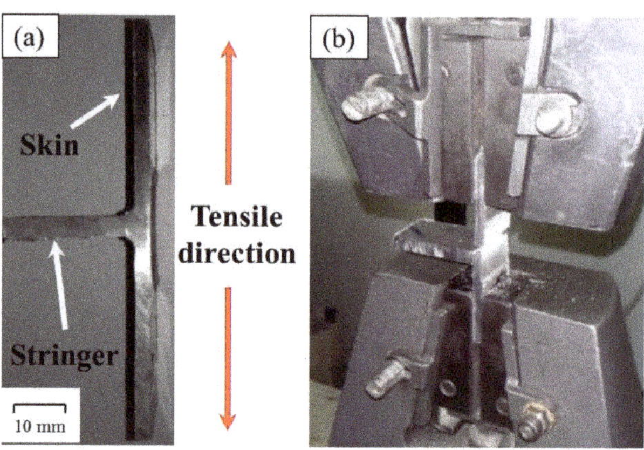

Figure 2. Photographs of tensile testing along skin direction (**a**) T-butt joint tensile test specimen and (**b**) after tensile test.

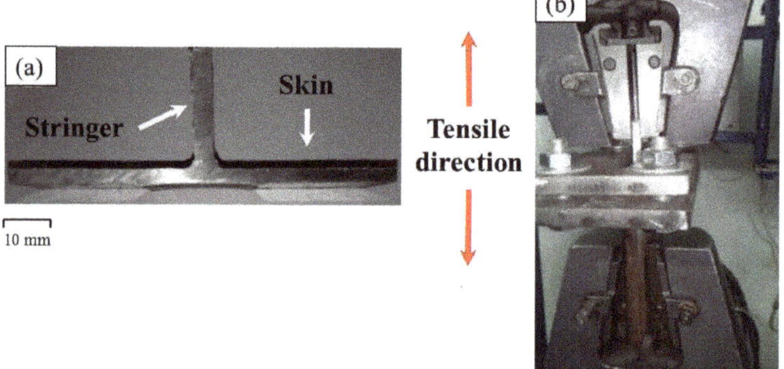

Figure 3. Photographs of tensile testing along stringer direction (**a**) T-butt joint tensile test specimen and (**b**) tensile test clamping system.

3. Results and Discussion

3.1. Joint Appearance and Macro-Examinations

Figure 4 shows the appearance of the T-butt joints friction stir welded at different rotational speeds of 400, 600 and 800 rpm. It can be observed that the top surfaces are defect free and also the exit holes are complete with no indications of any tunnel defects.

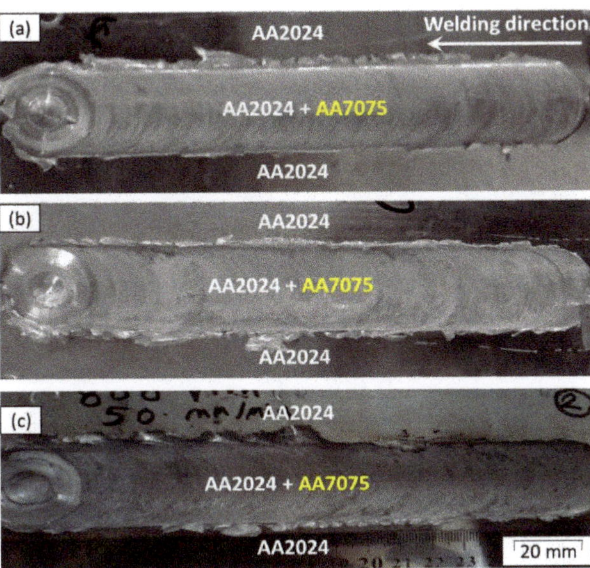

Figure 4. T-butt joint appearance of friction stir welded AA2024 and AA7075 at 50 mm/min welding travel speed and different rotational speeds (**a**) 400 rpm, (**b**) 600 rpm, and (**c**) 800 rpm.

The examination of the cross-section macrostructure of welding zone may be used to clarify the quality of friction stir welded joints. It has been reported that four different zones, namely stir zone (SZ) or nugget zone, thermo-mechanically affected zone (TMAZ), heat affected zone (HAZ) and base material (BM) can be identified in the macrostructure of welds [4]. Several variables like types of BM, thickness of work piece, welding parameters, and tool design play a significant role in the formation of friction stir welded joint features. Actually, the amount of heat input has a great effect on material flow, SZ shape, microstructure grain size and FSW defects. Hence, the quality of joints can be enhanced if the rotation and welding speeds of the FSW process can be carefully controlled. Typical transverse cross-section macrographs of the friction stir welded T-butt-joints of AA2024-T4 and AA7075-T6 are shown in Figure 5. It can be observed that the transverse cross sections showed sound T-butt joints at all the applied FSW parameters. However, the mixing process between the two alloys is significantly enhanced by increasing the rotation speed of the FSW tool such that at the low rotation speed is entirely pertains the whole stringer material (AA7075) within the SZ zone, by increasing the rotation speed the stringer material starts to get dispersed and mixed with the skin material (AA2024). This implies that increasing the rotation speed has resulted in a complex flow pattern that enhanced the two materials mixing and joining.

It is worth mentioning here that the dimensions of the SZ are controlled by the dimensions of the tool shoulder and pin. SZ is usually diffused to be little bit larger than the pin dimensions and has a like-conical form [10]. The widest region is beneath the shoulder diameter, which is ~32 mm and become narrower with the depth to be determined by the pin diameters (12 to 10 mm) and length (h = 3.8 mm) as it has been defined in the Figure 5 by the dashed lines.

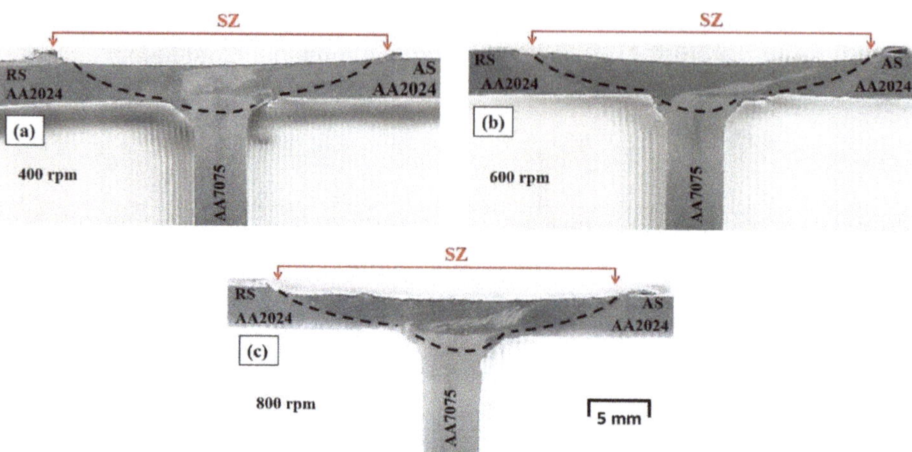

Figure 5. Macrographs of T-butt-joints of friction stir weldedAA2024 and AA7075 at different rotational speeds of (**a**) 400 rpm, (**b**) 600 rpm, and (**c**) 800 rpm.

3.2. OM Investigations

Optical microscopy investigations were performed on the transverse cross-sections of the FSWd T-joint specimen's perpendicular to the welding direction. The three zones (SZ, TMAZ, and HAZ) were arranged by SZ from the center line of the weld joint perpendicular to the welding direction in both sides ending HAZ in the BM. The width of each zone and its features are related to the heat input introduced to the work-piece during FSW process, as shown in the joints' macrographs in Figure 5. The variation in heat input in the present work came mainly from the applied different rotational speeds of 400, 600 and 800 rpm where the travel speed is kept constant at 50 mm/min, the heat input index is equal to (ω^2/v) [10]. The microstructures of the cross-sectional dissimilar friction stir welded T-butt joints of AA2024-T4 and AA7075-T6 Al alloys show new formation of the grain structures in the SZ (Figure 6). For friction stir welded T-joint at 400 rpm good mixing between skin and stringer can be clearly seen, as given in Figure 6a. The mixing between skin and stringer in the SZ has improved as the rotation welding speed increased to 600 rpm, Figure 6c. FSW at 600 rpm welding condition converts the initial elongated grains in as AA2024 and AA7075 aluminum alloys to fine equiaxed recrystallized grains in the weld SZ. Moreover, the coarse precipitates given by the lowest rotational speed of 400 rpm become finer and more dispersed. This change in microstructure is ascribed to high temperature and high stirring action by the pin inside skin and stringer which positioned between the two pieces forming the skin (Figure 1a,b). Remarkable coarser grains with very good mixing are observed at the highest rotational speed of 800 rpm, as shown in Figure 6e,f. However, the high heat input presented a chance to generate coarse precipitates in both SZ and TMAZ. Similar results of coarsening the precipitates at such conditions have been noticed in other works [1,27]. It can be concluded that, the best mixing between skin and stringer is obtained at rotational speeds of 600 and 800 rpm. The T-joint produced at 600 rpm rotational speed exhibits finer grains and precipitates in the SZ than the other two joints. Detailed EBSD grain size measurements for the parent alloys and for the friction stir welded material are presented in subchapter 3.5. In the TMAZ no recrystallization is noticed because the frictional heat and plastic deformation are not high enough to cause recrystallization Figure 6b,d,f.

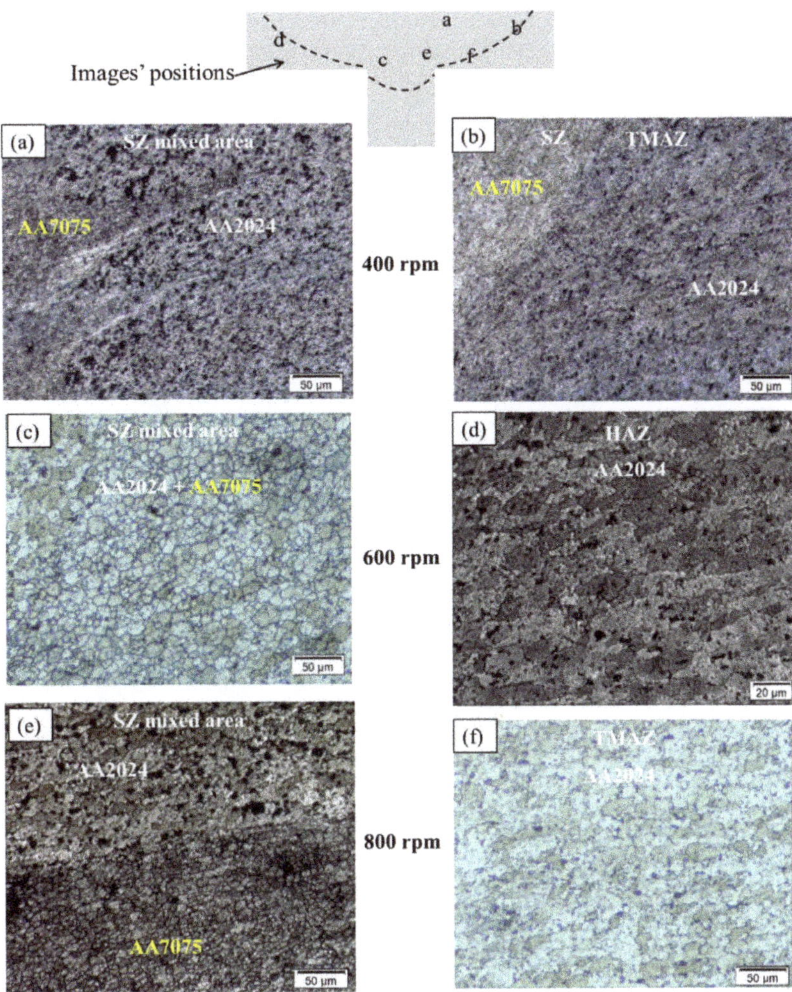

Figure 6. Optical microstructures of AA2024-AA7075 T-butt joint welded at different rotational speeds of (**a**,**b**) 400, (**c**,**d**) 600 and (**e**,**f**) 800 rpm. The positions of the different images are shown in the top schematic drawing.

3.3. Hardness Distribution

The hardness variation along the skin and the stringer of the friction stir welded T-butt joint may give a better prediction about the phenomena involved in this welding process and the causes of loss or increase in strength compared with the base material. The hardness distribution maps of the T-butt joints for all the applied rotational speeds indicate good material mixing at the nugget zone as shown in Figure 7, which is evidence of selecting a proper tool design with shoulder/pin diameter ratio (D/d). The D/d ratio in the current study was 3.20. This ratio agrees well within the value that reported by Saravanan et al. [28] who focused their work on studying the effect of the D/d ratio in the range of 1–4 on the microstructure and mechanical properties of the dissimilar butt-joint of AA2024-T6 and AA7075-T6 Al-alloys. They concluded that the butt joint friction stir welded using a D/d ratio of 3.00 exhibited higher mechanical properties when compared to the other welded joints [28].

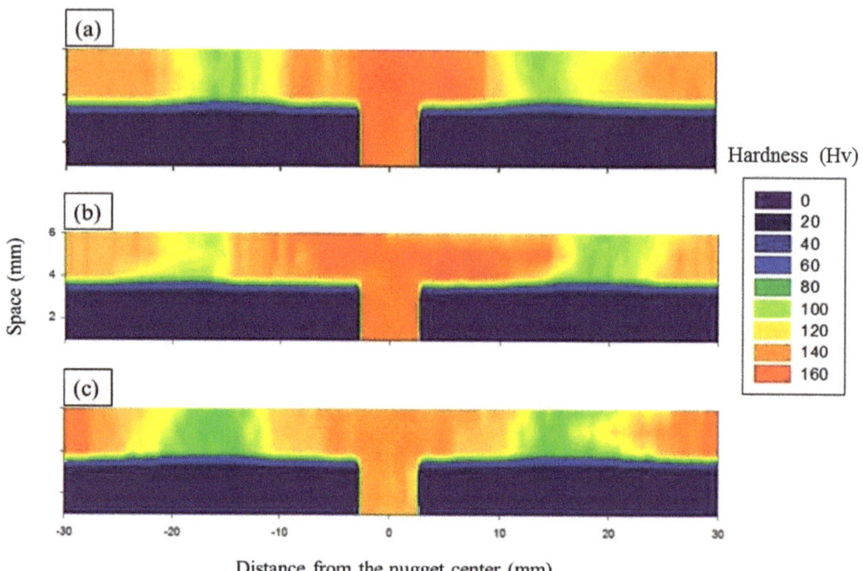

Figure 7. Hardness maps of the T-butt joint AA2024-T4 and AA7075-T6 friction stir welded at different rotational speeds of (**a**) 400 rpm, (**b**) 600 rpm and (c) 800 rpm.

The hardness maps show higher hardness at the SZ of the mixed alloys compared to the skin base material AA2024-T4 at the rotational speeds of 400 and 600 rpm, as shown in Figure 7a,b. This increase in the hardness at the SZ can be attributed to two reasons: (i) high proportion of the AA7075-T6 (high hardness alloy) in the SZ and (ii) the finer grain size (dynamically recrystallized) at SZ compared to the base material. According to Hall-Petch relation [29], smaller grain size leads to harder material property. The increasing amount of AA7075-T6 material in the microstructure of the stir zone seems to be also a reason for the higher hardness measured within the nugget zone of the T-butt joint. The highest rotational speed of 800 rpm leads to higher heat generation and slower cooling rate and then causes the formation of relatively coarse grains. And also, leads to the formation coarser precipitates during re-precipitation in the cooling cycle, thus result in lower hardness in the SZ and TMAZ compared to other joints welded at 400 and 600 rpm. The lower hardness observed in the HAZ of both advancing side (AS) and retreating side (RS) of the joints are results of coarsening and dissolution of strengthening precipitates [1–3,13,30]. No obvious differences are seen in the mean values of hardness of HAZ at both the AS and RS of all the welded joints. For the T-joint welded at 800 rpm, a wider softened area evidenced with slightly lower hardness in the SZ as well as in the HAZ of both sides. The slight lower hardness in the SZ and the wider softened region in this joint can be attributed to the high heat input experienced which affects the precipitates coarsening and/or dissolution. The softening behavior at HAZ with a minimum hardness ranging from 98 to 105 Hv is observed. The observed best joint hardness at a rotational speed of 600 rpm which gives proper mixing with adequate generated heat and cooling rate at travel speed of 50 mm/min.

The hardness profile of friction stir welded of heat-treatable AA2024 and AA7075 aluminum alloys depends strongly on the grain size structure, and the precipitates in terms of type, amount, morphology and distribution [4,5]. The difference in grain shape and size of the two base materials is surely important. Furthermore, the amount of solute atoms (Zn, Mg and Cu) of the mixed two materials is also an essential reason of the high hardness in the nugget zone.

3.4. SEM Investigation

As noted by optical examinations of the welded joints (Figure 8) and confirmed with the two modes SEM investigations (Figure 9) the initial coarse elongated grains of the as-received AA2024-T4 and AA7075-T6 base alloys are thermomechanically deformed during FSW and resulted in new recrystallized equiaxed fine grain structures. The microstructures display also fine precipitates decorated the grain boundaries of the fine grains of both AA2024 and AA7075, Figure 8a–f. In addition, coarse and fragmented precipitates are detected in the SZ of the T-joints as given in VCD-mode images Figure 8b,e respectively. Moreover, the friction stir welded AA7075 alloy shows finer recrystallize grain size than that of AA2024 alloy, Figure 8d.

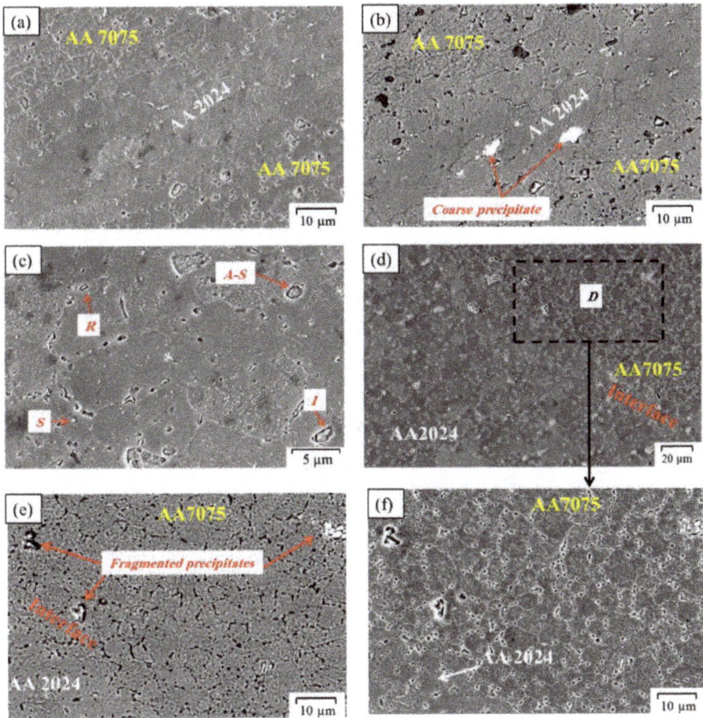

Figure 8. SEM microstructure of Friction stir welded T-joints at different rotational speeds: (**a**) ETD mode at 400 rpm, (**b**) in VCD mode, (**c**) ETD at 600 rpm, (**d**) low magnification in ETD at 800 rpm, (**e**) in VCD and (**f**) high magnification of the selected area (D) in (**d**).

The FSW and FSP lead to significant change in size and morphology of the precipitates. It can be effectively refined and redistributed both soluble and insoluble particles in precipitation strengthening non-ferrous alloys [31,32]. The precipitates in different regions of FSW zones are strongly function of the local thermo-mechanical cycles. The evolved microstructures in the SZ of the friction stir welded dissimilar T-joints at different rotational speeds are expected to be very complex. In fact, the precipitation phenomena need more research and still out of completely understood [33]. The current study is roughly an important attempt to evaluate morphologies and types of formed precipitates in the SZ of the Friction stir welded T-butt joints of AA2024-T4 and AA7075-T6. This attempt is extended to relate the precipitates influence on the recrystallization process during FSW. The precipitates display four types of morphologies: irregular, almost-spherical, spherical and rod-like and marked by symbols *I*, *A–S*, *S* and *R*, respectively. These morphologies are shown in the SZ of all the welded joints at the different rotational speeds (400–800 rpm) and typically observed as given for example in Figure 8c.

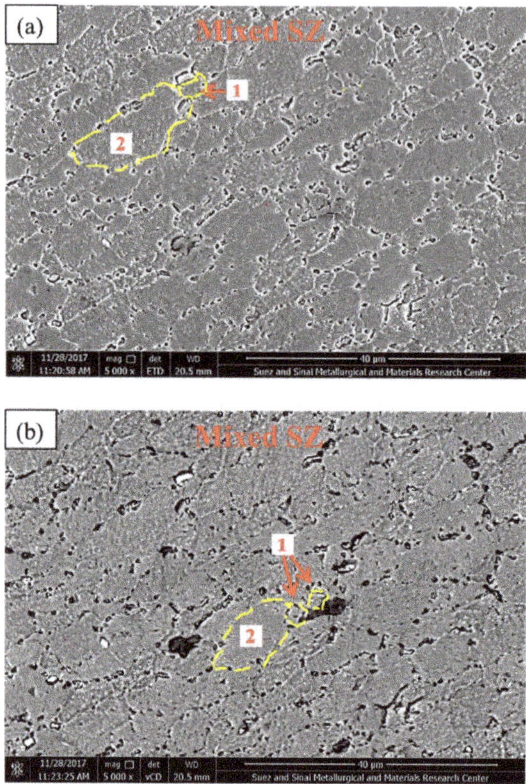

Figure 9. SEM images of T-joint at 400 rpm (**a**) ETD mode and (**b**) VCD mode.

Generally, the 2024 and 7075 Al-alloys have unstable nature of precipitates. This means that, the precipitates can coarsen and transformed into more stable precipitates, and/or undergo partial or complete dissolution during suffering high temperatures and may reappear in various morphologies, amounts and crystal structures during cooling.

Precipitates may accelerate or retard the recrystallization process depending on their size and volume fraction. Coarse particles can intensify the driving force of recrystallization and act as nucleation sites, which is known as particle stimulated nucleation (PSN). In contrast, relatively finer particle can retard recrystallization by the particle pinning of the grain boundaries, which is referred to as "Zener Pinning" [34]. Figure 9 shows two examination modes SEM microstructure of Friction stir welded sample at 400 rpm, where location 1 represents fine Dynamically recrystallized (DRX) grains adjacent to coarse particle. However, location 2 denotes coarse DRX grains at a distance from the coarse particles. PSN mechanism is observed to be more effective when the precipitates are present near the grain boundaries rather than within the grain, where the local strain gradient is relatively lower [34,35].

Coarse thermal stable and non-deformable particles interact with moving dislocations producing dislocation piled-ups or loops at the particle-matrix interface, which generates a local strain mismatch at the interface. This local strain field then eases the operation of different slip systems at the interface, which causes the surrounding matrix to rotate to fit the external matrix, creating the so-called particle deformation zone (PDZ) [35,36]. Further hot deformation, of such PDZs makes them favorable nucleation sites for recrystallization. Due to the high dislocation density in a PDZ, recovery takes place by the formation of new sub-grains, which then increases the misorientation by absorbing more dislocations. When this misorientation achieves from 10 to 15°, a likely recrystallization nuclei has evolved that may then grow into the surrounding matrix [37].

3.5. BM Grain Structure and Texture

For all the welded T-joint, the SZ is mixed regions of AA2024-T4 and AA7075-T6. Degree of mixing enhances as the plasticity of the two Al-alloys increases during the FSW. The plasticity of both alloys is related to the amount of heat input introduced to the work-piece. Actually, plasticity increases as the rotational welding speeds increase at the constant welding parameters. Selected areas of both Al-alloys in the SZ have been analyzed in terms of inverse pole figure (IPF) coloring maps and texture and compared with the as-received features of BMs. Figure 10 shows the IPF coloring maps with respect to the rolling direction (RD) and grain boundary (GB) maps of the BMs (a) AA2024-T4 and (b) AA7075-T6 Al alloys. It can be observed that the BMs are almost similar in terms of grain structure and low angle boundaries (LABs) distributions. The grain shapes are that of the typical rolled materials characterized by large and elongated grains. The GB maps are almost free of LABs, which indicate that both materials were in full-recovered state. The AA2024-T4 Al-alloy showed an average grain size of 47 μm and the AA7075-T6 Al-alloy revealed relatively finer average grain size of 43 μm, as obtained from the grain size distribution histogram shown in Figure 11a,b, respectively. The misorientation angle distribution displays very low density of LABs for the as-received materials, as shown in Figure 11a,b. In terms of texture the 101 and 111 pole figures (PFs) illustrated in Figure 12 fairly shows the rolled texture of the fcc metals although the number of grains obtained in the analyzed areas are limited number.

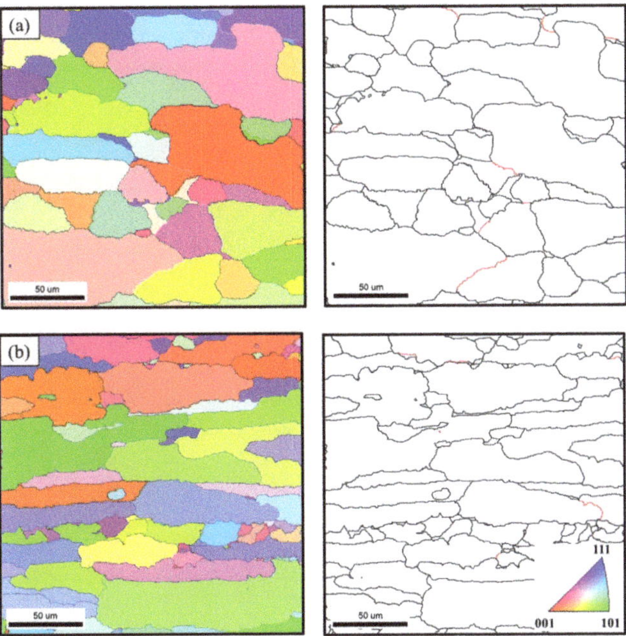

Figure 10. IPF coloring OIM maps relative to rolling direction (RD) and the grain boundary maps with high angle grain boundaries (HAGBs) > 15 in black lines and low angle boundaries (LAGBs) < 15 in red lines, (**a**) AA2024-T4 and (**b**) AA7075-T6.

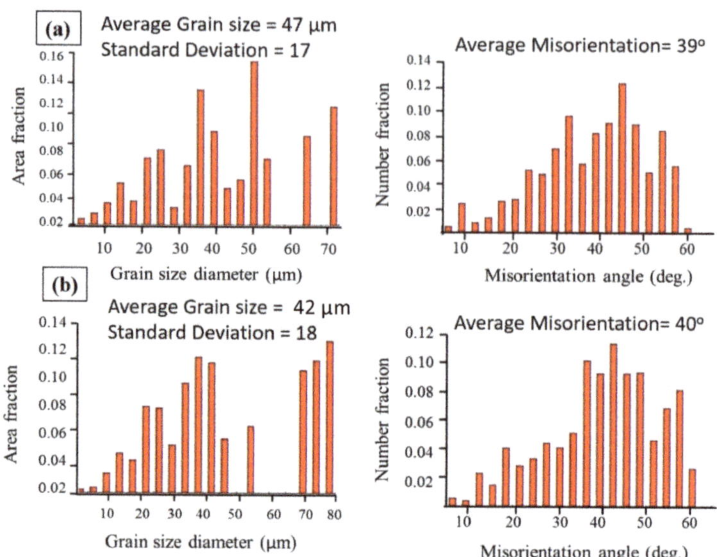

Figure 11. Grain size distribution and misorientation angle distribution of BMs, where (**a**) AA2024-T4and (**b**) AA7075-T6.

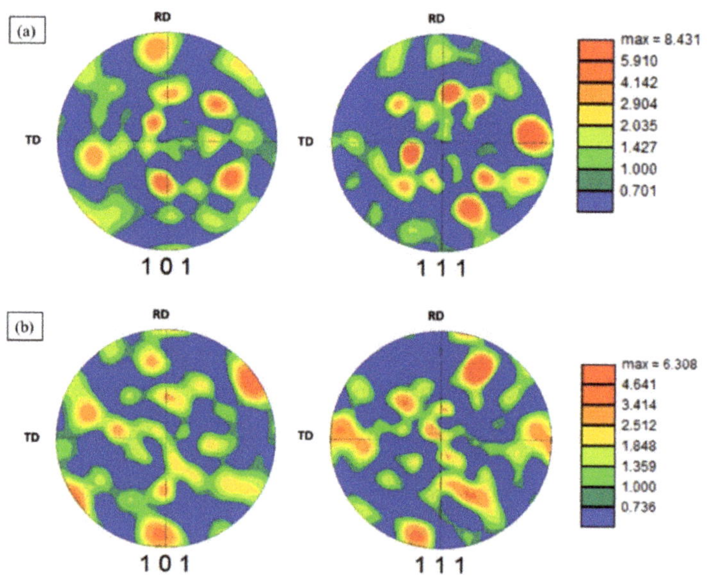

Figure 12. 101 and 111 pole figures (PFs) of BMs, where (**a**) AA2024-T4 and (**b**) AA7075-T6.

3.6. Grain Structure and Texture of Friction Stir Welded T-Butt Joints

The SZ of the T-butt joint Friction stir welded AA2024-T4 and AA7075-T6 has been investigated using EBSD by acquiring data at AA2024-T4 and AA7075-T6 regions. The IPF coloring maps with respect to the ND and their corresponding grain boundary maps are presented in Figure 13. It can be observed that the grain structure in both regions is dynamically recrystallized of equiaxed grains. Geometric dynamic recrystallization has been suggested by number of researchers to be the main recrystallization mechanism in the SZ of the welded aluminum [38,39]. In terms of grain size, it can be noted that in comparison to the BMs it is extremely fine, however a variation in the grain size can be observed between AA2024-T4 and AA7075-T6 regions, Figure 13a,b, respectively. The AA2024-T4 region exhibits coarser grain size relative to the AA7075-T6. The GB maps of the EBSD data obtained from the SZ of the T-butt Friction stir welded joint between AA2024-T4 and AA7075-T6 are presented in Figure 14. The maps are consisted of HAGB > 15° of the major density and a lower density of the LAB (5–15°). The examined SZ areas of AA2024-T4 and AA7075-T6 dominate mainly by HAGBs > 15° with a fraction of 0.801 and 0.924, respectively. While the value of LAGBs 5–15° displays with a fraction of 0.113 and 0.071, respectively.

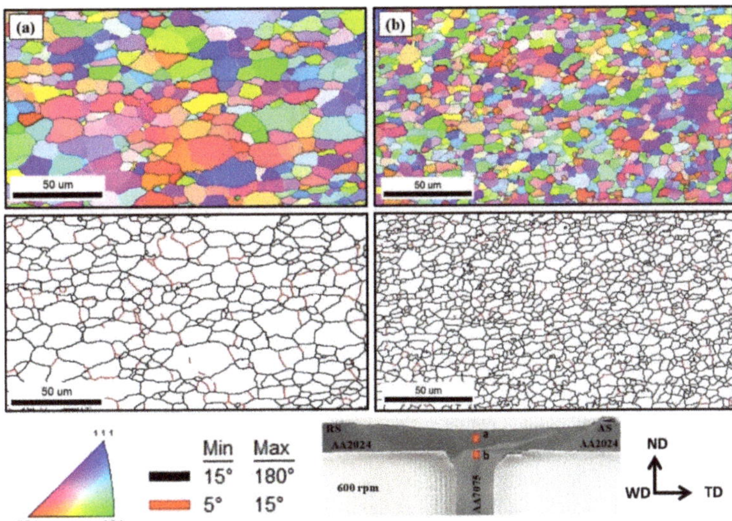

Figure 13. IPF coloring maps and their corresponding grain boundary maps for SZ of T-butt Friction stir welded joint AA2024-T4 and AA7075-T6 produced at 600 rpm, where (**a**) at AA2024-T4 region and (**b**) at AA7075-T6 region in the joint produced at 600 rpm. The positions at which the EBSD data collected are indicated with two red rectangles on the macrograph of the joint below the figure. The IPF maps key coloring, grain boundary legend and FSW axes are also presented below the figure.

This data is presented as misorientation angle distribution histogram for the two alloys in Figure 14a,b with the corresponding grain size distribution. The average grain size of the AA2024-T4 is about 11 μm while that of the AA7075-T6 is about 6 μm as given in Figure 14a,b, respectively. This variation can be attributed to the variation in the chemical composition of the two alloys and in consequently to the different types of precipitates in the two alloys. Ahmed et al. [2] in their study of similar and dissimilar Friction stir welded AA7075 and AA5083 reported a variation in the resulted grain size between the AA7075 and the AA5083 inside the same SZ of the dissimilar joints. In terms of the misorientation angle distribution which clearly showing an increase in the LABs density for the data obtained in the AA2024-T4. In terms of texture, the 101 and 111 PFs of this data also calculated and presented in Figure 15a,b. The two alloys show the simple shear texture with only about two times random. This is typical type of texture reported to be obtained in the SZ of the Friction stir welded aluminum alloys [27,40–42]. Crystallographic texture affects the mechanical properties of materials as it expresses the orientation of the crystallographic plans relative to the material reference axes [13]. This directly affects the alignment of the slip systems relative to the axis of the applied load. This why the isotropic materials that of random texture has the same mechanical properties in all direction while in contrast the anisotropic material that of strong texture its mechanical properties varied based on the direction of the applied load relative to the material reference axes [43–46].

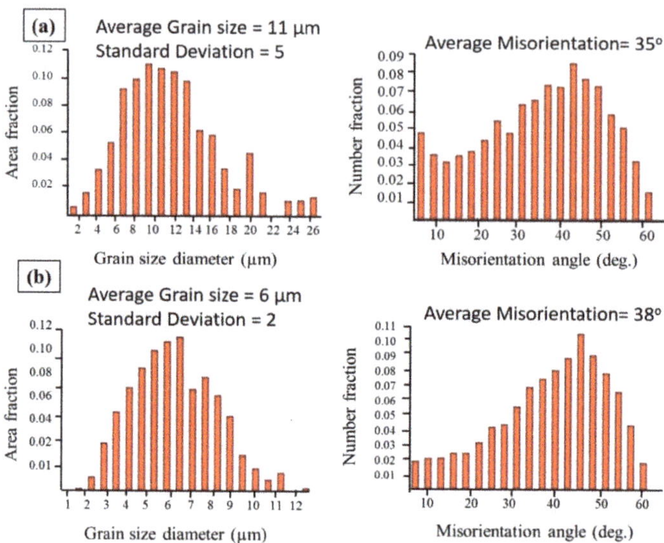

Figure 14. (**a**) Grain size and misorientation angle distributions of the dissimilar T-butt Friction stir welded joint, where (**a**) AA2024-T4 region and (**b**) AA7075-T6 region in the joint produced at 600 rpm.

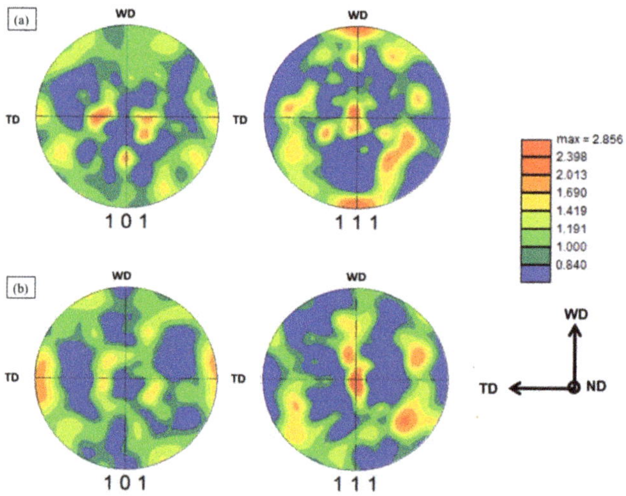

Figure 15. 101 and 111 PFs calculated from the data from SZ of T-butt Friction stir welded joint presented in Figure 13, where (**a**) AA2024-T4 and (**b**) AA7075-T6.

3.7. T-Butt Joints Tensile Properties

Figures 16 and 17 show photographs of typical fracture locations of Friction stir weldedAA2024-T4 and AA7075-T6 T-butt joints, which were pulled along skin and stringer, respectively. For the welded specimens at 400 rpm pulled along skin (Figure 17a) fracture occurs at the SZ in the RS. This may be attributed to reduction of hardness at the interface between AA2024-T4 skin and AA7075-T6 stringer (Figure 7a) for the T-joint as a result of insufficient heat input for good mixing in the SZ of the dissimilar aluminum alloys. For 600 rpm, fracture occurs at HAZ region in the RS of AA2024-T4 skin, as shown in Figure 17b. This in agreement with the distribution of the lowest value of hardness in hardness map (Figure 7b) of the joint. For 800 rpm, fracture occurs at TMAZ of RS of the skin, as shown in Figure 17c. This may be ascribed to the lowering hardness map (Figure 7c) in the weld joint as a result of higher heat input.

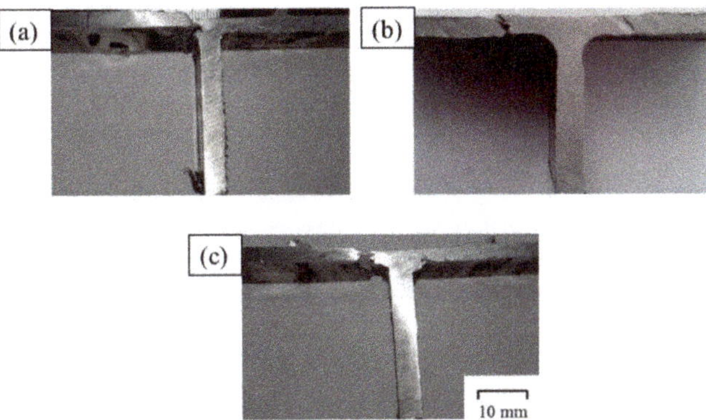

Figure 16. Photographs of fracture locations of T-butt welded specimens along skin. (**a**) 400 rpm, (**b**) 600 rpm, and (**c**) 800 rpm.

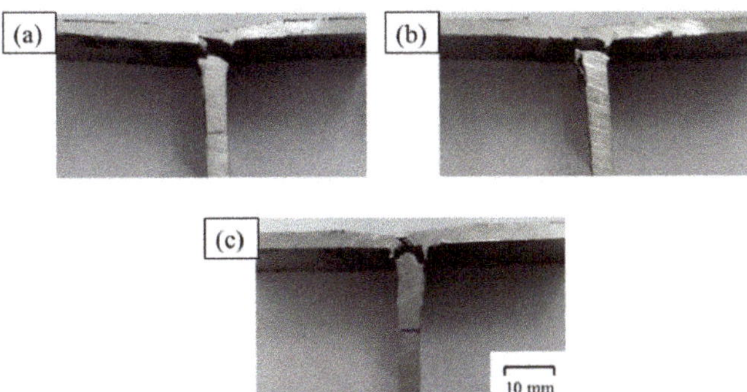

Figure 17. Photographs of fracture locations of T-butt welded specimens along stringer. (**a**) 400 rpm, (**b**) 600 rpm, and (**c**) 800 rpm.

Furthermore, the fracture locations of T-butt welded specimens along stringer are shown in Figure 17. For 400 rpm, Figure 17a, fracture occurs at center of the SZ. It coincides with the variation of hardness values in hardness map (Figure 7a) due to insufficient mixing of AA2024-T4 and AA7075-T6 in the SZ at the lowest rotational speed. For 600 rpm, Figure 17b, the fracture occurs at HAZ in stringer part. It is in good agreement with the lowest hardness values in hardness map (Figure 7b) of stringer AA7075-T6. For 800 rpm, Figure 17c, the fracture occurs at SZ. It coincides with the reduction of hardness values in SZ in hardness map compared to BM AA2024-T4 (Figure 7c).

Joint Efficiency

Joint efficiency of the welded material means the ratio of ultimate tensile strength (UTS) of the joint to the UTS of its base material. It gives an important perception for the joint quality and the mechanical property of the joint. Thus, tensile tests of the produced

T-butt joints were carried out along the skin and stringer directions. Figure 18 shows the joint efficiencies along the skin and stringer of the AA2024 and AA7075 T-butt joints produced using the welding parameters of 400, 600 and 800 rpm rotational speeds, and a constant travel speed of 50 mm/min. For the skin tensile test (perpendicular to welding direction), Figure 18a shows no significant difference in the efficiency values between joints welded at 400 and 600 rpm, while the rotational speed increases from 600 to 800 rpm the efficiency decreases from 83.40 % to reach 72.02 % of the base metal AA2024-T4. Tensile properties in the direction of stringer (perpendicular to welding direction) are given in Figure 18b. It can be seen that the joint efficiencies along stringer increases with increasing tool rotational speeds. The T-butt joints produced at 800 rpm displays the highest joint efficiency of 52.30% of the base metal AA2024-T4. It can be noted from Figure 18a,b, that the joint efficiencies of all produced T-butt joints along the stringer are lower than those of all joints along the skin. This trend is in good agreement with the results obtained by Cui et al. [37] for the similar AA6061-T4 T-butt joints (C-series). Furthermore, they reported that all T-joint efficiencies showed tendency to enhancement with increasing the rotational speeds at constant travel speed.

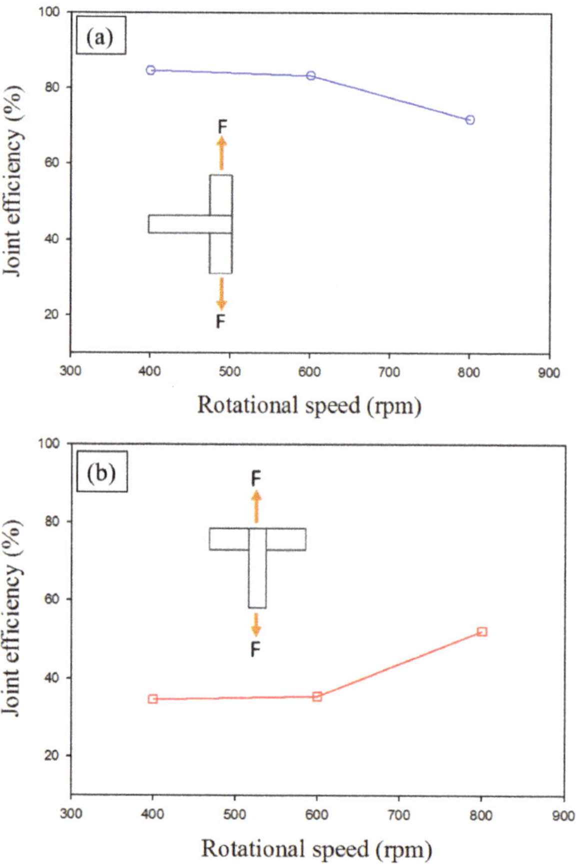

Figure 18. Joint efficiencies against rotational speeds for the T-butt joints of 2024-7075 Al alloys; (**a**) tensile tested along the skin and (**b**) along the stringer directions.

It can be noted that or the joint welded at 800 rpm the joint efficiency along the skin direction is the lowest and for the stringer direction is the highest as shown in

Figure 18a,b can be due to two reasons. First, the reduction in the hardness relative to the other two revolution speeds as shown in Figure 7, due to the increased heat input. Second, the reduced skin thickness due to the pressing effect of the shoulder (Figure 5), this is not the case for tension in the stinger direction. Moreover, at 800 rpm the pressing by the shoulder and the increased softening has generated larger fillet than the other two lower rotation speeds (400 and 600 rpm). This large fillet represents more support for the stringer leading to the highest joint efficiency.

The authors are still not satisfied with the attained joint efficiency especially in the skin direction, so that further welding trials should be performed with other welding parameters to overcome the reduction in the skin thickness due to the applied pressure by the tool shoulder.

4. Conclusions

1. Sound T-joints are successfully produced with three plates in T-butt configuration; (two AA2024 plates as skin and one AA7075 plate as stringer) at different rotation speeds and constant traverse speed using FSW tool of large diameter pin of 12 mm to enable one track FSW of the three pieces.
2. All T-butt joint specimens exhibit a W-shaped appearance hardness profiles along the center line of the skin of the welds and reaches the highest level at 600 rpm.
3. Significant grain refining in stir zone of FSW welds is accompanied with dynamic recrystallization compared with the as-received base metals. EDS analyses of the stir zones of the T-joints show four types of precipitates; $Al_6(Mn,Fe,Cu)$, Al_2Cu, Al_2CuMg and Al_7Cu_2Fe.
4. The differences between the AA2024-T4 and AA7075-T6 aluminum alloys in terms of chemical composition, starting grain structure and type of precipitates have strong impact on the recrystallization process, joint efficiency and failure mode of the Friction stir welded T-butt joints.
5. Fracture surfaces along the skin and stringer of all T-welds indicate that the joint failed through mixed modes of fracture, mainly ductile by locally deformed aluminum base and partially brittle via hard phase particles.

Author Contributions: Conceptualization, M.M.Z.A., M.M.E.-S.S., Z.A.Z., and R.M.R.; methodology, Z.A.Z., R.M.R., M.M.Z.A. and M.M.E.-S.S.; validation, S.A., N.A.A. and M.M.Z.A.; formal analysis, S.A. and M.M.Z.A.; investigation, M.M.E.-S.S., S.A. and M.M.Z.A.; writing original draft preparation, M.M.E.-S.S.; M.M.Z.A., R.M.R. and Z.A.Z.; writing—review and editing, S.A., M.M.Z.A. and N.A.A.; project administration, M.M.Z.A., N.A.A. and M.M.E.-S.S. All authors have read and agreed to the published version of the manuscript.

Funding: This research received no external funding.

Institutional Review Board Statement: Not applicable.

Informed Consent Statement: Not applicable.

Data Availability Statement: The data presented in this study are available on request from the corresponding author. The data are not publicly available due to the extremely large size.

Conflicts of Interest: The authors declare no conflict of interest.

References

1. Ahmed, M.M.Z.; Wynne, B.P.; Rainforth, W.M.; Threadgill, P.L. Microstructure, crystallographic texture and mechanical properties of friction stir welded AA2017A. *Mater. Charact.* **2012**, *64*, 107–117. [CrossRef]
2. Ahmed, M.M.Z.; Ataya, S.; El-Sayed Seleman, M.M.; Ammar, H.R.; Ahmed, E. Friction stir welding of similar and dissimilar AA7075 and AA5083. *J. Mater. Process. Technol.* **2017**, *242*, 77–91. [CrossRef]
3. Ahmed, M.M.Z.; Ahmed, E.; Hamada, A.S.; Khodir, S.A.; El-Sayed Seleman, M.M.; Wynne, B.P. Microstructure and mechanical properties evolution of friction stir spot welded high-Mn twinning-induced plasticity steel. *Mater. Des.* **2016**, *91*, 378–387. [CrossRef]
4. Threadgill, P.L. Terminology in friction stir welding. *Sci. Technol. Weld. Join.* **2007**, *12*, 357–360. [CrossRef]

5. Mishra, R.S.; Ma, Z.Y. Friction stir welding and processing. *Mater. Sci. Eng. R* **2005**, *50*, 1–78. [CrossRef]
6. Refat, M.; Elashery, A.; Toschi, S.; Ahmed, M.M.Z.; Morri, A.; Mahallawi, I.E.; Ceschini, L. Microstructure, Hardness and Impact Toughness of Heat-Treated Nanodispersed Surface and Friction Stir-Processed Aluminum Alloy AA7075. *Proc. J. Materials Eng. Perform.* **2016**, *25*, 5087–5101. [CrossRef]
7. Oliveira, J.P.; Barbosa, D.; Fernandes, F.M.B.; Miranda, R.M. Tungsten inert gas (TIG) welding of Ni-rich NiTi plates: Functional behavior. *Smart Mater. Struct.* **2016**, *25*, 03LT01. [CrossRef]
8. Costa, A.M.S.; Oliveira, J.P.; Pereira, V.F.; Nunes, C.A.; Ramirez, A.J.; Tschiptschin, A.P. Ni-based Mar-M247 superalloy as a friction stir processing tool. *J. Mater. Process. Technol.* **2018**, *262*, 605–614. [CrossRef]
9. Oliveira, J.P.; Duarte, J.F.; Inácio, P.; Schell, N.; Miranda, R.M.; Santos, T.G. Production of Al/NiTi composites by friction stir welding assisted by electrical current. *Mater. Des.* **2017**, *113*, 311–318. [CrossRef]
10. Ahmed, M.M.Z.; Ataya, S.; Seleman, M.M.E.; Mahdy, A.M.A.; Alsaleh, N.A.; Ahmed, E. Heat Input and Mechanical Properties Investigation of Friction Stir Welded AA5083/AA5754 and AA5083/AA7020. *Metals (Basel)* **2021**, 68. [CrossRef]
11. Boldsaikhan, E.; Fukada, S.; Fujimoto, M.; Kamimuki, K.; Okada, H. Refill friction stir spot welding of surface-treated aerospace aluminum alloys with faying-surface sealant. *J. Manuf. Process.* **2019**, *42*, 113–120. [CrossRef]
12. Li, S.; Chen, Y.; Zhou, X.; Kang, J.; Huang, Y.; Deng, H. High-strength titanium alloy/steel butt joint produced via friction stir welding. *Mater. Lett.* **2019**, *234*, 155–158. [CrossRef]
13. Ahmed, M.M.Z.; Wynne, B.P.; Rainforth, W.M.; Addison, A.; Martin, J.P.; Threadgill, P.L. Effect of Tool Geometry and Heat Input on the Hardness, Grain Structure, and Crystallographic Texture of Thick-Section Friction Stir-Welded Aluminium. *Metall. Mater. Trans. A* **2019**, *50*, 271–284. [CrossRef]
14. Ahmed, M.M.Z.; Wynne, B.P.; Martin, J.P. Effect of friction stir welding speed on mechanical properties and microstructure of nickel based super alloy Inconel 718. *Sci. Technol. Weld. Join.* **2013**, *18*, 680–687. [CrossRef]
15. Acerra, F.; Buffa, G.; Fratini, L.; Troiano, G. On the FSW of AA2024-T4 and AA7075-T6 T-joints: An industrial case study. *Int. J. Adv. Manuf. Technol.* **2010**, *48*, 1149–1157. [CrossRef]
16. Shinoda, T.; Suzuki, J. Observation of metal flow phenomenon of lap joints during friction stir welding. *Mater. Sci. Forum* **2004**, *449–452*, 421–424. [CrossRef]
17. Argade, G.R.; Shukla, S.; Liu, K.; Mishra, R.S. Friction stir lap welding of stainless steel and plain carbon steel to enhance corrosion properties. *J. Mater. Process. Technol.* **2018**, *259*, 259–269. [CrossRef]
18. Babu, S.; Janaki Ram, G.D.; Venkitakrishnan, P.V.; Reddy, G.M.; Rao, K.P. Microstructure and Mechanical Properties of Friction Stir Lap Welded Aluminum Alloy AA2014. *J. Mater. Sci. Technol.* **2012**, *28*, 414–426. [CrossRef]
19. Scialpi, A.; De Filippis, L.A.C.; Cavaliere, P. 39-Influence of shoulder geometry on microstructure and mechanical properties of friction stir welded 6082 aluminium alloy. *Mater. Des.* **2007**, *28*, 1124–1129. [CrossRef]
20. Arora, K.S.; Pandey, S.; Schaper, M.; Kumar, R. Microstructure Evolution during Friction Stir Welding of Aluminum Alloy AA2219. *J. Mater. Sci. Technol.* **2010**, *26*, 747–753. [CrossRef]
21. Sadeesh, P.; M, V.K.; Rajkumar, V.; Avinash, P.; Arivazhagan, N.; Ramkumar, K.; Narayanan, S. Studies on friction stir welding of AA 2024 and AA 6061 dissimilar metals. *Procedia Eng.* **2014**, *75*, 145–149. [CrossRef]
22. Manuel, N.; Galv, I.; Leal, R.M.; Costa, J.D. Nugget Formation and Mechanical Behaviour of Friction Stir Welds of Three Dissimilar Aluminum Alloys. *Materials* **2020**, *13*, 2664. [CrossRef] [PubMed]
23. Dursun, T.; Soutis, C. Recent developments in advanced aircraft aluminium alloys. *Mater. Des.* **2014**, *56*, 862–871. [CrossRef]
24. Tavares, S.M.O.; dos Santos, J.F.; de Castro, P.M.S.T. Friction stir welded joints of Al-Li Alloys for aeronautical applications: Butt-joints and tailor welded blanks. *Theor. Appl. Fract. Mech.* **2013**, *65*, 8–13. [CrossRef]
25. Dubourg, L.; Merati, A.; Gallant, M.; Jahazi, M. Manufacturing of aircraft panels by friction stir lap welding of 7075- T6 stringers on 2024-T3 skin: Process optimisation and mechanical properties. In Proceedings of the 7th International Friction Stir Welding Symposium; TWI: Osaka, Japan, 2008.
26. Cui, L.; Yang, X.; Zhou, G.; Xu, X.; Shen, Z. Characteristics of defects and tensile behaviors on friction stir welded AA6061-T4. *Mater. Sci. Eng. A* **2012**, *543*, 58–68. [CrossRef]
27. Ahmed, M.M.Z.; Wynne, B.P.; El-Sayed Seleman, M.M.; Rainforth, W.M. A comparison of crystallographic texture and grain structure development in aluminum generated by friction stir welding and high strain torsion. *Mater. Des.* **2016**, *103*, 259–267. [CrossRef]
28. Saravanan, V.; Rajakumar, S.; Banerjee, N.; Amuthakkannan, R. Effect of shoulder diameter to pin diameter ratio on microstructure and mechanical properties of dissimilar friction stir welded AA2024-T6 and AA7075-T6 aluminum alloy joints. *Int. J. Adv. Manuf. Technol.* **2016**, 3637–3645. [CrossRef]
29. Hansen, N. Hall—Petch relation and boundary strengthening. *Scr. Mater.* **2004**, *51*, 801–806. [CrossRef]
30. Ahmed, M.M.Z.; Wynne, B.P. Post Weld Heat Treatment of Friction Stir Welded AA2017. In *Proceedings of the Light Metals 2012*; TMS: Warrendale, PA, USA; pp. 12–14. [CrossRef]
31. Pasebani, S.; Charit, I.; Mishra, R.S. Effect of tool rotation rate on constituent particles in a friction stir processed 2024Al alloy. *Mater. Lett.* **2015**, *160*, 64–67. [CrossRef]
32. Palanivel, S.; Arora, A.; Doherty, K.J.; Mishra, R.S. A framework for shear driven dissolution of thermally stable particles during friction stir welding and processing. *Mater. Sci. Eng. A* **2016**, *678*, 308–314. [CrossRef]

33. Su, J.-Q.; Nelson, T.; Mishra, R.; Mahoney, M. Microstructural investigation of friction stir welded 7050-T651 aluminium. *Acta Mater.* **2003**, *51*, 713–729. [CrossRef]
34. Schlegel, S.M.; Hopkins, S.; Frary, M. Effect of grain boundary engineering on microstructural stability during annealing. *Scr. Mater.* **2009**, *61*, 88–91. [CrossRef]
35. Tangen, S.; Sjølstad, K.; Furu, T.; Nes, E. Effect of Concurrent Precipitation on Recrystallization and Evolution of the P-Texture Component in a Commercial Al-Mn Alloy. *Metall. Mater. Trans. A* **2010**, *41*, 2970–2983. [CrossRef]
36. Robson, J.D.; Henry, D.T.; Davis, B. Particle effects on recrystallization in magnesium—manganese alloys: Particle-stimulated nucleation. *Acta Mater.* **2009**, *57*, 2739–2747. [CrossRef]
37. Cai, B.; Adams, B.L.; Nelson, T.W. Relation between precipitate-free zone width and grain boundary type in Relation between precipitate-free zone width and grain boundary type in 7075-T7 Al alloy. *Acta Mater.* **2007**, *55*, 1543–1553. [CrossRef]
38. Mcnelley, T.R.; Swaminathan, S.; Su, J.Q. Recrystallization mechanisms during friction stir welding / processing of aluminum alloys. *Scr. Mater.* **2008**, *58*, 349–354. [CrossRef]
39. Etter, A.L.; Baudin, T.; Fredj, N.; Penelle, R. Recrystallization mechanisms in 5251 H14 and 5251 O aluminum friction stir welds. *Mater. Sci. Eng. A* **2007**, *445–446*, 94–99. [CrossRef]
40. Ahmed, M.Z.; Wynne, B.P.; Rainforth, W.M.; Martin, J.P. Crystallographic Texture Investigation of Thick Section Friction Stir Welded AA6082 and AA5083 Using EBSD. *Key Eng. Mater.* **2018**, *786*, 44–51. [CrossRef]
41. Ahmed, M.M.Z.; Wynne, B.P.; Rainforth, W.M.; Threadgill, P.L. Quantifying crystallographic texture in the probe-dominated region of thick-section friction-stir-welded aluminium. *Scr. Mater.* **2008**, *59*, 507–510. [CrossRef]
42. Ahmed, M.M.Z.; Wynne, B.P.; Rainforth, W.M.; Threadgill, P.L. Through-thickness crystallographic texture of stationary shoulder friction stir welded aluminium. *Scr. Mater.* **2011**, *64*, 45–48. [CrossRef]
43. Divinski, S.V.; Dnieprenko, V.N. Texture contribution to the anisotropy of physical properties. *J. Phys. Condens. Matter* **1994**, *6*, 8503–8512. [CrossRef]
44. Fouad, D.M.; El-Garaihy, W.H.; Ahmed, M.M.Z.; El-Sayed Seleman, M.M.; Salem, H.G. Influence of multi-channel spiral twist extrusion (MCSTE) processing on structural evolution, crystallographic texture and mechanical properties of AA1100. *Mater. Sci. Eng. A* **2018**, *737*, 166–175. [CrossRef]
45. Randle, V. Application of electron backscatter diffraction to grain boundary characterisation. *Int. Mater. Rev.* **2004**, *49*, 1–11. [CrossRef]
46. Rollett, A.D. Crystallographic Texture Change during Grain Growth. *JOM* **2004**, *56*, 63–68. [CrossRef]

Article

Grain Structure, Crystallographic Texture, and Hardening Behavior of Dissimilar Friction Stir Welded AA5083-O and AA5754-H14

Mohamed Mohamed Zaky Ahmed [1,2,*], Sabbah Ataya [2,3], Mohamed Mohamed El-Sayed Seleman [2], Tarek Allam [2,4], Naser Abdulrahman Alsaleh [3] and Essam Ahmed [2]

1 Department of Mechanical Engineering, College of Engineering at Al Kharj, Prince Sattam Bin Abdulaziz University, Al Kharj 16273, Saudi Arabia
2 Department of Metallurgical and Materials Engineering, Faculty of Petroleum and Mining Engineering, Suez University, Suez 43512, Egypt; sabbah.ataya@suezuniv.edu.eg (S.A.); mohamed.elnagar@suezuniv.edu.eg (M.M.E.-S.S.); tarek.allam@iehk.rwth-aachen.de (T.A.); essam.ahmed@suezuniv.edu.eg (E.A.)
3 Department of Mechanical Engineering, College of Engineering, Al Imam Mohammad Ibn Saud Islamic University, Riyadh 11432, Saudi Arabia; naalsaleh@imamu.edu.sa
4 Steel Institute (IEHK), RWTH Aachen University, D-52056 Aachen, Germany
* Correspondence: moh.ahmed@psau.edu.sa; Tel.: +966-011-588-1200

Abstract: This work investigated the effect of friction stir welding (FSW) tool rotation rate and welding speed on the grain structure evolution in the nugget zone through the thickness of the 10 mm thick AA5083/AA5754 weldments. Three joints were produced at different combinations of FSW parameters. The grain structure and texture were investigated using electron backscattering diffraction (EBSD). In addition, both the hardness and tensile properties were investigated. It was found that the grain size varied through the thickness in the nugget (NG), which was reduced from the top to the base in all welds. Reducing the rotation rate from 600 rpm to 400 rpm at a constant welding speed of 60 mm/min reduced the average grain size from 33 µm to 25 µm at the top and from 19 µm to 12 µm at the base. On the other hand, the increase of the welding speed from 20 mm/min to 60 mm/min had no obvious effect on the average grain size. This implied that the rotation rate was more effective in grain size reduction than the welding speed. The texture was the mainly simple shear texture that required some rotations to obtain the ideal simple shear texture. The hardness distribution, mapped for the nugget zone, and the parent alloys indicated a diffused softened welding zone. The heating effect of the pressure and rotation of the pin shoulder and the heat input parameter (ω/v) on the hardness value of the nugget zone were dominating. Tensile stress-strain curves of the base alloys and that of the FSWed joints were evaluated and presented. Moreover, the true stress-true strain curves were determined and described by the empirical formula after Ludwik, and then the materials strengthening parameters were determined. The tensile specimens of the welded joint at a revolution speed of 400 rpm and travel speed of 60 mm/min possessed the highest strain hardening parameter (n = 0.494).

Keywords: friction stir welding; dissimilar welding; aluminum; mechanical properties; microstructure; texture; fracture

Citation: Ahmed, M.M.Z.; Ataya, S.; Seleman, M.M.E.; Allam, T.; Alsaleh, N.A.; Ahmed, E. Grain Structure, Crystallographic Texture, and Hardening Behavior of Dissimilar Friction Stir Welded AA5083-O and AA5754-H14. *Metals* **2021**, *11*, 181. https://doi.org/10.3390/met11020181

Academic Editors: João Pedro Oliveira and Zhi Zeng
Received: 31 December 2020
Accepted: 18 January 2021
Published: 20 January 2021

1. Introduction

Aluminum alloys have remained the prime selection in producing various components in many industries like aerospace, automotive, and shipbuilding because of their perfect strength to weight ratio [1–5]. AA5000-series alloys are characterized by a good strength-to-weight ratio and an appropriate corrosion resistance. However, they are difficult to join by conventional fusion welding techniques because of their dendritic structure, which seriously weakens the mechanical properties. Solid-state welding processes are appropriate joining for either similar or dissimilar aluminum alloys [6]. Resistance spot welding is considered one of the dominant solid-state welding processes in automotive constructions [7,8]. However, the use of a continuous welding line process instead of weld spots leads to higher structural stiffness and better crash performance [9]. Friction stir welding (FSW) of the AA5000 series represents a promising technique to obtain defect-free and sound joints, either in similar [10] and dissimilar [11–13] welding combinations. FSW can also be used effectively for the welding of different types of materials [14–17], and the same principle of FSW can be used for the development of metal matrix composites [18–22]. In FSW, a non-consumable rotating tool induces a stirring action until the tool shoulder contacts the top surface of the sheets with a given plunge depth, generating a large amount of frictional heat [23]. As the tool moves along the welding line, the blanks are joined through a solid-state process, owing to the severe plastic strain and the metal mixing across the weld. The weld zone undergoes a solid-state process promoted by the frictional heat between the wear-resistant welding tool and the materials to be joined. The plasticized zone is further extruded from the tool advancing side to the retreating side during its steady traversing along the joint line [24]. FSW process parameters influence the final joint quality and performance, including traverse welding speed; tool rotational speed, geometry, and shape; blank thickness; heat input; applied force; tilt angle; specimen preparation; sheet-rolling direction; plates/sheets metallurgical history. It has been demonstrated that, among process parameters, the tool rotational speed and traverse welding speed have a strong effect on heat generation, heat dissipation, and cooling rate. Hence, the microstructure and texture, and mechanical properties evolution of the FSW joints are significantly affected by traverse welding speed and tool rotational speed values [6,11,24–26]. For this reason, an accurate choice of the FSW process parameters and of the tool material and geometry is required. In fact, the joint mechanical properties can be optimized by increasing the tool rotational speed or by decreasing the traverse welding speed [27,28]. The excessive agglomerations and joints defects are produced when the high strength aluminum alloy on the advancing side (AS) of AA5052/AA5J32 is placed because of material flow limitation [29]. Both material flow and joint quality are more dependent on the FSW conditions and their effects on heat input and temperature distribution in weld nugget, regardless of base material (BM) placement [30]. During FSW, the heat generation is controlled by tool rotation and welding speed due to the material plastic flow [30–32]. However, very high rotation speeds lead to macroscopic defects because of the excessive heat input [24,33]. Due to FSW, three different metallurgical zones are usually recognized, namely, nugget zone (NZ), thermomechanically affected zone (TMAZ), and heat-affected zone (HAZ) [34]. In the NZ, the metal is in direct contact with the pin being continuously stirred during the passage of the rotating tool, thus creating the necessary strong bond between the two metals under the welding. Fast thermomechanical heating (peak temperature may reach 0.6 to 0.95 TM) and cooling occur, and they favor the occurrence of dynamic recrystallization (DRX) phenomena, generating fine grain structures in the form of onion rings [34,35]. From a microstructural viewpoint, the NZ is generally characterized by a fine or even very-fine equiaxed grained structure, as mentioned in [34]. In the TMAZ, the microstructure experiences a significant grain morphology and size modification. Because of the insufficient deformation strain, DRX does not occur in the TMAZ. In the third zone, HAZ, the materials are subjected to thermal cycles with no plastic deformation, and the microstructure has the same grain structures as the parent material (BM) [6,25]. The transients and gradients in strain, strain rate, and temperature are inherent in the thermomechanical cycles of FSW, which control and shape the characteristic microstructural zones of a typical FSW joint. During FSW, material flows in a complex, vortex-like pattern around the pin from the advancing side to the retreating side [14]. The high stacking fault energy metallic materials, such as aluminum, enhance the dynamic recovery (DRV) to occur during the hot working process [36,37]. As the DRV rate is increased, low-angle grain boundaries (LABs) are formed to minimize the dislocation forest/multiplication by the rearrangement of most of the dislocations. In DRX, new, dislocation-free grains form at high energy sites, such as prior grain boundaries, deformation band interfaces, or boundaries of newly recrystallized grains [38,39]. All the herein mentioned mechanisms of formation for sub-grains and grains (TMAZ) and recrystallized fine grains (NZ) are always also dependent on the material's initial metallurgical conditions and are subject to different FSW process and tool parameters. Thus, the aim of this work was to examine the effect of FSW tool rotation rate and the welding speed on the grain structure, texture, and mechanical properties of AA5083/AA5754. In this work, three FSWed AA5083/AA5754 joints (J1: 600 rpm and 60 mm/min, J2: 400 rpm and 60 mm/min, and J3: 400 rpm and 20 mm/min) were produced. Through the thickness of the produced joints, the grain structure and texture were investigated using EBSD. In addition, both the hardness distribution and tensile properties measurements were investigated. A full description of materials and experimental procedures is in Section 2. The results and discussion are presented in Section 3. The conclusion drawn from this work is in Section 4.

2. Materials and Methods

The materials under investigation are the aluminum alloys AA5083-O (AlMg4.5Mn0.6) and AA5754-H14 (AlMg3.1). More details on the full chemistry of both AA5083-O and AA5754-H14 are found in our previous work [11]. The temper designated "O" in AA5083-O state means in the annealed condition, which is applied to increase subsequent alloy workability. While "H14" in AA5754-H14 states that strain hardened-1/2 hard condition. The hardness values of AA5083-O and AA5754-H14 are 68 and 74 HV, respectively; the hardness values were measured using 1 Kg load and averaged out of 10 measurements. Vickers hardness tester machine (HWDV-75, TTS Unlimited, Osaka, Japan) was used. The alloys were supplied in the form of rolled plates of 10 mm thick. The FSW butt joints were designed to be 200 mm $\times$ 110 mm on each side. Figure 1 represents a schematic for the FSW process, showing all the basic elements and the movement direction. Friction stir welding was carried to produce three different joints between the two aluminum alloys with FSW rotation rates and welding traverse speeds combination as follows: J1 (600 rpm-60 mm/min), J2 (400 rpm-60 mm/min), and J3 (400 rpm-20 mm/min). The FSW tool used was made from the H13 tool steel that was heat treated and tempered to 54 HRC (hardness Rockwell C) hardness. The joints after production were section perpendicular to the welding direction (WD) and prepared to read the optical macrographs. For the EBSD investigation, samples from the top and bottom of the weld Nugget were cut. These samples were then mechanically polished and subsequently electropolished with a solution of 30% nitric acid in methanol for 60 s at 14 V and -15 °C. FEI Quanta FEG 250 Field Emission Gun Scanning Electron Microscope (FEGSEM), FEI company (Hillsboro, OR, USA), equipped with a Hikari EBSD camera controlled by EDAX-OIM7.3 (EDAX Inc. Mahwah, NJ, USA) analysis software, was used for EBSD data acquisition and post processing. To evaluate the changes in the mechanical properties due to the FSW process, the base alloys, as well as the FSWed joints, were tested using tensile and hardness testing. Vickers macro-hardness tests were performed on the transverse cross-sections with an interspacing distance of 2 mm using a test load of 1 kg force and a dwell time of 15 s. The tensile test properties of the welded stir zone and transverse flat tensile specimens were used. Tensile samples were machined perpendicular to the FSW direction to the dimensions, as shown in Figure 2. The specimen's dimensions agree with the DIN EN10002-1 2001(D) standards. Tensile tests were carried out at room temperature and at a quasi-static strain rate of ε 0.001 s^{-1} using the tensile testing machine Instron Type 4210, Norwood, MA, USA.

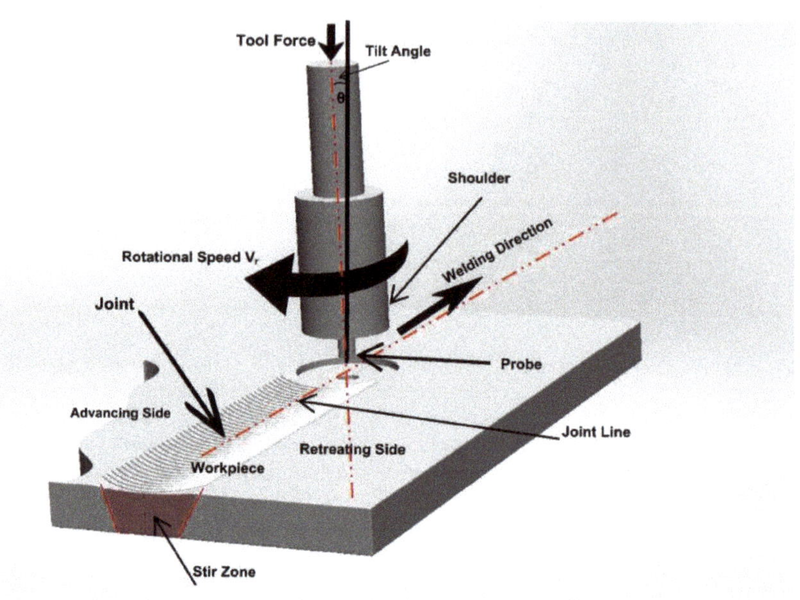

Figure 1. Schematic representation of the friction stir welding process.

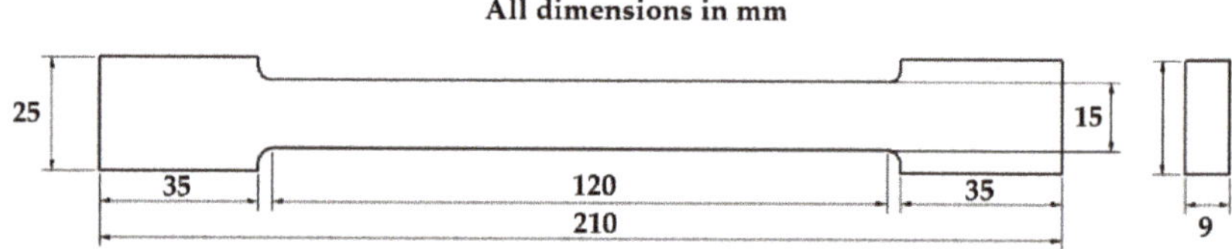

Figure 2. Tensile test specimen dimensions.

3. Results and Discussion

3.1. Microstructural Features of the Base Aluminum Alloys

Figure 3 shows the inverse pole figure coloring maps (a, b), their corresponding grain boundary maps with high angle boundaries (HABs) > 15° in black lines and low angle boundaries (LABs) from 5° to <15° in red lines, and the grain size distribution charts for the as-received aluminum alloys AA5754 and AA5083. The microstructures of the AA5754 and AA5083 Al-alloys in the as-received conditions revealed a recrystallized grain structure. The presented maps of both alloys showed random and fully recrystallized grain structures without pronounced textures, as indicated by the color-code legend of grain orientations and the low density of low angle grain boundaries. The average grain diameters of AA5754 and AA5083 alloys were measured to be 82.3 with a standard deviation of 29 and 93.5 µm with a standard deviation of 34, respectively, as can be seen from the corresponding distribution of grain diameters. The results of EBSD measurements of the as-received conditions demonstrated insignificant differences in the initial grain structure of the base alloys.

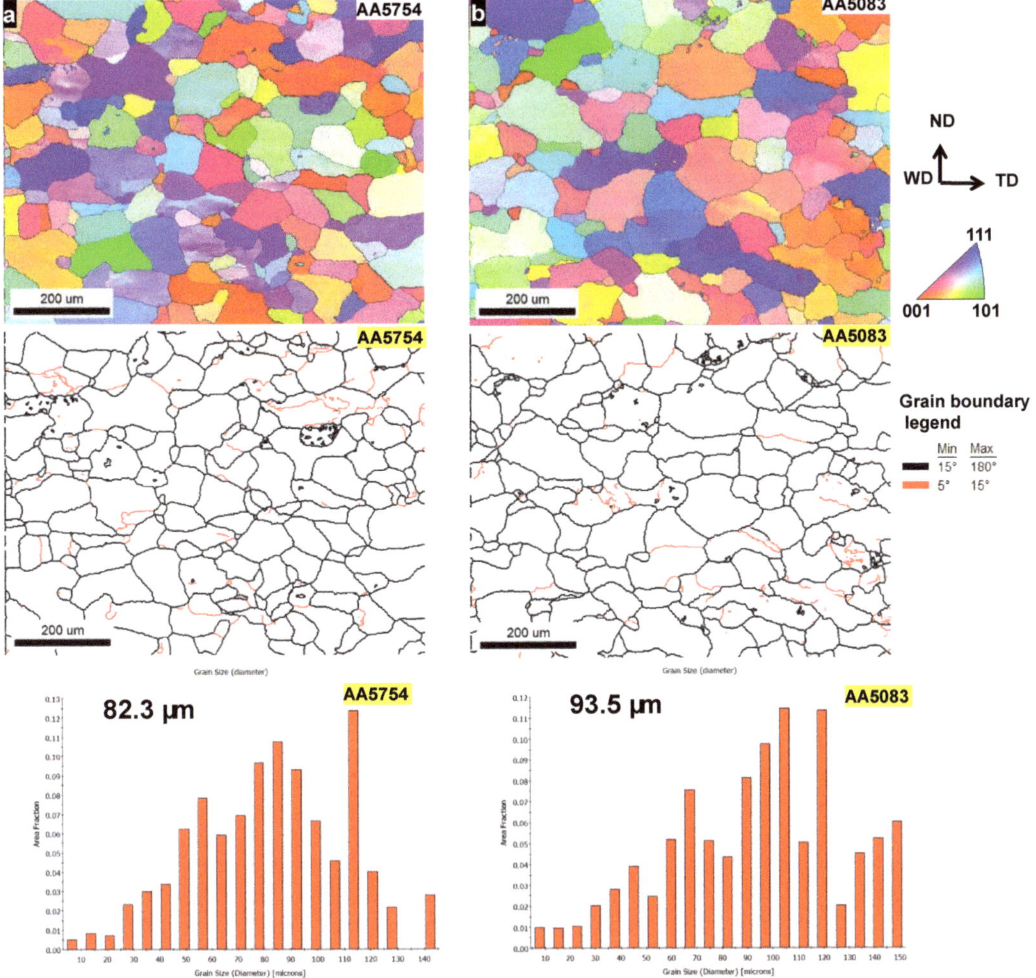

Figure 3. Microstructural characteristics of the AA5754 and AA5083 Al-alloys in the as-received condition. EBSD orientation map, corresponding grain boundaries map, and grain size distribution histograms for (**a**) AA5754 and (**b**) AA5083 Al-alloys.

3.2. Microstructural Features of the FSWed Dissimilar AA5083-AA5754 Joints

Figure 4 shows a collage of the macro- and micrographs that reveal the main characteristics of the grain structures for the dissimilar AA5754-AA5083 joints welded using different combinations of rotation rates and travel speeds of the FSW tool. The macrographs in the middle point out to the locations (Figure 4), where the EBSD measurements were performed using 1 µm step size. Two locations (one at the top and the other at the bottom) were investigated in the NG zone, almost along the vertical centerline for each joint. Generally, the top locations in the NG zones (a, b, and c) showed always larger grain structures than those developed at the corresponding bottom locations (d, e, and f). This can be attributed to the high heat generated at the top surface due to the effect of both the shoulder and the pin, while near the bottom of the NG is only affected by the pin with lower heat experienced [12,24,32,40,41]. Furthermore, it is clear that the grain sizes in the NG zones showed a dependency on the rotation and travel speeds as well. As can be seen, the grain sizes of J1 manufactured at 600 rpm–60 mm/min (Figure 4a,d) were coarser than their counterparts of J2 manufactured at 400 rpm–60 mm/min (Figure 4b,e), indicating a grain refining effect induced by the decrease in the rotation rate of the tool from 600 to 400 rpm. On the other hand, the decrease in the welding traverse speed from 60 mm/min to 20 mm/min at a constant rotation rate of 400 rpm had not resulted in a significant effect on the grain structure and the average grain size, as can be observed from Figure 4c,f. The variation in grain sizes from the top to the bottom locations through the thickness in the NG zones can be explained by the higher heat experienced at the top regions of the joints due to the friction-induced heat caused by the contact between the work-piece and the tool shoulder and pin during FSW, while the bottom regions are only affected by the pin and accordingly experience a lower heat [42,43]. Another factor that promotes a variation in heat from the top to the bottom of the NG zones is the thick section of the welded plates, which contribute to a higher cooling capacity during FSW [44]. It is also expected that the variation in heat from the top to the bottom through the thickness of NG zones can be affected by the rotation and travel speeds. Accordingly, higher heat input is excepted for the higher rotation speed and slower travel speed, which reflects the grain structure evolution in J1 that experiences the highest heat input (coarse grain structure) and in J2 that is exposed to the lowest heat input (finer grain structure). The obtained results here are in agreement with that reported in work conducted by Ahmed et al. [24] for the FSW of the thick section AA6082. They reported a significant reduction in the grain size towards the bottom part of the weld NG, which they attributed to the lower heat input experienced at the lower part due to the only pin effect relative to the top part of the NG, which was affected by both the pin and the shoulder of the tool. Besides, there was a significant reduction in the grain size by decreasing the heat input through the reduction of the tool rotation rate. The grain-size distributions represented in grain diameter based on the measured grain areas in the NG zones of J1, J2, and J3 are shown in Figure 5. The same data-sets represented in Figure 3 were utilized to calculate the grain-size distributions at the top locations (a, b, and c) and at the bottom locations (d, e, and f) for J1, J2, and J3, respectively. It was remarked that the average measured grain diameters in the NG zones at the top locations varied from 33, 25, to 24.5 μm, and at the bottom locations, changed from 19, 12, to 11.8 μm for J1, J2, and J3, respectively. Obviously, the grain sizes in the NG zones at the bottom locations were more than two times finer than those counterparts at the top locations. It should be noted here that the effect of reducing the tool rotation rate was more effective in controlling the grain size than increasing the traverse speed. Reducing the tool rotation rate from 600 rpm to 400 rpm resulted in a reduction of the average grain size at the top from 33 μm to 25 μm and at the bottom from 19 μm to 12 μm. On the other hand, decreasing the traverse speed from 60 mm/min to 20 mm/min almost did not affect the grain size parameters. In both cases, the average grain size was almost similar at the top locations, about 25 μm, and at the bottom locations, about 12 μm. In terms of grain orientation of the maps presented in Figure 4 and obtained at the top and the bottom locations of the NG from each weld, it could be considered randomly orientated with mixed <001> red, <101>green, and <111> blue orientations. It should be mentioned here that the data presented in Figure 4 is the as-collected data in which there was a difference between the FSW reference frame (TD, ND, WD) and the actual shear reference frame (θ, z, r), as quantitatively determined in a detailed study by Ahmed et al. [45,46] for the methodology to be applied to align the FSW reference frame with the shear reference frame to obtain the real FSW texture and orientations. Figure 6 shows the inverse pole figure (IPF) coloring maps with their corresponding (111) pole figures for the same data presented in Figure 4 after applying the required rotations to align the FSW reference frame with the shear reference frame. Now the IPF maps (Figure 6a–f) were dominated by the <111> blue orientations due to the alignment of the <111> poles with shear plan normal (r). In terms of texture, it could be observed from the (111) pole figures (PFs) that the texture was strong texture with up to 10 times random and was mainly of simple shear texture. The (111) PF of the J1 joint (Figure 6a,d) had the strongest texture with 10 times random at the top and 7 times random at the bottom of the NG. This could be attributed to the high amount of deformation experienced due to the high tool rotation rate (600 rpm) and the fast welding speed (60 mm/min). The (111) PF of the J2 joint (Figure 6b,e) had slightly relatively less strong texture with 6 and 5 times random at the top and bottom of the NG, respectively. The (111) PF of the J3 joint (Figure 6c,f) showed strong texture with 7 times random at the top and only 3 times at the bottom. This indicates the effect of the FSW parameters on the strength of the texture components. In all cases, the textures were of the simple shear, which is the main type of texture reported in the NG of FSWed aluminum alloys [45,46].

3.3. Mechanical Properties

Vickers macro-hardness distribution profiles on the transverse cross-sections of the joints produced by FSW are shown in Figure 7. Figure 7a–c show the hardness maps for the three joints (J1, J2, and J3); it can be noted that the FSW-affected zones were diffusing and extended to a width of 22 mm at the bottom of the butt joint and increased to reach around 40 mm at the upper surface due to the effect of the friction and the pressure applied by the rotating shoulder to the surface of the joint.

The conical shape of the SZ and HAZ was more obvious at the joints with low ω/v values of 10 (J1: 600 rpm and 60 mm/min). The FSW nuggets showed the lowest hardness values due to the heat input concentrated in these regions, causing softening of the stirred regions of the joined materials. At both applied rotational speeds (400 and 600 rpm), the lower hardness region took place in the upper surface of the joints at the lowest travel speed (20 mm/min) and then appeared in the lower half of the cross-section at the highest travel speed (60 mm/min). This statement confirmed the softening effect of the friction and pressure of the pin shoulder on the upper surface of specimens [47].

Figure 8 represents the engineering tensile stress-strain curves of the base alloys AA5083 and AA5754 and the FSWed dissimilar joints. Flow behavior of the Al–Mg alloys of the series AA5XXX have been investigated at quasi-static and high strain rate ranges [48,49] and showed similar serration in the flow curves, which are related to the so-called Portevin-Le Chatelier effect [49–51]. This effect is due to successive pinning and unpinning of the moving dislocations by the solute atoms. The base aluminum alloys show typical stress-strain curves with moderate hardening, followed by a wide plastic strain range up to the ultimate tensile stress, followed by a slow decrease of stress value up to fracture.

Table 1 includes the tensile properties of the tested specimens of the welded joints compared with that of the base alloys. The tensile sample of the FSWed joint at the revolution of 400 rpm and travel speed of 60 mm/min (J2: 400-60) showed similar behavior to the base materials, except that the short plastic strain range was lower than the base alloys. This showed higher tensile stress than the base alloy AA5083 from the beginning of the plastic strain region till its ultimate tensile stress value (224 MPa) and decreased till fracture at a total elongation of 23%. Relating the ultimate tensile value of this joint to the ultimate tensile value of the base alloy AA5083 resulted in a welding efficiency of 96%. The other two tensile samples of the FSWed joints ((J1: 600-60) and (J3: 400-20)) were early fractured at strains of 5.5% and 4.3%, respectively, before reaching the ultimate tensile value. This behavior was due to the presence of some welding defects, such as tunnels or pores [52]. However, the yield stress of the tensile sample taken from these joints (J1: 600-60 and J3: 400-20) was comparable with the yield stress of the base alloy AA5083. The increased strength and the soundness of the sample (J2: 400-60) were related to the lowest heat input value, as shown in Table 1, where its heat index value was one third that of the sample J3: 400-20 and one half of the sample J1: 600-60.

To describe the flow behavior of the tensile stress-strain curves ($\sigma - \varepsilon$) of the materials under investigation, the engineering curves were transferred to the true stress-true strain ($\sigma_f - \varphi$) up to the ultimate point by these formulas: true stress $\sigma_f = \sigma\,(1+\varepsilon)$ and true strain $\varphi = ln\,(1+\varepsilon)$.

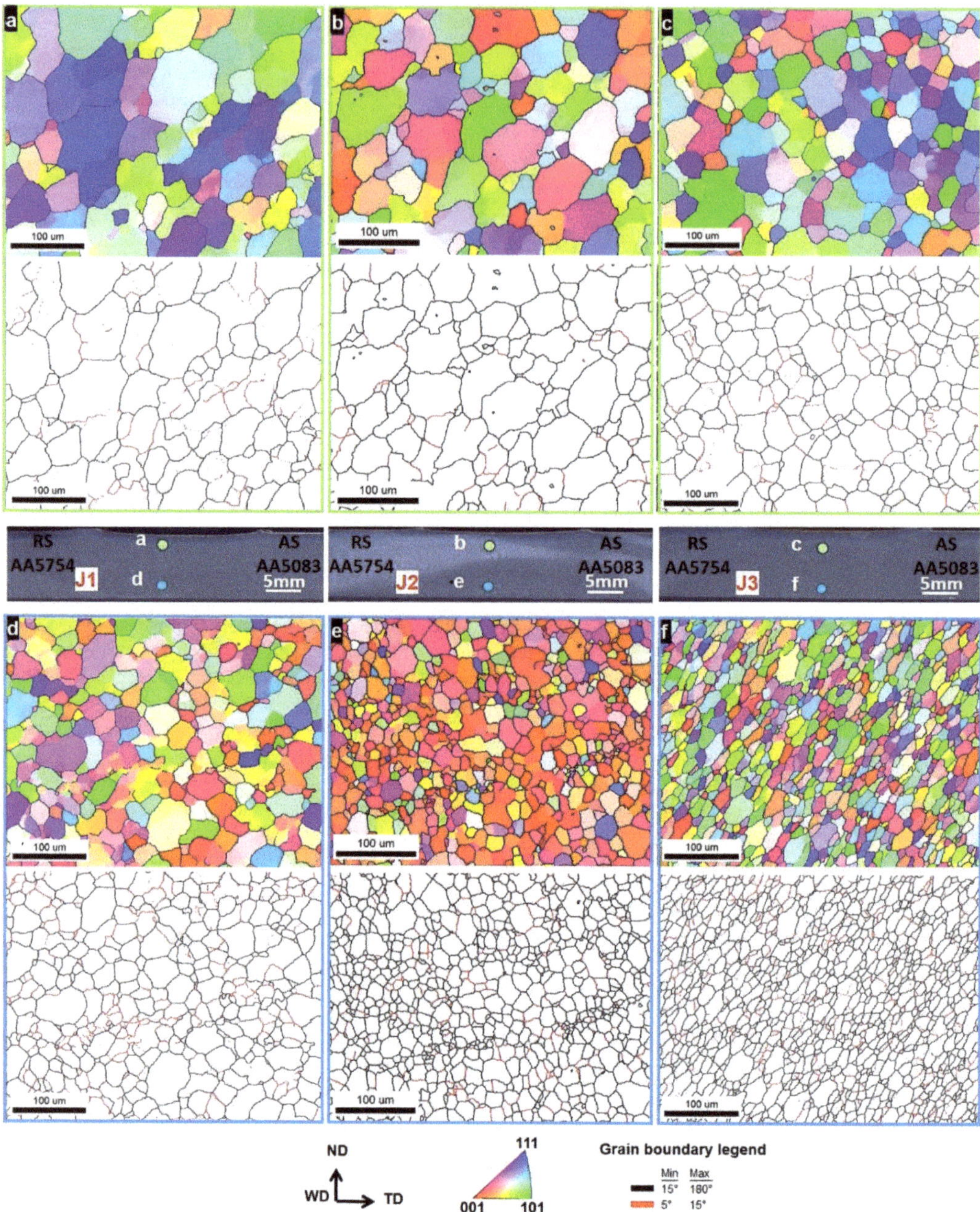

Figure 4. Macrostructure representing three different FSWed AA5754-AA5083 joints J1, J2, and J3 prepared by applying different combinations of rotation and travel speeds (rpm-mm/min) of 600-60, 400-60, and 400-20, respectively. EBSD measurements were performed at the centerline in the NG zone for each joint at the corresponding specified top locations (**a–c**) and bottom locations (**d–f**) for J1, J2, and J3, respectively. The inverse pole figure coloring (IPF) maps and their corresponding grain boundary (GB) maps are represented for all the denoted locations.

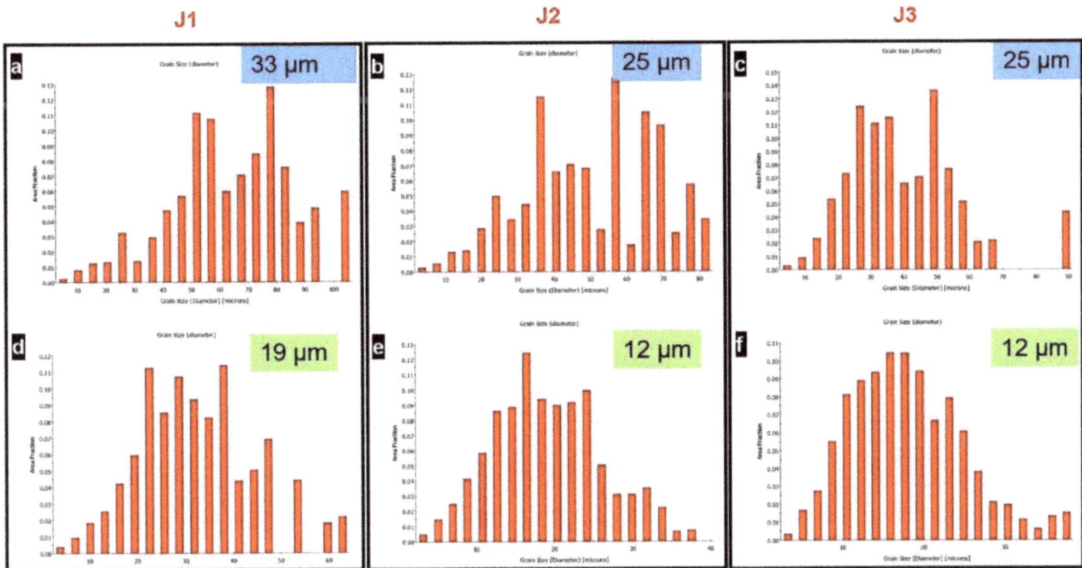

Figure 5. Distribution of grain diameters for the different dissimilar friction stir welded (FSWed) joints prepared using different rotation and travel speeds. (**a–c**) at the top locations and (**d–f**) at the bottom locations in the NG zones of J1, J2, and J3, respectively.

Table 1. Friction stir welding conditions and tensile properties.

State	Welding Conditions		Tensile Properties		
	#	Heat Index ω^2/v	σ_{UTS} (MPa)	Total Strain (%)	Welding Efficiency (%)
Base AA5754	-	-	251	28.50	–
Base AA5083	-	-	233	34	–
AA5083-AA5754	J1	6000	178	5.50	77
	J2	2666	224	23	96
	J3	8000	153	4.30	66

There are many published models describing the flow behavior of metallic materials [53–55]. The description model can be selected depending on the suitability for the specific material and the test conditions. The model simplicity for application, represented in the low number of model parameters, is a factor helping the spread of some models. The flow curves of the tested samples were described using the empirical formula relating the flow stress (σ_f) and true strain (φ) after Ludwik [56]:

$$\sigma_f = \sigma_o + k\,(\varphi)^n \tag{1}$$

where initial flow stress (σ_o) is the flow stress at the plastic strain of $\varphi = 0$, k is a material parameter, and n is the material strengthening parameter.

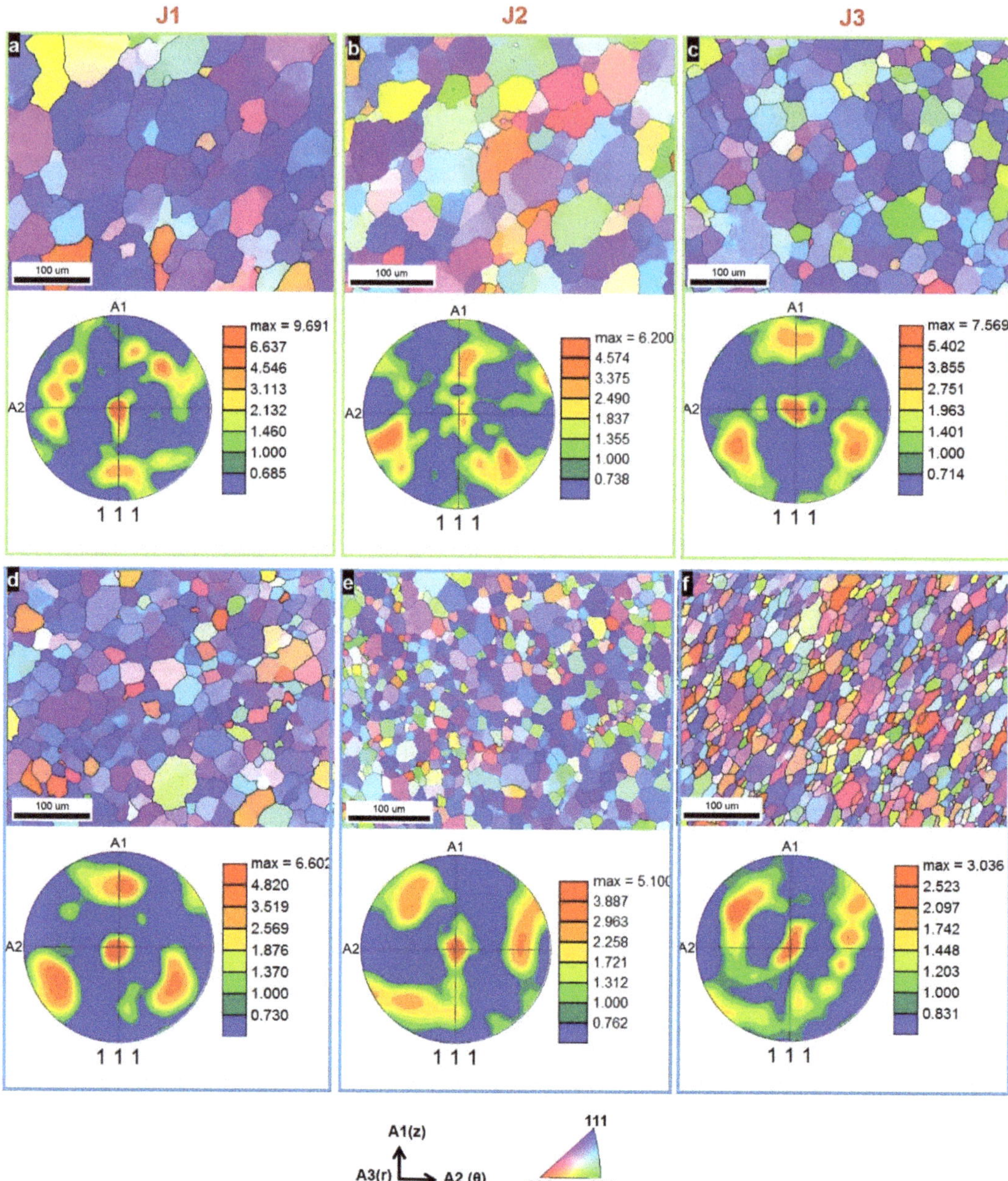

Figure 6. IPF coloring maps with their corresponding (111) pole figures for the same data presented in Figure 4 after applying the required rotations to align the FSW reference frame with the shear reference frame. (**a–c**) are the IPF maps after rotation and their corresponding (111) pole figures for the EBSD data obtained at the top locations given in Figure 4. (**d–f**) are the IPF maps after rotation and their corresponding (111) pole figures for the EBSD data obtained at the bottom locations given in Figure 4.

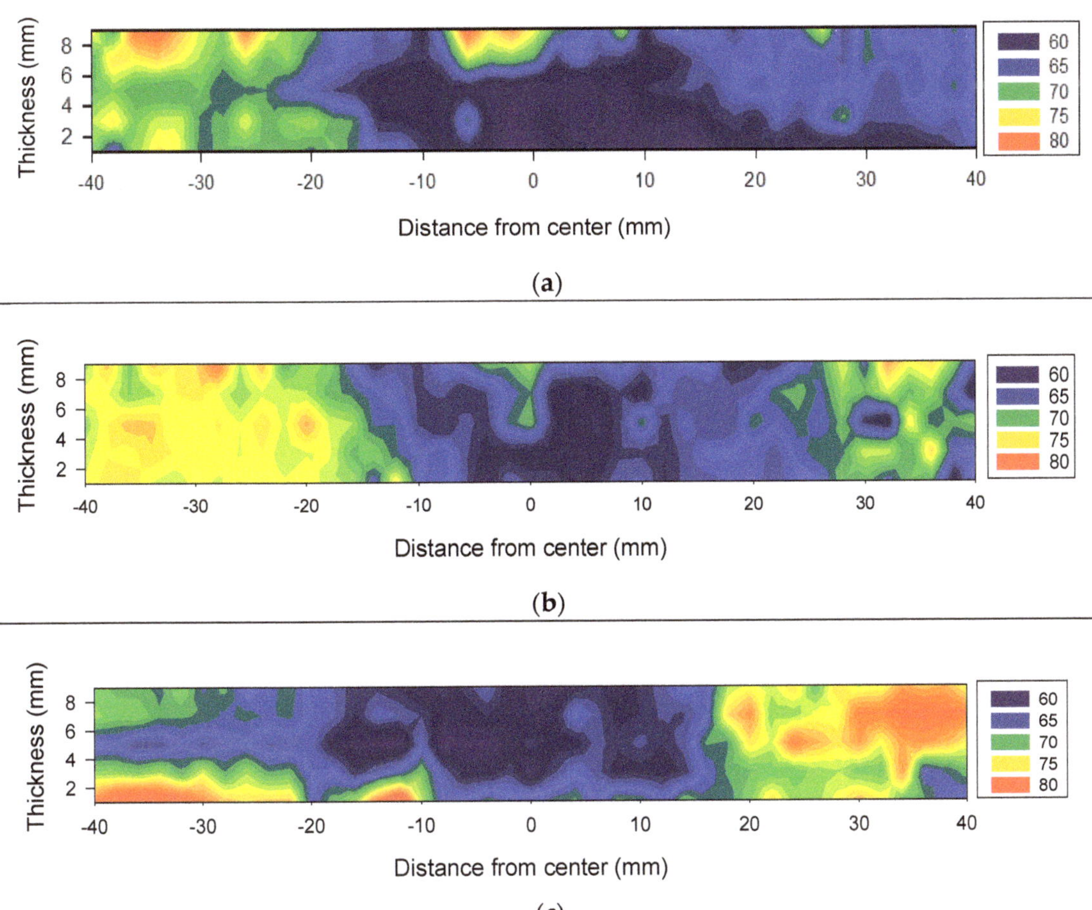

Figure 7. Hardness distribution maps over the cross-section of the FSWed joints (**a**) J1: 600 rpm–60 mm/min, (**b**) J2: 400 rpm 60 mm/min, and (**c**) J3: 400 rpm and 20 mm/min.

Figure 9 shows the description of the plastic flow curves of the base alloys AA5754 and AA5083 and the FSWed joints using the Ludwik formula. It can be seen that the selected empirical model described the curves very well. The materials parameters (k and n) of the base alloys were relatively low due to the combination of the flow curve of the higher strengthening rate region at the beginning of the flow curve and the moderate hardening in the steady-state region up to the end of the flow curve. The samples of the joints welded at the conditions 400-20 and 600-60 showed higher strengthening parameter (n) and higher material parameter (k) than the base alloys due to the early fracture of the samples, leading to shortening of the flow curves, especially the lower strengthening rate region at the end of the curve. However, the FSWed joint using pin revolution of 400 rpm and a travel speed of 60 mm/min showed the highest strengthening parameter (n = 0.494) with a moderate k value of 413. The tensile flow parameters of the flow curves are summarized in Table 2. In the three joints (J1, J2, and J3), the fracture mechanism was ductile mode with very clear dimple features, and it is fully characterized in [9].

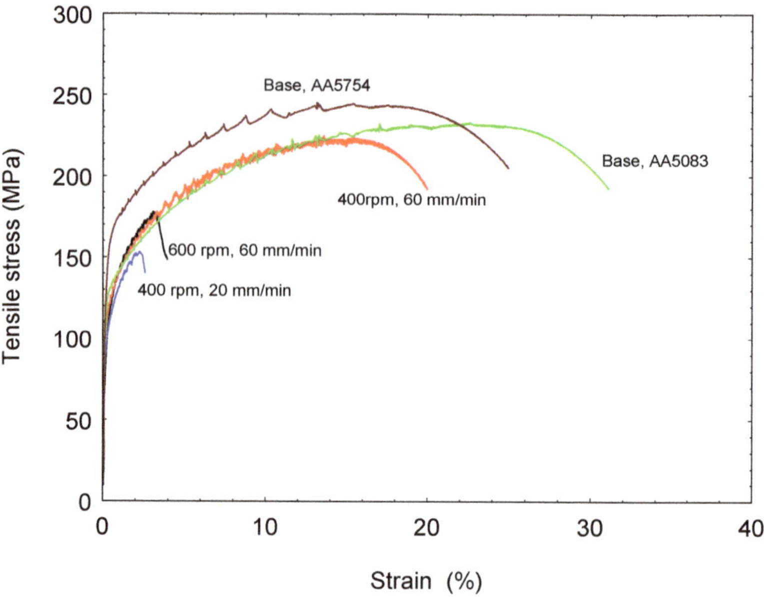

Figure 8. Engineering tensile stress-strain curves of the base alloys AA5083 and AA5754 and the FSWed dissimilar joints at the conditions 400 rpm/20 mm/min, 400 rpm/60 mm/min, and at 600 rpm/60 mm/min.

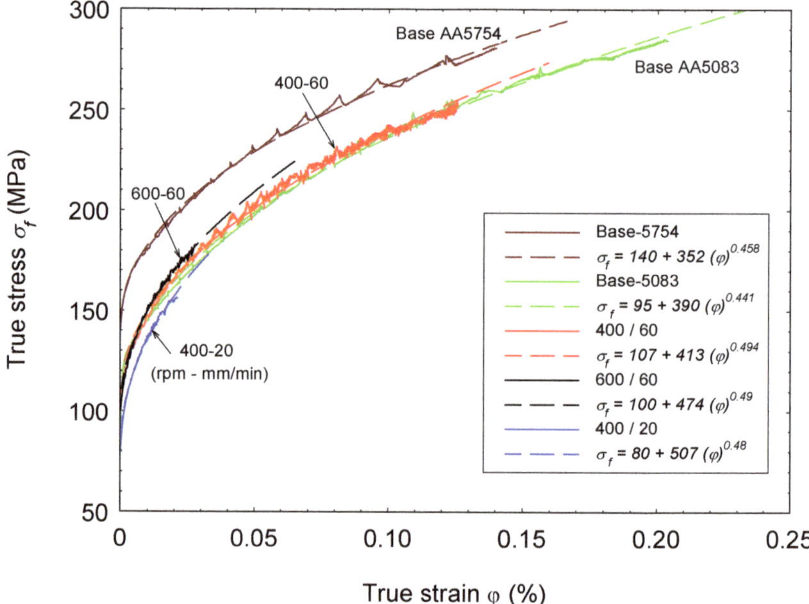

Figure 9. Description of the true tensile stress-strain curves of the base alloys AA5083 and AA5754 and the FSWed dissimilar joints using Ludwik formula.

Table 2. Tensile flow parameters of the flow curves.

State	ω (rpm)	v (mm/min)	σ_o (MPa)	k (MPa)	n (–)
Base 5754	–	–	140	352	0.458
Base 5083	–	–	95	390	0.441
FSWed	600	60	100	474	0.490
	400	60	107	413	0.494
	400	20	80	507	0.48

4. Conclusions

- Microstructure observations using EBSD revealed a significant grain refinement effect for the rotation rate than that of the welding speed during dissimilar FSW of AA5754-AA5083 joints. The average grain size reduced from 19 μm to 12 μm by the reduction of the rotation rate from 600 rpm to 400 rpm at a constant welding speed of 60 mm/min, while almost similar average grain size (12 μm) was obtained by the reduction of welding speed from 60 mm/min to 20 mm/min at a constant rotation rate of 400 rpm.
- The combination of the lowest applied tool rotation rate of 400 rpm and welding speed of 20 mm/min promoted a significant grain structure refinement, attributable to a decreased heat input compared with other welded joints at 400 rpm-40 mm/min and 600 rpm-60 mm/min.
- The generally observed fine grain structure in the bottom region of nugget zones for all joints was explained by the thickness-induced high cooling capacity, preventing grain growth, besides being the bottom region affected by the pin not by the shoulder and pin together as the case in the top regions.
- Hardness distribution maps revealed the softening of the nugget zone. The increased heat generated by the pin shoulder made the upper region of the nugget zone more soft than the lower zone.
- Tensile flow curves of the tested materials were well described using the Ludwik formula, and the materials parameters were sensitive to the hardening effect resulting from the FSW-ing process. The FSWed joint (400-60) showed the highest strengthening parameter ($n = 0.494$) with a moderate k value of 413 MPa.

Author Contributions: Conceptualization, S.A., E.A., and M.M.Z.A.; methodology, T.A., E.A., and M.M.E.-S.S.; validation, S.A., N.A.A., and E.A.; formal analysis, S.A. and M.M.Z.A.; investigation, M.M.E.-S.S., S.A., and E.A.; writing original draft preparation, T.A., S.A., and E.A.; writing—review and editing, T.A., M.M.Z.A., and N.A.A.; project administration, M.M.Z.A. and M.M.E.-S.S. All authors have read and agreed to the published version of the manuscript.

Funding: This research received no external funding.

Data Availability Statement: The data presented in this study are available on request from the corresponding author. The data are not publicly available due to the extremely large size.

Acknowledgments: The authors acknowledge the financial support rendered by the Science and Technology Development Fund (STDF), Ministry of Higher Education and Scientific Research, Egypt.

Conflicts of Interest: The authors declare no conflict of interest.

Abbreviations

AA	Aluminum alloy
AS	Advancing side
BM	Base material
DRV	Dynamic recovery
DRX	Dynamic recrystallization
EBSD	Electron backscatter diffraction
EDAX	Energy dispersive analysis of X-rays
EDX	Energy-dispersive X-ray
FSW	Friction stir welding
FSWed	Friction stir welded
HABs	High angle grain boundaries
HAZ	Heat affected zone
HI	Heat input, J/mm
HV	Hardness Vickers
HRC	Hardnedd Rockwell C
IPF	Inverse pole figure
k	Material parameter
LABs	Low angle grain boundaries
Mg	Magnesium
Mn	Manganese
n	Strengthening parameter
ND	Normal direction
NG	Nugget zone
rpm	Revolution per minute
RS	Retreating side
SEM	Scanning electron microscope
SZ	Stirred zone
TD	Transverse direction
TMAZ	Thermomechanical affected zone
v	Welding speed, mm/min
WD	Welding direction
WN	Welding nugget
ε	Engineering strain
σ	Engineering stress
$\sigma_{0.2\%}$:	0.2 offset yield stress, MPa
σ_f:	Flow stress, MPa
σ_{UTS}:	Ultimate tensile strength, MPa
φ	True strain
ω	Rotational speed, rpm

References

1. Sahu, P.K.; Pal, S.; Pal, S.K.; Jain, R. Influence of Plate Position, Tool Offset and Tool Rotational Speed on Mechanical Properties and Microstructures of Dissimilar Al/Cu Friction Stir Welding Joints. *J. Mater. Process. Technol.* **2016**, *235*, 55–67. [CrossRef]
2. Mabuwa, S.; Msomi, V. Review on Friction Stir Processed Tig and Friction Stirwelded Dissimilar Alloy Joints. *Metals* **2020**, *10*, 142. [CrossRef]
3. Singh, V.P.; Patel, S.K.; Ranjan, A.; Kuriachen, B. Recent Research Progress in Solid State Friction-Stir Welding of Aluminium–Magnesium Alloys: A Critical Review. *J. Mater. Res. Technol.* **2020**, *9*, 6217–6256. [CrossRef]
4. Mehta, K.P. A review on friction-based joining of dissimilar aluminum-steel joints. *J. Mater. Res.* **2019**, *34*, 78–96. [CrossRef]
5. Ahmed, M.M.Z.; Seleman, M.M.E.; Zidan, Z.A.; Ramadan, R.M.; Ataya, S.; Alsaleh, N.A. Microstructure and Mechanical Properties of Dissimilar Friction Stir Welded AA2024-T4/AA7075-T6 T-Butt Joints. *Metals* **2021**, *11*, 128. [CrossRef]
6. Cabibbo, M.; Forcellese, A.; Santecchia, E.; Paoletti, C.; Spigarelli, S.; Simoncini, M. New Approaches to Friction Stir Welding of Aluminum Light-Alloys. *Metals* **2020**, *10*, 233. [CrossRef]
7. Oliveira, J.P.; Ponder, K.; Brizes, E.; Abke, T.; Edwards, P.; Ramirez, A.J. Combining Resistance Spot Welding and Friction Element Welding for Dissimilar Joining of Aluminum to High Strength Steels. *J. Mater. Process. Technol.* **2019**, *273*, 116192. [CrossRef]
8. Ahmed, M.M.Z.; Ahmed, E.; Hamada, A.S.; Khodir, S.A.; Seleman, M.E.-S.; Wynne, B.P. Microstructure and Mechanical Properties Evolution of Friction Stir Spot Welded High-Mn Twinning-Induced Plasticity Steel. *Mater. Des.* **2016**, *91*. [CrossRef]

9. Merklein, M.; Johannes, M.; Lechner, M.; Kuppert, A. A Review on Tailored Blanks—Production, Applications and Evaluation. *J. Mater. Process. Technol.* **2014**, *214*, 151–164. [CrossRef]
10. El Rayes, M.M.; Soliman, M.S.; Abbas, A.T.; Pimenov, D.Y.; Erdakov, I.N.; Abdel-mawla, M.M. Effect of Feed Rate in FSW on the Mechanical and Microstructural Properties of AA5754 Joints. *Adv. Mater. Sci. Eng.* **2019**, *2019*, 4156176. [CrossRef]
11. Ahmed, M.M.Z.; Ataya, S.; Seleman, M.M.E.; Mahdy, A.M.A.; Alsaleh, N.A.; Ahmed, E. Heat Input and Mechanical Properties Investigation of Friction Stir Welded AA5083/AA5754 and AA5083/AA7020. *Metals* **2021**, *11*, 68. [CrossRef]
12. Ahmed, M.M.Z.; Ataya, S.; Seleman, M.E.-S.; Ammar, H.R.; Ahmed, E. Friction Stir Welding of Similar and Dissimilar AA7075 and AA5083. *J. Mater. Process. Technol.* **2017**, *242*, 77–91. [CrossRef]
13. Sangalli, G.; Lemos, G.V.B.; Martinazzi, D.; De Lima Lessa, C.R.; Beskow, A.B.; Reguly, A. Towards Qualification of Friction Stir Welding to AA5083-O and AA5052-O Aluminum Alloys. *Mater. Res.* **2019**. [CrossRef]
14. Ahmed, M.M.Z.; Seleman, M.E.-S.; Shazly, M.; Attallah, M.M.; Ahmed, E. Microstructural Development and Mechanical Properties of Friction Stir Welded Ferritic Stainless Steel AISI 409. *J. Mater. Eng. Perform.* **2019**, *28*, 6391–6406. [CrossRef]
15. Ahmed, M.M.Z.; Wynne, B.P.; Martin, J.P. Effect of Friction Stir Welding Speed on Mechanical Properties and Microstructure of Nickel Based Super Alloy Inconel 718. *Sci. Technol. Weld. Join.* **2013**, *18*, 680–687. [CrossRef]
16. Hamada, A.S.; Järvenpää, A.; Ahmed, M.M.Z.; Jaskari, M.; Wynne, B.P.; Porter, D.A.; Karjalainen, L.P. The Microstructural Evolution of Friction Stir Welded AA6082-T6 Aluminum Alloy during Cyclic Deformation. *Mater. Sci. Eng. A* **2015**, *642*, 366–376. [CrossRef]
17. Khodir, S.A.; Ahmed, M.M.Z.; Ahmed, E.; Mohamed, S.M.R.; Abdel-Aleem, H. Effect of Intermetallic Compound Phases on the Mechanical Properties of the Dissimilar Al/Cu Friction Stir Welded Joints. *J. Mater. Eng. Perform.* **2016**, *25*, 4637–4648. [CrossRef]
18. Refat, M.; Elashery, A.; Toschi, S.; Ahmed, M.M.Z.; Morri, A.; El-Mahallawi, I.; Ceschini, L. Microstructure, Hardness and Impact Toughness of Heat-Treated Nanodispersed Surface and Friction Stir-Processed Aluminum Alloy AA7075. *J. Mater. Eng. Perform.* **2016**, *25*, 5087–5101. [CrossRef]
19. Hoziefa, W.; Toschi, S.; Ahmed, M.M.Z.; Morri, A.; Mahdy, A.A.; Seleman, M.E.-S.; El-Mahallawi, I.; Ceschini, L.; Atlam, A. Influence of Friction Stir Processing on the Microstructure and Mechanical Properties of a Compocast AA2024-Al_2O_3 Nanocomposite. *Mater. Des.* **2016**, *106*, 273–284. [CrossRef]
20. Zayed, E.M.; El-Tayeb, N.S.M.; Ahmed, M.M.Z.; Rashad, R.M. Development and Characterization of AA5083 Reinforced with SiC and Al_2O_3 Particles by Friction Stir Processing. In *Engineering Design Applications*; Springer: Cham, Switzerland, 2019; Volume 92.
21. Oliveira, J.P.; Duarte, J.F.; Inácio, P.; Schell, N.; Miranda, R.M.; Santos, T.G. Production of Al / NiTi Composites by Friction Stir Welding Assisted by Electrical Current. *JMADE* **2017**, *113*, 311–318. [CrossRef]
22. Tonelli, L.; Refat, M.; Toschi, S.; Ahmed, M.M.Z.; Ahmed, E.; Morri, A.; El-Mahallawi, I.; Ceschini, L. Production of AlSi12CuNiMg/Al_2O_3 Micro/Nanodispersed Surface Composites Using Friction Stir Processing for Automotive Applications BT—Friction Stir Welding and Processing X. In *Friction Stir Welding and Processing X*; Hovanski, Y., Mishra, R., Sato, Y., Upadhyay, P., Yan, D., Eds.; Springer International Publishing: Cham, Switzerland, 2019; pp. 233–242.
23. Costa, A.M.S.; Oliveira, J.P.; Pereira, V.F.; Nunes, C.A.; Ramirez, A.J.; Tschiptschin, A.P. Ni-Based Mar-M247 Superalloy as a Friction Stir Processing Tool. *J. Mater. Process. Technol.* **2018**, *262*, 605–614. [CrossRef]
24. Ahmed, M.M.Z.Z.; Wynne, B.P.; Rainforth, W.M.; Addison, A.; Martin, J.P.; Threadgill, P.L. Effect of Tool Geometry and Heat Input on the Hardness, Grain Structure, and Crystallographic Texture of Thick-Section Friction Stir-Welded Aluminium. *Metall. Mater. Trans. A Phys. Metall. Mater. Sci.* **2019**, *50*, 271–284. [CrossRef]
25. Cabibbo, M.; Forcellese, A.; El Mehtedi, M.; Simoncini, M. Double Side Friction Stir Welding of AA6082 Sheets: Microstructure and Nanoindentation Characterization. *Mater. Sci. Eng. A* **2014**, *590*, 209–217. [CrossRef]
26. Su, H.; Wu, C.S.; Bachmann, M.; Rethmeier, M. Numerical Modeling for the Effect of Pin Profiles on Thermal and Material Flow Characteristics in Friction Stir Welding. *Mater. Des.* **2015**, *77*, 114–125. [CrossRef]
27. Sarkari Khorrami, M.; Kazeminezhad, M.; Kokabi, A.H. Microstructure Evolutions after Friction Stir Welding of Severely Deformed Aluminum Sheets. *Mater. Des.* **2012**, *40*, 364–372. [CrossRef]
28. Kumbhar, N.T.; Sahoo, S.K.; Samajdar, I.; Dey, G.K.; Bhanumurthy, K. Microstructure and Microtextural Studies of Friction Stir Welded Aluminium Alloy 5052. *Mater. Des.* **2011**, *32*, 1657–1666. [CrossRef]
29. Kim, N.K.; Kim, B.C.; An, Y.G.; Jung, B.H.; Song, S.W.; Kang, C.Y. The Effect of Material Arrangement on Mechanical Properties in Friction Stir Welded Dissimilar A5052/A5J32 Aluminum Alloys. *Met. Mater. Int.* **2009**, *15*, 671–675. [CrossRef]
30. Mastanaiah, P.; Sharma, A.; Reddy, G.M. Dissimilar Friction Stir Welds in AA2219-AA5083 Aluminium Alloys: Effect of Process Parameters on Material Inter-Mixing, Defect Formation, and Mechanical Properties. *Trans. Indian Inst. Met.* **2016**, *67*, 1397–1415. [CrossRef]
31. Kalemba-Rec, I.; Kopyściański, M.; Miara, D.; Krasnowski, K. Effect of Process Parameters on Mechanical Properties of Friction Stir Welded Dissimilar 7075-T651 and 5083-H111 Aluminum Alloys. *Int. J. Adv. Manuf. Technol.* **2018**, *97*, 2767–2779. [CrossRef]
32. Essa, A.R.S.; Ahmed, M.M.Z.; Mohamed, A.Y.A.; El-Nikhaily, A.E. An Analytical Model of Heat Generation for Eccentric Cylindrical Pin in Friction Stir Welding. *J. Mater. Res. Technol.* **2016**, *5*, 234–240. [CrossRef]
33. Kasman, Ş.; Yenier, Z. Analyzing Dissimilar Friction Stir Welding of AA5754/AA7075. *Int. J. Adv. Manuf. Technol.* **2014**, *70*, 145–156. [CrossRef]
34. Threadgill, P.L. Terminology in Friction Stir Welding. *Sci. Technol. Weld. Join.* **2007**, *12*, 357–360. [CrossRef]

35. Threadgill, P.L.; Leonard, A.J.; Shercliff, H.R.; Withers, P.J. Friction Stir Welding of Aluminium Alloys. *Int. Mater. Rev.* **2009**, *54*, 49–93. [CrossRef]
36. Rollett, A.; Humphreys, F.; Rohrer, G.S.; Hatherly, M. *Recrystallization and Related Annealing Phenomena*, 2nd ed.; Elsevier: Amsterdam, The Netherlands, 2004; ISBN 9780080441641.
37. McQueen, H.J.; Blum, W. Dynamic Recovery: Sufficient Mechanism in the Hot Deformation of Al (<99.99). *Mater. Sci. Eng. A* **2000**, *290*, 95–107. [CrossRef]
38. Su, J.-Q.Q.; Nelson, T.W.; Sterling, C.J. Microstructure Evolution during FSW/FSP of High Strength Aluminum Alloys. *Mater. Sci. Eng. A* **2005**, *405*, 277–286. [CrossRef]
39. Fonda, R.W.; Bingert, J.F.; Colligan, K.J. Development of Grain Structure during Friction Stir Welding. *Scr. Mater.* **2004**, *51*, 243–248. [CrossRef]
40. Ahmed, M.M.Z.; Elnaml, A.; Shazly, M.; Seleman, M.M.E. The Effect of Top Surface Lubrication on the Friction Stir Welding of Polycarbonate Sheets. *Intern. Polym. Process.* **2021**, 1–9. [CrossRef]
41. Ahmed, M.M.Z.; Wynne, B.P.; Rainforth, W.M.; Threadgill, P.L. Microstructure, Crystallographic Texture and Mechanical Properties of Friction Stir Welded AA2017A. *Mater. Charact.* **2012**, *64*, 107–117. [CrossRef]
42. Zhang, F.; Su, X.; Chen, Z.; Nie, Z. Effect of Welding Parameters on Microstructure and Mechanical Properties of Friction Stir Welded Joints of a Super High Strength Al-Zn-Mg-Cu Aluminum Alloy. *Mater. Des.* **2015**, *67*, 483–491. [CrossRef]
43. Cao, X.; Jahazi, M. Effect of Welding Speed on the Quality of Friction Stir Welded Butt Joints of a Magnesium Alloy. *Mater. Des.* **2009**, *30*, 2033–2042. [CrossRef]
44. Bagheri, B.; Abbasi, M.; Dadaei, M. Effect of Water Cooling and Vibration on the Performances of Friction-Stir-Welded AA5083 Aluminum Joints. *Metallogr. Microstruct. Anal.* **2020**, *9*, 33–46. [CrossRef]
45. Ahmed, M.M.Z.; Wynne, B.P.; Rainforth, W.M.; Threadgill, P.L. Quantifying Crystallographic Texture in the Probe-Dominated Region of Thick-Section Friction-Stir-Welded Aluminium. *Scr. Mater.* **2008**, *59*, 507–510. [CrossRef]
46. Ahmed, M.M.Z.; Wynne, B.P.; Rainforth, W.M.; Threadgill, P.L. Through-Thickness Crystallographic Texture of Stationary Shoulder Friction Stir Welded Aluminium. *Scr. Mater.* **2011**, *64*, 45–48. [CrossRef]
47. Zhao, Y.H.; Lin, S.B.; Qu, F.X.; Wu, L. Influence of Pin Geometry on Material Flow in Friction Stir Welding Process. *Mater. Sci. Technol.* **2006**, *22*, 45–50. [CrossRef]
48. Sajuri, Z.; Mohamad Selamat, N.F.; Baghdadi, A.H.; Rajabi, A.; Omar, M.Z.; Kokabi, A.H.; Syarif, J. Cold-Rolling Strain Hardening Effect on the Microstructure, Serration-Flow Behaviour and Dislocation Density of Friction Stir Welded AA5083. *Metals* **2020**, *10*, 70. [CrossRef]
49. Bintu, A.; Vincze, G.; Picu, R.C.; Lopes, A.B. Effect of Symmetric and Asymmetric Rolling on the Mechanical Properties of AA5182. *Mater. Des.* **2016**, *100*, 151–156. [CrossRef]
50. Gabrielli, F.; Forcellese, A.; El Mehtedi, M.; Simoncini, M. Mechanical Properties and Formability of Cold Rolled Friction Stir Welded Sheets in AA5754 for Automotive Applications. *Procedia Eng.* **2017**, *183*, 245–250. [CrossRef]
51. Hollomon, J.H. Tensile Deformation. *Trans. AIME* **1945**, *126*, 268–290.
52. Yamada, H.; Kami, T.; Mori, R.; Kudo, T.; Okada, M. Strain Rate Dependence of Material Strength in AA5xxx Series Aluminum Alloys and Evaluation of Their Constitutive Equation. *Metals* **2018**, *8*, 576. [CrossRef]
53. Swift, H.W. Plastic Instability under Plane Stress. *J. Mech. Phys. Solids* **1952**, *1*, 1–18. [CrossRef]
54. Meckings, H.; Koccks, U.F. Kinetics of Flow and Strain Hardening. *Acta Metall.* **1981**, *29*, 1865–1875. [CrossRef]
55. Estrin, Y.; Mecking, H. A unified phenomenological description of work hardening and creep based on one-parameter models. *Acta Met.* **1984**, *32*, 57–70. [CrossRef]
56. Ludwik, P. Fließvorgänge Bei Einfachen Beanspruchungen. In *Elemente der Technologischen Mechanik*; Springer: Berlin/Heidelberg, Germany, 1909. [CrossRef]

metals

MDPI

Article

Weldability Evaluation of Alloy 718 Investment Castings with Different Si Contents and Thermal Stories and Hot Cracking Mechanism in Their Laser Beam Welds

Pedro Álvarez [1,*], Alberto Cobos [1], Lexuri Vázquez [1], Noelia Ruiz [2], Pedro Pablo Rodríguez [2], Ana Magaña [3], Andrea Niklas [3] and Fernando Santos [3]

1 LORTEK Technological Centre, Basque Research and Technology Alliance (BRTA), 20240 Ordizia, Spain; acobos@lortek.es (A.C.); lvazquez@lortek.es (L.V.)
2 EIPC RESEARCH CENTER, AIE, Torrekua 3, 20600 Eibar, Spain; nruiz@eipc.es (N.R.); prodriguez@eipc.es (P.P.R.)
3 Fundación AZTERLAN, Basque Research and Technology Alliance (BRTA), Aliendalde Auzunea 6, 48200 Durango, Spain; amagana@azterlan.es (A.M.); aniklas@azterlan.es (A.N.); fsantos@azterlan.es (F.S.)
* Correspondence: palvarez@lortek.es; Tel.: +34-943-88-23-03

Abstract: In this work, weldability and hot cracking susceptibility of five alloy 718 investment castings in laser beam welding (LBW) were investigated. Influence of chemical composition, with varying Si contents from 0.05 to 0.17 wt %, solidification rate, and pre-weld heat treatment were studied by carrying out three different weldability tests, i.e., hot ductility, Varestraint, and bead-on-plate tests, after hot isostatic pressing (HIP) and solution annealing treatment. Onset of hot ductility drop was directly related to the presence of residual Laves phase, whereas the hot ductility recovery behaviour was connected to the Si content and γ grain size. LBW Varestraint tests gave rise to enhanced fusion zone (FZ) cracking with much more reduced heat-affected zone (HAZ) cracking that was mostly independent of Si content and residual Laves phase. Microstructural characterisation of bead-on-plate welding samples showed that HAZ cracking susceptibility was closely related to welding morphology. Multiple HAZ cracks were detected in nail or mushroom welding shapes, typical in keyhole mode LBW, irrespective of the chemical composition and thermal story of castings. In all LBW welds, Laves phase with a composition similar to the eutectic of the pseudo-binary equilibrium diagram of alloy 718 was formed in the FZ. The composition of this regenerated Laves phase matched with the continuous Laves phase film observed along HAZ cracks. This was strong evidence of backfilling mechanism, which is described as wetting and infiltration of terminal liquid along γ grain boundaries of parent material. The current results suggest that this cracking mechanism was activated in three-point intersections resulting from perpendicular crossing of columnar grain boundaries with fusion line and was enhanced by nail or mushroom weld shapes and narrow and columnar γ grain characteristics of castings. Neither Varestraint nor hot ductility weldability tests can reproduce this particular cracking mechanism.

Keywords: investment casting; alloy 718; hot cracking mechanism; Varestraint test; laser beam welding

Citation: Álvarez, P.; Cobos, A.; Vázquez, L.; Ruiz, N.; Rodríguez, P.P.; Magaña, A.; Niklas, A.; Santos, F. Weldability Evaluation of Alloy 718 Investment Castings with Different Si Contents and Thermal Stories and Hot Cracking Mechanism in Their Laser Beam Welds. *Metals* **2021**, *11*, 402. https://doi.org/10.3390/met11030402

Academic Editors: João Pedro Oliveir and Zhi Zeng

Received: 3 December 2020
Accepted: 9 February 2021
Published: 1 March 2021

1. Introduction

Alloy 718 was developed almost 60 years ago [1] and it has been the most widely used Ni-based superalloy to date. Being a precipitation-strengthened superalloy, it has been broadly used for the manufacturing of both land-based energy and aircraft turbine components, showing an outstanding performance at working temperatures up to 700 °C under high structural loading and corrosive conditions [2].

While other precipitation-hardened Ni superalloys have relatively high amounts of gamma prime (γ') former elements, i.e., Al and Ti, alloy 718 is based on the addition of Nb, which forms metastable gamma double-prime (γ'') precipitates of Ni_3Nb. The precipitation kinetics of γ'' is slower compared to γ' (Ni_3Al, Ni_3Ti and $Ni_3(Ti, Al)$), which contributes to improve castability, hot working, and weldability. The improvement is basically due to the fact that alloy 718 remains in a softer state during these manufacturing processes, avoiding the build-up of internal stresses. In terms of weldability, the sluggish precipitation kinetics of alloy 718 minimises strain age cracking (SAC) after welding and during post-welding heat treatment [2].

Alloy 718 investment castings are usually melted and poured inside vacuum furnaces and subsequently heat treated by hot isostatic pressing (HIP) to ensure highest performance. Intermetallic phases such as NbC and Nb-rich Laves phase can be found in alloy 718 castings due to segregation of chemical elements during slow solidification [3,4]. These secondary phases solidify at low temperatures (e.g., γ/Laves eutectic at temperatures down to 1180 °C) and they are usually concentrated along grain boundaries. Melting of Laves phases that are formed during the terminal solidification has been identified as the origin of the higher cracking susceptibility of castings during welding in comparison with wrought alloy 718 [4,5]. The reason for this is that Laves phase is readily melted upon heating in contrast with constitutional liquation of NbC that requires a dissolution reaction to form a liquid [2,6]. The latter is the predominant liquation mechanism identified in welds of wrought parts [2,6,7].

Incipient melting of Laves phase gives rise to a liquid which is distributed along grain boundaries, drastically reducing the strength of the material and its capability to withstand stresses. Therefore, sophisticated thermal treatments have been developed with the aim of reducing the amount of Laves phases and consequently heat-affected zone (HAZ) liquation cracking susceptibility of alloy 718 castings [4,8,9]. The aim of these treatments is to solubilise deleterious Laves phase and reduce compositional gradients in the as-cast microstructure. In fact, the concentration of several residual elements such as B, P, and S in grain boundaries can promote HAZ liquation cracking to a higher degree by decreasing even more the initial melting temperature and modifying the wetting characteristics of the intergranular liquid [4,10–12].

Formation of γ/Laves eutectic can also cause fusion zone (FZ) cracking in alloy 718 welds. This eutectic solidifies at a much lower temperature than the bulk matrix and wi-dens the solidification temperature range [2,4]. Wider solidification temperature range is directly associated with higher hot cracking susceptibility due to a longer coexistence of solid and liquid phases.

In order to study hot cracking susceptibility of Ni superalloys, researchers have defined and implemented different weldability assessment trials [2]. They are usually classified into three categories.

Self-restrained or representative tests use the inherent strain of the welding to induce cracking. They try to reproduce real joint configuration and residual stress levels. The drawback of this type of test is that it does not give quantitative values of cracking susceptibility and the result only indicates if the weld cracks pop up or not. Sometimes circular welding paths are applied to induce higher residual stresses.

In simulative tests, either a tension or bending deformation is externally applied du-ring welding. The Varestraint test is probably the best known test in this category, entai-ling the application of bending deformation along the longitudinal direction of the weld. The development of the Varestraint testing method from its origin was thoroughly reviewed by Andersson et al. [13]. Deformation enhances hot cracking and its extension, i.e., number and length of cracks, which can be determined at different strain levels. In this way, cracking susceptibility of different materials can be compared [14,15]. However, in real welding applications, residual stresses are predominant in contrast with plastic strain.

Finally, both strength and ductility are directly measured at high temperatures in hot ductility tests. In these tests, the temperature at which the material loses complete its strength (nil strength temperature (NST)) is determined by heating up testing samples under a constant tensile load. Additionally, nil ductility temperature (NDT), correspon-ding to the peak heating temperature at which the area reduction of the broken surface is 0%, and the ductility recovery temperature (DRT), at which 5% of area reduction is reco-vered after cooling down from a temperature close to NST, are computed. NDT and DRT are determined from on-cooling curves and the strain is only applied when the testing temperature has been reached. Hot ductility behaviour is linked to hot cracking susceptibility since cracks are generated when the material cannot accommodate stresses and strains induced during welding [2,9].

In a recent paper [16], the current authors compared the hot cracking susceptibility of wrought and investment casting alloy 718 by Varestraint test while applying pulsed and continuous tungsten inert gas (TIG) welding and LBW. It was concluded that hot cracking was enhanced in LBW samples due to extended centre line fusion zone (FZ) cracking showing a fishbone-like cracking pattern. Minor influence of pulsation mode and grain size was observed and, in fact, casting samples with grain sizes 30 times coarser showed slightly better performance than wrought material. It must be noted that Laves phases were not observed in investment casting samples and only some traces of needle-like delta (δ) phase and Mo sulphide were detected in base material.

Pulsed current can refine solidification microstructure and reduce the amount of Laves phase and Nb segregations in TIG welds according to [17,18]. Moreover, Bai et al. [19] recently investigated the potential benefits of combining high-frequency micro-vibration and LBW. Under particular vibration frequencies, the length of the liquation cracks in HAZ was reduced, but not completely avoided. The authors performed bead-on-plate tests and obtained weld cross sections with nail or mushroom shape. LBW has also been studied in alloy 718 parts by other researchers [20–25]. These investigations targeted the influence of LBW parameters and energy input on porosity, microstructure, and mechanical properties of these welds. A deep analysis of HAZ cracking susceptibility of this alloy during LBW was performed in particular by [20,22]. Impact of weld shape morphology, grain size, pre-weld heat treatment, and boron segregation were investigated.

In this work, weldability and hot cracking susceptibility in laser beam welds of five alloy 718 investment castings were investigated. Influence of chemical composition, Si content, solidification rate, and pre-weld heat treatment was studied by carrying out three different weldability tests. Cracking behaviour was compared, and results were completed with the detailed microstructural analysis after welding tests. Additionally, influence of pre-weld heat treatment on microstructure of parent material is discussed. An analysis about the correlation of weldability assessment test results, i.e., Varestraint and hot ductility, with cracking trend observed in real bead-on-plate LBW trials was performed. Fundamentals of the mechanism that triggers HAZ cracking in bead-on-plate test are explained.

2. Materials and Methods

2.1. Investment Casting and Chemical Composition of Alloy 718 Casting Heats

Investment casting moulds were manufactured, incorporating 20 test samples as flat plates of 150 × 50 × 10 mm in each mould. Moulds were covered with a ceramic shell composed by three primary layers and five backup layers, dewaxed, preheated, poured, and later demoulded, as described in Figure 1.

Casting samples were cast at industrial facilities under vacuum conditions. The mould preheating temperature was always 1150 °C, whereas the molten material was poured at 1450 °C.

The chemical composition of the different castings is shown in Table 1. The content of each alloying element was determined in cast samples using the following analytic techniques. C and S contents were measured by combustion and infrared absorption; N by inert gas fusion and thermal conductivity; O by inert gas fusion and infrared absorption; and finally Si, Mn, P, Fe, Cr, Mo, Ti, Al, and Nb contents were determined by spark atomic emission spectrometry. Co and B contents were not measured in casted samples, but they were taken from the chemical composition of the ingots employed as raw material for castings averaging 0.11 wt % and 0.002 wt %, respectively.

(a) Mould in wax

(b) Mould with shell

(c) Mould without shell

Figure 1. Investment casting moulds in different manufacturing steps.

Table 1. Chemical composition of alloy 718 casting heats in weight percentage.

Ref.	Ni	C	Si	Mn	P	S	Fe	Cr	Mo	Ti	Al	Nb + Ta
O	51.9	0.047	0.051	<0.050	<0.010	<0.005	21.1	17.8	3.02	0.89	0.47	4.75
E	52.1	0.049	0.11	0.037	<0.010	<0.005	20.4	17.6	2.91	0.98	0.59	4.92
P	51.7	0.038	0.17	<0.050	<0.010	<0.005	21.1	17.7	3.02	0.89	0.46	4.85
N/NP	52.5	0.058	0.12	0.038	<0.010	<0.005	20.3	17.7	2.88	0.77	0.47	4.88

Note that mould O (low Si content) was manufactured by using high purity ingots as raw material. In mould P (high Si content), Si was intentionally added during vacuum melting, and chemically adjusted ingots (28 Kg in total) were manufactured in a first melting step. Mould E corresponded to conventional chemical composition and casting process, whereas the cooling rate during solidification was reduced in moulds N and NP by incorporating 1 ceramic blanket over cast parts. The cooling rates between 800 and 500 °C were determined and they were 0.52 °C/s and 1.65 °C/s for the moulds with blanket (ref. N/NP) and without blanket (ref. O/E/P), respectively.

Once parts were shot blasted, they were submitted to a heat treatment process, according to GKN V.AC:9922 standard and comprising HIP and solution annealing thermal cycles. An additional solubilisation pre-HIP treatment consisting of solubilisation at 1052 °C for 2 h followed by air cooling was applied to slowly cooled mould NP. The goal of this treatment was to dissolve Laves phases before HIP treatment.

2.2. Weldability Assessment Trials

Three different weldability assessment trials were carried out with the 5 investigated alloy 718 casting heats after completing heat treatments described in the previous section, thus, in the solution annealing state. LBW Varestraint tests were carried out in a testing device (Figure 2) that was fully designed and manufactured at LORTEK [15,16]. Performance of the test bench complied with general requirements of ISO/TR 17643-1 "Destructive tests on welds in metallic materials—Hot cracking tests for weldments—Arc welding processes—Part 3: Externally loaded tests" [26]. LBW Varestraint tests were performed on 3.2 mm thickness samples that were electric discharge machined (EDM) from 10 mm thickness casting plates. External surfaces were milled before testing. LBW was applied on the surface of testing samples without adding any filler metal by TRUDISK 6002 disk laser from TRUMPF company, Ditzingen, Germany. The laser beam was guided through

400 μm diameter fibre to a TRUMPF BEO D70 laser welding head (200 mm focus length and 200 mm collimation length). LBW Varestraint tests were completed at 0.5 m/min welding speed, 2300 W continuous mode power, and 0.8 mm diameter spot size. Tests were performed in a closed chamber filled by argon gas to avoid surface oxidation and provide good shielding conditions.

(a)

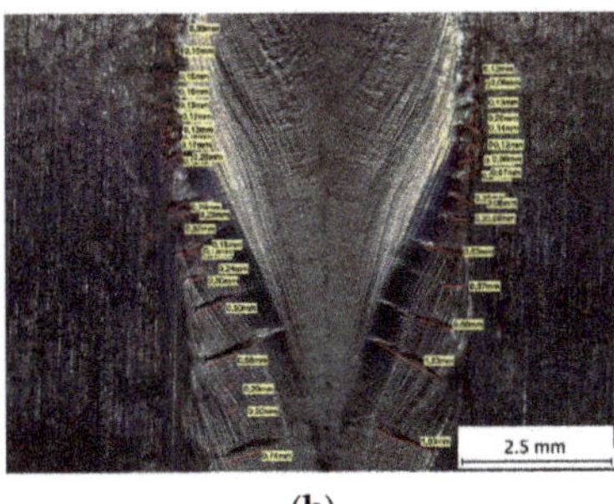

(b)

Figure 2. (**a**) LBW Varestraint testing device and (**b**) measurement of total crack lengths (TCL). Labels of individual cracks in fusion zone (FZ) and heat-affected zone (HAZ) are displayed.

Different augmented strains (ε) were applied during LBW by bending the samples along their length at 150 mm/s stroke rate and employing several interchangeable die blocks with radii varying from 20 to 320 mm. The induced augmented strains which are calculated by the following equation were in the range from 0.5 to 8%.

$$\varepsilon = \frac{t}{2 \cdot R} \times 100, \tag{1}$$

where ε is the resulting augmented strain as a percentage, t is the thickness of the sample in millimetres, and R is the radius of the die block in millimetres. Two expendable support plates of 304 stainless steel were positioned in both sides of the testing samples to avoid kinking. Both FZ and HAZ cracking susceptibility were studied by determining total crack lengths (TCL) in these two zones of the welds [2,13,14].

Note that in this case, cracks were measured on the surface of the testing samples using magnification lenses (up to 150×) and after cleaning the surface of the welds by soft manual polishing and oxalic acid electroetching to avoid reflections.

Additionally, hot ductility tests of 5 casting heats were carried out in Gleeble 3800D thermomechanical simulator (DYNAMIC SYSTEMS INC., Austin, USA) owned by West University in Sweden. Here, 6 mm (−0.025 mm, +0.01 mm) diameter cylindrical shape samples were finely turned from 10 mm thickness as-cast plates. Hot ductility testing setup and guidelines included in Gleeble Users Training 2010 handbook were applied. These are comparable to the testing specifications included in procedure B of [26], with minimum differences in samples length. NST temperature was only determined in mould E, concentrating the overall weldability assessment of 5 alloys on on-heating and on-cooling tests. A heating rate of 111 °C/s from room to testing temperatures was employed in on-heating tests, whereas samples were heated up to 1195 °C at the same heating rate and subsequently cooled down at 50 °C/s to each testing temperature in on-cooling trials. Temperature profile was recorded with K-type thermocouple welded to the surface of testing samples in the area between clamps. The samples were pulled to fracture at 55 mm/s stroke rate. Percentages of area reduction from initial 28.3 mm^2 (i.e., 6 mm diameter) were measured to determine ductility at the different testing temperatures.

Finally, self-restrained bead-on-plate LBW trials were carried out in casting plates. In these representative tests, the parent materials were remelted by scanning the surface with a laser beam with the same energy distribution, shielding conditions and parameters employed in LBW Varestraint tests. In this case, trials were carried out in samples with less than 3 mm thickness (between 2.6 and 2.9 mm) and 9 mm thickness, the latter resulting from the surface grinding of casting samples. Cross-sections of bead-on-plate samples were metallographically characterised to detect cracks in FZ and HAZ.

2.3. Metallographic Characterisation

As-cast, heat-treated, and bead-on-plate welding samples were characterised by optical microscopy (OM) in a LEICA MEF4 microscope (LEICA MICROSYSTEMS GmbH, Wetzlar, Germany) and field emission scanning electron microscopy (FESEM) with a ZEISS Ultra Plus microscope (CARL ZEISS AG, Oberkochen, Germany). Energy-dispersive X-ray (EDX) spectroscopy analysis was conducted in the FESEM microscope to determine local chemical composition of precipitates and phases. Average chemical composition of the Laves phases was determined through EDX analysis of 5 phases for each casting heat. For the metallographic analysis, cross sections were prepared by grinding and polishing using standard procedures. The area percentage of carbides was determined by image analysis of 5 images obtained by OM at 100× (HAZ and BM) and 500x (FZ) using Leica application suite V4.2. The area percentage of Laves phases + carbides was determined through SEM at 500× in the heat-affected zone (HAZ) and in the base metal (BM), and at 2500× in the fusion zone (FZ). Finally, the Laves phase area percentage was obtained by subtracting the carbide area percentage obtained by OM from the area percentage of Laves phases + carbides obtained by SEM.

The microsegregation degree of alloying elements was evaluated by the segregation ratio (SR), which can be calculated using the following formula [27]:

$$SR = C_{i;IR}/C_{i;DC} \quad (2)$$

where $C_{i,IR}$ is the maximum concentration of element i in interdendritic region, and $C_{i,DC}$ is the minimum concentration of element i in dendrite core. For SR < 1, solute elements tend to segregate to dendrite core during solidification; when SR > 1, the alloying elements partition towards the interdendritic region. If SR values are close to 1, the corresponding elements do not favourably segregate to any region. The concentration of each element was determined by EDX analysing 8 points in 3 different regions; an example is shown in Figure 3.

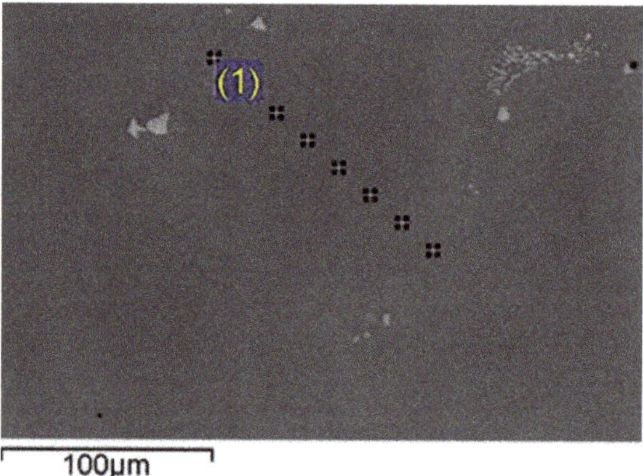

Figure 3. Location of EDX analysis points for determining the segregation ratio.

Grain size was revealed by etching with Kalling 2 reagent. Dimensions of at least 10 grains were analysed for each casting heat in cross-sections. The grain size was determined from the horizontal and vertical mean intersection lengths between 2 grain boundaries of grains.

The solidification path and evolution of Laves phases were studied by Thermo-Calc software (using TCNI10 and MOBNI4 databases). Scheil simulation, considering back diffusion of alloying elements in the primary phase, was performed at a high cooling rate of 100 °C/s, similar to the expected welding cooling rates, for alloys with different Si contents (moulds O, E, and P).

3. Results

3.1. As-Cast Microstructure

The secondary phases observed in the five moulds or casting heats in as-cast condition are shown in Figure 4. The as-cast microstructure of the five moulds consisted of a γ matrix, Laves phases, Nb carbides (NbC), and smaller quantities of TiNb carbonitrides (TiNbCN). In the moulds with slow solidification rate, additional δ and γ'' phases were observed in the segregated interdendritic region. Laves phases and carbides show similar colour and also morphologies when analysed by SEM; thus, for identification it is necessary to perform an EDX analysis. Therefore, for the quantification of the Laves phase area percentage included in Table 2, first the area percentage of Laves phases + carbides was determined by SEM images. Second, the area fraction of carbides was evaluated by optical microscopy, which revealed in the unetched state only the presence of carbides. Finally, as it is also explained in Section 2.3, the Laves phase area percentage was obtained by subtracting the area percentage of carbides obtained by OM from the area percentage obtained by SEM (Laves + carbides). It can be observed that, as the Si content of the alloy increased from 0.051 wt % (mould O) to 0.17 wt % (mould P), the area percentage of Laves phases increased from 2.2% to 3.5% and the Si content in the Laves phases increased from 0.28 to 1.29 wt %.

High segregation of alloying elements to the interdendritic spaces, particularly for Nb and Ti but also Mo, was observed in the as-cast state of every mould. A slight depletion in Fe and Cr was observed in those interdendritic spaces (Table 3).

The characteristics of the γ grains are depicted in Table 4. It can be observed that the grains of moulds fabricated with normal solidification rate (moulds O, E, P) showed a highly columnar grain morphology and an aspect ratio between 3.0 and 3.2, while the slowly solidified moulds presented a coarser grain size and aspect ratio below 2.0. Mean width was measured as the mean horizontal distance between grain boundaries, whereas mean length corresponded to distance along sample thickness. Morphology and grain size of γ grains in as-cast samples is shown in Figure 5.

3.2. Microstructure after Heat Treatment and before Welding Trials

Heat treatment did not modify the as-cast grain size and morphology observed in Figure 5, however, segregation and amount of Laves phase were significantly reduced in comparison with as-cast condition as can be concluded by comparing Tables 3 and 5. The characteristics of the Laves phases after heat treatment (HIP+S) are shown in Table 2, together with the ones of as-cast samples. No Laves phases were detected in the low and standard Si-bearing alloys (moulds O and E), but small area percentages between 0.14 and 0.35% remained in high-Si alloy (mould P) and slowly solidified alloy with (mould NP) and without pre-HIP (mould N) heat treatments. It is also worthwhile mentioning that after heat treatment, the Laves phase composition was enriched in Mo and Si.

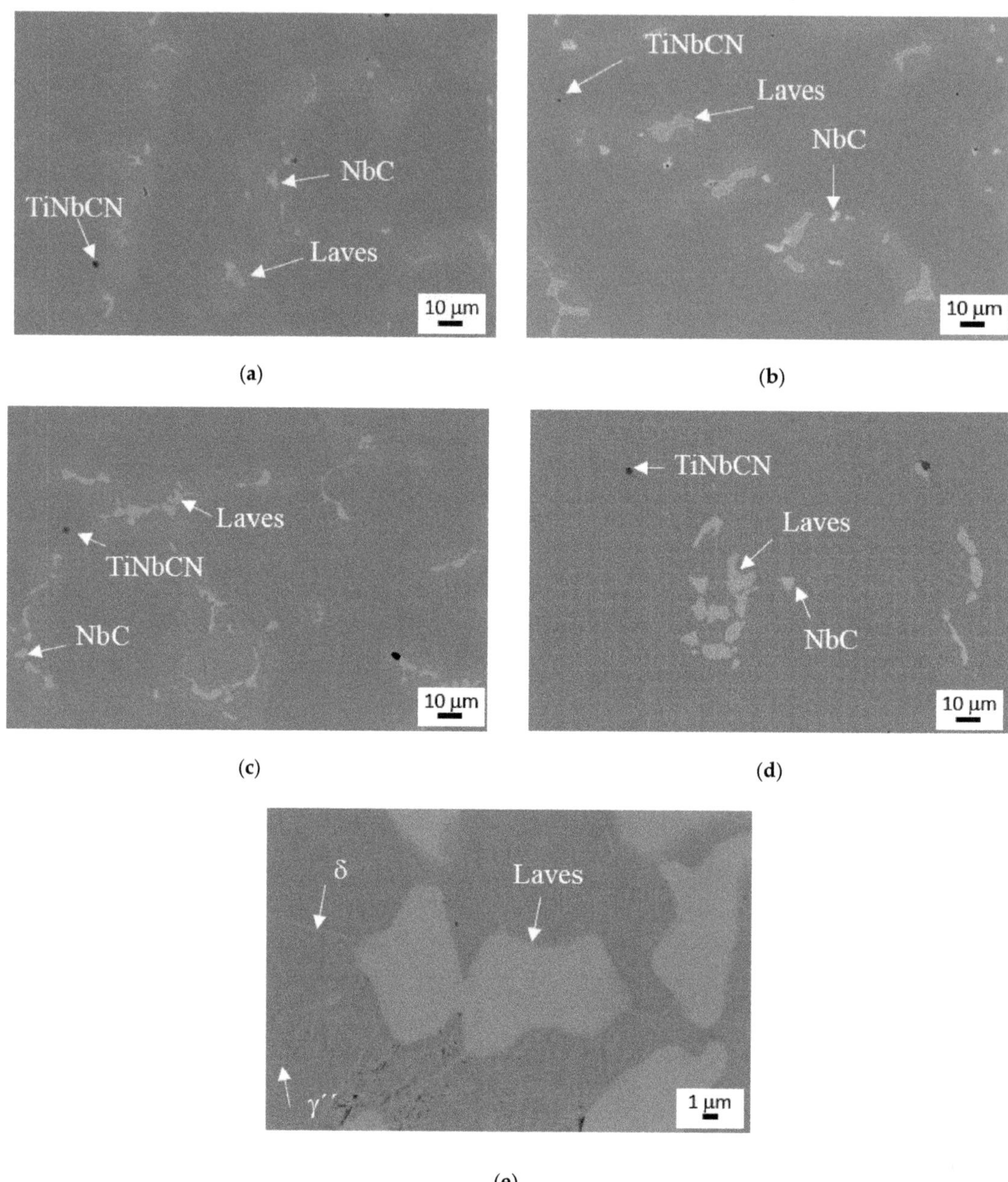

Figure 4. Secondary phases present in the microstructure of different moulds in the as-cast state: (**a**) O, (**b**) E, (**c**) P, and (**d**) NP; (**e**) NP: magnification of (**d**) showing the presence of Laves, δ phase, and γ″ in the interdendritic region.

Table 2. Area percentage and chemical composition of Laves phases (in wt %) as-cast and after heat treatment obtained by / EDX) analysis (* Laves phase was not detected).

Mould	State	Area % Laves	Al	Si	Ti	Cr	Fe	Ni	Nb	Mo
O	As-cast	2.20	0.13 ± 0.05	0.28 ± 0.06	0.91 ± 0.07	11.86 ± 0.05	11.98 ± 0.24	34.02 ± 0.64	33.90 ± 0.50	7.67 ± 0.50
	HIP + S *	0	-	-	-	-	-	-	-	-
E	As-cast	2.40	0.15 ± 0.03	0.78 ± 0.09	1.02 ± 0.06	11.57 ± 0.46	11.63 ± 0.40	34.94 ± 0.48	32.49 ± 0.79	7.42 ± 0.38
	HIP + S *	0	-	-	-	-	-	-	-	-
P	As-cast	3.50	0.13 ± 0.05	1.29 ± 0.08	0.95 ± 0.05	11.15 ± 0.41	11.95 ± 0.36	34.63 ± 0.43	32.11 ± 0.70	7.80 ± 0.34
	HIP + S	0.35	0.08 ± 0.06	2.02 ± 0.07	0.59 ± 0.09	11.43 ± 0.07	12.82 ± 0.24	29.76 ± 0.33	30.90 ± 0.76	12.39 ± 0.53
N	As-cast	2.60	0.17 ± 0.02	0.91 ± 0.11	0.81 ± 0.08	11.19 ± 0.21	11.10 ± 0.21	34.54 ± 0.26	33.94 ± 0.35	7.40 ± 0.16
	HIP + S	0.19	0.09 ± 0.06	1.71 ± 0.07	0.52 ± 0.11	11.48 ± 0.24	11.89 ± 0.24	30.15 ± 0.037	30.85 ± 0.38	13.99 ± 0.69
NP	Pre-HIP	2.10	0.17 ± 0.06	1.19 ± 0.12	0.67 ± 0.14	10.90 ± 0.70	11.29 ± 0.39	32.59 ± 0.55	34.10 ± 0.97	9.09 ± 0.49
	Pre-HIP + HIP + S	0.14	0.16 ± 0.05	1.76 ± 0.11	0.53 ± 0.05	10.98 ± 0. 30	11.83 ± 0.20	30.97 ± 0.71	31.28 ± 0.88	12.62 ± 0.93

Table 3. Segregation ratio (SR) of alloying elements in the as-cast state.

Mould	Ti	Nb	Mo	Fe	Cr
O	2.28	3.70	1.49	0.76	0.86
E	2.37	3.75	1.25	0.79	0.85
P	2.57	3.95	1.40	0.78	0.88
N	2.04	2.73	1.37	0.78	0.82

Table 4. Grain size (mean with and length with standard deviation) in the as-cast state.

Mould	Mean Width (mm)	Mean Length (mm)	Aspect Ratio	Morphology
O	1.1 ± 0.35	3.4 ± 0.92	3.1	Columnar
E	1.0 ± 0.46	3.2 ± 0.97	3.2	Columnar
P	1.1 ± 0.33	3.3 ± 0.78	3.0	Columnar
N/NP	2.1 ± 0.34	3.4 ± 0.71	1.5	Coarse, slightly columnar

3.3. LBW Varestraint Weldability Test Results

Figure 6 shows the hot cracking behaviour observed in the five casting heats that was determined by LBW Varestraint test. Moulds O (low Si), P (high Si), and E (standard Si), which were cast without ceramic blanket and therefore solidified at quicker cooling rates, were tested at augmented strain levels from 1 to 8%. Slowly cooled heats (moulds N and NP) were only tested up to 4% augmented strain. Results show that in all casting heats, FZ cracking was much more prominent than HAZ cracking. In fact, TCL measured in FZ was more than 5–6 times longer than in HAZ. At 8% augmented strains, TCL determined in the HAZ ranged between 2.5 and 5 mm, whereas for the same testing conditions, TCL in FZ was between 20 and 30 mm. Therefore, it was observed that LBW Varestraint test mainly enhanced FZ cracking.

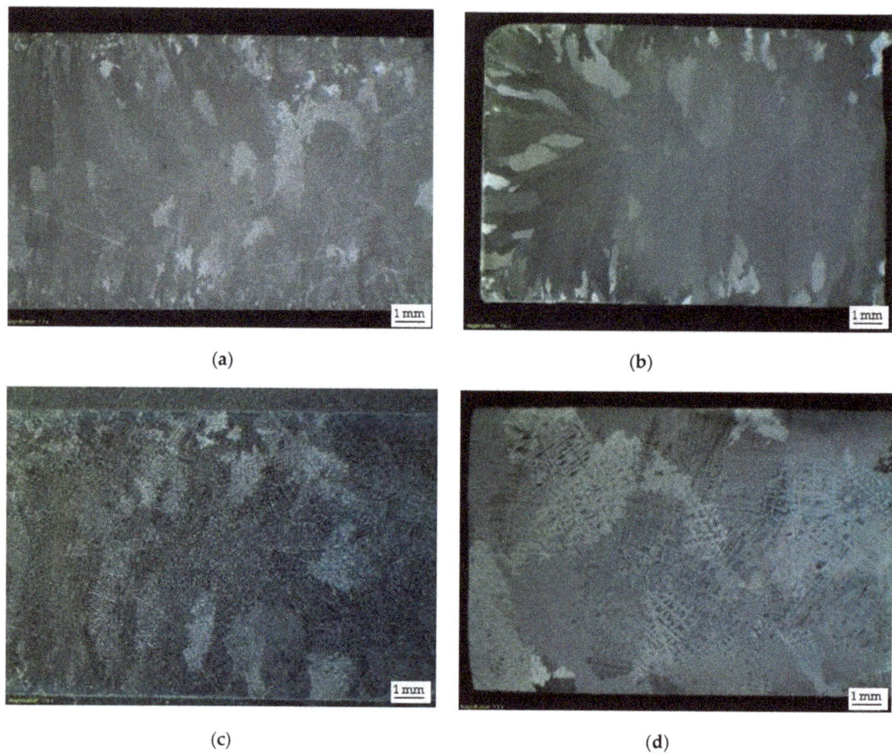

Figure 5. Grain size of different moulds in the as-cast state: (**a**) O, (**b**) E, (**c**) P, and (**d**) NP.

Table 5. Segregation ratio (SR) of alloying elements after heat treatment.

Mould	Ti	Nb	Mo	Fe	Cr
O	1.29	1.37	1.28	0.89	0.93
E	1.27	1.16	1.32	0.94	0.95
P	1.20	1.34	1.28	0.93	0.94
N	1.13	1.52	1.52	0.98	0.99
NP	1.33	1.35	1.23	0.92	0.94

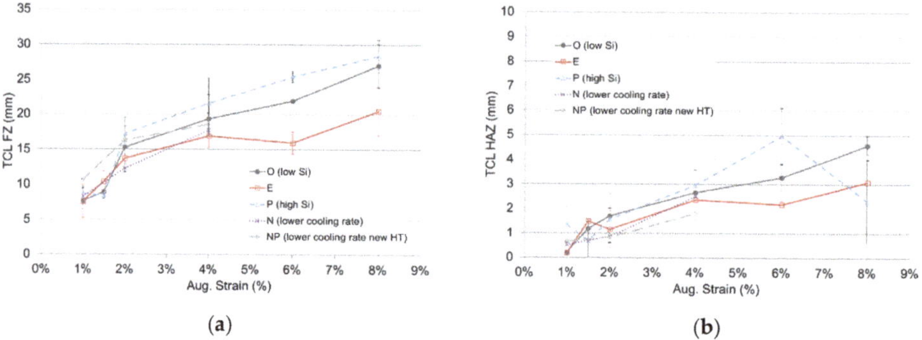

Figure 6. (**a**) FZ and (**b**) HAZ cracking response of five alloy 718 casting heats in LBW Varestraint test.

Mould P (high Si content) showed a comparatively higher cracking susceptibility than the rest of the heat levels. Nevertheless, mould O (low Si) was ranked second in terms of trend towards FZ and HAZ cracking. Mould E with intermediate Si content and NP showed slightly lower TCL values, whereas behaviour of mould N was in between.

3.4. Hot Ductility Weldability Test Results

Figure 7 describes the area reduction percentage determined in on-heating and on-cooling hot ductility tests for the five casting heats. Continuous lines correspond to the grade 3 polynomial fitting of experimentally determined on-heating test results, whereas dashed lines depict on-cooling behaviour. Looking at on-heating curves, the ductility at temperatures between 950 and 1000 °C was higher than 60% for the five moulds, showing comparatively higher values in the case of mould E and O (low Si content). Between 1000 and 1050 °C, ductility started to drop, and at 1150 °C it was already below 2%, which meant that the capability to deform without breaking had been completely lost. Again, at intermediate 1100 °C, moulds E and O (low Si content) showed comparatively better performance in terms of ductility, which was the reason why fitting curves were slightly displaced towards higher temperatures. This means that the onset of ductility drop in these two moulds was delayed to some extent.

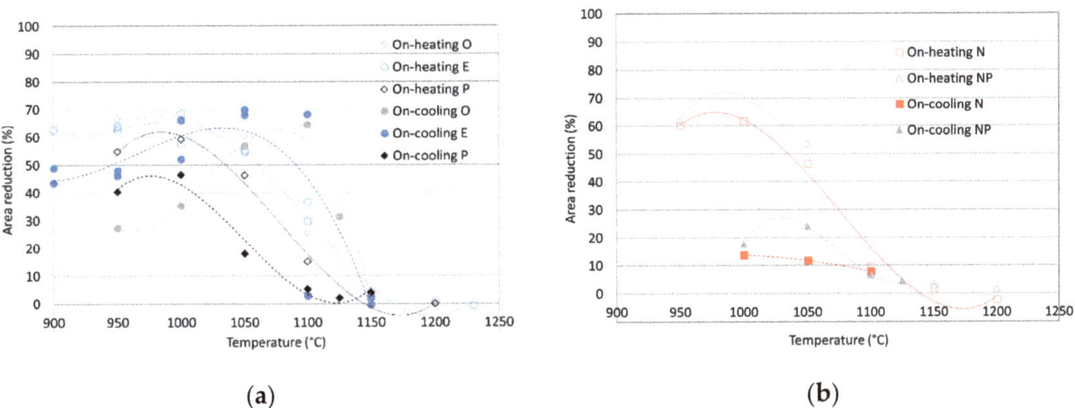

Figure 7. On-heating and on-cooling curves in terms of area reduction percentage and testing temperature for moulds E, O, and P (**a**), and N and NP (**b**).

Differences between casting heats were much more evident in on-cooling curves. Mould O (low Si content) presented remarkable ductility recovery behaviour, quickly reaching an area reduction value of 64% after testing at 1100 °C. Note that the thermal sequence of on-cooling hot ductility tests involved fast heating to peak temperature of 1195 °C and cooling down to the corresponding test temperature. Peak temperature was selected after defining NST in mould E samples that reached 1263.5 °C ± 5.8 °C. It was decided to limit peak temperature in on-cooling test to 1195 °C in order to ensure repeatable and stable ductility recovery behaviour.

After reaching ductility values or original parent material, mould O showed a ductility drop at lower testing temperatures down to 27% at 950 °C. This drop was not so remarkable either in mould E (from 69% to 46%) or high Si P (from 46% to 40%). Thus, it is quite clear that the ductility recovery rate is strongly related to the Si content of the alloys.

Moulds N and NP with slower cooling rates in the casting process depicted a completely different ductility recovery performance during on-cooling tests. In these two heats, restored ductility values did not surpass 14% and 24%, respectively, and the slope of the curves was drastically reduced. Calculated DRT and brittle temperature range (BTR) values are included in Table 6. BTR is the difference between peak temperature employed in on-cooling tests and determined DRT, at which 5% of area reduction is recovered. BTR is a parameter widely used to conclude on the hot cracking susceptibility of superalloys [2,9,26].

Table 6. Weldability parameters obtained from hot ductility tests.

Mould	Peak Temperature (°C)	DRT (°C)	BTR (°C)	Max. Ductility after Cooling	Ductility Recovery Rate
O	1195	1150	45	64% at 1100 °C	Very fast
E	1195	1145	50	69% at 1050 °C	Fast
P	1195	1090	105	46% at 1000 °C	Intermediate
N	1195	1110	85	14% at 1000 °C	Very low
NP	1195	1110	85	24% at 1050 °C	Very low

3.5. Bead-on-Plate Weldability Test Results

Figure 8 shows an image of the bead-on-plate tests performed on casting plates from five heats. Both straight and circular weld paths were applied. Circular welds were 15 mm in diameter and the goal was to modify the self-constraint condition. As described above, bead-on-plate tests were carried out both in 9 mm and less than 3 mm thickness plates, after surface grinding and EDM cutting, respectively, employing similar welding parameters as in LBW Varestraint tests. In the case of thinner samples, laser power was decreased to 2050 W to avoid excessive root overhang.

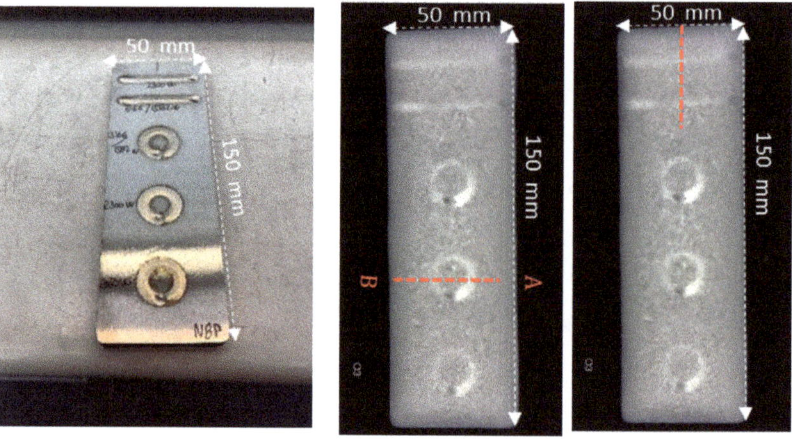

Figure 8. Bead-on-plate trials and X-ray digital images showing cuts for cross-section metallographic examination.

Every examined weld was free of cracks in less than 3 mm thickness samples, showing a "bowl-like" welding shape (Figure 9), that is, a shape without characteristic nail head of keyhole mode LBW. However, several cracks were detected in the HAZ in 9 mm thickness samples (Figure 10) welded with similar process parameters (2300 W), following both straight and circular welding paths. Note that in this case, welding resembled "nail or mushroom shape" usually observed in keyhole mode LBW.

Total number of cracks and TCL determined in each cross section were included in Figure 11. For circular welds, average values determined from cross sections A and B were represented. It is clearly shown that straight welds gave rise to longer and higher number of cracks with minor influence of alloy composition.

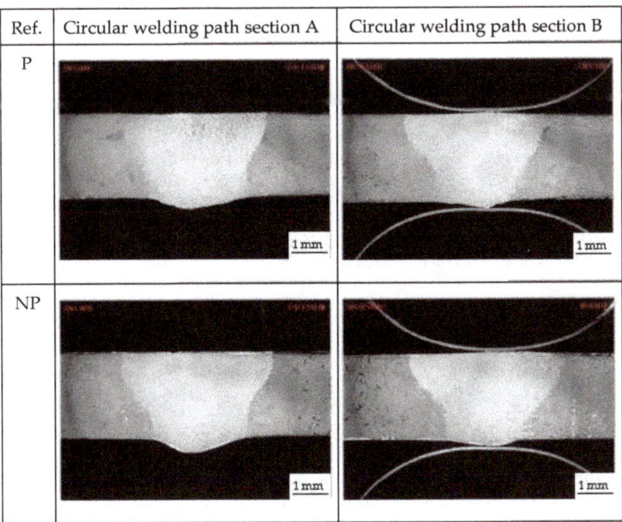

Figure 9. Cross-section of bead-on-plate tests of moulds P and NP with less than 3 mm thickness. Laser power 2050 W.

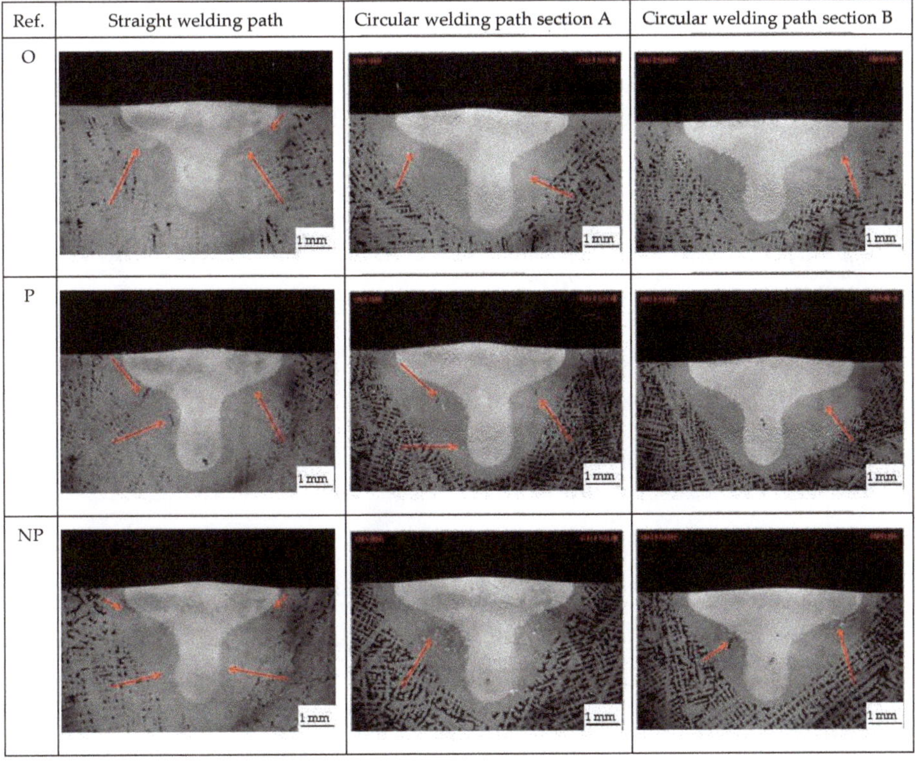

Figure 10. Cross-section of bead-on-plate tests of moulds O, P, and NP with 9 mm thickness. Laser power 2330 W.

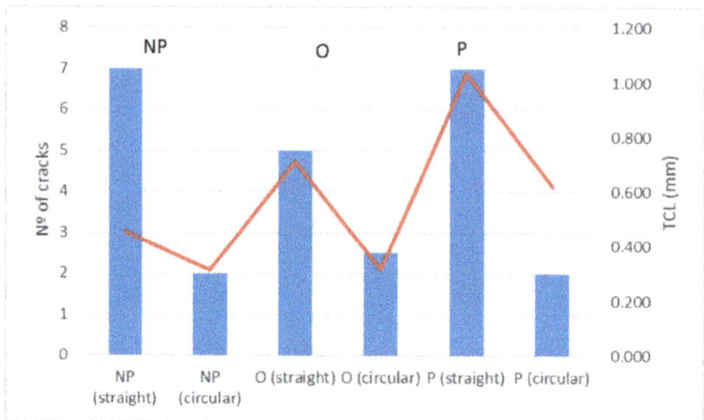

Figure 11. Number of cracks (blue bars) and TCL (red line) in bead-on-plate cross-sections of moulds O, P, and NP with 9 mm thickness. Laser power 2330 W.

3.6. Microstructural Characterisation of Bead-on-Plate Welding Samples

The composition of Laves phases in FZ and HAZ/BM (base metal) was analysed by EDX in both 9 and less than 3 mm thickness plates; corresponding values are included in Table 7. It can be observed that Laves phases in FZ had much lower Si, Nb, and Mo contents with respect to the HAZ and BM, and, on the contrary, they were enriched in Ti, Cr, Fe, and Ni elements. The Si content of Laves phase in FZ increased with the Si weight percentage of the alloy, with 0.22% being the lowest in mould O heat. In FZ the area percentage of Laves phases increased with increasing Si content of the alloy. Surprisingly, the area percentage of Laves phases of 3 mm plates of both moulds (P and NP) were lower than in the 9 mm thickness plates. The 9 mm thickness plates evacuated heat more quickly by thermal conduction, and therefore comparatively higher solidification rates and less Laves phases were expected.

Table 7. Area percentage and mean chemical composition of Laves phase (in wt %) and standard deviation in FZ and HAZ/BM of bead-on-plate welding samples.

Mould	Location	Area % Laves	Al	Si	Ti	Cr	Fe	Ni	Nb	Mo
					9 mm					
O	FZ	1.78	0.41	0.22	1.51	14.02	14.80	43.52	20.53	4.99
	HAZ/BM	-	-	-	-	-	-	-	-	-
P	FZ	4.13	0.40	0.65	1.68	13.44	14.19	44.12	20.51	5.02
	HAZ/BM	0.20	-	1.98	0.53	11.16	12.55	30.70	30.33	12.74
NP	FZ	2.41	0.56	0.47	1.57	13.53	13.42	43.65	22.11	4.68
	HAZ/BM	0.32	-	1.55	0.57	11.64	12.24	31.20	29.14	13.66
					Less than 3.0 mm plate					
P	FZ	2.63	0.51	0.33	1.36	14.34	15.17	43.29	20.17	4.82
	HAZ/BM	0.20	0.14	2.01	0.5	10.96	12.43	30. 92	30.42	12.58
NP	FZ	1.50	0.48	0.49	1.46	13.61	13.88	43.07	22.07	4.94
	HAZ/BM	0.30	0.40	1.43	0.81	11.94	12.03	35.76	26.73	11.17

Laves phase was not detected in HAZ and BM of the low Si O alloy. In the high Si alloy (mould P) and the standard Si alloy with low solidification rate and pre-HIP (NP), small quantities of Laves phases in HAZ/BM were still observed, which were higher in the latter. This indicates that the time and temperature of additional heat treatment before HIP were not enough to completely dissolve these phases.

As described above, cracks in HAZ were only observed in the 9 mm thickness plates along the grain boundaries. Higher magnification SEM images are displayed in Figure 12.

This micrograph corresponded to mould P (high Si) and it was representative of the rest of the observed cracks in 9 mm thickness samples. These cracks were partially filled with a continuous Laves phase film, which was observed either at the tip of the crack or at the edges of the open crack. The chemical composition of this Laves film matched exactly with the Laves phase composition in FZ (Table 8).

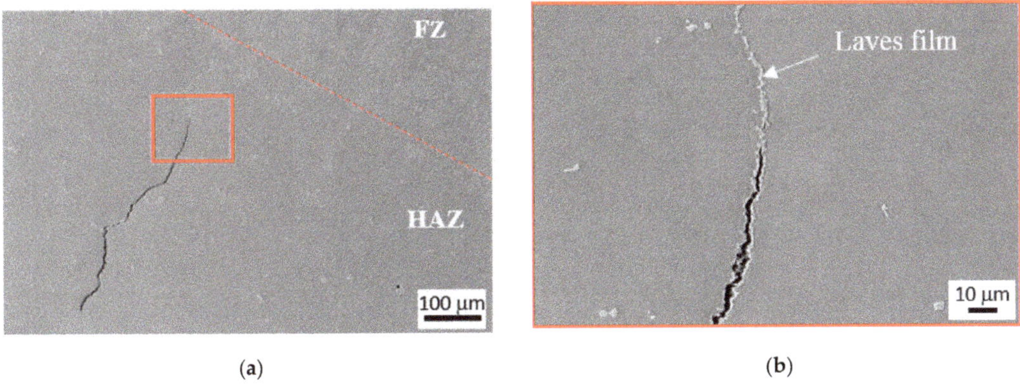

Figure 12. (**a**) Grain boundary cracking in HAZ and (**b**) higher magnification of squared zone. Mould P.

Table 8. Composition of Laves phases (in wt %) at different locations in 9 mm thickness sample of Mould P.

Laves Phase	Al	Si	Ti	Cr	Fe	Ni	Nb	Mo
Crack-HAZ	0.83	0.73	1.45	12.93	12.94	42.36	23.68	5.08
FZ	0.40	0.65	1.68	13.44	14.19	44.12	20.51	5.02
HAZ	-	1.98	0.53	11.16	12.55	30.70	30.33	12.74

Besides cracks, there was also evidence of Laves phase liquation observed in the HAZ of moulds P and NP for both less than 3 mm and 9 mm thickness plates. These phases were located in the interdendritic region, also showing the presence of carbides and δ phase (Figure 13).

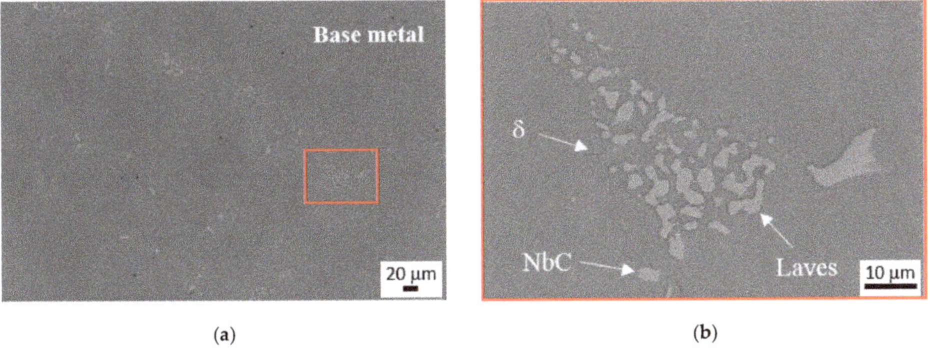

Figure 13. *Cont.*

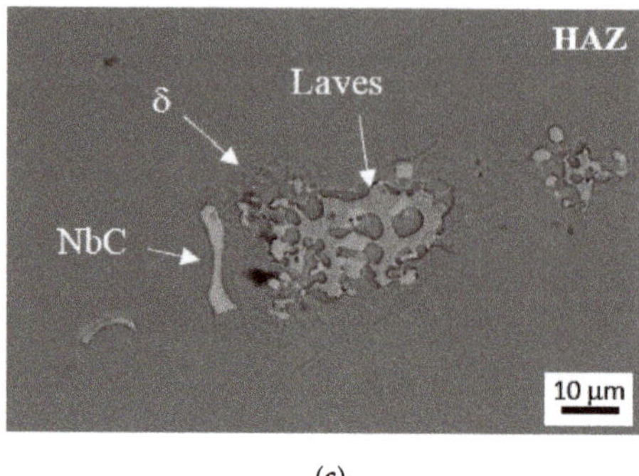

(c)

Figure 13. (**a**) Phases in the interdendritic region. (**b**) Magnification of the squared zone showing the presence of δ phase, carbides, and Laves phases in base metal (BM) in the interdendritic segregation area. (**c**) Segregation area with evidence of Laves and δ phase liquation in HAZ (NP- less than 3 mm plate).

4. Discussion

4.1. Influence of Chemical Composition, Investment Casting Conditions, and Heat Treatment on Microstructure and Weldability

Moulds E and O with higher casting solidification rates (1.65 °C/s) and standard and low Si contents, respectively, did not present any Laves phases in solution annealing state after complete HIP + solubilisation annealing heat treatment. Area percentage of Laves phase observed in moulds N, NP, and P in the solution-annealed condition ranged from 0.14 to 0.35%, without significant differences between moulds N and NP and being the residual content of Laves phase higher in mould P, that is, the heat with higher Si content and faster solidification rate. Note that this mould also had the highest amount of Laves phase in the as-cast condition (Table 2).

In moulds P, N, and NP containing residual Laves phase in the solution annealing state, the onset of ductility drop in on-heating hot ductility tests was triggered at lower temperatures, and corresponding fitting curves were shifted to the left side of the chart. This was observed in curves displayed in Figures 7a and 14, with the latter superposing the on-heating curves of the five studied heats. Note that dashed fitting lines corresponded to those three moulds. Therefore, this early ductility drop was associated with incipient melting of Laves phase. It is worth mentioning that melting of Laves phase is a fast event which does not require any significant reaction time to form the liquid as in constitutional liquation of NbC [2,6], and, consequently, it readily melts upon heating at very fast rates, as in current hot ductility tests (111 °C/s heating rate).

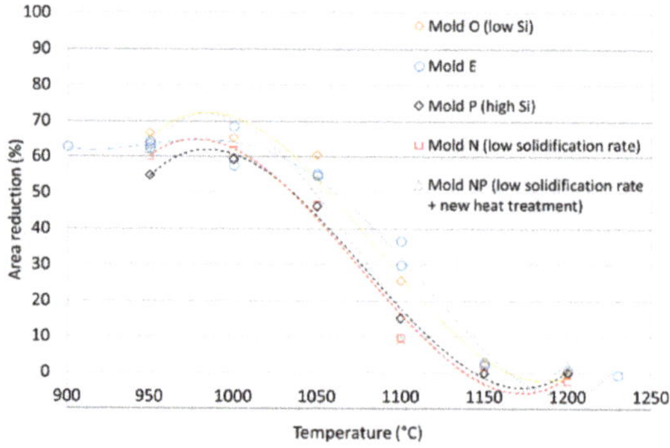

Figure 14. Comparison of on-heating curves in terms of area reduction percentage and testing temperature for moulds E, O, P, N, and NP.

Effective dissolution of Laves phase in alloy 718 castings prior to welding and high temperature use is pursued in industrial manufacturing processes because this phase impairs both weldability and mechanical properties [4,5]. This has been particularly shown in castings with volume fractions of secondary particles (including Laves and NbC) higher than the residual values reported in this work.

The chemical composition of remaining Laves phase was remarkably modified in moulds P, N, and NP after HIP and solubilisation treatment. The considerable enrichment in Mo could be explained by differences in diffusivities of solute elements in the austenite matrix. The dissolution kinetics of Laves phase in IN718 has been studied by [28]. In this work, the authors concluded that the back-diffusion of molybdenum in austenite is the controlling micro-mechanism for dissolution of the undesirable Laves phase. On the basis of the Johnson–Mehl–Avrami–Kolmogorov (JMAK) analysis at different temperatures, the authors determined the activation energy of 274.5 kJ/mol for the dissolution of Laves phase, which was close to the activation energy for diffusion of Mo in Ni (288 kJ/mol). This was also supported by the diffusion calculations shown in Table 9, which shows lower diffusivity values for Mo in Ni than for Nb and Ti.

Table 9. Parameters for diffusivity calculation of Nb, Ti, and Mo in Ni and calculated diffusivities.

Element	Q (kJ/mol)	Do (m^2/s)	$D_{1050\ °C}$ (m^2/s)	$D_{1100\ °C}$ (m^2/s)	$D_{1150\ °C}$ (m^2/s)	Reference
Nb	202	1.0×10^{-6}	1.1×10^{-14}	2.1×10^{-14}	4.0×10^{-14}	[29]
Ti	257	86×10^{-6}	0.6×10^{-14}	1.4×10^{-14}	3.2×10^{-14}	[30]
Mo	288	300×10^{-6}	0.1×10^{-14}	0.8×10^{-14}	0.8×10^{-14}	[31]

It is also interesting to note that the chemical composition of remaining Laves phase in these three heats was comparable with only minor differences in Si content.

HIP and solubilisation treatment did not modify original grain size and morphology resulting from investment casting. Consequently, different cooling conditions during casting yielded different grain sizes and aspect ratios which remained in the base material employed for the weldability tests. Thus, moulds O, E, and P cooled without ceramic blanket had highly columnar grains elongated along plate thickness. On the contrary, HIP and subsequent solution annealing heat treatment was effective in reducing segregation in interdendritic regions, leading to comparable segregation ratios in the five heats (Table 5).

4.2. Correlation between Weldability Assessment Trials

In LBW Varestraint test, which is an externally loaded weldability test, cracks were mainly induced on the surface and in FZ since the strain was applied while the melt pool was solidifying and it was forced to pull away when the material did not have minimum strength and ductility to accommodate residual deformations. FZ cracking was highly enhanced in LBW Varestraint test, featuring an elongated V-shape solidification line due to the LBW parameters and energy density which was required to achieve full penetration and minimum weld width requested by industrial quality standards [16]. In this study, the section of Varestraint testing samples had to be reduced to 3.2 mm to deform them by bending due to test bench capability and to adjust them to representative welding applications.

Minimum differences in HAZ cracking were determined between heats with somehow better performance of NP and E. However, the relatively high scattering of TCL va-lues made the comparison between alloys difficult.

Clear differences between alloys were observed in on-cooling hot ductility Gleeble tests. In these tests, only incipient melting of low melting point phases took place. This melting was enough to wet grain boundaries, leading to a brittle fracture without any area reduction at the test temperatures close to NDT. By reducing on-cooling testing temperature, we were able to study ductility recovery behaviour. On-cooling hot ductility results showed that moulds O and E (without Laves phase in BM at the beginning of the welding test) had shorter BTR and fast ductility recovery rates reaching the previous values before later ductility drop (Table 6). In both moulds, the ductility was effectively recovered at temperatures above 1050 °C (Figure 7). Moulds N and NP presented higher BTR (85°) and very low recovery rate and ductility recovery capability. Coarser γ grain sizes in these samples could be the reason for this behaviour, since large grain sizes promote continuity of liquid resulting from incipient melting at grain boundaries and reduce extension of interfacial area between solid-state γ grains [2].

Mould P with higher Si content in its chemical composition had the largest BTR (105 °C) and intermediate recovery rate and capability. Longer BTR can be related to both the higher amount of residual Laves phase in the microstructure and its greater Si content. Both factors will contribute to increasing the volume fraction of intergranular liquid formed at the peak temperature reached in on-cooling test. Consequently, the temperature must be decreased to a lower point to allow full resolidification of the Si-enriched liquid. It must be mentioned that the effect of Si content on 718 alloys has been previously investigated, concluding that HAZ cracking trend is favoured if high Si contents are combined with high Mn or C contents [2,4].

Hot ductility tests give an insight about the response of material to liquation and subsequent resolidification. However, only a very limited amount of material is melted as opposed to real welding applications in which a relatively high amount of material is melted in the FZ and significant microstructural changes occur in this zone.

Indeed, this was observed in the FZ microstructural characterisation of bead-on-plate samples. Laves phases were regenerated in FZ, whose chemical composition was completely different from the original Laves phase in HAZ and BM. FZ Laves phase in 9 mm plates were particularly depleted in Nb, Si, and Mo (Table 7). Surprisingly, the number of Laves phases in the less than 3 mm plates were much lower than in the 9 mm plates with faster welding cooling rates.

Whereas 3 mm plates were free of cracks, remarkable microfissures were identified in HAZ along grain boundaries in 9 mm bead-on-plate samples for the five investigated heats. As described previously, these cracks were decorated by continuous film, whose composition matched with Laves phase of FZ (Nb (20.5–22.1 wt %), Mo (4.7–5.0 wt %), and Ti (1.5–1.7 wt %)). Composition of this Laves phase film is quite independent of chemical composition, with only minor deviations in Si content. Laves phase with very similar Nb concentrations in LBW welds were reported by Odabaçi et al. [25].

As can be observed in Figures 12 and 15, corresponding to cross-sections of 9 mm thickness bead-on-plate samples of moulds P and O, respectively, the Laves film was also extended through FZ. This is strong evidence of backfilling mechanism. The current results demonstrate that in 9 mm bead-on-plate samples, the terminal liquid which remains in the melt pool at the end of the solidification process was diffused along grain boundaries, giving rise to a continuous Laves phase film that caused HAZ microfissuration.

Looking at the micrographs, we concluded that this diffusion was completed along base material γ grain boundaries which are perpendicular to fusion line and delimit three-point intersections. Similar crack morphologies have been reported by Bai et al. [19] in LBW. The area below the nail head is a particular critical point where these microfissures pop up. It is worth mentioning that grain morphology was quite columnar, with grain boundaries elongated along the thickness, particularly in samples cast without ceramic blanket. Therefore, at the nail head area where the fusion line was quite horizontal, there was a high probability of having grain boundaries in the parent material intersecting fusion line, leading to three-point intersection, which is a critical point with high stresses during melt pool solidification [2]. Moreover, the elongated and straight morphology of vertical grain boundaries would be favourable for the described diffusion-based backfilling mechanism. Evidence of comparable backfilling mechanism which promotes HAZ cracking has been recently reported in Haynes® 282® casting alloy [32]. In this case, cracking was exacerbated by diffusion of B. Other authors cited by [2] also concluded that segregation of B has a detrimental effect of HAZ liquation and FZ solidification cracking. However, this is not an influencing parameter in this work since ingots with the same lean B content of 0.002 wt % were used as raw material for the five heats.

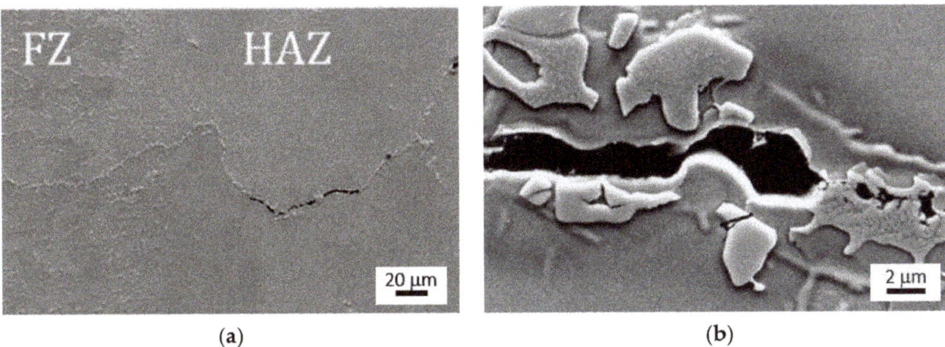

(a) **(b)**

Figure 15. (**a**) Film along crack in HAZ from 9 mm thickness sample mould O cross-section. (**b**) Enlargement of crack tip.

Straight bead-on-plate welds were more prone to HAZ cracking phenomena since both total number of cracks observed in cross-sections and TCL of those cracks were comparatively higher than in circular welds (Figure 11). Bead-on-plate samples from NP mould showed a consistently lower TCL than equivalent samples from moulds O and P. Therefore, it can be concluded that the lower aspect ratio of these grains due to slower solidification rates during initial casting reduced the probability of having deleterious three-point intersections along the length of fusion line.

No cracks were observed in less than 3 mm thickness samples welded with comparable LBW parameters. In this case, two remarkable differences were observed when looking at the cross-section of these samples. On one hand, melt pool or FZ had a "bowl shape" without nail head, typical in keyhole mode LBW, which reduces the risk of perpendicular intersection with columnar grains of parent material in comparison with "nail or mushroom" shape observed in 9 mm thickness plates. On the other hand, percentage of Laves phases in FZ was much lower than in 9 mm plates, as observed from Table 7. Both factors were critical to avoid formation of HAZ microfissures observed in thicker plates.

The percentage of Nb content of Laves phase observed in FZ of bead-on-plate samples was close to the eutectic point of the pseudo-binary diagram of alloy 718 (Figure 16a). This means that the terminal liquid would solidify as $L \rightarrow \gamma$ + Laves eutectic. It must be noted that most Laves phases observed at high magnification had a eutectic microstructure composed of γ and Laves phase (Figure 16b). This solidification path would be associated with larger volumes of terminal liquid that solidified at lower temperatures (down to 1180 °C) [4], since stepped liquid to solid transformation is hampered. If the amount of terminal liquid and its coexistence temperature range are increased, then one should expect a higher grain boundary wetting and infiltration risk in the three-point intersections, particularly if those grain boundaries are long and straight, as in the case of columnar microstructures obtained in the investment casting process.

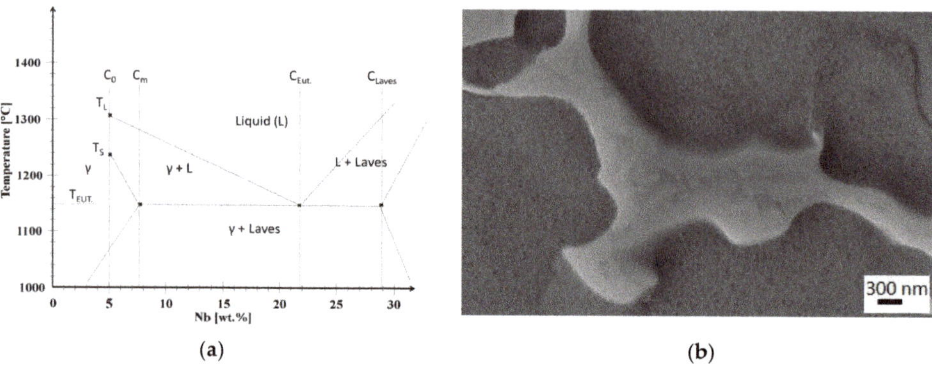

(a) (b)

Figure 16. (**a**) Pseudo-binary phase diagram for alloy 718 [33][1] and (**b**) eutectic Laves phase in FZ (mould NP). [1] Reproduced from [33], with permission from J. Andersson, 2021.

In order to analyse the solidification path and steps, we carried out thermodynamic and diffusion-based simulations with the three heats with different Si contents. Figure 17a shows the calculated solidification path of mould E determined using Scheil simulation, which considers back diffusion of elements in the primary phase at high cooling rate of 100 °C/s, representative of welding processes. The simulation predicted the formation of NbC carbides and Laves phase at the final stage of solidification, significantly reducing the solidus temperature compared to solidification under equilibrium condition. In Figure 17b, we can see that at the end of solidification, the liquid enriched in Nb, Mo, Ti, and Si, which allowed the formation of Carbides and Laves phases, lowering the solidus temperature. Table 10 depicts the solidus and liquidus temperatures of the heats with different Si contents. It can be observed that the solidus temperature decreased with increasing Si content and the solidification range increased. As observed experimentally, the percentage of Laves phase and its Si content increased in the as-cast material through increasing the Si content of the alloy. However, in LBW, the cooling rate can be extremely fast [25] and this leads to the limitation of Nb segregation. Therefore, the resulting terminal interdendritic liquid will have a composition close to the eutectic point, and therefore it will have long persistence and relatively high volume.

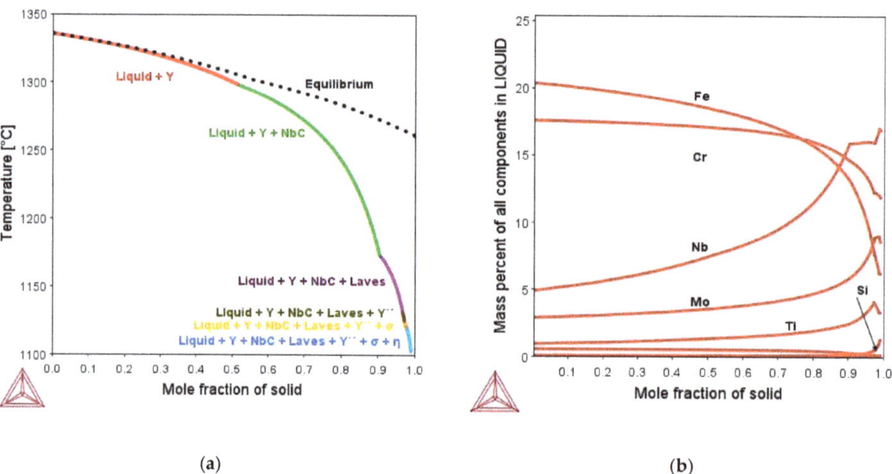

(a) (b)

Figure 17. Thermo-Calc simulation of alloy E (standard Si content): (**a**) solidification path and (**b**) microsegregation prediction using the Scheil model.

Table 10. Solidus and liquidus temperatures determined by equilibrium and Scheil simulation.

Ref.	Si in Alloy (wt %)	Si in Laves (wt %)	Laves wt %	Equilibrium Simulation			Scheil Simulation		
				Tsol (°C)	Tliq (°C)	ΔT (°C)	Tsol (°C)	Tliq (°C)	ΔT (°C)
O	0.051	0.03	1.71	1213.6	1339.8	126.2	1110.1	1339.8	229.8
E	0.110	0.13	1.78	1212.0	1335.8	123.8	1103.5	1335.8	232.4
P	0.170	0.43	1.91	1209.2	1338.0	128.8	1084.0	1338.0	254.1

5. Conclusions

In this work, influence of Si content, solidification rate, and pre-weld heat treatment on as-cast microstructure, weldability, and hot cracking susceptibility of alloy 718 investment castings were investigated. The following conclusions can be drawn about the impact of these factors and the selection of the different weldability tests which have been employed:

- Microstructural analysis of as-cast samples showed differences between heats in terms of amount and chemical composition of Laves phase, grain size, and aspect ratio. Shape of γ grains mainly depended on cooling rate.
- After HIP and solution annealing heat treatment, residual contents (less than 0.35% in area) of Laves phase were only observed in the samples with higher Si content and slower solidification rate, i.e., moulds P, N, and NP. The application of additional pre-HIP cycle to slowly solidified casting at 1052 °C for 2 h was not enough to completely remove the Laves phase.
- Onset of hot ductility drop in on-heating hot ductility test was directly related to the presence of residual Laves phase, whereas the hot ductility recovery behaviour was connected to the Si content and parent material grain size. Coarser grain size was associated with very slow recovery rate and very limited ductility recovery capability due to longer liquid continuity after incipient melting. In parallel, higher Si content reduced DRT and enlarged BTR.
- LBW Varestraint tests gave rise to enhanced fusion zone (FZ) cracking with much reduced heat-affected zone (HAZ) cracking on the surface. Both TCL FZ and TCL HAZ were mostly independent of Si content and presence of residual Laves phase.
- In all LBW welds, Laves phase was formed again in the FZ. The chemical composition of regenerated Laves phase has a composition similar to the eutectic of the pseudo-binary equilibrium diagram of alloy 718, and this suggests a long persistence of terminal liquid during the welding solidification. FZ Laves phase had eutectic morpho-logy.
- The composition of the regenerated FZ Laves phase matched with the continuous Laves phase film observed along HAZ microfissures in LBW bead-on-plate samples with nail or mushroom shapes which are characteristic in keyhole mode LBW.
- The observed HAZ cracking can be explained by the following hot cracking mechanism: backfilling and infiltration of terminal liquid along parent material γ grain boundaries in three point intersections resulting from perpendicular crossing of columnar grain boundaries with fusion line. This cracking mechanism was enhanced by both nail or mushroom weld shapes and narrow and columnar grain sizes of castings.
- The described cracking mechanism did not depend on the Si content, the effective dissolution of Laves phase, or homogenization of segregation gradients with proper heat treatments before welding, since the formation of detrimental Laves phases happens in the final solidification of melt pool after welding, which takes place at very fast cooling rates and limits Nb segregation in comparison with slow cooling condition.
- Neither Varestraint nor hot ductility weldability tests can reproduce this particular cracking mechanism, which is activated inside the samples and requires remelting of significant amount of material to form the melt pool.

Author Contributions: P.Á.: Investigation in Varestraint, hot ductility, and bead-on-plate tests; methodology; formal analysis; and writing—original draft preparation. L.V.: Investigation in Varestraint, hot ductility, and bead-on-plate tests; methodology; data curation; and visualisation. A.C.: Investigation in Varestraint, hot ductility, and bead-on-plate tests; microstructural investigation; and visualisation. N.R.: Methodology, formal analysis, Varestraint investigation, and data curation. P.P.R.: Conceptualisation, methodology, investigation in investment casting, and writing—review. A.N.: Methodology, formal analysis, thermodynamic simulation, microstructural investigation, data curation, and writing—original draft preparation. A.M.: Methodology, formal analysis, microstructural investigation, and data curation. F.S.: Conceptualisation, methodology, supervision, writing—review and editing, project administration, and funding acquisition. All authors have read and agreed to the published version of the manuscript.

Funding: This research was performed under the framework of HiperTURB project, which was funded by Clean Sky 2 Joint Undertaking under the European Union's Horizon 2020 research and innovation program, grant agreement no. 755561.

Institutional Review Board Statement: Not applicable.

Informed Consent Statement: Not applicable.

Data Availability Statement: Data available on request due to restrictions. The data presented in this study are available on request from the corresponding author. The data are not publicly available due to some IPR and confidentiality issues.

Acknowledgments: Bengt Pettersson and Vikström Fredrik from GKN Aerospace company in Trollhättan (Sweden) are gratefully acknowledged for their technical support and fruitful discussions. Additionally, the authors would like to thank the support received from Kejll Hurtig and Joel Andersson (Department of Engineering Science at West University) in carrying out hot ductility tests.

Conflicts of Interest: The authors declare no conflict of interest.

References

1. Eiselstein, H.L. Age-Hardenable Nickel Alloy. United States Patent No. 3,046,108, 24 July 1962.
2. Lippold, J.C.; Kiser, S.D.; DuPont, J.N. *Welding Metallurgy and Weldability of Nickel-Base Alloys*; John Wiley & Sons, Inc.: Hoboken, NJ, USA, 2009. [CrossRef]
3. Schirra, J.J.; Caless, R.H.; Hatala, R.W. The effect of Laves phase on the mechanical properties of wrought and cast+ HIP Inconel 718. *Superalloys* **1991**, *718*, 375–388. [CrossRef]
4. Muralidharan, B.G.; Shankar, V.; Gill, T.P.S. *Weldability of Inconel 718—A Review*; Indira Gandhi Centre for Atomic Research: Kalpakkam, India, 1996.
5. Woo, I.; Nishimoto, K.; Tanaka, K.; Shirai, M. Effect of grain size on heat affected zone cracking susceptibility. Study of weldability of Inconel 718 cast alloy (2nd report). *Weld. Int.* **2000**, *14*, 514–522. [CrossRef]
6. Baeslack, W.A.; Nelson, D.E. Morphology of weld heat-affected zone liquation in cast alloy 718. *Metallography* **1986**, *19*, 371–379. [CrossRef]
7. Thompson, G.R.; Genculu, S. Microstructural Evolution in the HAZ of Inconel 718 and Correlation with the Hot Ductility Test. *Weld. Res. Suppl.* **1983**, *62*, 337–345.
8. Singh, S.; Andersson, J. Hot cracking in cast alloy 718. *Sci. Technol. Weld. Join.* **2018**, *1718*, 568–574. [CrossRef]
9. Singh, S.; Hanning, F.; Andersson, J. Influence of Hot Isostatic Pressing on the Hot Ductility of Cast Alloy 718: The Effect of Niobium and Minor Elements on the Liquation Mechanism. *Metall. Mater. Trans. A* **2020**, *51*, 6248–6257. [CrossRef]
10. Guo, H.; Chaturvedi, M.C.; Richards, N.L. Effect of boron concentration and grain size on weld heat affected zone microfissuring in Inconel 718 base superalloys. *Sci. Technol. Weld. Join.* **1999**, *4*, 257–264. [CrossRef]
11. Richards, N.L.; Chaturvedi, M.C. Effect of minor elements on weldability of nickel base superalloys. *Int. Mater. Rev.* **2000**, *45*, 109–129. [CrossRef]
12. Huang, X.; Chaturvedi, M.C.; Richards, N.L. An Investigation of Microstructure and HAZ Microfissuring of Cast Alloy 718. *Superalloys* **1994**. [CrossRef]
13. Andersson, J.; Jacobsson, J.; Lundin, C. A historical perspective on Varestraint testing and the importance of testing parameters. In *Cracking Phenomena in Welds IV*; Springer International Publishing: New York, NY, USA, 2016; pp. 3–23. [CrossRef]
14. Andersson, J. Weldability of Precipitation Hardening Superalloys—Influence of Microstructure. Ph.D. Thesis, Chalmers University of Technology, Gothenburg, Sweden, 2011.
15. Alvarez, P.; Vázquez, L.; García-Riesco, P.M.; Rodríguez, P.P.; Magaña, A.; Santos, F. A Simplified Varestraint Test for Analyzing Weldability of Fe-Ni Based Superalloys. In Proceedings of the 9th International Symposium on Superalloy 718 & Derivatives: Energy, Aerospace, and Industrial Applications, Pittsburgh, PA, USA, 13 May 2018; pp. 849–865. [CrossRef]

16. Alvarez, P.; Vázquez, L.; Ruiz, N.; Rodríguez, P.; Magaña, A.; Niklas, A.; Santos, F. Comparison of hot cracking susceptibility of TIG and laser beam welded alloy 718 by varestraint testing. *Metals* **2019**, *9*, 985. [CrossRef]
17. Ram, G.D.J.; Reddy, A.V.; Rao, K.P.; Reddy, G.M. Control of Laves phase in Inconel 718 GTA welds with current pulsing. *Sci. Technol. Weld. Join.* **2004**, *9*, 390–398. [CrossRef]
18. Tharappel, J.T.; Babu, J. Welding processes for Inconel 718- A brief review. *IOP Conf. Ser. Mater. Sci. Eng.* **2018**. [CrossRef]
19. Bai, Y.; Lu, Q.; Ren, X.; Yan, H.; Zhang, P. Study of Inconel 718 Welded by Bead-On-Plate Laser Welding under High-Frequency Micro-Vibration Condition. *Metals* **2019**, *9*, 1335. [CrossRef]
20. Woo, I. The Factors Affecting HAZ Crack Susceptibility in the Laser Weld. Study on Weldability of Cast Alloy 718 (Report 4). *Q. J. Jpn. Weld. Soc.* **2001**, 308–316. [CrossRef]
21. Allen, C.; Shaw-edwards, R.; Nijdam, T.; Allen, C.; Shaw-edwards, R. Nickel-containing superalloy laser weld qualities and properties. *J. Laser Appl.* **2017**, *27*, S29001. [CrossRef]
22. Oshobe, O.E. Fiber Laser Welding of Nickel-Based Superalloy Inconel 718. Ph.D. Thesis, University of Manitoba, Fort Garry, MB, Canada, 2012.
23. Khan, A.; Hilton, P.; Blackburn, J.; Allen, C. Meeting weld quality criteria when laser welding Ni-based alloy 718. In Proceedings of the International Congress on Applications of Lasers & Electro-Optics, Anaheim, CA, USA, 23–27 September 2012; pp. 549–557. [CrossRef]
24. Kuo, T.Y.; Jeng, S.L. Porosity reduction in Nd–YAG laser welding of stainless steel and inconel alloy by using a pulsed wave. *J. Phys. D Appl. Phys.* **2005**, *38*, 722–728. [CrossRef]
25. Odabaçi, A.; Ünlü, N.I.; Göller, G.I.; Eruslu, M.N. A study on laser beam welding (LBW) technique: Effect of heat input on the microstructural evolution of superalloy Inconel 718. *Metall. Mater. Trans. A* **2010**, *41*, 2357–2365. [CrossRef]
26. International Organization for Standardization. *ISO/TR 17641-3:2005(E) Destructive Tests on Welds in Metallic Materials—Hot Cracking Tests for Weldments—Arc Welding Processes—Part 3: Externally Loaded Tests*; International Organization for Standardization (ISO): Geneva, Switzerland, 2005.
27. Zhang, Y.; Li, J. Characterization of the microstructure evolution and microsegregation in a Ni-based superalloy under super-high thermal gradient directional solidification. *Mater. Trans.* **2012**, *53*, 1910–1914. [CrossRef]
28. Rafiei, M.; Mirzadeh, H.; Malekan, M.; Sohrabi, M.J. Homogenization kinetics of a typical nickel-based superalloy. *J. Alloys Compd.* **2019**, *793*, 277–282. [CrossRef]
29. Patil, R.V.; Kale, G.B. Chemical diffusion of niobium in nickel. *J. Nucl. Mater* **1996**, *230*, 57–60. [CrossRef]
30. Jung, S.B.; Yamane, T.; Minamino, Y.; Hirao, K.; Araki, H.; Saji, S. Interdiffusion andits size effect in nickel solid solutions of Ni-Co, Ni-Cr and Ni-Ti systems. *J. Mater. Sci. Lett.* **1992**, *11*, 1333–1337. [CrossRef]
31. Swalin, R.A.; Martin, A.; Olson, R. Diffusion of magnesium, silicon, and molybdenum in nickel. *JOM* **1957**, *9*, 936–939. [CrossRef]
32. Singh, S.; Andersson, J. Heat-Affected-Zone Liquation Cracking in Welded Cast Haynes®282®. *Metals* **2020**, *10*, 29. [CrossRef]
33. Andersson, J.; Raza, S.; Eliasson, A.; Surreddi, K.B. Solidification of alloy 718, ATI 718Plus® and waspaloy. In Proceedings of the 8th International Symposium on Superalloy 718 and Derivatives, TMS (The Minerals, Metals & Materials Society), Pittsburgh, PA, USA, 28 September–1 October 2014; pp. 145–156.

MDPI
St. Alban-Anlage 66
4052 Basel
Switzerland
Tel. +41 61 683 77 34
Fax +41 61 302 89 18
www.mdpi.com

Metals Editorial Office
E-mail: metals@mdpi.com
www.mdpi.com/journal/metals

www.ingramcontent.com/pod-product-compliance
Lightning Source LLC
LaVergne TN
LVHW071619170726
843515LV00009B/2435
* 9 7 8 3 0 3 6 5 5 6 7 5 8 *